# Classical Groups, Derangements and Primes

A classical theorem of Jordan states that every nontrivial finite transitive permutation group contains a derangement. This existence result has interesting and unexpected applications in many areas of mathematics, including graph theory, number theory and topology. Various generalisations have been studied in recent years, with a particular focus on the existence of derangements with special properties.

Written for academic researchers and postgraduate students working in related areas of algebra, this introduction to the finite classical groups features a comprehensive account of the conjugacy and geometry of elements of prime order. The development is tailored towards the study of derangements in finite primitive classical groups; the basic problem is to determine when such a group $G$ contains a derangement of order $r$, where $r$ is a given prime divisor of the degree of $G$. This involves a detailed analysis of the conjugacy classes and subgroup structure of the finite classical groups.

Australian Mathematical Society Lecture Series: 25

# Classical Groups, Derangements and Primes

TIMOTHY C. BURNESS
*University of Bristol*

MICHAEL GIUDICI
*The University of Western Australia*

CAMBRIDGE
UNIVERSITY PRESS

# CAMBRIDGE
## UNIVERSITY PRESS

University Printing House, Cambridge CB2 8BS, United Kingdom

Cambridge University Press is part of the University of Cambridge.

It furthers the University's mission by disseminating knowledge in the pursuit of education, learning and research at the highest international levels of excellence.

www.cambridge.org
Information on this title: www.cambridge.org/9781107629448

© Timothy C. Burness and Michael Giudici 2016

First published 2016

*A catalogue record for this publication is available from the British Library*

*Library of Congress Cataloguing in Publication data*
Burness, Timothy C., 1979–
Classical groups, derangements, and primes / Timothy C. Burness, University of Bristol,
Michael Giudici, The University of Western Australia
pages    cm. – (Australian Mathematical Society lecture series ; 25)
Includes bibliographical references and index.
ISBN 978-1-107-62944-8
1. Logic, Symbolic and mathematical.   2. Group theory.   3. Algebra.   4. Numbers,
Prime.   I. Giudici, Michael, 1976–   II. Title.
QA9.B856   2016
512.7–dc23
2015013719

ISBN 978-1-107-62944-8 Paperback

*Dedicated to the memory of our friend and colleague*
*Ákos Seress, 1958–2013*

# Contents

# Tables

# Preface

The theory of permutation groups is a classical area of algebra, which arises naturally in the study of symmetry in a vast range of mathematical and physical systems. Originating in the early nineteenth century, permutation group theory continues to be a very active area of current research, with far-reaching applications across the sciences and beyond. In the last thirty years, the subject has been revolutionised by the *Classification of Finite Simple Groups* (CFSG), a truly remarkable theorem that is widely recognised as one of the greatest achievements of twentieth century mathematics. This has led to many interesting problems, and the development of powerful new techniques to solve them.

Let $G$ be a transitive permutation group on a finite set $\Omega$ of size at least 2. By a classical theorem of Jordan [82], $G$ contains an element that acts fixed-point-freely on $\Omega$; such elements are called *derangements*. The existence of derangements has interesting and unexpected applications in many other areas of mathematics, such as graph theory, number theory and topology (we will briefly discuss some of these applications in Chapter 1).

The study of derangements can be traced all the way back to the origins of probability theory over three centuries ago. Indeed, in 1708 Pierre de Montmort [106] studied the proportion of derangements in the symmetric group $S_n$ in its natural action on $n$ points. Montmort obtained a precise formula, which shows that this proportion tends to the constant $1/e$ as $n$ tends to infinity. This work played an important role in his pioneering mathematical analysis of various games of chance that were popular in the salons and gambling dens of early eighteenth century Paris. As we will see in Chapter 1, a wide range of related problems concerning derangements has been studied in more recent years.

One of the main motivations for our work stems from a theorem of Fein, Kantor and Schacher that provides a powerful extension of Jordan's aforementioned existence result. The main theorem in [52] states that every transitive

group $G$ as above contains a derangement of *prime power* order. It is interesting to note that the only known proof of this theorem requires CFSG. Naturally,
we can ask whether or not $G$ always contains a derangement of *prime* order.
It turns out that this is false in general, although examples with no such elements appear to be somewhat rare, which explains why the permutation groups
with this property are called *elusive* groups. For example, a theorem of Giudici
[61] implies that the smallest Mathieu group $M_{11}$ is the only almost simple
primitive elusive group (with respect to its 3-transitive action on 12 points).

This observation raises several natural and intriguing questions. Clearly, if $r$
is a prime number then $G$ contains a derangement $x$ of order $r$ only if $r$ divides
$|\Omega|$ (every cycle in the disjoint cycle decomposition of $x$ has to have length $r$).
This leads naturally to a *local* notion of elusivity: for a prime $r$ we say that $G$
is *r-elusive* if $r$ divides $|\Omega|$ and $G$ does not contain a derangement of order $r$.
In particular, $G$ is elusive if and only if $G$ is $r$-elusive for every prime divisor $r$
of $|\Omega|$. Given a non-elusive group $G$, we can ask whether or not $G$ contains a
derangement of order $r$ for *every* prime divisor $r$ of $|\Omega|$, or whether $G$ contains
a derangement of order 2 (when $|\Omega|$ is even), or a derangement of order $r$ for
the largest prime divisor $r$ of $|\Omega|$, and so on.

In this book we will address questions of this nature for a particularly
important family of finite permutation groups. Recall that a transitive group
$G$ as above is *primitive* if $\Omega$ is indecomposable, in the sense that there are
no $G$-invariant partitions of $\Omega$ (except for the two trivial partitions $\{\Omega\}$ and
$\{\{\alpha\} \mid \alpha \in \Omega\}$). The primitive permutation groups are the basic building
blocks of all finite permutation groups, and they play a central role in permutation group theory. The structure of such a group is described by the
O'Nan–Scott Theorem, which classifies the finite primitive permutation groups
according to their socle and the action of a point stabiliser (see [94], for example). This theorem can often be used to reduce a general problem concerning
primitive groups $G$ to the so-called *almost simple* case, where

$$T \leqslant G \leqslant \mathrm{Aut}(T)$$

for a nonabelian simple group $T$ (the socle of $G$). At this point, CFSG can be
invoked to describe the possibilities for $T$ (and thus $G$), and this can be combined with detailed information on the subgroup structure, conjugacy classes
and representation theory of almost simple groups. When applicable, this
reduction strategy is an extremely powerful tool in permutation group theory.

In [23], the O'Nan–Scott Theorem is used to reduce the problem of determining the $r$-elusive primitive permutation groups to the almost simple case.
Now, according to CFSG, there are three possibilities for the socle $T$ of an
almost simple group:

(i)   $T$ is an alternating group $A_n$ of degree $n \geqslant 5$;
(ii)  $T$ is one of 26 sporadic simple groups;
(iii) $T$ is a simple group of Lie type.

All the $r$-elusive primitive groups arising in cases (i) and (ii) are determined in [23], so it remains to deal with the almost simple groups of Lie type, which are either *classical* or *exceptional*. Recall that the classical groups arise naturally from groups of invertible matrices defined over finite fields, and the exceptional groups can be constructed from the exceptional simple Lie algebras over $\mathbb{C}$, of type $E_6$, $E_7$, $E_8$, $F_4$ and $G_2$.

The purpose of this book is to provide a detailed analysis of derangements of prime order in primitive almost simple classical groups. In a strong sense, 'most' almost simple groups are classical, so our work is a major contribution to the project initiated in [23]. A summary of our main results will be presented in Chapter 1 (see Section 1.5), with more detailed statements available later in the text. In order to do this, we require detailed information on the subgroup structure and conjugacy classes of elements (of prime order) in finite almost simple classical groups. Indeed, observe that if $G$ is a primitive permutation group with point stabiliser $H$, then $H$ is a maximal subgroup of $G$, and an element $x \in G$ is a derangement if and only if the conjugacy class of $x$ in $G$ fails to meet $H$.

The study of the subgroup structure of an almost simple classical group $G$, and in particular its maximal subgroups, can be traced all the way back to Galois and his letter to Chevalier written on the eve of his fatal duel in 1832 [59]. More recently, in particular post-CFSG, there have been great advances in this area.

The main result is a theorem of Aschbacher from 1984. In [3], eight collections of 'natural', or *geometric*, subgroups of $G$ are defined in terms of the underlying geometry of the natural (projective) module $V$ for the socle of $G$. These subgroup collections include the stabilisers of appropriate subspaces and direct sum decompositions of $V$, for example. Given a subgroup $H$ of $G$, Aschbacher proves that either $H$ is contained in a member of one of these geometric collections, or $H$ is almost simple and the socle of $H$ acts absolutely irreducibly on $V$ (and satisfies several additional conditions). Detailed information on the subgroups in the geometric collections (in terms of existence, structure, maximality and conjugacy) is given by Kleidman and Liebeck [86], and further details for the low-dimensional classical groups can be found in the recent book [13] by Bray, Holt and Roney-Dougal.

In this book, our analysis of derangements for geometric actions of classical groups is organised according to the subgroup collections arising in Aschbacher's theorem. Rather different methods are needed to study the remaining *non-geometric* actions of classical groups, and a detailed analysis of derangements in this situation will be given in a future paper.

In Chapters 2 and 3 we aim to provide the reader with an accessible introduction to the finite classical groups and their underlying geometry, with a particular focus on the conjugacy classes of elements of prime order. Our treatment is inevitably tailored towards the application to derangements, and we make no attempt to give a comprehensive introduction.

We start with a discussion of forms, standard bases and automorphisms, and we describe some specific classical group embeddings that will be useful later. A brief description of the subgroup collections arising in Aschbacher's theorem is provided in Section 2.6. Chapter 3 is dedicated to the study of conjugacy classes of elements of prime order in almost simple classical groups. Our main aim is to bring together a wide range of results on conjugacy classes that are somewhat scattered through the literature in this area. In particular, we provide a detailed analysis of involutions (both semisimple, unipotent and outer automorphisms), which complements important earlier work of Aschbacher and Seitz [5], and Gorenstein, Lyons and Solomon [67]. Of course, we require this information for the application to derangements given in Chapters 4, 5 and 6, but we hope that our treatment will be useful more generally, in a wide range of problems concerning finite permutation groups. For example, this sort of information has been essential in recent work on fixed point ratios and bases for primitive permutation groups (see [17, 18, 19, 20] and [21, 25, 26, 27, 28]), in the classification of almost simple extremely primitive groups and $\frac{3}{2}$-transitive groups (see [6, 29, 30]), and in a wide range of problems concerning the generation and random generation of finite simple groups (see [14, 24, 70], for example). In Chapter 5 we also determine the precise structure of some specific geometric maximal subgroups of classical groups, extending the analysis given in [86] (see Sections 5.3.2 and 5.5.1).

We hope that this book will be useful to graduate students and researchers who work with finite classical groups. Indeed, the background material in Chapters 2 and 3 will be accessible to graduate students with a basic understanding of linear algebra and group theory. In particular, we hope that our detailed treatment of conjugacy classes of elements of prime order will serve as a useful general reference. Our study of derangements of prime order in almost simple classical groups provides an immediate application, building naturally on the material presented in Chapters 2 and 3. We anticipate that the content of Chapters 4, 5 and 6 will appeal to researchers working in permutation group theory.

Finally, some words on the organisation of this book. Chapter 1 serves as a general introduction. Here we provide a brief survey of earlier work on derangements, we describe several applications and we present a summary of our main results on derangements in finite classical groups. The next two chapters provide a brief introduction to the finite classical groups, giving the necessary background information that we will need for the application to derangements investigated in Chapters 4, 5 and 6.

Our analysis of derangements of prime order in finite classical groups begins in Chapter 4, where we handle the so-called *subspace actions*. Guided by Aschbacher's structure theorem, the remaining geometric subgroup collections are studied in Chapter 5, together with a small additional collection of *novelty* subgroups that arises when $G$ has socle $\mathrm{Sp}_4(q)'$ ($q$ even) or $\mathrm{P}\Omega_8^+(q)$. Finally, in Chapter 6 we use our earlier work to present detailed results on derangements in the low-dimensional classical groups, including both geometric and non-geometric actions.

We also include two appendices. In Appendix A we record several number-theoretical results that will be needed in our analysis of conjugacy classes and derangements. Finally, in Appendix B we present various tables that conveniently summarise some of the information on conjugacy classes in finite classical groups discussed in Chapter 3.

*Tim Burness and Michael Giudici, March 2015*

# Acknowledgements

Burness was supported by Engineering and Physical Sciences Research Council grant EP/I019545/1 and the Royal Society of London. He thanks the Centre for the Mathematics of Symmetry and Computation at the University of Western Australia for their generous hospitality during several visits.

Giudici was supported by Australian Research Council Discovery Project grant DP0770915, and he thanks the Schools of Mathematics at the University of Southampton and the University of Bristol for their generous hospitality.

Both authors thank Robert Guralnick and Martin Liebeck for their encouragement and helpful comments. They also thank Martin Liebeck and Gary Seitz for providing advanced access to their book [96] on unipotent and nilpotent classes, and they thank John Bray, Derek Holt and Colva Roney-Dougal for providing numerous details (pre-publication) from their book [13] on the subgroup structure of the low-dimensional finite classical groups.

# Notational conventions

*Group-theoretic notation*

Let $G$ and $H$ be groups, $n$ a positive integer, $p$ a prime number.

| | |
|---|---|
| $|G|$ | order of $G$ |
| $G'$ | derived subgroup of $G$ |
| $Z(G)$ | centre of $G$ |
| $\mathrm{soc}(G)$ | socle of $G$ (subgroup of $G$ generated by its minimal normal subgroups) |
| $G^{\infty}$ | last term in the derived series of $G$ |
| $G^n$ | direct product of $n$ copies of $G$ |
| $H \leqslant G$ | $H$ is a subgroup of $G$ |
| $H < G$ | $H$ is a proper subgroup of $G$ |
| $|G:H|$ | index of a subgroup $H$ of $G$ |
| $G.H$ | an extension of $G$ by $H$ |
| $G{:}H$ | a split extension of $G$ by $H$ |
| $G \wr H$ | wreath product of $G$ and $H$, $H \leqslant S_n$ |
| $N_G(H)$ | normaliser in $G$ of $H$ |
| $C_G(x)$ | centraliser in $G$ of $x$ |
| $x^G$ | $G$-conjugacy class of $x$ |
| $|x|$ | order of $x$ |
| $C_n$ or just $n$ | cyclic group of order $n$ |
| $C_p^n$ or just $p^n$ | elementary abelian group of order $p^n$ |
| $D_n$ | dihedral group of order $n$ |
| $S_n$ | symmetric group of degree $n$ |
| $A_n$ | alternating group of degree $n$ |
| $[n]$ | unspecified soluble group of order $n$ |

## Other notation

| | |
|---|---|
| $\mathbb{F}_q$ | field of size $q$ |
| $\mathbb{F}^\times$ | nonzero elements of a field $\mathbb{F}$ |
| $\overline{\mathbb{F}}$ | algebraic closure of a field $\mathbb{F}$ |
| $\mathbb{E}/\mathbb{F}$ | $\mathbb{E}$ is a field extension of $\mathbb{F}$ |
| $\mathscr{S}_r$ | set of all nontrivial $r$th roots of unity in a field |
| $D(Q)$ | discriminant of a quadratic form $Q$ |
| $\Phi(r,q)$ | smallest $i \in \mathbb{N}$ such that $r$ divides $q^i - 1$ |
| $\delta_{i,j}$ | Kronecker delta |
| $(a_1,\ldots,a_t)$ | greatest common divisor of integers $a_1,\ldots,a_t$ |
| $[a_1,\ldots,a_t]$ | least common multiple of integers $a_1,\ldots,a_t$ |
| $(a)_b$ | highest power of $b$ dividing $a$, for integers $a,b$ |
| $a'$ | largest odd divisor of the integer $a$ |
| $\mathbb{N}_0$ | $\mathbb{N} \cup \{0\}$ |
| $A^\mathsf{T}$ | transpose of a matrix $A$ |
| $\lfloor x \rfloor$ | largest integer $n \leqslant x$ ($x \in \mathbb{R}$) |
| $x - \varepsilon$ | $x - \varepsilon 1$ with $\varepsilon = \pm$ ($x \in \mathbb{R}$) |

## Classical group notation

Our notation for classical groups is fairly standard, and we closely follow the notation used by Kleidman and Liebeck [86]. In particular, we write

$$\mathrm{PSL}_n^+(q) = \mathrm{PSL}_n(q), \ \ \mathrm{PSL}_n^-(q) = \mathrm{PSU}_n(q)$$

for the projective special linear and unitary groups of dimension $n$ over $\mathbb{F}_q$, respectively. Specific notation for orthogonal groups will be defined in Section 2.5. We write $x = [x_1,\ldots,x_k]$ to denote a block-diagonal matrix $x \in \mathrm{GL}_n(q)$ with blocks $x_1,\ldots,x_k$. Furthermore, if the $x_i$ are all equal then we write $x = [x_1^k]$. We use $I_n$ for the identity matrix in $\mathrm{GL}_n(q)$, and we write $J_n$ for the standard (lower-triangular) unipotent Jordan block of size $n$ in $\mathrm{GL}_n(q)$. Further notation will be introduced as and when needed in the text.

# 1
# Introduction

The study of derangements in transitive permutation groups has a long and rich history, which can be traced all the way back to the origins of probability theory in the early eighteenth century. In 1708, the French mathematician Pierre de Montmort wrote one of the first highly influential books on probability, entitled *Essay d'Analyse sur les Jeux de Hazard* [106], in which he presents a systematic combinatorial analysis of games of chance that were popular at the time. Through studying the card game *treize* (and variations), he calculates the proportion of derangements in the symmetric group $S_{13}$ in its natural action on 13 points, and he proposes the general formula

$$\frac{1}{2!} - \frac{1}{3!} + \cdots + \frac{(-1)^n}{n!}$$

for the natural action of $S_n$. In a second edition, published in 1713, he reports on his correspondence with Nicolaus Bernoulli, who proved the above formula using the inclusion-exclusion principle (see [117] for further details). In particular, it follows that the proportion of derangements in $S_n$ tends to $1/e$ as $n$ tends to infinity.

In the context of permutation group theory, derangements have been widely studied since the days of Jordan in the nineteenth century, finding a range of interesting applications and connections in diverse areas such as graph theory, number theory and topology. In more recent years, following the Classification of Finite Simple Groups, the subject has been reinvigorated and our understanding of derangements has advanced greatly. As we shall see, many new results on the proportion of derangements in various families of groups have been obtained, and there has been a focus on studying the existence of derangements with special properties.

In the first three sections of this introductory chapter we will briefly survey some of these results and applications, focusing in particular on derangements

1

of prime order. Given a fixed prime number $r$, we will see that the problem of determining the existence of a derangement of order $r$ in a finite transitive permutation group $G$ can essentially be reduced to the case where $G$ is a primitive almost simple group of Lie type. In this book, we aim to provide a detailed analysis of derangements of prime order in classical groups; the basic problem is introduced in Section 1.4, and we present a brief summary of our main results in Section 1.5 (with more detailed results given later in the text).

## 1.1 Derangements

We start by recalling some basic notions. We refer the reader to the books by Cameron [35], Dixon and Mortimer [48] and Wielandt [120] for excellent introductions to the theory of permutation groups.

Let $G$ be a permutation group on a set $\Omega$, so $G$ is a subgroup of $\mathrm{Sym}(\Omega)$, the group of all permutations of $\Omega$. We will use exponential notation for group actions, so $\alpha^g$ denotes the image of $\alpha \in \Omega$ under the permutation $g \in G$. The cardinality of $\Omega$ is called the *degree* of $G$.

We say that $G$ is *transitive* on $\Omega$ if for all $\alpha, \beta \in \Omega$ there exists an element $g \in G$ such that $\alpha^g = \beta$. The *stabiliser in $G$ of* $\alpha$, denoted by $G_\alpha$, is the subgroup of $G$ consisting of all the permutations that fix $\alpha$. The familiar Orbit-Stabiliser Theorem implies that if $G$ is transitive then $\Omega$ can be identified with the set of (right) cosets of $G_\alpha$ in $G$. Moreover, the action of $G$ on $\Omega$ is equivalent to the natural action of $G$ on this set of cosets by right multiplication.

Given a subgroup $H$ of $G$, we will write $H^g$ to denote the conjugate subgroup $g^{-1}Hg = \{g^{-1}hg \mid h \in H\}$. It is easy to see that $G_{\alpha^g} = (G_\alpha)^g$ for all $\alpha \in \Omega$, $g \in G$. In particular, if $G$ is transitive then $G_\alpha$ and $G_\beta$ are conjugate subgroups for all $\alpha, \beta \in \Omega$.

The notion of primitivity is a fundamental indecomposability condition in permutation group theory. We say that a transitive group $G$ is *imprimitive* if $\Omega$ admits a nontrivial $G$-invariant partition (there are two trivial partitions, namely $\{\Omega\}$ and $\{\{\alpha\} \mid \alpha \in \Omega\}$), and *primitive* otherwise. Equivalently, $G$ is primitive if and only if $G_\alpha$ is a maximal subgroup of $G$. The finite primitive groups are the basic building blocks of all finite permutation groups.

Notice that if $N$ is a normal subgroup of $G$, then the set of orbits of $N$ on $\Omega$ forms a $G$-invariant partition of $\Omega$. Thus, if $G$ is primitive, every nontrivial normal subgroup of $G$ is transitive. We can generalise the notion of primitivity by defining a group to be *quasiprimitive* if every nontrivial normal subgroup is transitive.

**Definition 1.1.1** Let $G$ be a group acting on a set $\Omega$. An element of $G$ is a *derangement* (or *fixed-point-free*) if it fixes no point of $\Omega$. We write $\Delta(G)$ for the set of derangements in $G$. In addition, if $G$ is finite then $\delta(G) = |\Delta(G)|/|G|$ denotes the proportion of derangements in $G$.

Note that if $G$ is transitive with point stabiliser $H$ then

$$\Delta(G) = G \setminus \bigcup_{g \in G} H^g \qquad (1.1.1)$$

so an element $x \in G$ is a derangement if and only if $x^G \cap H$ is empty, where $x^G = \{g^{-1}xg \mid g \in G\}$ is the conjugacy class of $x$ in $G$. We also observe that $\Delta(G)$ is a normal subset of $G$.

Let $G$ be a finite group acting transitively on a set $\Omega$ with $|\Omega| \geqslant 2$. By the Orbit-Counting Lemma we have

$$\frac{1}{|G|} \sum_{x \in G} |\mathrm{fix}_{\Omega}(x)| = 1$$

where $\mathrm{fix}_{\Omega}(x) = \{\alpha \in \Omega \mid \alpha^x = \alpha\}$ is the set of fixed points of $x$ on $\Omega$. Since $|\mathrm{fix}_{\Omega}(1)| = |\Omega| \geqslant 2$, there must be an element $x \in G$ with $|\mathrm{fix}_{\Omega}(x)| = 0$ and thus $G$ contains a derangement. This is a theorem of Jordan, which dates from 1872 (see [82]).

**Theorem 1.1.2** *Let $G$ be a finite group acting transitively on a set $\Omega$ with $|\Omega| \geqslant 2$. Then $G$ contains a derangement.*

In particular, every nontrivial finite transitive permutation group contains a derangement. In view of (1.1.1), Jordan's theorem is equivalent to the fact that

$$G \neq \bigcup_{g \in G} H^g \qquad (1.1.2)$$

for every proper subgroup $H$ of a finite group $G$.

It is easy to see that Jordan's theorem does *not* extend to transitive actions of infinite groups:

(i) Let $\mathrm{FSym}(\Omega)$ be the *finitary symmetric group* on an infinite set $\Omega$; it comprises the permutations of $\Omega$ with finite support (that is, the permutations that move only finitely many elements of $\Omega$). Clearly, this transitive group does not contain any derangements.

(ii) Let $V$ be an $n$-dimensional vector space over $\mathbb{C}$ and let $G = \mathrm{GL}(V)$ be the general linear group of all invertible linear transformations of $V$. Let $\Omega$ be the set of complete flags of $V$, that is, the set of subspace chains

$${0} = V_0 \subset V_1 \subset V_2 \subset \cdots \subset V_{n-1} \subset V_n = V$$

where each $V_i$ is an $i$-dimensional subspace of $V$. The natural action of $G$ on $V$ induces a transitive action of $G$ on $\Omega$. For each $x \in G$ there is a basis of $V$ in which $x$ is represented by a lower-triangular matrix (take the Jordan canonical form of $x$, for example), so $x$ fixes a complete flag and thus $G$ has no derangements.

(iii) More generally, consider a connected algebraic group $G$ over an algebraically closed field $K$ of characteristic $p \geqslant 0$, and let $B$ be a Borel subgroup of $G$. Then every element of $G$ belongs to a conjugate of $B$, so $G$ has no derangements in its transitive action on the flag variety $G/B$. In fact, by a theorem of Fulman and Guralnick [55, Theorem 2.4], if $G$ is a simple algebraic group acting on a coset variety $G/H$, then $G$ contains no derangements if and only if one of the following holds:

(a) $H$ contains a Borel subgroup of $G$;

(b) $G = \mathrm{Sp}_n(K)$, $H = \mathrm{O}_n(K)$ and $p = 2$;

(c) $G = G_2(K)$, $H = \mathrm{SL}_3(K).2$ and $p = 2$.

Moreover, if $G$ is simple then [55, Lemma 2.2] implies that $\Delta(G)$ is a dense subset of $G$ (with respect to the Zariski topology) if and only if $H$ does not contain a maximal torus of $G$.

As observed by Serre, Jordan's theorem has some interesting applications in number theory and topology (see Serre's paper [113] for further details).

(i) *A number-theoretic application.* Let $f \in \mathbb{Z}[x]$ be an irreducible polynomial over $\mathbb{Q}$ with degree $n \geqslant 2$. Then $f$ has no roots modulo $p$ for infinitely many primes $p$.

(ii) *A topological application.* Let $f : T \to S$ be a finite covering of a topological space $S$, where $f$ has degree $n \geqslant 2$ (so that $|f^{-1}(s)| = n$ for all $s \in S$) and $T$ is path-connected and non-empty. Then there exists a continuous map $\varphi : \mathbb{S}_1 \to S$ from the circle $\mathbb{S}_1$ that cannot be lifted to the covering $T$.

In view of Jordan's theorem, two natural questions arise:

**Question 1.** *How abundant are derangements in transitive groups?*

**Question 2.** *Can we find derangements with special properties, such as a prescribed order?*

Both of these questions have been widely investigated in recent years, and in the next two sections we will highlight some of the main results.

**Remark 1.1.3** We will focus on Questions 1 and 2 above. However, there are many other interesting topics concerning derangements that we will not discuss. Here are some examples:

(i) *Normal coverings.* Let $G$ be a finite group and recall that if $H$ is a proper subgroup of $G$ then $\bigcup_{g \in G} H^g$ is a proper subset of $G$ (see (1.1.2)). A collection of proper subgroups $\{H_1, \ldots, H_t\}$ is a *normal covering* of $G$ if

$$G = \bigcup_{i=1}^{t} \bigcup_{g \in G} H_i^g$$

and we define $\gamma(G)$ to be the minimal size of a normal covering of $G$. By Jordan's theorem, $\gamma(G) \geqslant 2$, and this invariant has been investigated in several recent papers (see [15, 16, 42], for example). The connection to derangements is transparent: if $\{H_1, \ldots, H_t\}$ is a normal covering then each $x \in G$ has fixed points on the set of cosets $G/H_i$, for some $i$.

(ii) *Algorithms.* Given a set of generators for a subgroup $G \leqslant S_n$, it is easy to determine whether or not $G$ is transitive. If $G$ is transitive and $n \geqslant 2$, then Jordan's theorem implies that $G$ contains a derangement, and there are efficient randomised algorithms to find a derangement in $G$. In a recent paper, Arvind [2] has presented the first elementary *deterministic* polynomial-time algorithm for finding a derangement.

(iii) *Thompson's question.* A finite transitive permutation group $G \leqslant \mathrm{Sym}(\Omega)$ is *Frobenius* if $|G_\alpha| > 1$ and $G_\alpha \cap G_\beta = 1$ for all distinct $\alpha, \beta \in \Omega$. By a theorem of Frobenius, $\{1\} \cup \Delta(G)$ is a normal transitive subgroup and thus $\Delta(G)$ is a transitive subset of $G$. The following, more general question, has been posed by J. G. Thompson.

**Question.** *Let $G \leqslant \mathrm{Sym}(\Omega)$ be a finite primitive permutation group. Is $\Delta(G)$ a transitive subset of $G$?*

This is Problem 8.75 in the Kourovka Notebook [84]. It is easy to see that the primitivity condition here is essential; there are imprimitive groups $G$ such that $\Delta(G)$ is intransitive. For instance, take the natural action of the alternating group $A_4$ on the set of 2-element subsets of $\{1, 2, 3, 4\}$.

## 1.2 Counting derangements

Let $G$ be a transitive permutation group on a finite set $\Omega$ with $|\Omega| = n \geqslant 2$. Recall that $\Delta(G)$ is the set of derangements in $G$, and $\delta(G) = |\Delta(G)|/|G|$ is the proportion of derangements. In general, it is difficult to compute $\delta(G)$

precisely. Of course, Jordan's theorem (Theorem 1.1.2) implies that $\delta(G) > 0$, and stronger lower bounds have been obtained in recent years. In [37], for example, Cameron and Cohen use the Orbit-Counting Lemma to show that $\delta(G) \geqslant 1/n$, with equality if and only if $G$ is *sharply* 2-*transitive*, that is, either $(G, n) = (S_2, 2)$, or $G$ is a Frobenius group of order $n(n-1)$, with $n$ a prime power. This has been extended by Guralnick and Wan (see [73, Theorem 1.3]).

**Theorem 1.2.1** *Let $G$ be a transitive permutation group of degree $n \geqslant 2$. Then one of the following holds:*

(i) $\delta(G) \geqslant 2/n$;
(ii) $G$ *is a Frobenius group of order $n(n-1)$ with $n$ a prime power;*
(iii) $G = S_n$ *and $n \in \{2, 4, 5\}$.*

It is worth noting that this strengthening of the lower bound on $\delta(G)$ from $1/n$ to $2/n$ requires the classification of the finite 2-transitive groups, which in turn relies on the Classification of Finite Simple Groups. As explained in [73], Theorem 1.2.1 has interesting applications in the study of algebraic curves over finite fields.

Inspired by Montmort's formula

$$\delta(S_n) = \frac{1}{2!} - \frac{1}{3!} + \cdots + \frac{(-1)^n}{n!}$$

(with respect to the natural action of $S_n$), it is natural to consider the asymptotic behaviour of $\delta(G)$ when $G$ belongs to an interesting infinite family of groups. From the above formula, we immediately deduce that $\delta(S_n)$ tends to $1/e$ as $n$ tends to infinity. Similarly, we find that $\delta(A_n) \geqslant 1/3$ and $\delta(\mathrm{PSL}_2(q)) \geqslant 1/3$ for all $n, q \geqslant 5$, with respect to their natural actions of degree $n$ and $q + 1$ (see [12, Corollary 2.6 and Lemma 2.8]). In these two examples, we observe that $G$ belongs to an infinite family of finite simple groups, and $\delta(G)$ is bounded away from zero by an absolute constant.

In fact, a deep theorem of Fulman and Guralnick [55, 56, 57, 58] shows that this is true for *any* transitive simple group.

**Theorem 1.2.2** *There exists an absolute constant $\varepsilon > 0$ such that $\delta(G) \geqslant \varepsilon$ for any transitive finite simple group $G$.*

This theorem confirms a conjecture of Boston *et al.* [12] and Shalev. The asymptotic nature of the proof does not yield an explicit constant, although [57, Theorem 1.1] states that $\varepsilon \geqslant 0.016$ with at most finitely many exceptions. It is speculated in [12, p. 3274] that the optimal bound is $\varepsilon = 2/7$, which is realised by the standard actions of $\mathrm{PSL}_3(2)$ and $\mathrm{PSL}_3(4)$, of degree

7 and 21, respectively. In fact, it is easy to check that the action of the Tits group $G = {}^2F_4(2)'$ on the set of cosets of a maximal subgroup $2^2.[2^8].S_3$ yields $\delta(G) = 89/325 < 2/7$, and we expect $89/325$ to be the optimal constant in Theorem 1.2.2.

Fulman and Guralnick also establish strong asymptotic results. For instance, they show that apart from some known exceptions, $\delta(G)$ tends to 1 as $|G|$ tends to infinity (the exceptions include $G = A_n$ acting on the set of $k$-element subsets of $\{1,\ldots,n\}$ with $k$ bounded, for example). Further information on the limiting behaviour of the proportion of derangements in the natural action of $S_n$ or $A_n$ on $k$-sets is given by Diaconis, Fulman and Guralnick [44, Section 4], together with an interesting application to card shuffling.

As explained in [55, Section 6], one can show that the above theorem of Fulman and Guralnick does *not* extend to almost simple groups. For example, let $p$ and $r$ be primes such that $r$ and $|\mathrm{PGL}_2(p)| = p(p^2 - 1)$ are coprime, and set $G = \mathrm{PGL}_2(p^r){:}\langle\phi\rangle$ and $\Omega = \phi^G$, where $\phi$ is a field automorphism of $\mathrm{PGL}_2(p^r)$ of order $r$. By [71, Corollary 3.7], the triple $(G,\mathrm{PGL}_2(p^r),\Omega)$ is *exceptional* and thus [71, Lemma 3.3] implies that every element in a coset $\mathrm{PGL}_2(p^r)\phi^i$ (with $1 \leqslant i < r$) has a unique fixed point on $\Omega$. Therefore

$$\delta(G) \leqslant \frac{|\mathrm{PGL}_2(p^r)|}{|G|} = \frac{1}{r}$$

and thus $\delta(G)$ tends to 0 as $r$ tends to infinity.

It is worth noting that Theorem 1.2.2 indicates that the proportion of derangements in simple primitive groups behaves rather differently to the proportion of derangements in more general primitive groups. Indeed, by a theorem of Boston *et al.* [12, Theorem 5.11], the set

$$\{\delta(G) \mid G \text{ is a finite primitive group}\}$$

is dense in the open interval $(0,1)$.

In a slightly different direction, if $G$ is a transitive permutation group of degree $n \geqslant 2$, then $\Delta(G)$ is a normal subset of $G$ and we can consider the number of conjugacy classes in $\Delta(G)$, which we denote by $\kappa(G)$. Of course, Jordan's theorem implies that $\kappa(G) \geqslant 1$. In [31], the finite primitive permutation groups with $\kappa(G) = 1$ are determined (it turns out that $G$ is either sharply 2-transitive, or $(G,n) = (A_5,6)$ or $(\mathrm{PSL}_2(8){:}3,28)$), and this result is used to study the structure of finite groups with a nonlinear irreducible complex character that vanishes on a unique conjugacy class. We refer the reader to [31] for more details and further results.

An extension of the main theorem of [31] from primitive to transitive groups has recently been obtained by Guralnick [69]. He shows that every transitive group $G$ with $\kappa(G) = 1$ is primitive, so no additional examples arise.

## 1.3 Derangements of prescribed order

In addition to counting the number of derangements in a finite permutation group, it is also natural to ask whether or not we can find derangements with special properties, such as a specific order.

### 1.3.1 Prime powers

The strongest result in this direction is the following theorem of Fein, Kantor and Schacher [52], which concerns the existence of derangements of prime power order.

**Theorem 1.3.1** *Every nontrivial finite transitive permutation group contains a derangement of prime power order.*

This theorem was initially motivated by an important number-theoretic application, which provides another illustration of the utility of derangements in other areas of mathematics. Here we give a brief outline (see [52] and [87, Chapter III] for more details; also see [68] for further applications in this direction).

Let $K$ be a field and let $A$ be a central simple algebra (CSA) over $K$, so $A$ is a simple finite-dimensional associative $K$-algebra with centre $K$. By the Artin–Wedderburn theorem, $A$ is isomorphic to a matrix algebra $M_n(D)$ for some positive integer $n$ and division algebra $D$. Under the *Brauer equivalence*, two CSAs $A$ and $A'$ over $K$ are equivalent if $A \cong M_n(D)$ and $A' \cong M_m(D)$ for some $n$ and $m$, and the set of equivalence classes forms an abelian group under tensor product. This is called the *Brauer group* of $K$, denoted $\mathscr{B}(K)$.

Let $L/K$ be a field extension. The inclusion $K \subseteq L$ induces a group homomorphism $\mathscr{B}(K) \to \mathscr{B}(L)$, and the *relative Brauer group* $\mathscr{B}(L/K)$ is the kernel of this map. The connection to derangements arises from the remarkable observation that Theorem 1.3.1 is equivalent to the fact that $\mathscr{B}(L/K)$ is infinite for any nontrivial finite extension of global fields (where a *global field* is a finite extension of $\mathbb{Q}$, or a finite extension of $\mathbb{F}_q(t)$, the function field in one variable over a finite field $\mathbb{F}_q$).

In order to justify this equivalence, as explained in [52, Section 3], there is a reduction to the case where $L/K$ is separable, and by a further reduction one can assume that $L = K(\alpha)$. Let $E$ be a Galois closure of $L/K$, let $\Omega$ be the set of roots in $E$ of the minimal polynomial of $\alpha$ over $K$, and let $G$ be the Galois group $\mathrm{Gal}(E/K)$. Then $G$ acts transitively on $\Omega$, and [52, Corollary 3] states that $\mathscr{B}(L/K)$ is infinite if and only if $G$ contains a derangement of prime power order. More precisely, if $r$ is a prime divisor of $|\Omega|$ then the $r$-torsion subgroup of $\mathcal{B}(L/K)$ is infinite if and only if $G$ contains a derangement of $r$-power order.

Although the existence of derangements in Theorem 1.1.2 is an easy corollary of the Orbit-Counting Lemma, the extension to prime powers in Theorem 1.3.1 appears to require the full force of the Classification of Finite Simple Groups. The basic strategy is as follows. First observe that if $G \leqslant \mathrm{Sym}(\Omega)$ is an imprimitive permutation group and every $x \in G$ of prime power order fixes a point, then $x$ must also fix the set that contains this point in an appropriate $G$-invariant partition of $\Omega$. Hence the primitive group induced by $G$ on a maximal $G$-invariant partition also has no derangements of prime power order, so the existence problem is reduced to the primitive case. We now consider a minimal counterexample $G$. If $N$ is a nontrivial normal subgroup of $G$, then $N$ acts transitively on $\Omega$ (by the primitivity of $G$), so the minimality of $G$ implies that $N = G$ and thus $G$ is simple. The proof now proceeds by working through the list of finite simple groups provided by the Classification. It would be very interesting to know if there exists a Classification-free proof of Theorem 1.3.1.

**Remark 1.3.2** The finite primitive permutation groups with the property that *every* derangement has $r$-power order, for some fixed prime $r$, are investigated in [32]. The groups that arise are almost simple or affine, and the almost simple groups with this extremal property are determined in [32, Theorem 2].

### 1.3.2 Isbell's Conjecture

Let $G$ be a finite transitive permutation group. Although Theorem 1.3.1 guarantees the existence in $G$ of a derangement of prime power order, the proof does not provide any information about the primes involved. However, there are some interesting conjectures in this direction. For example, it is conjectured that if a particular prime power dominates the degree of $G$, then $G$ contains a derangement that has order a power of that prime. This is known as *Isbell's Conjecture*.

**Conjecture 1.3.3** *Let $p$ be a prime. There is a function $f(p,b)$ with the property that if $G$ is a transitive permutation group of degree $n = p^{a}b$ with $(p,b) = 1$ and $a \geqslant f(p,b)$, then $G$ contains a derangement of $p$-power order.*

The special case $p = 2$ arises naturally in the study of $n$-player games, and the conjecture dates back to work of Isbell on this topic in the late 1950s [77,

78, 79]. The formulation of the conjecture stated above is due to Cameron, Frankl and Kantor [38, p. 150].

Following [78], let us briefly explain the connection to $n$-player games. A *fair game* (or *homogeneous game*) is a method for resolving binary questions without giving any individual player an advantage. If such a game has $n$ players, then it can be modelled mathematically as a family $\mathscr{W}$ of subsets of a set $X$ of size $n$, called *winning sets*, with the following four properties:

(a) If $A \subseteq B \subseteq X$ and $A \in \mathscr{W}$ then $B \in \mathscr{W}$.
(b) If $A \in \mathscr{W}$ then $X \setminus A \notin \mathscr{W}$.
(c) If $A \notin \mathscr{W}$ then $X \setminus A \in \mathscr{W}$.
(d) If $G \leqslant \mathrm{Sym}(X)$ is the setwise stabiliser of $\mathscr{W}$, then $G$ is transitive on $X$.

For example, if $n$ is odd then 'majority rules', where $\mathscr{W}$ is the set of all subsets of $X$ of size at least $n/2$, is a fair game.

We claim that the existence of a fair game with $n$ players is equivalent to the existence of a transitive permutation group of degree $n$ with no derangements of 2-power order (see [77, Lemma 1]).

To see this, suppose that $\mathscr{W}$ is a fair game with $n$ players and associated group $G$. Clearly, if $n$ is odd then $G$ has no derangements of 2-power order, so let us assume that $n$ is even. A derangement in $G$ of 2-power order would map some subset $A$ of size $n/2$ to its complement, but this is ruled out by (b) and (c) above.

Conversely, suppose $G \leqslant \mathrm{Sym}(X)$ is a transitive permutation group of degree $n$ with no derangements of 2-power order. As noted above, if $n$ is odd then $G$ preserves the fair game 'majority rules', so let us assume that $n$ is even. Consider the action of $G$ on the set of subsets of $X$ of size $n/2$, and suppose that $G$ contains an element $g$ that maps such a subset to its complement. Then $g$ is a derangement. Moreover, if the cycles of $g$ have length $n_1, \ldots, n_k$, then $g^m$ is a derangement of 2-power order, where $m = [n_1', \ldots, n_k']$ is the least common multiple of the $n_i'$, and $n_i'$ is the largest odd divisor of $n_i$. This is a contradiction. Therefore, the orbits of $G$ on the set of subsets of size $n/2$ can be labelled

$$\mathscr{O}_1, \ldots, \mathscr{O}_\ell, \mathscr{O}_1^c, \ldots, \mathscr{O}_\ell^c$$

where $\mathscr{O}_i^c = \{X \setminus A \mid A \in \mathscr{O}_i\}$. Then

$$\mathscr{W} = \{A \subseteq X \mid B \subseteq A \text{ for some } B \in \mathscr{O}_i, \ 1 \leqslant i \leqslant \ell\}$$

is preserved by $G$ and so it models a fair game with $n$ players. This justifies the claim.

Isbell's Conjecture remains an open problem, although some progress has been made in special cases. For example, Bereczky [8] has shown that if $n = p^a b$, where $p$ is an odd prime, $a \geqslant 1$ and $p + 1 < b < \frac{3}{2}(p+1)$, then $G$ contains a derangement of $p$-power order. An even more general version of Isbell's Conjecture, due to Cameron [35, p. 176], was refuted by Crestani and Spiga [43] for $p \geqslant 5$, and more recently by Spiga [114] for $p = 3$.

### 1.3.3 Semiregular elements

In view of Theorem 1.3.1, it is natural to ask whether or not every nontrivial finite transitive permutation group contains a derangement of *prime* order. In fact, it is not too difficult to see that there are transitive groups with no such elements, but examples appear to be somewhat rare. Following [39], we say that a transitive permutation group is *elusive* if it does not contain a derangement of prime order. For instance, the 3-transitive action of the smallest Mathieu group $M_{11}$ on 12 points is elusive since $M_{11}$ has a unique conjugacy class of involutions, and also a unique class of elements of order 3 (and moreover, the point stabiliser $PSL_2(11)$ contains elements of order 2 and 3).

The first construction of an elusive group was given by Fein, Kantor and Schacher in [52]. Let $p$ be a Mersenne prime, let $G$ be the group

$$AGL_1(p^2) = \{x \mapsto ax + b \mid a, b \in \mathbb{F}_{p^2}, a \neq 0\}$$

of affine transformations of $\mathbb{F}_{p^2}$ and let $H$ be the subgroup of transformations with $a, b \in \mathbb{F}_p$. Then the natural action of $G$ on the set of cosets of $H$ gives a transitive permutation group of degree $p(p + 1)$ with the property that all elements of order 2 and $p$ have fixed points. Therefore, $G$ is elusive. Generalisations of this construction are given in [39], producing elusive groups of degree $p^m(p + 1)$ for all Mersenne primes $p$ and positive integers $m$. In particular, this family of examples shows that the natural extension of Isbell's Conjecture, from prime-powers to primes, is false.

A nontrivial permutation is said to be *semiregular* if all of its cycles have the same length. Clearly, a derangement of prime order is semiregular, and since any power of a semiregular element is either trivial or semiregular, the existence of a semiregular element is equivalent to the existence of a derangement of prime order.

Determining the existence of semiregular elements is a classical problem with a long history. For example, Burnside [33, p. 343] showed that if $G$ is a primitive permutation group of degree $p^a$, where $p$ is a prime and $a > 1$, and $G$ contains a cycle of length $p^a$, then $G$ is 2-transitive. This was later extended by Schur [112], who proved that any primitive permutation group of composite degree $n$ containing an $n$-cycle is 2-transitive. The complete list of such 2-transitive groups was later independently determined by Jones [80] and Li [90], following earlier work of Feit [53]. These results have found a wide range of applications in combinatorics, including coding theory [9], Cayley graphs of cyclic groups (see [91], for example) and rotary embeddings of graphs on surfaces [92]. In a different direction, the existence of semiregular elements has also been used to study tame ramification in number fields [81].

### 1.3.4 The Polycirculant Conjecture

The notion of a semiregular permutation arises naturally in graph theory. In order to describe the connection, let us recall some standard terminology. A *digraph* $\Gamma$ consists of a set $V\Gamma$ of vertices and a set $A\Gamma$ of ordered pairs of distinct elements of $V\Gamma$, called *arcs*. If $\Gamma$ has the property that $(u,v) \in A\Gamma$ if and only if $(v,u) \in A\Gamma$, then $\Gamma$ is called a *graph*. In this situation, the set of edges of $\Gamma$ is denoted by $E\Gamma = \{\{u,v\} \mid (u,v) \in A\Gamma\}$, and we say that $u$ is adjacent to $v$, denoted $u \sim v$, if $\{u,v\} \in E\Gamma$. An *automorphism* of a digraph $\Gamma$ is a permutation $g$ of $V\Gamma$ such that $(u,v) \in A\Gamma$ if and only if $(u^g,v^g) \in A\Gamma$. We denote the group of all automorphisms of $\Gamma$ by $\mathrm{Aut}(\Gamma)$. If $\mathrm{Aut}(\Gamma)$ acts transitively on $V\Gamma$ then we say that $\Gamma$ is *vertex-transitive*. Similarly, $\Gamma$ is *arc-transitive* if $\mathrm{Aut}(\Gamma)$ acts transitively on $A\Gamma$, and *edge-transitive* if $\mathrm{Aut}(\Gamma)$ acts transitively on $E\Gamma$.

In 1981, Marušič [101, Problem 2.4] asked the following question.

**Question.** *Does every finite vertex-transitive digraph admit a semiregular automorphism?*

Note that in Marušič's terminology in [101], a digraph $\Gamma$ is *galactic* if it has a semiregular automorphism. For any prime $p$, he showed that a transitive permutation group of $p$-power degree, or degree $mp$ with $m \leqslant p$, has a derangement of order $p$. The same question for graphs has subsequently been posed by both Leighton [89] and Jordan [83], in 1983 and 1988, respectively.

The existence of a semiregular automorphism is closely related to the notion of a *Cayley digraph*. Given a group $G$ and subset $S$ with $1 \notin S$, the Cayley digraph $\mathrm{Cay}(G,S)$ is the digraph with vertex set $G$ and the property that $(g,h)$ is an arc if and only if $hg^{-1} \in S$. Note that $\mathrm{Cay}(G,S)$ is connected if and only if $G = \langle S \rangle$. If $S$ is symmetric in the sense that $S = S^{-1} := \{s^{-1} \mid s \in S\}$, then $(g,h)$ is an arc if and only if $(h,g)$ is an arc, so in this situation we refer to the *Cayley graph* of $G$ with respect to $S$. The group $G$ acts on itself by right multiplication, mapping arcs to arcs, and so it induces a regular group of automorphisms of $\mathrm{Cay}(G,S)$ (in particular, $\mathrm{Cay}(G,S)$ is vertex-transitive). Sabidussi [110] showed that a digraph $\Gamma$ is a Cayley digraph if and only if the full automorphism group of $\Gamma$ contains a regular subgroup.

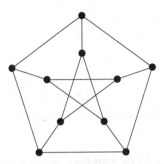

Figure 1.3.1 The Petersen graph

Clearly, every Cayley digraph admits a semiregular automorphism. However, not all vertex-transitive digraphs are Cayley digraphs. For example, it is easy to see that the familiar *Petersen graph* in Figure 1.3.1 is not a Cayley graph, but it visibly has a semiregular automorphism of order 5.

A Cayley digraph of a cyclic group is called a *circulant*. Notice that a circulant with $n$ vertices admits a semiregular automorphism of order $n$. Similarly, a digraph $\Gamma$ with $n$ vertices is called a *bicirculant* if it has a semiregular automorphism of order $n/2$; for example, the Petersen graph has this property (with $n = 10$). Bicirculants have been studied by various authors in recent years, see [100] for example, and the notion has been extended to tricirculants, etc.

The existence of a semiregular automorphism of a graph $\Gamma$ has other applications. For example, it facilitates a particularly succinct notation to describe the vertices and adjacency relation in $\Gamma$ (see [10, 54], for example). Indeed, if $g \in \text{Aut}(\Gamma)$ is semiregular, with $r$ cycles of length $m$, then we can represent the graph $\Gamma$ using only $r$ vertices, with each vertex corresponding to one of the cycles of $g$. A label $m|k$ on a vertex corresponding to the $m$-cycle of $g$ containing the vertices $v_1, v_2, \ldots, v_m$ of $\Gamma$ indicates that each $v_i$ is adjacent to $v_{i+k}$, where addition is calculated modulo $m$. Similarly, given two $m$-cycles $u_1, u_2, \ldots, u_m$ and $v_1, v_2, \ldots, v_m$, an unlabelled edge between the corresponding nodes signifies that $u_i$ is adjacent to $v_i$, while an edge labelled by a positive integer $k$ means that $u_i$ is adjacent to $v_{i+k}$. This representation of $\Gamma$ corresponds to the quotient of $\Gamma$ by the partition of $V\Gamma$ determined by the cycles of $g$. This is called the *Frucht notation* for $\Gamma$. For instance, the representation of the Petersen graph is given in Figure 1.3.2. This notation can be adjusted appropriately for digraphs.

Semiregular automorphisms of graphs have also been used to construct Hamiltonian paths and cycles. As explained in [1], this can be done by lifting such a path or cycle in the quotient graph corresponding to the semiregular

5|1 ●

5|2 ●

Figure 1.3.2  Frucht notation for the Petersen graph

automorphism. These automorphisms have also played a role in the enumeration of all vertex-transitive graphs with a small number of vertices (see [103], for example).

The existence of elusive permutation groups implies that there are transitive groups that do not contain semiregular elements. However, not every transitive permutation group is the full automorphism group of a digraph. For example, $G = M_{11}$ has a 2-transitive action on 12 points, so the only digraphs with 12 vertices that are preserved by $G$ are the complete graph and the edgeless graph on 12 vertices, both of which admit semiregular automorphisms.

In order to generalise Marušič's question, we need the notion of 2-closure. Let $G$ be a permutation group on a finite set $\Omega$. The 2-*closure* of $G$, denoted by $G^{(2)}$, is the largest subgroup of $\mathrm{Sym}(\Omega)$ that preserves the orbits of $G$ on $\Omega \times \Omega$. For instance, if $G$ is 2-transitive then

$$\{(\alpha, \alpha) \mid \alpha \in \Omega\}, \ \ \{(\alpha, \beta) \mid \alpha, \beta \in \Omega, \alpha \neq \beta\}$$

are the orbits of $G$ on $\Omega \times \Omega$, so $G^{(2)} = \mathrm{Sym}(\Omega)$. We say that $G$ is 2-*closed* if $G = G^{(2)}$. Note that the automorphism group $\mathrm{Aut}(\Gamma)$ of a finite digraph $\Gamma$ is 2-closed: any permutation that fixes the orbits of $\mathrm{Aut}(\Gamma)$ on ordered pairs of vertices also fixes $A\Gamma$ setwise, and is therefore an automorphism of $\Gamma$. However, not every 2-closed group is the full automorphism group of a digraph. For example, the regular action of the Klein 4-group $C_2 \times C_2$ on four points is 2-closed, but it is not the full automorphism group of any digraph.

In 1997, Klin [34, Problem 282 (BCC15.12)] extended Marušič's question to 2-closed groups. This is now known as the *Polycirculant Conjecture*.

**Conjecture 1.3.4** *Every nontrivial finite transitive 2-closed permutation group contains a derangement of prime order.*

One obvious way to attack this conjecture is to determine all elusive groups and show that none are 2-closed. Although elusive groups have been much studied in recent years (see [39, 51, 62] for some specific constructions), a complete classification remains out of reach. However, the following result of Giudici [61] classifies the quasiprimitive elusive groups.

**Theorem 1.3.5** *Let $G \leqslant \mathrm{Sym}(\Omega)$ be an elusive permutation group with a transitive minimal normal subgroup. Then $G = \mathrm{M}_{11} \wr K$ acting with its product action on $\Omega = \Delta^k$ for some $k \geqslant 1$, where $K$ is a transitive subgroup of $S_k$ and $|\Delta| = 12$.*

The proof of Theorem 1.3.5 relies on the list of pairs $(G,H)$ given in [95, Table 10.7], where $G$ is a simple group and $H$ is a maximal subgroup of $G$ with the property that $|G|$ and $|H|$ have the same set of prime divisors.

None of the groups arising in Theorem 1.3.5 are 2-closed and so every minimal normal subgroup of a counterexample to the Polycirculant Conjecture must be intransitive.

Further progress in this direction has been made by Giudici and Xu in [65], where all the elusive biquasiprimitive permutation groups are determined. (A transitive permutation group is *biquasiprimitive* if it contains a nontrivial intransitive normal subgroup, and all nontrivial normal subgroups have at most two orbits.) As a corollary, it follows that every locally quasiprimitive graph has a semiregular automorphism. (A graph $\Gamma$ with automorphism group $G$ is *locally quasiprimitive* if for all vertices $v \in V\Gamma$, the stabiliser $G_v$ acts quasiprimitively on the set of vertices adjacent to $v$.) This family of graphs includes all arc-transitive graphs of prime valency, and all 2-arc transitive graphs.

Another approach to the Polycirculant Conjecture, and also the original question of Marušič, is to show that digraphs with additional properties must admit a semiregular automorphism. For instance, Marušič and Scapellato [102] showed that every vertex-transitive graph of valency 3, or with $2p^2$ vertices ($p$ a prime), has a semiregular automorphism. Similarly, all vertex-transitive graphs of valency 4 [50], or with a square-free number of vertices [49], also have semiregular automorphisms. In fact, [64] reveals that any vertex-transitive group of automorphisms of a connected graph of valency at most 4 contains a semiregular element, and any vertex-transitive digraph of out-valency at most 3 admits a semiregular automorphism. By the main theorem of [88], all distance-transitive graphs have a semiregular automorphism. This remains an active area of current research.

### 1.3.5 Derangements of prime order

Let $G$ be a transitive permutation group on a finite set $\Omega$. Notice that $G$ contains a derangement of prime order $r$ only if $r$ divides $|\Omega|$. One of the main aims of this book is to initiate a quantitative study of derangements of prime order, motivated by the following basic question:

**Question.** *Let r be a prime divisor of* $|\Omega|$*. Does G contain a derangement of order r?*

This question leads us naturally to the following *local* notion of elusivity, which was introduced in [23].

**Definition 1.3.6** Let $G$ be a transitive permutation group on a finite set $\Omega$ and let $r$ be a prime divisor of $|\Omega|$. Then $G$ is *r-elusive* if it does not contain a derangement of order $r$.

In this terminology, $G$ is elusive if and only if it is $r$-elusive for every prime divisor $r$ of $|\Omega|$. Similarly, we say that $G$ is *strongly r-elusive* if $r$ divides $|\Omega|$ and $G$ does not contain a derangement of $r$-power order.

Recall that $G$ is primitive if a point stabiliser $G_\alpha$ is a maximal subgroup of $G$. The existence of a core-free maximal subgroup imposes restrictions on the abstract structure of $G$ (for instance, it implies that $G$ has at most two minimal normal subgroups). This is formalised in the statement of the O'Nan–Scott Theorem, which describes the structure of a finite primitive permutation group. This important theorem divides primitive groups into a certain number of classes according to the structure of the *socle* (the subgroup generated by the minimal normal subgroups) and the action of a point stabiliser. The precise number of classes depends on how fine a subdivision is required; for example, see [35, Section 4.5] for a subdivision into four classes, and [94] and [109, Section 6] for more refined subdivisions. Roughly speaking, the theorem states that a primitive group $G \leqslant \mathrm{Sym}(\Omega)$ either preserves some natural structure on $\Omega$, for example a product structure, or the structure of an affine space, or there is a nonabelian simple group $T$ such that

$$T \leqslant G \leqslant \mathrm{Aut}(T)$$

In the latter case, $G$ is an *almost simple* group.

In many situations, the O'Nan–Scott Theorem can be used to reduce a general problem concerning primitive groups to the almost simple case. At this point, the Classification of Finite Simple Groups can be invoked to describe the possibilities for $T$ (and thus $G$), and the vast literature on finite simple groups (in particular, information on their subgroup structure, conjugacy classes and representations) can be brought to bear on the problem.

For example, in order to determine all the $r$-elusive primitive permutation groups, the O'Nan–Scott Theorem was used in [23] to establish the following reduction theorem (see [23, Theorem 2.1]).

**Theorem 1.3.7** *Let $G \leqslant \mathrm{Sym}(\Omega)$ be a finite primitive permutation group with socle $N$. Let $r$ be a prime divisor of $|\Omega|$. Then one of the following holds:*

(i) *$G$ is almost simple;*
(ii) *$N$ contains a derangement of order $r$;*
(iii) *$G \leqslant H \wr S_k$ acting with its product action on $\Omega = \Delta^k$ for some $k \geqslant 2$, where $H \leqslant \mathrm{Sym}(\Delta)$ is primitive, almost simple and the socle of $H$ is $r$-elusive.*

In view of Theorem 1.3.7, we may focus our attention on the almost simple primitive groups. Let $G$ be such a group, with socle $T$. By the Classification of Finite Simple Groups, there are four cases to consider:

(a) $T$ is an alternating group $A_n$ of degree $n \geqslant 5$;
(b) $T$ is one of 26 sporadic simple groups;
(c) $T$ is a simple classical group;
(d) $T$ is a simple group of exceptional Lie type.

By Theorem 1.3.5, the 3-transitive action of $M_{11}$ on 12 points is the only almost simple primitive elusive group.

All the $r$-elusive groups in cases (a) and (b) are determined in [23]. For example, in case (a) the main theorem is the following (see [23, Section 3]).

**Theorem 1.3.8** *Let $G = A_n$ or $S_n$ be an almost simple primitive permutation group on a set $\Omega$ with point stabiliser $H$. Let $r$ be a prime divisor of $|\Omega|$.*

(i) *If $H$ acts primitively on $\{1, \ldots, n\}$, then $G$ is $r$-elusive if and only if $r = 2$ and $(G, H) = (A_5, D_{10})$ or $(A_6, \mathrm{PSL}_2(5))$.*
(ii) *Let $\Omega$ be the set of partitions of $\{1, \ldots, n\}$ into $b$ parts of size $a$ with $a, b \geqslant 2$. Write $a \equiv \ell \pmod{r}$ and $b \equiv k \pmod{r}$ with $0 \leqslant \ell, k < r$. Then $G$ is $r$-elusive if and only if $r \leqslant a$ and one of the following holds:*
  (a) *$\ell = 0$;*
  (b) *$k = 0$ and $\ell = 1$;*
  (c) *$0 < k\ell < r$ and either $b < r$ or $(k + r)\ell \leqslant ka + r$.*
(iii) *Let $\Omega$ be the set of $k$-element subsets of $\{1, \ldots, n\}$ with $1 \leqslant k < n/2$. Write $n \equiv i \pmod{r}$ and $k \equiv j \pmod{r}$ with $0 \leqslant i, j < r$.*
  (a) *If $r$ is odd, then $G$ is $r$-elusive if and only if $r \leqslant k$ and $i \geqslant j$.*
  (b) *$G$ is 2-elusive if and only if $k$ is even, or $n$ is odd, or $G = A_n$ and $n/2$ is odd.*

Table 1.4.1 *The finite simple classical groups*

| Type | Notation | Conditions |
|------|----------|------------|
| Linear | $PSL_n(q)$ | $n \geqslant 2$, $(n,q) \neq (2,2), (2,3)$ |
| Unitary | $PSU_n(q)$ | $n \geqslant 3$, $(n,q) \neq (3,2)$ |
| Symplectic | $PSp_n(q)'$ | $n \geqslant 4$ even |
| Orthogonal | $\left\{\begin{array}{l} \Omega_n(q) \\ P\Omega_n^{\pm}(q) \end{array}\right.$ | $nq$ odd, $n \geqslant 7$ <br> $n \geqslant 8$ even |

Further results are established in [23]. For example, the conjugacy classes of derangements of prime order are determined for almost all primitive actions of almost simple sporadic groups (including the *Baby Monster* sporadic group for example, and almost all primitive actions of the *Monster*). In addition, the strongly $r$-elusive primitive actions of the almost simple groups with socle an alternating or sporadic group are determined in [23]. We also show that if $r$ is the largest prime divisor of $|\Omega|$, then such a group $G$ contains a derangement of order $r$, unless $G = M_{11}$, $|\Omega| = 12$ and $r = 3$ (see [23, Corollary 1.2]).

In view of Theorem 1.3.7, and the work in [23], the challenge now is to extend the study of $r$-elusivity to almost simple groups of Lie type. Here we will focus on classical groups; derangements of prime order in almost simple groups of exceptional Lie type will be investigated in future work.

## 1.4 Derangements in classical groups

Let $G \leqslant \mathrm{Sym}(\Omega)$ be a primitive almost simple classical group over $\mathbb{F}_q$ with socle $T$ and natural (projective) module $V$ of dimension $n$. Let $H = G_\alpha$ be a point stabiliser. The possibilities for $T$ (up to isomorphism) are listed in Table 1.4.1. These groups will be formally introduced in Chapter 2, where the notation and given conditions will be explained. In Chapter 2 we will also describe the associated geometries and automorphisms of the classical groups.

We are interested in the following problem:

**Problem 1.4.1** *For each prime divisor $r$ of $|\Omega|$, determine whether $T$ is $r$-elusive, that is, determine whether or not $T$ contains a derangement of order $r$.*

Let $x$ be an element of $T$. Recall that

(i) $H$ is a maximal subgroup of $G$ such that $G = HT$, and
(ii) $x$ is a derangement if and only if $x^G \cap H$ is empty.

Therefore, in order to attack Problem 1.4.1 we require detailed information on the subgroup structure of $G$ (in order to determine the possibilities for $H$). We also need a description of the $G$-classes of elements of prime order in $T$, and we need to study the fusion of the $H$-classes of such elements in $G$ (to determine whether $x^G \cap H$ is non-empty for all $x \in T$ of a given prime order).

In view of (ii) above, our aim in Chapter 3 is to bring together a range of results on conjugacy classes of elements of prime order in the finite classical groups. Most of these results can be found in the literature, in one form or another, but it is desirable to have a single reference for this important information. Indeed, a detailed description of conjugacy classes is essential for our application to derangements, and more generally we expect that the content of Chapter 3 will be useful in many other problems involving finite classical groups.

Let $V$ be the natural $T$-module, let $x \in T$ be an element of prime order $r$, and write $q = p^f$ where $p$ is a prime. (Here $V$ is an $n$-dimensional vector space over $\mathbb{F}_{q^u}$, where $u = 2$ if $T$ is a unitary group, otherwise $u = 1$.) Let $\mathbb{F}$ be the algebraic closure of $\mathbb{F}_q$ and set $\overline{V} = V \otimes \mathbb{F}$. Since $x \in \mathrm{PGL}(V)$, we may define $\hat{x} \in \mathrm{GL}(\overline{V})$ to be a preimage of $x$.

In order to describe the conjugacy class of $x$ in $G$, we distinguish the cases $r = p$ and $r \neq p$. In the former case, $x$ is a *unipotent* element; 1 is the only eigenvalue of $\hat{x}$ on $\overline{V}$, and the $G$-class of $x$ is essentially determined by the Jordan block structure of $\hat{x}$ on $\overline{V}$. If $r \neq p$ then $x$ is *semisimple*; here $\hat{x} \in \mathrm{GL}(\overline{V})$ is diagonalisable and the $G$-class of $x$ can typically be described in terms of the multiset of eigenvalues of $\hat{x}$ on $\overline{V}$. In both cases we will discuss class representatives, and we will provide information on the centraliser $C_G(x)$ and the type of subspace decompositions of $V$ fixed by $x$. Some of these results are conveniently summarised in the tables in Appendix B. We will also discuss the conjugacy classes of outer automorphisms of $T$ of prime order.

The case $r = 2$ requires special attention. Indeed, our treatment of semisimple involutions is one of the main features of Chapter 3. This detailed analysis is needed for the application to derangements, and more generally it is designed to complement the extensive information in [67, Table 4.5.1] by Gorenstein, Lyons and Solomon.

The main theorem on the subgroup structure of finite classical groups is due to Aschbacher. In [3], Aschbacher introduces eight *geometric* families of subgroups of $G$, denoted by $\mathscr{C}_i$ ($1 \leqslant i \leqslant 8$), which are defined in terms of the underlying geometry of the natural $T$-module $V$. For example, these collections include the stabilisers of suitable subspaces of $V$, and the stabilisers of appropriate direct sum and tensor product decompositions of $V$. Essentially, Aschbacher's main theorem states that if $H$ is a maximal subgroup of $G$ with

Table 1.4.2 *Aschbacher's subgroup collections*

| Collection | Description |
|---|---|
| $\mathscr{C}_1$ | Stabilisers of subspaces, or pairs of subspaces, of $V$ |
| $\mathscr{C}_2$ | Stabilisers of decompositions $V = \bigoplus_{i=1}^{t} V_i$, where $\dim V_i = a$ |
| $\mathscr{C}_3$ | Stabilisers of prime degree extension fields of $\mathbb{F}_q$ |
| $\mathscr{C}_4$ | Stabilisers of decompositions $V = V_1 \otimes V_2$ |
| $\mathscr{C}_5$ | Stabilisers of prime index subfields of $\mathbb{F}_q$ |
| $\mathscr{C}_6$ | Normalisers of symplectic-type $r$-groups, $r \neq p$ |
| $\mathscr{C}_7$ | Stabilisers of decompositions $V = \bigotimes_{i=1}^{t} V_i$, where $\dim V_i = a$ |
| $\mathscr{C}_8$ | Stabilisers of nondegenerate forms on $V$ |
| $\mathscr{S}$ | Almost simple absolutely irreducible subgroups |
| $\mathscr{N}$ | Novelty subgroups ($T = \mathrm{P}\Omega_8^{+}(q)$ or $\mathrm{Sp}_4(q)'$ ($p = 2$), only) |

$HT = G$, then either $H$ is contained in one of the $\mathscr{C}_i$ collections, or $H$ is almost simple and the socle of $H$ acts absolutely irreducibly on $V$. Following [86], we use $\mathscr{S}$ to denote the latter collection of *non-geometric* subgroups. It turns out that a small additional subgroup collection (denoted by $\mathscr{N}$) arises when $T = \mathrm{Sp}_4(q)'$ (with $p = 2$) or $\mathrm{P}\Omega_8^{+}(q)$, due to the existence of certain exceptional automorphisms. A brief description of these subgroup collections is presented in Table 1.4.2, and we refer the reader to Section 2.6 for further details.

Our study of derangements in finite classical groups is organised in terms of the subgroup collections in Table 1.4.2. The *subspace actions* corresponding to subgroups in $\mathscr{C}_1$ require special attention, and they are handled first in Chapter 4. Here we need to determine whether a given prime order element in $T$ fixes an appropriate subspace (or pair of subspaces) of $V$. In order to answer this question, we need the detailed information on conjugacy class representatives recorded in Chapter 3 (in particular, we need to understand the subspace decompositions of $V$ fixed by such elements). The remaining geometric subgroup collections $\mathscr{C}_i$, with $2 \leqslant i \leqslant 8$, are handled in Chapter 5, together with the small collection of *novelty* subgroups denoted by $\mathscr{N}$. In Chapter 6, we present detailed results on the $r$-elusivity of the low-dimensional almost simple classical groups.

Rather different techniques are required to deal with the non-geometric actions corresponding to the subgroups in the collection $\mathscr{S}$. If $H$ is such a subgroup of $G$, with (simple) socle $S$, then there exists an absolutely irreducible representation $\rho : \hat{S} \to \mathrm{GL}(V)$, where $\hat{S}$ is a covering group of $S$. However, it is not easy to use this representation-theoretic description of the embedding of $H$ in $G$ to study the fusion of $H$-classes in $G$ (of course, even the dimensions of the irreducible $\mathbb{F}_q \hat{S}$-modules are not known, in general). Therefore, a somewhat different approach is required, and we will study the $r$-elusivity of $\mathscr{S}$-actions of finite classical groups in a separate paper.

## 1.5 Main results

We are now in a position to discuss some of our main results on derangements of prime order in almost simple classical groups. More detailed results will be presented in Chapters 4, 5 and 6; some of these statements are necessarily somewhat involved, with reference to a number of tables, so in this section we will state simplified versions. We also give precise references for the more detailed statements that can be found later in the text.

As before, let $G \leqslant \mathrm{Sym}(\Omega)$ be a primitive almost simple classical group over $\mathbb{F}_q$ with socle $T$ and natural (projective) module $V$ of dimension $n$. Let $H = G_\alpha$ be a point stabiliser, and define the subgroup collections as in Table 1.4.2. Write $q = p^f$, where $p$ is a prime, and assume that $H \notin \mathscr{S}$.

**Theorem 1.5.1** *Let $r$ be a prime divisor of $|\Omega|$.*

  (i) *If $H \in \mathscr{C}_1 \cup \mathscr{C}_2$, $r = p > 2$ and $T$ is $r$-elusive, then $(G,H)$ belongs to a known list of cases.*
 (ii) *In all other cases, $T$ is $r$-elusive if and only if $(G,H,r)$ belongs to a known list of cases.*

**Remark 1.5.2** Some comments on the statement of Theorem 1.5.1.

 (a) For $H \in \mathscr{C}_1 \cup \mathscr{C}_2$, we refer the reader to the theorems referenced in Table 1.5.1, which provide detailed results. For example, if $H \in \mathscr{C}_1$ and $r \neq p$ is odd, then $T$ is $r$-elusive if and only if $(G,H,r)$ is one of the cases recorded in Theorem 4.1.6. As indicated in Theorems 4.1.4 and 5.2.1, in some (but not all) cases we are able to present necessary and sufficient conditions for $r$-elusivity when $r = p > 2$, which typically depend on number-theoretic properties of partitions of $n$.
 (b) If $H \in \mathscr{C}_i$ (with $3 \leqslant i \leqslant 8$), then the precise conditions for $r$-elusivity are stated in Theorem 5.$i$.1 and the specific cases that arise are listed in Table 5.$i$.2 (for $i \neq 7$). We find that the collection of primes $r$ for which $T$ is $r$-elusive is rather restricted:
   • If $H \in \mathscr{C}_4 \cup \mathscr{C}_7$ then $T$ is $r$-elusive only if $r = 2$.
   • If $H \in \mathscr{C}_3$ and $T$ is $r$-elusive then either $r = 2$, or $T = \mathrm{PSL}_n^\varepsilon(q)$, $H$ is of type $\mathrm{GL}_{n/k}^\varepsilon(q^k)$ and $r = k$.
   • If $H \in \mathscr{C}_5$ is a subfield subgroup over $\mathbb{F}_{q_0}$, where $q = q_0^k$, then $T$ is $r$-elusive only if $r \in \{2, 3, 5, k, p\}$.
   • If $H \in \mathscr{C}_6$ then $T$ is $r$-elusive only if $r \leqslant 3$, or if $r$ is a Mersenne or Fermat prime.
   • If $H \in \mathscr{C}_8$ then $T$ is $r$-elusive only if $r \in \{2, 3, 5, p\}$.

Table 1.5.1 *References for $\mathscr{C}_i$-actions, $i = 1, 2$*

|                   | $r = p > 2$ | $r \neq p, r > 2$ | $r = 2$ |
|-------------------|-------------|-------------------|---------|
| $\mathscr{C}_1$   | 4.1.4       | 4.1.6             | 4.1.7   |
| $\mathscr{C}_2$   | 5.2.1       | 5.2.3             | 5.2.5   |

(c) By definition, if $T$ is not $r$-elusive then $T$ contains a derangement of order $r$. In this situation, specific derangements are usually identified in the proofs of the main theorems.

In the next theorem, we highlight the special case $r = 2$.

**Theorem 1.5.3** *$T$ is 2-elusive if and only if $|\Omega|$ is even and $(G, H)$ is one of the cases in Table 4.1.3 (for $H \in \mathscr{C}_1$) or Table 5.1.2 (in all other cases).*

We say that $T$ is $2'$-elusive if $|\Omega|$ is divisible by an odd prime, but $T$ does not contain a derangement of odd prime order.

**Theorem 1.5.4** *Let $G$ be a primitive almost simple classical group with socle $T$. Then $T$ is not $2'$-elusive.*

This is a special case of the main theorem of [22], which describes the structure of quasiprimitive and biquasiprimitive groups that are $2'$-elusive. In particular, if $G$ is a primitive almost simple group with socle $T$ and point stabiliser $H$, then $T$ is $2'$-elusive if and only if $(G, H)$ is one of the following (in terms of the Atlas [41] notation):

$$(M_{11}, \mathrm{PSL}_2(11)), \; (^2F_4(2)', \mathrm{PSL}_2(25)), \; (^2F_4(2), \mathrm{PSL}_2(25).2_3)$$

Our final theorem concerns the $r$-elusivity of the low-dimensional classical groups with $n \leqslant 5$.

**Theorem 1.5.5** *Let $G$ be a primitive almost simple classical group with socle $T$ and point stabiliser $H$, where*

$$T \in \{\mathrm{PSL}_2(q), \mathrm{PSL}_3^\varepsilon(q), \mathrm{PSL}_4^\varepsilon(q), \mathrm{PSp}_4(q)', \mathrm{PSL}_5^\varepsilon(q)\} \qquad (1.5.1)$$

*Let $r$ be a prime. Then $T$ is $r$-elusive if and only if $(G, H, r)$ is one of the cases recorded in Tables 6.4.1–6.4.8.*

The proof of Theorem 1.5.5 is given in Chapter 6. For $H \notin \mathscr{S}$, this is a corollary of Theorem 1.5.1, noting that it is straightforward to determine necessary and sufficient conditions in part (i) when $n$ is small. A complete list of the subgroups in $\mathscr{S}$ is given in [13, Chapter 8] (also see Table 6.3.1), and we study each possibility in turn, working with the corresponding irreducible representation (and its character) to investigate the fusion of $H$-classes in $G$.

**Corollary 1.5.6** *Let $G$ be a primitive almost simple classical group over $\mathbb{F}_q$ with point stabiliser $H$ and socle $T$ as in (1.5.1). Let $r > 5$ be a prime. Then $T$ is $r$-elusive only if one of the following holds:*

(i) *$H \in \mathscr{C}_5$ is a subfield subgroup over $\mathbb{F}_{q_0}$, where $q = q_0^k$ and $r \in \{k, p\}$.*

(ii) *$T = \mathrm{PSL}_n(q)$, $n \in \{3, 5\}$ and $H$ is a $\mathscr{C}_8$-subgroup of type $\mathrm{GU}_n(q_0)$, where $q = q_0^2$ and $r = p$.*

(iii) *$H \in \mathscr{S}$ has socle $S$ and $(T, S, r)$ is one of the following:*

  (a) *$T = \mathrm{PSL}_5^\varepsilon(q)$, $S = \mathrm{PSL}_2(11)$ and $r = 11$;*

  (b) *$T = \mathrm{PSL}_5(3)$, $S = \mathrm{M}_{11}$ and $r = 11$;*

  (c) *$T = \mathrm{PSL}_4^\varepsilon(q)$, $S = A_7$ or $\mathrm{PSL}_2(7)$, and $r = 7$;*

  (d) *$T = \mathrm{PSL}_3^\varepsilon(q)$, $S = \mathrm{PSL}_2(7)$ and $r = 7$.*

# 2

# Finite classical groups

In this chapter we provide a brief introduction to the finite classical groups. Our treatment is tailored towards the application to derangements explained in Chapter 1, and we focus on forms, standard bases and automorphisms. We also describe some specific classical group embeddings (or *constructions*, as we call them), which will be useful later. Finally, in Section 2.6 we briefly discuss Aschbacher's theorem [3] on the subgroup structure of finite classical groups, which plays an essential role in our analysis of derangements in primitive almost simple classical groups.

There are a number of excellent general references on finite classical groups, which provide a more comprehensive treatment. In particular, we refer the reader to Dieudonné [47] and Taylor [118]. Other good references include Chapter 2 in [86], Chapter 1 in [13], Chapter 3 in [122], Chapter 7 in [4], and Cameron's lecture notes on classical groups [36]. Sections 27 and 28 in the recent book by Malle and Testerman [99] provide an accessible account of Aschbacher's subgroup structure theorem (also see Chapter 2 in [13]).

## 2.1 Linear groups

Let $V$ be an $n$-dimensional vector space over the finite field $\mathbb{F}_q$ with $q = p^f$ for a prime number $p$. The *general linear group* $\mathrm{GL}(V)$ is the group of all invertible linear transformations of $V$. The centre $Z$ of $\mathrm{GL}(V)$ is the subgroup of all scalar transformations $v \mapsto \lambda v$ with $\lambda \in \mathbb{F}_q^\times$. The group $\mathrm{GL}(V)$ acts naturally on the set of all 1-dimensional subspaces of $V$, and the kernel of this action is $Z$. We define $\mathrm{PGL}(V) = \mathrm{GL}(V)/Z$, the *projective general linear group*, which is isomorphic to the permutation group induced by $\mathrm{GL}(V)$ in this action.

By fixing a basis for $V$ we can represent each element of $\mathrm{GL}(V)$ by an invertible $n \times n$ matrix with entries in $\mathbb{F}_q$. We denote the group of all such matrices by $GL_n(q)$. In this representation of $GL(V)$, the centre $Z$ consists of all scalar matrices $\lambda I_n$, where $I_n$ is the $n \times n$ identity matrix and $\lambda \in \mathbb{F}_q^\times$ is a nonzero scalar. We denote $GL_n(q)/Z$ by $PGL_n(q)$.

The subgroup of $GL_n(q)$ comprising the matrices of determinant 1 is the *special linear group* $SL_n(q)$; this is the kernel of the determinant homomorphism $\det: GL_n(q) \to \mathbb{F}_q^\times$. We also define the *projective special linear group*

$$PSL_n(q) = SL_n(q)Z/Z \cong SL_n(q)/(SL_n(q) \cap Z)$$

We will sometimes denote $PSL_n(q)$ by $PSL_n^+(q)$, and similarly $GL_n(q)$ by $GL_n^+(q)$, and $PGL_n(q)$ by $PGL_n^+(q)$. Membership in $PSL_n(q)$ is given by the following lemma.

**Lemma 2.1.1** *Let* $g \in PGL_n(q)$ *such that* $g = AZ$ *with* $A \in GL_n(q)$. *Then* $g \in PSL_n(q)$ *if and only if* $\det(A) = \lambda^n$ *for some* $\lambda \in \mathbb{F}_q^\times$.

*Proof*  Clearly, $g \in PSL_n(q)$ if and only if the coset $AZ$ contains a matrix with determinant 1, that is, if and only if

$$\det(\lambda A) = \lambda^n \det(A) = 1$$

for some $\lambda \in \mathbb{F}_q^\times$. The result follows.                              $\square$

The next result is well known (see [122, Section 3.3], for example).

**Theorem 2.1.2** *Suppose* $n \geqslant 2$ *and* $(n,q) \neq (2,2),(2,3)$. *Set* $d = (n, q-1)$. *Then* $PSL_n(q)$ *is a simple group of order*

$$\frac{1}{d} q^{\frac{1}{2}n(n-1)} \prod_{i=2}^{n} (q^i - 1)$$

*The exceptions* $PSL_2(2) \cong S_3$ *and* $PSL_2(3) \cong A_4$ *are soluble.*

Let $T = PSL_n(q)$. Elements of $PGL_n(q) \setminus T$ act on $T$ by conjugation; these are the nontrivial *diagonal automorphisms* of $T$. We define $\mathrm{Inndiag}(T)$ to be the group of *inner-diagonal automorphisms* of $T$, that is, the subgroup of $\mathrm{Aut}(T)$ generated by the inner and diagonal automorphisms. Evidently,

$$\mathrm{Inndiag}(T) = PGL_n(q) = \langle T, \delta \rangle$$

where $\delta = [\mu, I_{n-1}]Z \in PGL_n(q)$ and $\mu$ is a primitive element of $\mathbb{F}_q$ (here the notation $[\mu, I_{n-1}]$ represents a diagonal matrix with diagonal entries $\mu$ and 1, the latter with multiplicity $n-1$).

A map $g : V \to V$ is a *semilinear transformation* if there exists a field automorphism $\sigma$ of $\mathbb{F}_q$ such that for all $u, v \in V$ and $\lambda \in \mathbb{F}_q$

$$(u + v)^g = u^g + v^g$$
$$(\lambda v)^g = \lambda^\sigma v^g$$

Every linear map is semilinear. The *general semilinear group* $\Gamma L(V)$ is the group of all invertible semilinear transformations of $V$.

Let $\{v_1, v_2, \ldots, v_n\}$ be a basis for $V$, and recall that $\mathrm{Aut}(\mathbb{F}_q) = \langle \sigma_p \rangle \cong C_f$, where $\sigma_p : \mathbb{F}_q \to \mathbb{F}_q$ is the Frobenius map $\lambda \mapsto \lambda^p$. Using $\sigma_p$, we define the map $\phi : V \to V$ by

$$\phi : \sum_i \lambda_i v_i \mapsto \sum_i \lambda_i^p v_i \qquad (2.1.1)$$

Then $\phi$ is an invertible semilinear transformation of $V$, and for $(a_{ij}) \in \mathrm{GL}_n(q)$ we have $(a_{ij})^\phi = (a_{ij}^p)$. Therefore $\phi$ normalises $\mathrm{GL}_n(q)$ and we have $\Gamma L(V) = \langle \mathrm{GL}(V), \phi \rangle$, or equivalently, $\Gamma L_n(q) = \langle \mathrm{GL}_n(q), \phi \rangle$. Since $\phi$ normalises $Z$, it induces an automorphism of $\mathrm{PGL}(V)$ and we define the *projective semilinear group* $\mathrm{P\Gamma L}(V) = \langle \mathrm{PGL}(V), \phi \rangle$, or equivalently, $\mathrm{P\Gamma L}_n(q) = \langle \mathrm{PGL}_n(q), \phi \rangle$.

Let

$$\iota : \quad \begin{array}{ccc} \mathrm{GL}_n(q) & \to & \mathrm{GL}_n(q) \\ A & \mapsto & A^{-\mathsf{T}} \end{array}$$

be the *inverse-transpose* automorphism of $\mathrm{GL}_n(q)$, where $A^{-\mathsf{T}}$ is the matrix $(A^{-1})^{\mathsf{T}}$. Note that $\iota$ induces a permutation of the set of subspaces of $V$, mapping an $m$-dimensional subspace $U$ to the $(n - m)$-dimensional subspace

$$\{v \in V \mid uv^{\mathsf{T}} = 0 \text{ for all } u \in U\}$$

If $g = AZ \in \mathrm{PGL}_n(q)$ then $g^\iota = A^\iota Z$, so $\iota$ induces an automorphism of the projective group $\mathrm{PGL}_n(q)$. Note that if $n = 2$ then this automorphism coincides with the inner automorphism given by conjugation by

$$\begin{pmatrix} 0 & 1 \\ -1 & 0 \end{pmatrix}$$

The automorphism groups of the finite simple classical groups are the subject of Dieudonné's 1951 memoir [46] (see [40, Chapter 12] for a textbook reference). As a special case, we see that $\mathrm{Aut}(\mathrm{PSL}_n(q)) = \langle \mathrm{P\Gamma L}_n(q), \iota \rangle$ and

$$\mathrm{Out}(\mathrm{PSL}_n(q)) = C_{(n,q-1)} {:} (C_f \times C_a)$$

where $a = 2$ if $n \geqslant 3$, otherwise $a = 1$.

Following [67, Definition 2.5.13], an automorphism of $T$ is a *field automorphism* if it is $\mathrm{Aut}(T)$-conjugate to $\phi^i$ for some $1 \leqslant i < f$. More precisely, given the structure of Aut($T$), these are the $\mathrm{PGL}_n(q)$-conjugates of $\phi^i$, and all such elements lie in $\mathrm{P\Gamma L}_n(q)$. Notice that the conjugates of the map $\phi$ are obtained by varying the basis of $V$ used to define $\phi$ in (2.1.1).

The *graph automorphisms* of $T$ are the Aut($T$)-conjugates of elements of the form $x\iota$ with $x \in \mathrm{PGL}_n(q)$. In fact, in view of the structure of Aut($T$), it follows that every graph automorphism is of the form $x\iota$ with $x \in \mathrm{PGL}_n(q)$. Similarly, the *graph-field automorphisms* of $T$ are the $\mathrm{PGL}_n(q)$-conjugates of elements of the form $\phi^i\iota$ for some $1 \leqslant i < f$ (so every graph-field automorphism is of the form $x\phi^i\iota$ with $x \in \mathrm{PGL}_n(q)$). Note that the various conjugates of $\phi^i\iota$ arise by varying the basis used to define $\phi$ and $\iota$.

From the structure of the automorphism group, it is clear that every automorphism of $\mathrm{PSL}_n(q)$ is the product of an inner, a diagonal, a field and a graph automorphism. As we will see later in this chapter, the definition of these automorphisms can be appropriately extended to any finite classical group, and more generally to any simple group of Lie type. By a theorem of Steinberg [115, Theorem 30], every automorphism of a simple group of Lie type can be expressed as a product of such automorphisms.

**Theorem 2.1.3** *Let $T$ be a simple group of Lie type. Then every automorphism of $T$ is the product of an inner, a diagonal, a field and a graph automorphism of $T$.*

We end this section by recording a basic construction.

**Construction 2.1.4** *The embedding $\mathrm{GL}_{n/r}(q^r) \leqslant \mathrm{GL}_n(q)$, $r \geqslant 1$.*

Let $r$ be a positive integer and let $V$ be an $\frac{n}{r}$-dimensional vector space over $\mathbb{F}_{q^r}$. Since $\mathbb{F}_q$ is a subfield of $\mathbb{F}_{q^r}$, we may view $V$ as an $n$-dimensional vector space over $\mathbb{F}_q$. Moreover, any $\mathbb{F}_{q^r}$-linear transformation of $V$ is also $\mathbb{F}_q$-linear, and this yields an embedding of $\mathrm{GL}_{n/r}(q^r)$ in $\mathrm{GL}_n(q)$. In addition, note that if $\phi$ is the standard field automorphism of $\mathrm{GL}_{n/r}(q^r)$ of order $r$, then $\phi$ is $\mathbb{F}_q$-linear and thus $\mathrm{GL}_{n/r}(q^r).\langle\phi\rangle$ embeds in $\mathrm{GL}_n(q)$.                  □

## 2.2 Forms

In this section we provide a brief introduction to the theory of sesquilinear and quadratic forms, which is needed in order to define the remaining classical groups. For a more thorough account, we refer the reader to [4, Chapter 7], [13, Chapter 2], [36], [86, Chapter 2], [118] and [122, Section 3.4].

### 2.2.1 Sesquilinear forms

Let $V$ be an $n$-dimensional vector space over $\mathbb{F}_q$ and let $\sigma$ be an automorphism of $\mathbb{F}_q$. A $\sigma$-*sesquilinear form* on $V$ is a map $B : V \times V \to \mathbb{F}_q$ such that for all $u, v, w \in V$ and $\lambda, \mu \in \mathbb{F}_q$

$$B(u+v,w) = B(u,w) + B(v,w)$$
$$B(u,v+w) = B(u,v) + B(u,w)$$
$$B(\lambda u, \mu v) = \lambda \mu^\sigma B(u,v)$$

so $B$ is linear in the first coordinate and semilinear in the second. In the special case $\sigma = 1$, the form $B$ is *bilinear*. In addition, $B$ is *reflexive* if $B(u,v) = 0$ implies that $B(v,u) = 0$. The *radical* of a reflexive form $B$ is the subspace

$$\mathrm{rad}(B) = \{ u \in V \mid B(v,u) = 0 \text{ for all } v \in V \}$$

and we say that $B$ is *nondegenerate* if $\mathrm{rad}(B) = \{0\}$.

Given a basis $\beta = \{v_1, \ldots, v_n\}$ of $V$, the *Gram matrix* $J_\beta$ of a $\sigma$-sesquilinear form $B$ is the $n \times n$ matrix defined by $(J_\beta)_{ij} = B(v_i, v_j)$. Moreover, for all $u, v \in V$ we have $B(u,v) = u J_\beta v^{\sigma\mathsf{T}}$, where $\sigma$ induces the map $\sum_i \lambda_i v_i \mapsto \sum_i \lambda_i^\sigma v_i$ on $V$. Note that $B$ is nondegenerate if and only if $J_\beta$ is invertible. We say that $\beta$ is an *orthonormal basis* if $B(v_i, v_j) = \delta_{i,j}$ for all $i, j$, that is, if $J_\beta$ is the identity matrix $I_n$.

### 2.2.2 Symmetric and alternating forms

A bilinear form $B$ on $V$ is *symmetric* if $B(u,v) = B(v,u)$ for all $u, v \in V$, and *skew-symmetric* if $B(u,v) = -B(v,u)$ for all $u, v \in V$. In addition, $B$ is *alternating* if $B(v,v) = 0$ for all $v \in V$. Note that every alternating form is skew-symmetric, but the converse holds if and only if $q$ is odd. A familiar example of a nondegenerate symmetric form is given by the scalar product

$$B(u,v) = a_1 b_1 + a_2 b_2 + \cdots + a_n b_n$$

where $u = (a_1, \ldots, a_n)$ and $v = (b_1, \ldots, b_n)$, or equivalently, $B(u,v) = u J v^\mathsf{T}$ with $J = I_n$. Similarly, if $n$ is even,

$$B(u,v) = a_1 b_2 - b_1 a_2 + a_3 b_4 - b_3 a_4 + \cdots + a_{n-1} b_n - b_{n-1} a_n$$

is a nondegenerate alternating form. Here $B(u,v) = u J v^\mathsf{T}$ and $J$ is the block-diagonal matrix

$$J = \begin{pmatrix} 0 & 1 & & & & \\ -1 & 0 & & & & \\ & & \ddots & & & \\ & & & & 0 & 1 \\ & & & & -1 & 0 \end{pmatrix}$$

The next result is [86, Proposition 2.4.1]. The given basis of $V$ is called a *standard*, or *symplectic*, basis. In this situation, we say that $V$ is a *symplectic space*.

**Proposition 2.2.1** *Let $V$ be an $n$-dimensional vector space over $\mathbb{F}_q$ equipped with a nondegenerate alternating form $B$. Then $n$ is even and $V$ has a basis*

$$\{e_1,\ldots,e_{n/2},f_1,\ldots,f_{n/2}\}$$

*such that $B(e_i,e_j) = B(f_i,f_j) = 0$ and $B(e_i,f_j) = \delta_{i,j}$ for all $i,j$.*

### 2.2.3 Hermitian forms

Let $V$ be an $n$-dimensional vector space over $\mathbb{F}_{q^2}$. A map $B : V \times V \to \mathbb{F}_{q^2}$ is called an *hermitian form* (or a *unitary form*) if

$$B(\lambda u + \mu v, w) = \lambda B(u,w) + \mu B(v,w)$$
$$B(u,v) = B(v,u)^q$$

for all $u,v,w \in V$ and $\lambda,\mu \in \mathbb{F}_{q^2}$. Note that an hermitian form is $\sigma$-sesquilinear, where $\sigma$ is the involutory field automorphism of $\mathbb{F}_{q^2}$. Also note that $B(u,u) \in \mathbb{F}_q$ for all $u \in V$. An example of a nondegenerate hermitian form on $V$ is given by

$$B(u,v) = a_1 b_1^q + a_2 b_2^q + \cdots + a_n b_n^q$$

where $u = (a_1,\ldots,a_n)$ and $v = (b_1,\ldots,b_n)$. If $V$ is equipped with a nondegenerate hermitian form then we say that $V$ is an *hermitian space*. Note that if $J$ is the Gram matrix of an hermitian form on $V$ (with respect to a fixed basis), then $J = (j_{ik})$ is an *hermitian matrix* in the sense that $J = (J^\sigma)^{\mathsf{T}}$, where $J^\sigma = (j_{ik}^\sigma)$.

In Proposition 2.2.2 we present two special bases for hermitian spaces (see [86, Propositions 2.3.1 and 2.3.2]). The basis given in part (ii) is called a *standard*, or *hermitian*, basis of $V$.

**Proposition 2.2.2** *Let $V$ be an $n$-dimensional vector space over $\mathbb{F}_{q^2}$, equipped with a nondegenerate hermitian form $B$.*

(i) *V has an orthonormal basis.*
(ii) *V has a basis*

$$\begin{cases} \{e_1,\ldots,e_{n/2},f_1,\ldots,f_{n/2}\} & n \text{ even} \\ \{e_1,\ldots,e_{(n-1)/2},f_1,\ldots,f_{(n-1)/2},x\} & n \text{ odd} \end{cases}$$

*where $B(e_i,e_j) = B(f_i,f_j) = 0$, $B(e_i,f_j) = \delta_{i,j}$ and $B(e_i,x) = B(f_i,x) = 0$ for all $i,j$, and $B(x,x) = 1$.*

The next result is an easy exercise.

**Lemma 2.2.3** *Let $V$ be an even-dimensional hermitian space over $\mathbb{F}_{q^2}$ with hermitian basis $\{e_1,\ldots,e_{n/2},f_1,\ldots,f_{n/2}\}$. Let $d = 2m+1$ be an odd integer such that $1 \leqslant d < n$ and fix a scalar $\lambda \in \mathbb{F}_{q^2}$ such that $\lambda + \lambda^q \neq 0$. Then*

$$\langle e_1,\ldots,e_m,f_1,\ldots,f_m,e_{m+1}+\lambda f_{m+1} \rangle$$

*is a nondegenerate $d$-space.*

### 2.2.4 Polarities and the Birkhoff–von Neumann Theorem

Let $V$ be a finite-dimensional vector space over a finite field such that $\dim V \geqslant 3$. Let $\mathscr{P}(V)$ be the projective space associated to $V$, that is, the set of all subspaces of $V$, equipped with the partial order of inclusion. A *collineation* of $\mathscr{P}(V)$ is an inclusion-preserving permutation of $\mathscr{P}(V)$. Similarly, a *duality* of $\mathscr{P}(V)$ is an inclusion-reversing permutation of $\mathscr{P}(V)$, and a *polarity* of $\mathscr{P}(V)$ is a duality of order two. The Fundamental Theorem of Projective Geometry (see [118, Theorem 3.1], for example) asserts that the group of all collineations of $\mathscr{P}(V)$ is the projective semilinear group $\mathrm{P\Gamma L}(V)$.

Let $B$ be a reflexive sesquilinear form on $V$ and define

$$W^\perp = \{v \in V \mid B(w,v) = 0 \text{ for all } w \in W\}$$

for each subspace $W$ of $V$. We often refer to $W^\perp$ as the *orthogonal complement* of $W$. Given a vector $v \in V$, we will usually write $v^\perp$ to denote $\langle v \rangle^\perp$. Note that $V^\perp = \mathrm{rad}(B)$.

If $B$ is nondegenerate then $\dim W + \dim W^\perp = \dim V$ and the map

$$\perp : \mathscr{P}(V) \to \mathscr{P}(V)$$

is a duality. Moreover, since $B$ is reflexive, $(W^\perp)^\perp = W$ for every subspace $W$, so this map is a polarity. In fact, the following theorem, due to Birkhoff and von Neumann, asserts that *every* polarity of $\mathscr{P}(V)$ arises in this way from an appropriate reflexive sesquilinear form on $V$ (see [118, Theorem 7.1]).

**Theorem 2.2.4** *If* $\dim V \geqslant 3$ *then every polarity of* $\mathscr{P}(V)$ *arises from a nondegenerate reflexive sesquilinear form on* $V$ *that is either symmetric, alternating or hermitian.*

A subspace $W$ of $V$ is *nondegenerate* if $W \cap W^{\perp} = \{0\}$, that is, the restriction of $B$ to $W \times W$ is nondegenerate. Similarly, $W$ is *totally isotropic* if $W \subseteq W^{\perp}$, that is, $B(v,w) = 0$ for all $v,w \in W$ (see Remark 2.2.6 to follow). Finally, two subspaces $U, W$ of $V$ are *orthogonal* if either $U \subseteq W^{\perp}$ or $W \subseteq U^{\perp}$. In this situation, if $U + W = U \oplus W$ then we write $U + W = U \perp W$ to denote the orthogonality of $U$ and $W$. This notation naturally extends to a direct sum $U_1 \perp \ldots \perp U_t$ of mutually orthogonal subspaces.

### 2.2.5 Quadratic forms

Let $V$ be an $n$-dimensional vector space over $\mathbb{F}_q$. A map $Q : V \to \mathbb{F}_q$ is a *quadratic form* on $V$ if the following two conditions are satisfied:

(i) $Q(\lambda u) = \lambda^2 Q(u)$ for all $u \in V$ and $\lambda \in \mathbb{F}_q$, and
(ii) the map $B_Q : V \times V \to \mathbb{F}_q$ defined by

$$B_Q(u,v) = Q(u+v) - Q(u) - Q(v) \qquad (2.2.1)$$

is a bilinear form.

We call $B_Q$ the *associated bilinear form* of $Q$, and we say that $Q$ *polarises* to $B_Q$. Note that $B_Q$ is symmetric. Following [86, p. 13], we define a quadratic form $Q$ to be *nondegenerate* if and only if the associated bilinear form $B_Q$ is nondegenerate in the usual sense. In terms of a fixed basis of $V$, the *Gram matrix* of $Q$ is defined to be the Gram matrix of $B_Q$, as described in Section 2.2.1. If $V$ is equipped with a nondegenerate quadratic form then we say that $V$ is an *orthogonal space*.

The conditions (i) and (ii) above imply that

$$B_Q(u,u) = Q(2u) - Q(u) - Q(u) = 2Q(u)$$

for all $u \in V$. Therefore, if $q$ is odd then $Q(u) = \frac{1}{2}B_Q(u,u)$ and we can recover $Q$ from $B_Q$. However, if $q$ is even then $B_Q$ is an alternating form and we cannot recover $Q$ in this situation. Notice that if $q$ is even and $V$ is equipped with a nondegenerate quadratic form, then Proposition 2.2.1 implies that $n$ is even. (If $n$ is odd and $q$ is even, then we can define a *nonsingular* quadratic form on $V$; see Remark 2.5.1.) According to the next lemma, if $q$ is even then it is possible to recover $Q$ if we have additional information.

**Lemma 2.2.5** *Let* $V$ *be a vector space over* $\mathbb{F}_q$, *with* $q$ *even, equipped with an alternating form* $B$. *Let* $\{v_1, \ldots, v_n\}$ *be a basis for* $V$ *and fix scalars* $\lambda_i \in \mathbb{F}_q$,

$1 \leqslant i \leqslant n$. Then there is a unique quadratic form $Q$ on $V$ that polarises to $B$ with the property that $Q(v_i) = \lambda_i$ for all $i$.

*Proof*  For $u = \sum_i a_i v_i \in V$ we define a map $Q : V \to \mathbb{F}_q$ by setting

$$Q(u) = \sum_i a_i^2 \lambda_i + \sum_{i<j} a_i a_j B(v_i, v_j)$$

Then $Q$ is a quadratic form on $V$ that polarises to $B$ with $Q(v_i) = \lambda_i$ for all $i$. If $\overline{Q}$ is another quadratic form on $V$ with the same properties, then (2.2.1) implies that

$$\overline{Q}(u) = \sum_i a_i^2 \overline{Q}(v_i) + \sum_{i<j} a_i a_j B(v_i, v_j) = Q(u)$$

and thus $\overline{Q} = Q$.                                                                 $\square$

Let $u, v \in V$, where $u = (a_1, \ldots, a_n)$ and $v = (b_1, \ldots, b_n)$. If $n$ is odd then

$$Q(u) = a_1 a_2 + a_3 a_4 + \cdots + a_{n-2} a_{n-1} + a_n^2$$

defines a quadratic form on $V$, with associated bilinear form

$$B_Q(u, v) = a_1 b_2 + b_1 a_2 + \cdots + a_{n-2} b_{n-1} + b_{n-2} a_{n-1} + 2 a_n b_n$$

Note that $Q$ is nondegenerate if and only if $q$ is odd.
    Similarly, if $n$ is even then

$$Q_1(u) = a_1 a_2 + a_3 a_4 + \cdots + a_{n-1} a_n \tag{2.2.2}$$

is a nondegenerate quadratic form, with associated bilinear form

$$B_{Q_1}(u, v) = a_1 b_2 + b_1 a_2 + \cdots + a_{n-1} b_n + b_{n-1} a_n$$

In addition, for each scalar $\zeta \in \mathbb{F}_q$ we can define a quadratic form

$$Q_2(u) = a_1 a_2 + a_3 a_4 + \cdots + a_{n-3} a_{n-2} + a_{n-1}^2 + a_{n-1} a_n + \zeta a_n^2 \tag{2.2.3}$$

Here the associated bilinear form is

$$B_{Q_2}(u, v) = a_1 b_2 + b_1 a_2 + \cdots + a_{n-3} b_{n-2} + b_{n-3} a_{n-2}$$
$$+ 2 a_{n-1} b_{n-1} + a_{n-1} b_n + b_{n-1} a_n + 2\zeta a_n b_n$$

and we observe that $Q_2$ is nondegenerate if and only if $4\zeta \neq 1$. Also note that $B_{Q_1}$ and $B_{Q_2}$ are identical when $q$ is even.

Let $V$ be a vector space over $\mathbb{F}_q$ equipped with a quadratic form $Q$. A subspace $W$ of $V$ is *nondegenerate* if the restriction of $Q$ to $W$ is nondegenerate. A vector $v \in V$ is *singular* if $Q(v) = 0$, otherwise $v$ is *nonsingular*. A subspace $W$ of $V$ is *anisotropic* if all nonzero vectors in $W$ are nonsingular.

Similarly, $W$ is *totally singular* if $Q(w) = 0$ for all $w \in W$. Note that if $q$ is odd then a subspace is totally singular with respect to $Q$ if and only if it is totally isotropic with respect to $B_Q$. However, if $q$ is even then totally singular subspaces are totally isotropic, but not all totally isotropic subspaces are totally singular. For instance, since $B_Q$ is an alternating form, every 1-dimensional subspace is totally isotropic, but not all of these subspaces are totally singular. Finally, note that in an orthogonal space $(V, Q)$ we define the orthogonal complement of a subspace in terms of the bilinear form $B_Q$.

**Remark 2.2.6** Let $V$ be a vector space and let $B$ be a nondegenerate sesquilinear form on $V$. Recall that a subspace $W$ of $V$ is totally isotropic if $B(u, v) = 0$ for all $u, v \in W$. In some situations, it will also be convenient to use the term *totally singular* to describe such a subspace. This harmless abuse of terminology is adopted by Kleidman and Liebeck (see [86, p.16]), and we will use it for the remainder of this book.

The next result describes some special bases for orthogonal spaces; see [86, Proposition 2.5.3] for proofs. We distinguish three cases; in each case the given basis is called a *standard* basis. More precisely, the basis is *hyperbolic* in case (i), *elliptic* in case (ii) and *parabolic* in case (iii).

**Proposition 2.2.7** *Let $V$ be an $n$-dimensional vector space over $\mathbb{F}_q$ equipped with a nondegenerate quadratic form $Q$ and associated symmetric bilinear form $B_Q = B$. Then $V$ has exactly one of the following bases:*

(i) *$n$ even:* $\{e_1, \ldots, e_{n/2}, f_1, \ldots, f_{n/2}\}$ *where*

$$Q(e_i) = Q(f_i) = 0, \quad B(e_i, f_j) = \delta_{i,j}$$

*for all $i, j$.*

(ii) *$n$ even:* $\{e_1, \ldots, e_{n/2-1}, f_1, \ldots, f_{n/2-1}, x, y\}$ *where $Q(e_i) = Q(f_i) = 0$,*

$$B(e_i, x) = B(e_i, y) = B(f_i, x) = B(f_i, y) = 0, \quad B(e_i, f_j) = \delta_{i,j}$$

*for all $i, j$, $Q(x) = 1$, $B(x, y) = 1$ and $Q(y) = \zeta$ where $\mathbf{x}^2 + \mathbf{x} + \zeta \in \mathbb{F}_q[\mathbf{x}]$ is irreducible.*

(iii) *$n$ odd:* $\{e_1, \ldots, e_{(n-1)/2}, f_1, \ldots, f_{(n-1)/2}, x\}$ *where $Q(e_i) = Q(f_i) = 0$,*

$$B(e_i, x) = B(f_i, x) = 0, \quad B(e_i, f_j) = \delta_{i,j}$$

*for all $i, j$, and $Q(x) \neq 0$.*

If $V$ is equipped with a nondegenerate quadratic form $Q$ that admits a hyperbolic basis then we say that $Q$ is a *hyperbolic* quadratic form. Similarly, we refer to *elliptic* and *parabolic* quadratic forms. Since $B_Q$ is an alternating form when $q$ is even, it follows that $q$ is always odd in the parabolic case. If the restriction of a quadratic form to a nondegenerate subspace $W$ is hyperbolic, then we say that $W$ is a *hyperbolic* subspace. We define *elliptic* and *parabolic* subspaces similarly. By Proposition 2.2.7, every nondegenerate subspace of $V$ is of one of these three types. As a convenient shorthand, we say that a nondegenerate subspace $W$ has *type* $\varepsilon$, where $\varepsilon = +$ if $W$ is hyperbolic, $\varepsilon = -$ in the elliptic case, and $\varepsilon = \circ$ if $W$ is parabolic. Notice that a nondegenerate 2-space is elliptic if and only if it is anisotropic. For later use, it will be convenient to define a product on $\{+,-\}$ by setting

$$++ = -- = +, \ +- = -+ = - \qquad (2.2.4)$$

**Remark 2.2.8** If $q$ is odd and $Q$ is elliptic then it is often more convenient to work with the slightly modified basis

$$\{e_1,\ldots,e_{n/2-1},f_1,\ldots,f_{n/2-1},x',y'\}$$

where $Q(x') = 1$, $B(x',y') = 0$ and $Q(y') = \alpha$ with $-\alpha \in \mathbb{F}_q$ a nonsquare (that is, the polynomial $\mathbf{x}^2 + \alpha \in \mathbb{F}_q[\mathbf{x}]$ is irreducible).

We close this initial discussion of quadratic forms by introducing some additional notation and terminology that will be useful later.

Let $V$ be a vector space equipped with a sesquilinear or quadratic form $f$, and let $W$ be a subspace of $V$. If $f$ is a sesquilinear form, then $f|_W$ denotes the restriction of $f$ to $W \times W$, and it denotes the restriction of $f$ to $W$ when $f$ is a quadratic form. In addition, we define the *polar space* associated to $V$ to be the set of totally singular subspaces of $V$ (see Remark 2.2.6).

### 2.2.6 Isometries and similarities

Let $V$ and $V'$ be vector spaces over a finite field $\mathbb{F}$ equipped with respective sesquilinear forms $B$ and $B'$. An *isometry* from $V$ to $V'$ is an invertible linear map $g : V \to V'$ such that for all $u, v \in V$

$$B'(u^g, v^g) = B(u, v)$$

If such a map exists then $V$ and $V'$, and also $B$ and $B'$, are said to be *isometric*. More generally, subspaces $W \subseteq V$ and $W' \subseteq V'$ are *isometric* if the restrictions $B|_W$ and $B'|_{W'}$ are isometric. For instance, Proposition 2.2.1 implies that any two nondegenerate alternating forms on $V$ are isometric, and similarly Proposition 2.2.2 reveals that any two nondegenerate hermitian forms on $V$ are isometric.

An *isometry* of $B$ is an isometry $g : V \to V$ (in this situation, we say that $g$ *preserves* $B$). Fix a basis $\beta$ for $V$ and let $J_\beta$ be the corresponding Gram matrix of $B$. In terms of matrices, an invertible linear map $g : V \to V$ is an isometry of $B$ if and only if

$$AJ_\beta A^{\varphi\mathsf{T}} = J_\beta \qquad (2.2.5)$$

where $A$ is the matrix representing $g$ with respect to $\beta$, and $\varphi = 1$ unless $B$ is an hermitian form and $\mathbb{F} = \mathbb{F}_{q^2}$, in which case $\varphi$ is the map

$$\begin{array}{rcl} \varphi : \ \mathrm{GL}_n(q^2) & \to & \mathrm{GL}_n(q^2) \\ (a_{ij}) & \mapsto & (a_{ij}^q) \end{array} \qquad (2.2.6)$$

where $n$ denotes the dimension of $V$. The isometries of $B$ form a group under composition, which is called the *isometry group* of $B$. All the classical groups can be defined in terms of the isometry groups of suitable sesquilinear (or quadratic) forms. We will discuss the groups that arise in this way in Sections 2.3–2.5.

A *similarity* from $V$ to $V'$ is an invertible linear map $g : V \to V'$ such that there exists a scalar $\lambda \in \mathbb{F}^\times$ with

$$B'(u^g, v^g) = \lambda B(u, v)$$

for all $u, v \in V$. If such a map exists, then $V$ and $V'$ are said to be *similar*. This notion of similarity extends naturally to subspaces and the sesquilinear forms themselves.

A *semisimilarity* from $V$ to $V'$ is an invertible semilinear map $g : V \to V'$ with the property that there is a scalar $\lambda \in \mathbb{F}^\times$ such that

$$B'(u^g, v^g) = \lambda B(u, v)^\sigma$$

for all $u, v \in V$, where $\sigma \in \mathrm{Aut}(\mathbb{F})$ is the field automorphism defining $g$. We define a semisimilarity of a sesquilinear form on $V$ in an analogous manner.

Isometries, similarities and semisimilarities can also be defined for spaces equipped with a quadratic form. Given vector spaces $V$ and $V'$ over $\mathbb{F}_q$, with respective quadratic forms $Q$ and $Q'$, an *isometry* from $V$ to $V'$ is an invertible linear map $g : V \to V'$ such that

$$Q'(u^g) = Q(u)$$

for all $u \in V$. We can define similarities and semisimilarities between $Q$ and $Q'$ analogously. An *isometry* of $Q$ is an isometry from $V$ to $V$, and the *isometry group of* $Q$ is the set of isometries of $Q$ under composition. Note that the isometry group of $Q$ is a subgroup of the isometry group of the associated bilinear form $B_Q$; these groups are equal if and only if $q$ is odd. A *similarity* of $Q$ is an invertible linear map $g : V \to V$ for which there is a scalar $\tau(g) \in \mathbb{F}_q^\times$ such that

$$Q(u^g) = \tau(g)Q(u) \qquad (2.2.7)$$

for all $u \in V$. Semisimilarities of $Q$ are defined in a similar fashion.

Let $V$ be an $n$-dimensional vector space over $\mathbb{F}_q$. If $n$ is even then there are exactly two isometry types of nondegenerate quadratic forms on $V$; namely, the hyperbolic and elliptic forms introduced in Proposition 2.2.7.

Now assume $n$ and $q$ are both odd. Here there are two isometry types, but only one nondegenerate quadratic form up to similarity (in particular, the two isometry types give rise to isomorphic isometry groups).

To see this, let $Q$ be a nondegenerate quadratic form on $V$ and let $x \in V$ be a nonsingular vector, say $Q(x) = \alpha$. Since $Q(\lambda x) = \lambda^2 \alpha$ for all $\lambda \in \mathbb{F}_q^{\times}$, it follows that

$$\{Q(v) \mid v \in \langle x \rangle, v \neq 0\}$$

is either the set of all squares or the set of all nonsquares in $\mathbb{F}_q^{\times}$. In particular, if $Q_1$ and $Q_2$ are nondegenerate quadratic forms on $U = \langle x \rangle$ such that $Q_1(x)$ is a square and $Q_2(x)$ is a nonsquare, then $Q_1$ and $Q_2$ are not isometric. However, $Q_1(x) = \lambda Q_2(x)$ for some nonsquare scalar $\lambda \in \mathbb{F}_q$, so the identity map on $U$ is a similarity between $Q_1$ and $Q_2$. In general, if $x$ is the nonsingular vector in the parabolic basis for $V$ in Proposition 2.2.7(iii), then the isometry type of $Q$ can be identified by determining whether $Q(x)$ is a square or nonsquare.

For any $n$, if $q$ is odd then the two isometry types can be distinguished via the *discriminant*, which is defined as follows. Let $\beta$ be a basis for $V$, let $Q$ be a nondegenerate quadratic form on $V$, and let $J_\beta$ be the corresponding Gram matrix of the associated bilinear form $B_Q$. The determinant of $J_\beta$ is either a square or nonsquare in $\mathbb{F}_q$, and this is independent of the choice of basis $\beta$. Thus we define the discriminant $D(Q) \in \{\square, \boxtimes\}$ by setting

$$D(Q) = \begin{cases} \square & \det(J_\beta) \in \mathbb{F}_q \text{ is a square} \\ \boxtimes & \text{otherwise} \end{cases}$$

and we define a product on these symbols in the obvious way, so that

$$\square\square = \boxtimes\boxtimes = \square, \quad \square\boxtimes = \boxtimes\square = \boxtimes \tag{2.2.8}$$

and thus $\{\square, \boxtimes\}$ is a group with identity element $\square$. (Note that if $q$ is even then every element of $\mathbb{F}_q$ is a square, so the discriminant is not a useful notion in this situation.) The discriminant distinguishes the two isometry classes of quadratic forms in both the even-dimensional and odd-dimensional cases. More precisely, if $n$ is even then the next lemma describes how the discriminant distinguishes the forms (see [86, Proposition 2.5.10]).

**Lemma 2.2.9** *Let $Q$ be a nondegenerate quadratic form on an $n$-dimensional vector space over $\mathbb{F}_q$, where $n$ is even and $q$ is odd.*

  (i) *If $Q$ is hyperbolic, then $D(Q) = \square$ if and only if $n(q-1)/4$ is even.*
  (ii) *If $Q$ is elliptic, then $D(Q) = \square$ if and only if $n(q-1)/4$ is odd.*

**Remark 2.2.10** As previously noted, if $nq$ is odd then there are two distinct nondegenerate quadratic forms on $V$ up to isometry, but a unique such form up to similarity. We will say that $V$ is a *square-parabolic* space if $D(Q) = \square$, and $V$ is a *nonsquare-parabolic* space if $D(Q) = \boxtimes$. Unless stated otherwise, if $nq$ is odd and $Q$ is a nondegenerate quadratic form on $V$, then we will always assume that $D(Q) = \square$.

We also record the following lemma on orthogonal decompositions (see [86, Proposition 2.5.11]). Here we use the products on $\{\square, \boxtimes\}$ and $\{+, -\}$ defined in (2.2.8) and (2.2.4), respectively.

**Lemma 2.2.11** *Let $Q$ be a nondegenerate quadratic form on a vector space $V$ over $\mathbb{F}_q$, and let $U, W$ be nondegenerate subspaces such that $V = U \perp W$.*

(i) $D(Q) = D(Q|_U)D(Q|_W)$.
(ii) *If $U$ and $W$ are even-dimensional and of type $\varepsilon_1$ and $\varepsilon_2$ respectively, then $V$ has type $\varepsilon_1 \varepsilon_2$.*

**Corollary 2.2.12** *Let $V$ be an $n$-dimensional vector space over $\mathbb{F}_q$ equipped with a nondegenerate quadratic form $Q$. Let $m < n$ be a positive integer and assume $m$ is even if $q$ is even. Then $V$ contains a nondegenerate $m$-space of each isometry type.*

*Proof* To see this, it suffices to show that we can embed any $m$-dimensional orthogonal space in an $n$-dimensional orthogonal space of each isometry type.

Let $U$ and $W$ be vector spaces over $\mathbb{F}_q$ equipped with nondegenerate quadratic forms $Q_U$ and $Q_W$, where $\dim U = m$ and $\dim W = n - m$. Set $V = U \oplus W$ and define a nondegenerate quadratic form $Q$ on $V$ by setting

$$Q(u+w) = Q_U(u) + Q_W(w)$$

(note that $B_Q(u,w) = 0$ for all $u \in U$, $w \in W$, so $V = U \perp W$ is an orthogonal decomposition). In view of Lemma 2.2.11, by varying the isometry type of $W$ we can obtain all possible isometry types for $V$. The result follows. $\square$

**Lemma 2.2.13** *Fix $\zeta \in \mathbb{F}_q$ such that $\mathbf{x}^2 + \mathbf{x} + \zeta \in \mathbb{F}_q[\mathbf{x}]$ is irreducible.*

(i) *Let $V$ be a 3-dimensional parabolic space over $\mathbb{F}_q$ with standard basis $\{e_1, f_1, x\}$ such that $Q(x) = 1$. Then $\langle e_1 + \zeta f_1, f_1 + x \rangle$ is an elliptic 2-space.*
(ii) *Let $V$ be a 4-dimensional hyperbolic space over $\mathbb{F}_q$ with standard basis $\{e_1, e_2, f_1, f_2\}$. Then*

$$V = \langle e_1 + f_1, e_1 + e_2 + \zeta f_2 \rangle \perp \langle e_1 - f_1 + f_2, e_2 - \zeta f_2 \rangle$$

*is an orthogonal decomposition of $V$ into elliptic 2-spaces. Moreover, if $q$ is odd and $-\alpha \in \mathbb{F}_q$ is a nonsquare then*

$$V = \langle e_1 + f_1, e_2 + \alpha f_2 \rangle \perp \langle e_1 - f_1, e_2 - \alpha f_2 \rangle$$

*is also an orthogonal decomposition into elliptic 2-spaces.*

*Proof*  This is an easy exercise.  □

**Lemma 2.2.14** *Let $V$ be an $n$-dimensional vector space over $\mathbb{F}_q$, where $q$ is odd. Let $Q$ be a nondegenerate quadratic form on $V$, and let $B_Q$ be the associated bilinear form. Then $V$ has a basis $\{v_1, \ldots, v_n\}$ such that*

$$B_Q(v_i, v_j) = 0, \quad B_Q(v_i, v_i) = \begin{cases} 1 & i \geqslant 2 \ or \ D(Q) = \square \\ \mu & otherwise \end{cases}$$

*for all $i \neq j$, where $\mathbb{F}_q^\times = \langle \mu \rangle$. In particular, $V$ has an orthonormal basis if $D(Q) = \square$.*

*Proof*  This is [86, Proposition 2.5.12].  □

### 2.2.7 Witt's Lemma and applications

The following result is known as *Witt's Lemma*. It is an important tool in the study of classical groups (see [4, Section 20] for a proof).

**Lemma 2.2.15** *Let $V$ be a vector space equipped with a nondegenerate reflexive sesquilinear or quadratic form, let $U, W$ be subspaces of $V$, and let $g : U \to W$ be an isometry. Then $g$ extends to an isometry of $V$.*

An immediate consequence of Witt's Lemma is that the isometry group of a nondegenerate reflexive sesquilinear or quadratic form on $V$ is transitive on the set of all subspaces of $V$ of a given isometry type. It follows that all maximal totally singular subspaces have the same dimension, which is at most $\frac{1}{2}\dim V$. In particular, if $Q$ is a nondegenerate quadratic form and $U$ is a maximal totally singular subspace of $V$, then

$$\dim U = \frac{1}{2}\dim V - \delta$$

where $\delta = 0$ if $Q$ is hyperbolic, $\delta = 1$ if $Q$ is elliptic and $\delta = 1/2$ if $Q$ is parabolic.

Let $\{v_1, \ldots, v_n\}$ be a basis of $V$, where $n$ is even, and consider the quadratic forms $Q_1$ and $Q_2$ defined in (2.2.2) and (2.2.3) in Section 2.2.5. In the definition of $Q_2$, choose $\zeta \in \mathbb{F}_q$ so that the polynomial $\mathbf{x}^2 + \mathbf{x} + \zeta \in \mathbb{F}_q[\mathbf{x}]$ is irreducible over $\mathbb{F}_q$ (note that the irreducibility of $\mathbf{x}^2 + \mathbf{x} + \zeta$ implies that $4\zeta \neq 1$, so $Q_2$ is nondegenerate). Now $\langle v_1, v_3, \ldots, v_{n-1} \rangle$ is a maximal totally singular subspace of $V$ with respect to $Q_1$, so $Q_1$ is a hyperbolic quadratic form. On the other hand, the irreducibility of $\mathbf{x}^2 + \mathbf{x} + \zeta$ implies that $\langle v_{n-1}, v_n \rangle$ is anisotropic for $Q_2$, so $\langle v_1, v_3, \ldots, v_{n-3} \rangle$ is a maximal totally singular subspace with respect to $Q_2$, and it follows that $Q_2$ is elliptic. In particular, $Q_1$ and $Q_2$ are not isometric, even though they have the same associated bilinear form when $q$ is even.

Let $B$ be a nondegenerate reflexive sesquilinear form on $V$, let $W$ be a subspace of $V$ and recall that $\dim V = \dim W + \dim W^\perp$ (see Section 2.2.4). If $g$ is a semisimilarity of $B$ that fixes $W$ (setwise) then it is easy to see that $g$ also fixes $W^\perp$. Now, if $W$ is nondegenerate then $W^\perp$ is a complementary nondegenerate subspace, so $V = W \perp W^\perp$. On the other hand, if $W$ is totally isotropic then $W \subseteq W^\perp$ and $B$ induces a reflexive sesquilinear form $\overline{B}$ on $W^\perp/W$ given by $\overline{B}(u_1 + W, u_2 + W) = B(u_1, u_2)$. This is well defined since

$$B(u_1 + w_1, u_2 + w_2) = B(u_1, u_2) + B(u_1, w_2) + B(w_1, u_2) + B(w_1, w_2)$$
$$= B(u_1, u_2)$$

for all $u_1, u_2 \in W^\perp$ and $w_1, w_2 \in W$. In addition, $\overline{B}$ is nondegenerate since $(W^\perp)^\perp = W$.

Similarly, if $W$ is a totally singular subspace with respect to a nondegenerate quadratic form $Q$, then we can define a nondegenerate quadratic form $\overline{Q}$ on $W^\perp/W$ by setting $\overline{Q}(u + W) = Q(u)$. Then any isometry of $Q$ that fixes $W$ also fixes $W^\perp$, and it induces an isometry on $W^\perp/W$ with respect to $\overline{Q}$.

**Lemma 2.2.16** *Let $f$ be a nondegenerate reflexive sesquilinear or quadratic form on a vector space $V$, let $W$ be a nondegenerate subspace of $V$ and let $G$ be the isometry group of $f$. Let $H$ be the (setwise) stabiliser of $W$ in $G$. Then $H \cong G_1 \times G_2$ fixes the orthogonal decomposition $V = W \perp W^\perp$, where $G_1$ and $G_2$ are the isometry groups of $f|_W$ and $f|_{W^\perp}$, respectively.*

*Proof* This is entirely straightforward. □

In the next two lemmas, set $\mathbb{F} = \mathbb{F}_{q^2}$ if $f$ is an hermitian form, and $\mathbb{F} = \mathbb{F}_q$ in all other cases.

**Lemma 2.2.17** *Let $V$ be an $n$-dimensional vector space over $\mathbb{F}$ equipped with a nondegenerate reflexive sesquilinear or quadratic form $f$, and let $G$ be the isometry group of $f$.*

(i) *Suppose $n$ is even and $W$ is a totally singular subspace of $V$ of dimension $n/2$. Let $H$ be the stabiliser of $W$ in $G$. Then with respect to a suitable basis for $V$,*

$$H = \left\{ \begin{pmatrix} A & 0 \\ * & (A^{\varphi\mathsf{T}})^{-1} \end{pmatrix} \mid A \in \mathrm{GL}(W) \right\}$$

*where $\varphi$ is the map defined in (2.2.6) if $f$ is hermitian, and $\varphi = 1$ in all other cases.*

(ii) *More generally, if $g \in G$ fixes a totally singular $m$-dimensional subspace $U$ then there exists a complementary totally singular $m$-space $U^*$ such that $U \oplus U^*$ is nondegenerate and $g$-invariant. Moreover, with respect to a suitable basis, the matrix of $g$ on $U \oplus U^*$ has the form*

$$\begin{pmatrix} A & 0 \\ * & (A^{\varphi\mathsf{T}})^{-1} \end{pmatrix}$$

*where $A \in \mathrm{GL}(U)$ represents the action of $g$ on $U$.*

*Proof* First consider (i). Here $f$ is either an alternating or hermitian sesquilinear form, or a hyperbolic quadratic form. By Witt's Lemma, $G$ acts transitively on the set of totally singular $\frac{n}{2}$-spaces, so we may assume that $W = \langle e_1, \ldots, e_{n/2} \rangle$ where the $e_i$ are basis vectors in an appropriate standard basis of $V$. Then with respect to this basis, the Gram matrix of $f$ is

$$\begin{pmatrix} 0 & I_{n/2} \\ \xi I_{n/2} & 0 \end{pmatrix}$$

where $\xi = -1$ if $f$ is alternating, otherwise $\xi = 1$. Therefore, (2.2.5) implies that each $g \in H$ has the form

$$\begin{pmatrix} A & 0 \\ B & (A^{\varphi\mathsf{T}})^{-1} \end{pmatrix} \quad \text{with } BA^{-1} + \xi(A^{\varphi\mathsf{T}})^{-1}B^{\varphi\mathsf{T}} = 0$$

Moreover, by Witt's Lemma, $A$ can be any element of $\mathrm{GL}(W)$. An entirely similar calculation establishes part (ii).                                       $\square$

A pair $\{U, U^*\}$ of totally singular subspaces as in Lemma 2.2.17(ii) is called a *dual pair* of subspaces.

**Lemma 2.2.18** *Let $V$ be an $n$-dimensional vector space over $\mathbb{F}$ equipped with a nondegenerate reflexive sesquilinear or quadratic form $f$. Then the isometry* *group of f acts transitively on the set of dual pairs of m-dimensional totally singular subspaces of V.*

*Proof* Let $\{U_1, U_1^*\}$ and $\{U_2, U_2^*\}$ be dual pairs of totally singular $m$-spaces. Fix a basis $\{e_{i1}, \ldots, e_{im}, f_{i1}, \ldots, f_{im}\}$ of $U_i \oplus U_i^*$ so that the restriction of $f$ to $U_i \oplus U_i^*$ has Gram matrix

$$\begin{pmatrix} 0 & I_m \\ \xi I_m & 0 \end{pmatrix}$$

where $\xi = -1$ if $f$ is alternating, otherwise $\xi = 1$. Then the linear map

$$g : U_1 \oplus U_1^* \to U_2 \oplus U_2^*$$

that sends each $e_{1j}$ to $e_{2j}$, and each $f_{1j}$ to $f_{2j}$, is an isometry. By Witt's Lemma, $g$ extends to an isometry of $V$ and the result follows.                    $\square$

### 2.2.8 Tensor products

Let $U$ and $W$ be vector spaces over $\mathbb{F}_q$ of dimensions $a$ and $b$ respectively, and consider the tensor product space $V = U \otimes W$. Note that every element of $V$ can be written as a sum of simple tensors $u \otimes w$ with $u \in U$ and $w \in W$. Given sesquilinear forms $B_1$ and $B_2$ on $U$ and $W$, respectively, we can define a sesquilinear form $B = B_1 \otimes B_2$ on $V$ by setting

$$B(u_1 \otimes w_1, u_2 \otimes w_2) = B_1(u_1, u_2)B_2(w_1, w_2) \qquad (2.2.9)$$

for all $u_1, u_2 \in U$ and $w_1, w_2 \in W$, and then extending additively. If both $B_1$ and $B_2$ are nondegenerate then so is $B_1 \otimes B_2$. For the cases we are interested in, the type of $B_1 \otimes B_2$ is recorded in Table 2.2.1.

Recall that if $q$ is odd then a nondegenerate symmetric bilinear form $B$ on $V$ gives rise to a nondegenerate quadratic form $Q$ that polarises to $B$ ($Q$ is defined by setting $Q(v) = \frac{1}{2}B(v, v)$ for all $v \in V$). Therefore, if $Q_1$ and $Q_2$ are quadratic forms on $U$ and $W$, respectively, then we can construct a quadratic

Table 2.2.1 *Sesquilinear forms on a tensor product*

| $B_1$ | $B_2$ | $B_1 \otimes B_2$ |
|---|---|---|
| symmetric | symmetric | symmetric |
| alternating | alternating | { symmetric if $q$ odd  alternating if $q$ even |
| symmetric | alternating | alternating |
| hermitian | hermitian | hermitian |

Table 2.2.2 *Quadratic forms on a tensor product, q odd*

| $Q_1$ | $Q_2$ | $Q_1 \otimes Q_2$ |
|---|---|---|
| hyperbolic | hyperbolic | hyperbolic |
| hyperbolic | elliptic | hyperbolic |
| hyperbolic | parabolic | hyperbolic |
| elliptic | elliptic | hyperbolic |
| elliptic | parabolic | elliptic |
| parabolic | parabolic | parabolic |

form $Q = Q_1 \otimes Q_2$ on $V = U \otimes W$ by tensoring the bilinear forms corresponding to $Q_1$ and $Q_2$ as above. The type of the quadratic form on $V$ constructed in this way is given in Table 2.2.2 (see [86, Lemma 4.4.2], for example). In particular, note that if $Q_1$ is hyperbolic then $Q$ is also hyperbolic – indeed, if $X$ is a totally singular subspace of $U$ of dimension $a/2$, then $X \otimes W$ is a totally singular subspace of $V$ of dimension $ab/2$. The same argument shows that if $q$ is odd then the tensor product of two nondegenerate alternating forms yields a symmetric bilinear form that corresponds to a hyperbolic quadratic form. In addition, if $q$ is odd and $D(Q_i)$ denotes the discriminant of $Q_i$ (see Section 2.2.6), then it is easy to see that

$$D(Q) = D(Q_1)^b D(Q_2)^a \qquad (2.2.10)$$

(this is [86, Lemma 4.4.1]), where the product on $\{\square, \boxtimes\}$ is defined in (2.2.8).

In the next lemma, we consider the case where $q$ is even.

**Lemma 2.2.19** *Let $U$ and $W$ be vector spaces over $\mathbb{F}_q$ equipped with nondegenerate alternating forms $B_1$ and $B_2$, respectively. Assume $q$ is even and let $B$ be the bilinear form on $V = U \otimes W$ defined in (2.2.9). Let $Q$ be the unique quadratic form on $V$ polarising to $B$ such that $Q(u \otimes w) = 0$ for all $u \in U$, $w \in W$. Then $Q$ is hyperbolic.*

*Proof* First note that the existence and uniqueness of $Q$ follows from Lemma 2.2.5. The definitions of $B$ and $Q$ imply that if $X$ is a totally isotropic subspace of $U$ then $X \otimes W$ is a totally singular subspace of $V$. Therefore, $V$ has a totally singular subspace of dimension $\frac{1}{2} \dim V$, whence $Q$ is hyperbolic. $\qquad \square$

Let $\{u_1, \ldots, u_a\}$ be a basis for $U$ and let $\{w_1, \ldots, w_b\}$ be a basis for $W$. Given $g \in \mathrm{GL}(U)$ and $h \in \mathrm{GL}(W)$, we define $g \otimes h \in \mathrm{GL}(V)$ by setting

$$(u \otimes w)^{g \otimes h} = u^g \otimes w^h \qquad (2.2.11)$$

for all $u \in U$ and $w \in W$, and extending linearly to $V = U \otimes W$. In particular, if $g \in \mathrm{GL}(U)$ is represented by the matrix $A$ (with respect to the above basis), then $g \otimes 1 \in \mathrm{GL}(V)$ is given by the block-diagonal matrix $[A^b]$ (with $b$ blocks) in terms of the basis

$$\{u_1 \otimes w_1, \ldots, u_a \otimes w_1, u_1 \otimes w_2, \ldots, u_a \otimes w_b\}$$

for $V$.

Note that if $g$ and $h$ are isometries of $B_1$ and $B_2$, respectively, then the element $g \otimes h$ preserves the form $B_1 \otimes B_2$. Now, if $q$ is odd then an isometry of a symmetric bilinear form is also an isometry of the corresponding quadratic form, so if $g$ and $h$ preserve quadratic forms $Q_1$ and $Q_2$, respectively, then $g \otimes h$ preserves $Q_1 \otimes Q_2$. If $q$ is even and the $B_i$ are alternating forms, then Lemma 2.2.5 implies that $g \otimes h$ also preserves the quadratic form $Q$ defined in Lemma 2.2.19.

## 2.3 Unitary groups

Let $q = p^f$ be a power of a prime $p$ and let $V$ be an $n$-dimensional vector space over $\mathbb{F}_{q^2}$. Let $B : V \times V \to \mathbb{F}_{q^2}$ be a nondegenerate hermitian form on $V$ and let $\mathrm{GU}(V) = \mathrm{GU}_n(q)$ be the isometry group of $B$, which is called the *general unitary group*. Let $\{v_1, \ldots, v_n\}$ be an orthonormal basis for $V$ with respect to $B$ (see Proposition 2.2.2(i)). For the remainder of this section, all matrices will be written with respect to this basis, unless specified otherwise.

Define $\phi : V \to V$ by

$$\phi : \sum_i \lambda_i v_i \mapsto \sum_i \lambda_i^p v_i \tag{2.3.1}$$

and note that the map $\varphi$ defined in (2.2.6) satisfies $\varphi = \phi^f$. Then $\varphi$ normalises $\mathrm{GL}_n(q^2)$. With respect to the above orthonormal basis for $V$, (2.2.5) implies that

$$\mathrm{GU}_n(q) = \{A \in \mathrm{GL}_n(q^2) \mid AA^{\varphi\mathsf{T}} = I_n\}$$

and we define the *special unitary group* by setting

$$\mathrm{SU}_n(q) = \{A \in \mathrm{GU}_n(q) \mid \det(A) = 1\}$$

Observe that $\det(A)^{q+1} = 1$ for all $A \in \mathrm{GU}_n(q)$.

Let $\Delta\mathrm{U}_n(q)$ be the group of all similarities of $B$, that is, all $A \in \mathrm{GL}_n(q^2)$ for which there exists a scalar $\lambda \in \mathbb{F}_{q^2}$ such that $B(uA, vA) = \lambda B(u, v)$ for all $u, v \in V$. In this situation we often write $\tau(A) = \lambda$. We use $\Gamma\mathrm{U}_n(q)$ to denote the group of all semisimilarities of $B$.

Let $Z$ be the group of all scalar matrices in $\mathrm{GL}_n(q^2)$ and set $\mathrm{P}\Delta\mathrm{U}_n(q) = \Delta\mathrm{U}_n(q)/Z$ and

$$\mathrm{PGU}_n(q) = \mathrm{GU}_n(q)Z/Z \cong \mathrm{GU}_n(q)/(\mathrm{GU}_n(q) \cap Z)$$
$$\mathrm{PSU}_n(q) = \mathrm{SU}_n(q)Z/Z \cong \mathrm{SU}_n(q)/(\mathrm{SU}_n(q) \cap Z)$$

Also set $\mathrm{P\Gamma U}_n(q) = \Gamma \mathrm{U}_n(q)/Z$. Sometimes we will use $\mathrm{PGL}_n^-(q)$ to denote $\mathrm{PGU}_n(q)$, and $\mathrm{PSL}_n^-(q)$ for $\mathrm{PSU}_n(q)$. Note that $\mathrm{P\Delta U}_n(q) = \mathrm{PGU}_n(q)$ since $\Delta \mathrm{U}_n(q) = \mathrm{GU}_n(q)Z$ (see [86, (2.3.3)]). Also observe that

$$\mathrm{GU}_n(q) \cap Z = \{\mu I_n \mid \mu^{q+1} = 1\} \cong C_{q+1}, \;\; \mathrm{SU}_n(q) \cap Z \cong C_{(n,q+1)}$$

Membership in $\mathrm{PSU}_n(q)$ is given by the following lemma.

**Lemma 2.3.1** *Let $g \in \mathrm{P\Delta U}_n(q)$ such that $g = AZ$ for some $A \in \Delta \mathrm{U}_n(q)$ with $\tau(A) = \lambda$.*

(i) *$g \in \mathrm{PSU}_n(q)$ if and only if there exists $\mu \in \mathbb{F}_{q^2}$ such that $\mu^{q+1} = \lambda$ and $\det(A) = \mu^n$.*

(ii) *If $A \in \mathrm{GU}_n(q)$, then $g \in \mathrm{PSU}_n(q)$ if and only if there exists $\mu \in \mathbb{F}_{q^2}$ such that $\mu^{q+1} = 1$ and $\det(A) = \mu^n$.*

*Proof*   Clearly, $g \in \mathrm{PSU}_n(q)$ if and only if the coset $AZ$ contains a matrix in $\mathrm{SU}_n(q)$, that is, if and only if there exists $\mu \in \mathbb{F}_{q^2}$ such that $\mu A \in \mathrm{SU}_n(q)$. Since

$$\tau(\mu A) = \mu^{q+1} \tau(A) = \mu^{q+1} \lambda$$

it follows that $\mu A \in \mathrm{SU}_n(q)$ if and only if $\mu^{q+1} = \lambda^{-1}$ and $\det(A) = (\mu^{-1})^n$. Therefore (i) holds, and part (ii) is a special case of (i).     $\square$

The next result is the unitary group analogue of Theorem 2.1.2 (see [122, Section 3.6]). Note that $\mathrm{PSU}_2(q) \cong \mathrm{PSL}_2(q)$.

**Theorem 2.3.2** *Suppose $n \geqslant 2$ and $(n,q) \neq (2,2),(2,3),(3,2)$. Then $\mathrm{PSU}_n(q)$ is a simple group of order*

$$\frac{1}{d} q^{\frac{1}{2}n(n-1)} \prod_{i=2}^{n} (q^i - (-1)^i)$$

*where $d = (n, q+1)$. The exceptions*

$$\mathrm{PSU}_2(2) \cong S_3, \;\; \mathrm{PSU}_2(3) \cong A_4, \;\; \mathrm{PSU}_3(2) \cong 3^2{:}Q_8$$

*are soluble.*

Let $T = \mathrm{PSU}_n(q)$. The elements in $\mathrm{PGU}_n(q) \setminus T$ are the nontrivial *diagonal automorphisms* of $T$, and as in the linear case, we define $\mathrm{Inndiag}(T) = \mathrm{PGU}_n(q)$ to be the subgroup of $\mathrm{Aut}(T)$ generated by the inner and diagonal automorphisms of $T$. If we set $\delta = [\mu, I_{n-1}]Z \in \mathrm{PGU}_n(q)$, where $\mu \in \mathbb{F}_{q^2}$ has order $q + 1$, then $\mathrm{PGU}_n(q) = \langle T, \delta \rangle$.

Next observe that $B(u^\phi, w^\phi) = B(u, w)^p$ for all $u, w \in V$, where $\phi$ is defined in (2.3.1). Therefore $\phi$ is a semisimilarity of $B$ and we have $\mathrm{P\Gamma U}_n(q) = \langle \mathrm{PGU}_n(q), \phi \rangle$. Let us also observe that the inverse-transpose map

$$
\iota : \begin{array}{ccc} \mathrm{GL}_n(q^2) & \to & \mathrm{GL}_n(q^2) \\ A & \mapsto & A^{-\mathsf{T}} \end{array}
$$

normalises $\mathrm{GU}_n(q)$ and $Z$, so it induces an automorphism of $\mathrm{PGU}_n(q)$. Note that if $A \in \mathrm{GU}_n(q)$ then $A^\phi = A^{-\mathsf{T}}$ and so $\phi$ and $\iota$ induce the same automorphism of $\mathrm{GU}_n(q)$ and $\mathrm{PGU}_n(q)$. If $n \geqslant 3$ then $\mathrm{Aut}(\mathrm{PSU}_n(q)) = \mathrm{P\Gamma U}_n(q)$ (see [46]) and

$$
\mathrm{Out}(\mathrm{PSU}_n(q)) = C_{(n, q+1)}{:}C_{2f}
$$

Recall that $\mathrm{PSU}_2(q) \cong \mathrm{PSL}_2(q)$, so $\mathrm{Aut}(\mathrm{PSU}_2(q)) \cong \mathrm{P\Gamma L}_2(q)$. However, note that $|\mathrm{P\Gamma U}_2(q)| = 2|\mathrm{P\Gamma L}_2(q)|$. Indeed,

$$
\begin{pmatrix} 0 & 1 \\ -1 & 0 \end{pmatrix} \varphi \in \Gamma\mathrm{U}_2(q)
$$

induces the trivial automorphism of $\mathrm{PSU}_2(q)$.

Following [67, Definition 2.5.13], the *field automorphisms* of $T$ are the $\mathrm{Aut}(T)$-conjugates of $\phi^i$, where $1 \leqslant i < 2f$ and $\phi^i$ has odd order. In view of the structure of $\mathrm{Aut}(T)$, these are the $\mathrm{PGU}_n(q)$-conjugates of $\phi^i$. Note that each conjugate of $\phi$ corresponds to a specific choice of orthonormal basis used to define $\phi$ in (2.3.1). Similarly, the *graph automorphisms* of $T$ are the elements of the form $x\phi^i$ where $x \in \mathrm{PGU}_n(q)$ and $\phi^i$ has even order. In particular, $\varphi$ is a graph automorphism. Note that $T$ does not admit any graph-field automorphisms.

As recorded in Theorem 2.1.3, every automorphism of $\mathrm{PSU}_n(q)$ is the product of an inner, a diagonal, a field and a graph automorphism.

We close this preliminary discussion of unitary groups by describing the construction of a subgroup $\mathrm{GU}_{n/r}(q^r) \leqslant \mathrm{GU}_n(q)$ for any odd integer $r$ dividing $n$. This will be useful later.

**Construction 2.3.3** *The embedding* $\mathrm{GU}_{n/r}(q^r) \leqslant \mathrm{GU}_n(q)$, $r$ *odd.*

Let $r$ be an odd integer and let $V$ be an $\frac{n}{r}$-dimensional vector space over $\mathbb{F}_{q^{2r}}$ equipped with a nondegenerate hermitian form $\overline{B} : V \times V \to \mathbb{F}_{q^{2r}}$. Recall that $V$ can also be viewed as an $n$-dimensional vector space over $\mathbb{F}_{q^2}$ and this yields a natural embedding of $\mathrm{GL}_{n/r}(q^{2r})$ in $\mathrm{GL}_n(q^2)$ (see Construction 2.1.4). The form $B = \mathrm{Tr} \circ \overline{B} : V \times V \to \mathbb{F}_{q^2}$ is nondegenerate and hermitian, where

$$\text{Tr}: \quad \mathbb{F}_{q^{2r}} \;\longrightarrow\; \mathbb{F}_{q^2}$$
$$\lambda \;\longmapsto\; \lambda + \lambda^{q^2} + \cdots + \lambda^{q^{2(r-1)}}$$

is the familiar trace map. (Note that if $r$ is even then the form $B$ defined in this way is symmetric since $\mathbb{F}_{q^2} \subseteq \mathbb{F}_{q^r}$.) Moreover, if $U$ is an $\mathbb{F}_{q^{2r}}$-subspace of $V$ that is totally isotropic (respectively nondegenerate) with respect to $\overline{B}$, then it is also totally isotropic (respectively nondegenerate) with respect to $B$, as an $\mathbb{F}_{q^2}$-space.

Suppose $g \in \mathrm{GL}_{n/r}(q^{2r})$ is a similarity of $\overline{B}$, say $g$ preserves $\overline{B}$ up to a scalar $\tau(g) \in \mathbb{F}_{q^{2r}}$. If $\tau(g)$ is in the subfield $\mathbb{F}_{q^2}$, then $B(u^g, v^g) = \tau(g)B(u,v)$ for all $u, v \in V$ and thus $g$ is a similarity of $B$. This yields an embedding of $\mathrm{GU}_{n/r}(q^r)$ in $\mathrm{GU}_n(q)$. Note that not all of $\Delta \mathrm{U}_{n/r}(q^r)$ embeds in $\Delta \mathrm{U}_n(q)$, since we require the condition $\tau(g) \in \mathbb{F}_{q^2}$ (see [86, Proposition 4.3.5(i)]).  □

## 2.4 Symplectic groups

Let $V$ be an $n$-dimensional vector space over $\mathbb{F}_q$, where $q = p^f$ for a prime $p$, and let $B$ be a nondegenerate alternating form on $V$. Recall that $n$ is even (see Proposition 2.2.1). The isometry group of $B$ is the *symplectic group*, denoted by $\mathrm{Sp}(V) = \mathrm{Sp}_n(q)$. The groups of similarities and semisimilarities of $B$ are denoted by $\mathrm{GSp}_n(q)$ and $\Gamma\mathrm{Sp}_n(q)$, respectively.

Let $Z$ be the group of all scalar matrices in $\mathrm{GL}_n(q)$ and define $\mathrm{PGSp}_n(q) = \mathrm{GSp}_n(q)/Z$ and

$$\mathrm{PSp}_n(q) = \mathrm{Sp}_n(q)Z/Z \cong \mathrm{Sp}_n(q)/(\mathrm{Sp}_n(q) \cap Z)$$

Note that $\mathrm{Sp}_n(q) \cap Z = \{\pm I_n\}$.

The next result is proved in [122, Section 3.5]. Note that $\mathrm{PSp}_2(q) = \mathrm{PSL}_2(q)$.

**Theorem 2.4.1** *Suppose $n \geqslant 2$ is even and $(n,q) \neq (2,2),(2,3),(4,2)$. Set $d = (2, q-1)$. Then $\mathrm{PSp}_n(q)$ is a simple group of order*

$$\frac{1}{d} q^{\frac{1}{4}n^2} \prod_{i=1}^{n/2} (q^{2i} - 1)$$

*In the exceptional cases, $\mathrm{PSp}_2(2) \cong S_3$, $\mathrm{PSp}_2(3) \cong A_4$ and $\mathrm{PSp}_4(2) \cong S_6$.*

Let $T = \mathrm{PSp}_n(q)$. The elements of $\mathrm{PGSp}_n(q) \setminus T$ are the nontrivial *diagonal automorphisms* of $T$, and $\mathrm{Inndiag}(T) = \mathrm{PGSp}_n(q)$ is the group of inner-diagonal automorphisms. Let $\mu$ be a primitive element of $\mathbb{F}_q$ and set $\delta = \hat{\delta}Z$ where $\hat{\delta} = [\mu I_{n/2}, I_{n/2}]$ with respect to the symplectic basis

$$\{e_1,\ldots,e_{n/2},f_1,\ldots,f_{n/2}\}$$

described in Proposition 2.2.1. Then $\hat{\delta}$ is a similarity of $B$ with $\tau(\hat{\delta}) = \mu$, and $\mathrm{PGSp}_n(q) = \langle T,\delta \rangle$. Note that if $q$ is even then

$$\mathrm{PGSp}_n(q) = \mathrm{PSp}_n(q) = \mathrm{Sp}_n(q)$$

Define $\phi : V \to V$ by

$$\phi : \sum_i(\lambda_i e_i + \mu_i f_i) \mapsto \sum_i(\lambda_i^p e_i + \mu_i^p f_i)$$

Then $B(u^\phi, w^\phi) = B(u,w)^p$ for all $u, w \in V$ and thus $\phi$ is a semisimilarity of $B$. Moreover, $\mathrm{P\Gamma Sp}_n(q) = \langle \mathrm{PGSp}_n(q), \phi \rangle$.

If $(n,p) \neq (4,2)$ then $\mathrm{Aut}(\mathrm{PSp}_n(q)) = \mathrm{P\Gamma Sp}_n(q)$ (see [46]). However, if $(n,p) = (4,2)$ then there is a duality of the corresponding polar space that interchanges the sets of totally isotropic 1-spaces and totally isotropic 2-spaces (see [118, p. 201–202], for example). This duality induces an automorphism $\rho$ of $T = \mathrm{PSp}_4(q)$ such that $\rho^2 = \phi$. In particular, $\langle \rho \rangle \cong C_{2f}$ and we have $\mathrm{Aut}(\mathrm{PSp}_4(q)) = \langle \mathrm{PSp}_4(q), \rho \rangle$. Note that if $q = 2$, then $\rho$ corresponds to the exceptional automorphism of $S_6 \cong \mathrm{PSp}_4(2)$. The map $\rho$ will be discussed further in Section 3.4.6.

In conclusion,

$$\mathrm{Out}(\mathrm{PSp}_n(q)) = C_{(2,q-1)}{:}C_{fa}$$

where $a = 2$ if $(n,p) = (4,2)$, otherwise $a = 1$.

Following [67, Definition 2.5.13], the elements of $\mathrm{P\Gamma Sp}_n(q)$ that are conjugate to $\phi^i$ for some $1 \leqslant i < f$ are the *field automorphisms* of $T$. Similarly, if $(n,p) = (4,2)$ then the conjugates of $\phi^i \rho$ (with $1 \leqslant i \leqslant f$) are the *graph-field automorphisms* of $T$. Note that if $f$ is odd then $T$ admits graph-field automorphisms of order 2. There are no graph automorphisms of $T$.

We end this introductory discussion of symplectic groups by describing two field extension embeddings.

**Construction 2.4.2** *The embedding* $\mathrm{Sp}_{n/2}(q^2) \leqslant \mathrm{Sp}_n(q)$, $n \equiv 0 \pmod 4$.

Let $V$ be an $\frac{n}{2}$-dimensional vector space over $\mathbb{F}_{q^2}$ and let $\overline{B}_1 : V \times V \to \mathbb{F}_{q^2}$ be a nondegenerate alternating form on $V$. Then $V$ is an $n$-dimensional vector space over $\mathbb{F}_q$, and $B_1 = \mathrm{Tr} \circ \overline{B}_1 : V \times V \to \mathbb{F}_q$ is a nondegenerate alternating form on this $n$-space, where $\mathrm{Tr} : \mathbb{F}_{q^2} \to \mathbb{F}_q$ is the trace map $\lambda \mapsto \lambda + \lambda^q$. Moreover, if $U$ is an $\mathbb{F}_{q^2}$-subspace of $V$ that is totally isotropic (respectively nondegenerate) with respect to $\overline{B}_1$, then it is also totally isotropic (respectively nondegenerate) with respect to $B_1$.

Let $g \in \mathrm{GL}_{n/2}(q^2)$ be a similarity of $\overline{B}_1$; say $g$ preserves $\overline{B}_1$ up to multiplication by $\tau(g) \in \mathbb{F}_{q^2}$. If $\tau(g) \in \mathbb{F}_q$ then $B_1(u^g, v^g) = \tau(g)B_1(u,v)$ for all $u, v \in V$ and so $g$ induces a similarity of $B_1$. This allows us to embed $\mathrm{Sp}_{n/2}(q^2)$ inside $\mathrm{Sp}_n(q)$. Note that not all of $\mathrm{GSp}_{n/2}(q^2)$ embeds in $\mathrm{GSp}_n(q)$ (since we require the condition $\tau(g) \in \mathbb{F}_q$).

For $q$ odd, it will be useful to identify a standard basis for $V$ with respect to this embedding. To do this, let $\{e_1, \ldots, e_{n/4}, f_1, \ldots, f_{n/4}\}$ be a symplectic basis for $V$ over $\mathbb{F}_{q^2}$ and let $\xi \in \mathbb{F}_{q^2}$ be a scalar of order $2(q-1)_2$, where $(q-1)_2$ denotes the largest 2-power dividing $q-1$. Then $\mathrm{Tr}(\xi) = 0$ and

$$\{e_1, \xi e_1, \ldots, e_{n/4}, \xi e_{n/4}, f_1, \xi f_1, \ldots, f_{n/4}, \xi f_{n/4}\}$$

is a basis for $V$ over $\mathbb{F}_q$. This is not a symplectic basis with respect to $B_1$, but with a minor modification we obtain the following symplectic basis:

$$\left\{ e_1, \xi e_1, \ldots, e_{n/4}, \xi e_{n/4}, \frac{1}{2}f_1, \frac{1}{2}\xi^{-1}f_1, \ldots, \frac{1}{2}f_{n/4}, \frac{1}{2}\xi^{-1}f_{n/4} \right\} \qquad \square$$

**Construction 2.4.3** *The embedding* $\mathrm{GU}_{n/2}(q) \leqslant \mathrm{Sp}_n(q)$.

Let $V$ be an $\frac{n}{2}$-dimensional vector space over $\mathbb{F}_{q^2}$, let $\overline{B}_2 : V \times V \to \mathbb{F}_{q^2}$ be a nondegenerate hermitian form and fix $\lambda \in \mathbb{F}_{q^2}$ such that $\mathrm{Tr}(\lambda) = 0$. We may view $V$ as an $n$-dimensional space over $\mathbb{F}_q$. Then

$$B_2 = \mathrm{Tr} \circ (\lambda \overline{B}_2) : V \times V \to \mathbb{F}_q$$

is a nondegenerate alternating form on this $n$-space, and it is easy to check that if $U$ is an $\mathbb{F}_{q^2}$-subspace of $V$ that is totally isotropic (respectively nondegenerate) with respect to $\overline{B}_2$, then it is also totally isotropic (respectively nondegenerate) with respect to $B_2$.

Let $g \in \mathrm{GL}_{n/2}(q^2)$ be a similarity of $\overline{B}_2$. If $\tau(g) \in \mathbb{F}_q$ then $B_2(u^g, v^g) = \tau(g)B_2(u,v)$ for all $u, v \in V$, so $g$ induces a similarity of $B_2$. This yields an embedding of $\mathrm{GU}_{n/2}(q)$ in $\mathrm{Sp}_n(q)$, and we note that not all of $\Delta \mathrm{U}_{n/2}(q)$ embeds in $\mathrm{GSp}_n(q)$.

Finally, let us give standard bases for $V$ with respect to this embedding when $q$ is odd. As in the previous construction, let $\xi \in \mathbb{F}_{q^2}$ be an element of order $2(q-1)_2$ and note that $\mathrm{Tr}(\xi) = 0$, so we can take $\lambda = \xi$ in the above definition of $B_2$. Let $\{v_1, \ldots, v_{n/2}\}$ be an orthonormal basis for $V$ over $\mathbb{F}_{q^2}$ with respect to $\overline{B}_2$. One can check that

$$\left\{ v_1, \ldots, v_{n/2}, -\frac{1}{2}\xi^{-1}v_1, \ldots, -\frac{1}{2}\xi^{-1}v_{n/2} \right\}$$

is a symplectic basis for $V$ with respect to $B_2$.

If $n/2$ is even and $\{e_1, \ldots, e_{n/4}, f_1, \ldots, f_{n/4}\}$ is an hermitian $\mathbb{F}_{q^2}$-basis for $V$ (with respect to $\overline{B}_2$) then

$$\left\{ e_1, \xi e_1, \ldots, e_{n/4}, \xi e_{n/4}, -\frac{1}{2}\xi^{-1}f_1, \frac{1}{2}\xi^{-2}f_1, \ldots, -\frac{1}{2}\xi^{-1}f_{n/4}, \frac{1}{2}\xi^{-2}f_{n/4} \right\}$$

is a symplectic basis over $\mathbb{F}_q$ with respect to $B_2$. Similarly, if $n/2$ is odd and $\{e_1, \ldots, e_{(n-2)/4}, f_1, \ldots, f_{(n-2)/4}, x\}$ is an hermitian $\mathbb{F}_{q^2}$-basis for $V$ then

$$\left\{ e_1, \xi e_1, \ldots, e_{(n-2)/4}, \xi e_{(n-2)/4}, x, -\frac{1}{2}\xi^{-1}f_1, \frac{1}{2}\xi^{-2}f_1, \ldots \right.$$

$$\left. \ldots, -\frac{1}{2}\xi^{-1}f_{(n-2)/4}, \frac{1}{2}\xi^{-2}f_{(n-2)/4}, -\frac{1}{2}\xi^{-1}x \right\}$$

is a symplectic basis over $\mathbb{F}_q$ with respect to $B_2$. $\qquad\square$

## 2.5 Orthogonal groups

Let $V$ be an $n$-dimensional vector space over $\mathbb{F}_q$, let $Q$ be a nondegenerate quadratic form on $V$ and let $B_Q$ be the associated bilinear form on $V$. The corresponding *orthogonal group* is defined to be the isometry group of $Q$, denoted by $\mathrm{O}(V, Q)$. We write $\mathrm{SO}(V, Q)$ for the subgroup of isometries with determinant 1. The group of similarities is denoted by $\mathrm{GO}(V, Q)$, and the group of semisimilarities by $\Gamma\mathrm{O}(V, Q)$.

As discussed in Section 2.2.6, when $n$ is even there are two isometry types of nondegenerate quadratic forms on $V$, namely the hyperbolic and elliptic forms, and we denote the corresponding isometry groups by $\mathrm{O}_n^+(q)$ and $\mathrm{O}_n^-(q)$ respectively. (Note that this use of signs is consistent with the earlier notion of nondegenerate spaces of type $\pm$.) If $nq$ is odd, then once again there are two isometry types, but only one similarity type, so the corresponding isometry groups are isomorphic. (Recall that if $n$ is odd and $q$ is even then every quadratic form on $V$ is degenerate, but we can still define an orthogonal group in this situation – see Remark 2.5.1.) We denote the isometry group of a parabolic quadratic form by $\mathrm{O}_n(q)$, or occasionally $\mathrm{O}_n^\circ(q)$.

For $\varepsilon \in \{\pm, \circ\}$ we define the groups $\mathrm{SO}_n^\varepsilon(q)$, $\mathrm{GO}_n^\varepsilon(q)$ and $\Gamma\mathrm{O}_n^\varepsilon(q)$ in the obvious way. With the exception of $\mathrm{SO}_4^+(2) \cong S_3 \wr S_2$, the group $\mathrm{SO}_n^\varepsilon(q)$ contains a unique subgroup of index two that we denote by $\Omega_n^\varepsilon(q)$ (see [4, 22.9]; we will say more about this subgroup shortly). Sometimes it will also be convenient to use the notation $\mathrm{O}^\varepsilon(V)$, $\mathrm{SO}^\varepsilon(V)$, $\Omega^\varepsilon(V)$, etc., or just $\mathrm{O}(V)$, $\mathrm{SO}(V)$ and so on, if the isometry type of $V$ is clear.

**Remark 2.5.1** Suppose $n \geqslant 3$ is odd and $q$ is even, and let $Q$ be a quadratic form on $V$ with associated symmetric bilinear form $B_Q$. Recall that the *radical* of $B_Q$, denoted $\mathrm{rad}(B_Q)$, is the subspace

$$\mathrm{rad}(B_Q) = V^{\perp} = \{u \in V \mid B_Q(u,v) = 0 \text{ for all } v \in V\}$$

Note that $B_Q$ is alternating (since $q$ is even), so Proposition 2.2.1 implies that $\mathrm{rad}(B_Q)$ is nontrivial and thus $Q$ is degenerate. The radical of $Q$, denoted $\mathrm{rad}(Q)$, is the subspace of singular vectors in $\mathrm{rad}(B_Q)$. We say that $Q$ is *nonsingular* if $\mathrm{rad}(Q) = \{0\}$. As explained in [122, Section 3.4.7], if $Q$ is nonsingular then $\mathrm{rad}(B_Q)$ is 1-dimensional and $B_Q$ induces a nondegenerate alternating form $\overline{B}_Q$ on the quotient space $V/\mathrm{rad}(B_Q)$. Moreover, in this situation the isometry group of $Q$ on $V$, denoted by $O_n(q)$, is isomorphic to the isometry group of $\overline{B}_Q$ on $V/\mathrm{rad}(B_Q)$. In other words,

$$O_n(q) \cong \mathrm{Sp}_{n-1}(q)$$

In particular, $O_n(q)$ is simple unless $(n,q) = (3,2)$ or $(5,2)$.

Recall that we can define a homomorphism $\tau : \mathrm{GO}_n^{\varepsilon}(q) \to \mathbb{F}_q^{\times}$ such that $Q(v^g) = \tau(g)Q(v)$ for all $v \in V$ (see (2.2.7)). The next lemma describes the image of $\tau$. In the statement, $(\mathbb{F}_q^{\times})^2$ denotes the set of all squares in $\mathbb{F}_q^{\times}$.

**Lemma 2.5.2**

$$\tau(\mathrm{GO}_n^{\varepsilon}(q)) = \begin{cases} \mathbb{F}_q^{\times} & n \text{ even} \\ (\mathbb{F}_q^{\times})^2 & nq \text{ odd} \end{cases}$$

*Proof* Since $\mathrm{GO}_n^{\varepsilon}(q)$ contains all scalars, it follows that the image of $\tau$ contains $(\mathbb{F}_q^{\times})^2$. In particular, if $q$ is even then $\tau(\mathrm{GO}_n^{\varepsilon}(q)) = \mathbb{F}_q^{\times}$ as claimed. For the remainder, we may assume that $q$ is odd.

First assume $n$ is even and $\varepsilon = +$. Let $\mu$ be a primitive element of $\mathbb{F}_q$ and set $g = [\mu I_{n/2}, I_{n/2}]$ with respect to a hyperbolic basis $\{e_1, \ldots, e_{n/2}, f_1, \ldots, f_{n/2}\}$. Then $g \in \mathrm{GO}_n^+(q)$ and $\tau(g) = \mu$, so $\tau$ is onto in this case.

Now assume that $\varepsilon = -$. First consider the case $n = 2$. Let $\{x', y'\}$ be a basis for $V$ as described in Remark 2.2.8, so $B_Q(x', y') = 0$, $Q(x') = 1$ and $Q(y') = \alpha$ with $-\alpha \in \mathbb{F}_q$ a nonsquare. Now

$$g = \begin{pmatrix} 0 & 1 \\ \alpha & 0 \end{pmatrix} \in \mathrm{GO}_2^-(q)$$

is a similarity with $\tau(g) = \alpha$. Therefore, if $\alpha$ is a nonsquare then $\tau(\mathrm{GO}_2^-(q)) = \mathbb{F}_q^{\times}$. Now assume $\alpha$ is a square. Since the set of all squares in $\mathbb{F}_q$ is not closed under addition, there exists $\beta \in \mathbb{F}_q$ such that $\gamma = 1 + \alpha\beta^2$ is a nonsquare. Then

$$g = \begin{pmatrix} 1 & \beta \\ -\alpha\beta & 1 \end{pmatrix} \in \mathrm{GO}_2^-(q)$$

and $\tau(g) = \gamma$, so we conclude that $\tau$ is surjective when $n = 2$. Now, if $n > 2$ we may write $V = U \perp W$, where $U$ is an elliptic 2-space and $W$ is a hyperbolic $(n-2)$-space. In terms of this decomposition, we can take a block-diagonal element $[g, h] \in \mathrm{GO}_n^-(q)$ such that $\tau(g) = \tau(h)$ is a nonsquare. Once again, we conclude that $\tau$ is onto.

Finally, suppose $nq$ is odd. First recall that the group of similarities of $Q$ coincides with the group of similarities of the associated bilinear form $B_Q$. If $J$ is the Gram matrix of $B_Q$ with respect to some fixed basis of $V$, then $A \in \mathrm{GL}_n(q)$ is a similarity of $B_Q$ with $\tau(A) = \lambda$ if and only if $AJA^\mathsf{T} = \lambda J$ (see (2.2.5)). Taking determinants, we deduce that $\lambda^n = \det(A)^2$ and thus $\lambda^n$ is an $\mathbb{F}_q$-square. Since $n$ is odd, it follows that $\lambda$ is an $\mathbb{F}_q$-square and we deduce that the image of $\tau$ is precisely $(\mathbb{F}_q^\times)^2$. $\qquad\square$

**Remark 2.5.3** Suppose $q$ is odd and let $Q$ and $Q'$ be similar nondegenerate quadratic forms on $V$. Fix a basis $\beta$ of $V$ and let $g : V \to V$ be a similarity such that $Q'(u^g) = \lambda Q(u)$ for all $u \in V$. Consider the Gram matrices $J_\beta$ and $J_{\beta g}$ of the corresponding bilinear forms $B_Q$ and $B_{Q'}$, respectively. Then $J_{\beta g} = \lambda J_\beta$ and thus $\det(J_{\beta g}) = \lambda^n \det(J_\beta)$. Therefore, if $n$ is even then $Q$ and $Q'$ have the same discriminant. However, if $n$ is odd and $\lambda$ is an $\mathbb{F}_q$-nonsquare, then $Q$ and $Q'$ have different discriminants.

Let $Q$ be a hyperbolic quadratic form on $V$ and let $\mathscr{U}$ denote the set of maximal totally singular subspaces of $V$ (so $\dim U = n/2$ for all $U \in \mathscr{U}$). We define a relation $\sim$ on $\mathscr{U}$ as follows:

$$U \sim W \text{ if and only if } \dim U - \dim(U \cap W) \text{ is even.}$$

**Proposition 2.5.4**

(i) *The relation $\sim$ is an equivalence relation on $\mathscr{U}$ with exactly two equivalence classes $\mathscr{U}_1, \mathscr{U}_2$.*

(ii) *The partition $\mathscr{U} = \mathscr{U}_1 \cup \mathscr{U}_2$ is $\mathrm{O}_n^+(q)$-invariant, and the setwise stabiliser of $\mathscr{U}_1$ is $\mathrm{SO}_n^+(q)$ for $q$ odd, and $\Omega_n^+(q)$ for $q$ even.*

(iii) *$\Omega_n^+(q)$ acts transitively on both $\mathscr{U}_1$ and $\mathscr{U}_2$.*

*Proof* Part (i) follows from [4, 22.13]. See [86, Lemma 2.5.8] for (ii) and (iii). Incidentally, note that (ii) allows us to define $\Omega_4^+(2) \cong S_3 \times S_3$.  □

Let $v \in V$ be a nonsingular vector and define the *reflection in* $v$ to be the invertible linear map

$$
\begin{aligned}
r_v: \quad V &\rightarrow \quad V \\
w &\mapsto \quad w - \frac{B_Q(v,w)}{Q(v)}v
\end{aligned}
\tag{2.5.1}
$$

Note that $r_v$ maps $v$ to $-v$, and fixes the hyperplane $v^\perp$ pointwise. Also note that $r_v$ is an isometry of $Q$. With the exception of $O_4^+(2)$, every element of $O_n^\varepsilon(q)$ can be written as a product of reflections (see [118, Theorems 11.39 and 11.41]). Excluding the case $O_4^+(2)$, let $H$ be the subgroup of elements in $O_n^\varepsilon(q)$ that can be written as the product of an even number of reflections. By [118, Theorem 11.44], $H = \mathrm{SO}_n^\varepsilon(q)$ if $q$ is odd, and $H = \Omega_n^\varepsilon(q)$ if both $n$ and $q$ are even.

We introduce the spinor norm in order to identify $\Omega_n^\varepsilon(q)$ when $q$ is odd. Let $x \in \mathrm{SO}_n^\varepsilon(q)$ and write $x = r_{v_1} r_{v_2} \cdots r_{v_k}$ as a product of reflections for some nonsingular vectors $v_i \in V$. The *spinor norm* of $x$, denoted $\theta(x)$, is defined by

$$
\theta(x) = \prod_{i=1}^k B_Q(v_i, v_i) \ (\mathrm{mod}\ (\mathbb{F}_q^\times)^2) \in \{\square, \boxtimes\}
$$

It can be shown that the map

$$
\theta: \mathrm{SO}_n^\varepsilon(q) \rightarrow \{\square, \boxtimes\}
$$

is a well defined surjective homomorphism with kernel $\Omega_n^\varepsilon(q)$ (see [4, p. 97–98] for a proof using Clifford algebras, and [118, p. 163–164] for an alternative argument). For future reference, we record this fact in the following lemma.

**Lemma 2.5.5** *Suppose* $x \in \mathrm{SO}_n^\varepsilon(q)$, *where* $q$ *is odd. Then* $x \in \Omega_n^\varepsilon(q)$ *if and only if the spinor norm of* $x$ *is a square.*

Using this characterisation of $\Omega_n^\varepsilon(q)$ in terms of the spinor norm, we can determine when certain elements of $O_n^\varepsilon(q)$ belong to $\Omega_n^\varepsilon(q)$.

**Lemma 2.5.6** *Suppose* $n$ *is even and* $q$ *is odd. Then* $-I_n \in \Omega_n^\varepsilon(q)$ *if and only if* $D(Q) = \square$.

*Proof* Let $\beta = \{v_1, \ldots, v_n\}$ be the orthogonal basis for $V$ provided by Lemma 2.2.14, and note that $-I_n = r_{v_1} r_{v_2} \cdots r_{v_n}$. If $D(Q) = \square$ then $\beta$ is orthonormal, so $\theta(-I_n) = \square$ and thus $-I_n \in \Omega_n^\varepsilon(q)$ by Lemma 2.5.5. Similarly, if $D(Q) = \boxtimes$ then $\theta(-I_n) = \boxtimes$ and so $-I_n \notin \Omega_n^\varepsilon(q)$. $\square$

The next result is essentially a special case of Lemma 2.2.17. For a proof see [86, Lemma 4.1.9].

**Lemma 2.5.7** *Suppose $n$ is even and let $W_1$ and $W_2$ be complementary totally singular subspaces of dimension $n/2$. Then with respect to a suitable basis of $W_1 \oplus W_2$, the stabiliser in $O_n^+(q)$ of $W_1$ and $W_2$ is the group of all block-diagonal matrices of the form*

$$\begin{pmatrix} A & 0 \\ 0 & A^{-\mathsf{T}} \end{pmatrix}$$

*where $A \in \mathrm{GL}_{n/2}(q)$. Moreover, such an element lies in $\Omega_n^+(q)$ if and only if $\det(A) \in \mathbb{F}_q$ is a square.*

**Lemma 2.5.8** *Suppose $n$ is even and let $W_1$ and $W_2$ be complementary totally singular subspaces of dimension $n/2$. Then $\Omega_n^+(q)$ contains an element interchanging $W_1$ and $W_2$ if and only if $n \equiv 0 \pmod 4$.*

*Proof* By Proposition 2.5.4, $W_1$ and $W_2$ lie in the same $\Omega_n^+(q)$-orbit on maximal totally singular subspaces if and only if $n/2$ is even, so we can assume $n/2$ is even. Write $W_1 = \langle u_1, \ldots, u_{n/2} \rangle$ and $W_2 = \langle v_1, \ldots, v_{n/2} \rangle$, where $B_Q(u_i, v_j) = \delta_{i,j}$, and let $x \in O_n^+(q)$ be the element interchanging $u_i$ and $v_i$ for each $i$. Since $W_1$ and $W_2$ are in the same $\Omega_n^+(q)$-orbit, Proposition 2.5.4 implies that $x \in \mathrm{SO}_n^+(q)$ if $q$ is odd, and $x \in \Omega_n^+(q)$ if $q$ is even, so we may assume that $q$ is odd. Now

$$x = r_{u_1 - v_1} r_{u_2 - v_2} \cdots r_{u_{n/2} - v_{n/2}}$$

and $B_Q(u_i - v_i, u_i - v_i) = -2$ for all $i$, so $\theta(x) = \square$ and thus $x \in \Omega_n^+(q)$ by Lemma 2.5.5. $\square$

Let $V$ be an $n$-dimensional vector space over $\mathbb{F}_q$ equipped with a nondegenerate quadratic form $Q$ of type $\varepsilon$ and associated bilinear form $B_Q$. Write $V = U \perp W$, where $U$ and $W$ are nondegenerate subspaces, and set $\Omega(V) = \Omega_n^\varepsilon(q)$. For $X \in \{\Omega, \mathrm{SO}, \mathrm{O}\}$, let $X(U)$ be the subgroup of $O_n^\varepsilon(q)$ comprising block-diagonal matrices of the form $[g, I_W]$ with $g \in X(U, Q|_U)$, and define $X(W)$ analogously.

**Lemma 2.5.9**

(i) $\Omega(U) \times \Omega(W) \leqslant \Omega(V)$.
(ii) $\mathrm{O}(U) \cap \Omega(V) = \Omega(U)$.

(iii) *If U and W are isometric, then the diagonal subgroup*

$$\{[g,g] \mid g \in O(U)\} \leqslant O(U) \times O(W)$$

*is contained in* $\Omega(V)$.

*Proof*  See [86, Lemma 4.1.1] for parts (i) and (ii). For (iii), let $g \in O(U)$ and set $x = [g,g]$, which is a block-diagonal matrix with respect to the decomposition $V = U \perp W$. Write $g = r_{u_1} \cdots r_{u_k}$ as a product of reflections for some nonsingular vectors $u_i \in U$. Since $U$ and $W$ are isometric, there are nonsingular vectors $w_i \in W$ such that $x = r_{u_1} \cdots r_{u_k} r_{w_1} \cdots r_{w_k}$ and $B_Q(u_i, u_i) = B_Q(w_i, w_i)$ for each $i$, so $\theta(x) = \square$ and thus $x \in \Omega(V)$ by Lemma 2.5.5.                        $\square$

By Witt's Lemma (see Lemma 2.2.15), the isometry group $O_n^{\varepsilon}(q)$ acts transitively on the set of nondegenerate $m$-dimensional subspaces of $V$ of a given isometry type, and also on the set of totally singular $m$-spaces for any $m$. In fact, the next lemma reveals that the subgroup $\Omega_n^{\varepsilon}(q)$ also acts transitively on these sets (with the exception of totally singular $\frac{n}{2}$-spaces when $\varepsilon = +$).

**Lemma 2.5.10**  *The group $\Omega_n^{\varepsilon}(q)$ acts transitively on each of the following sets of subspaces of V:*

(i) *nondegenerate m-spaces of a fixed isometry type;*
(ii) *totally singular m-spaces, $m < n/2$;*
(iii) *nonsingular 1-spaces, q even.*

*Proof*  First we will show that $SO_n^{\varepsilon}(q)$ acts transitively on the sets in (i) and (ii). This is clear if $q$ is even since $O_n^{\varepsilon}(q) = SO_n^{\varepsilon}(q)$, so let us assume $q$ is odd. Let $U$ be a nondegenerate $m$-dimensional subspace of $V$ and let $u \in U$ be a nonsingular vector. Since $|O_n^{\varepsilon}(q) : SO_n^{\varepsilon}(q)| = 2$ it follows that $O_n^{\varepsilon}(q) = \langle SO_n^{\varepsilon}(q), x \rangle$, where $x$ is the reflection $r_u$ (see (2.5.1)). But $x$ fixes $U$, so $U^{SO_n^{\varepsilon}(q)} = U^{O_n^{\varepsilon}(q)}$ and thus $SO_n^{\varepsilon}(q)$ acts transitively on the set of nondegenerate $m$-spaces of a fixed isometry type. To deduce transitivity on totally singular $m$-spaces with $m < n/2$, observe that there exists a nonsingular $v \in V$ such that $v^{\perp}$ contains a totally singular subspace of dimension $\lfloor (n-1)/2 \rfloor$. Then $O_n^{\varepsilon}(q) = \langle SO_n^{\varepsilon}(q), y \rangle$, where $y = r_v$ fixes a totally singular $m$-space for all $m < n/2$. The result follows.

Now let us turn to the transitivity of $\Omega_n^{\varepsilon}(q)$. Let $W$ be a nondegenerate $m$-space and let $y \in SO(W) \setminus \Omega(W)$. Since $|SO_n^{\varepsilon}(q) : \Omega_n^{\varepsilon}(q)| = 2$, Lemma 2.5.9(ii) implies that $SO_n^{\varepsilon}(q) = \langle \Omega_n^{\varepsilon}(q), z \rangle$ where $z$ centralises $W^{\perp}$ and induces $y$ on $W$. Thus $W^{SO_n^{\varepsilon}(q)} = W^{\Omega_n^{\varepsilon}(q)}$ and so $\Omega_n^{\varepsilon}(q)$ is transitive on the set of nondegenerate $m$-spaces of a given isometry type. Moreover, if $q$ is even we can choose $W$ so that $W^\perp$ contains a nonsingular 1-space, whence $\Omega_n^\varepsilon(q)$ is transitive on the set of nonsingular 1-spaces.

It remains to show that $\Omega_n^\varepsilon(q)$ is transitive in case (ii). Note that $n \geqslant 3$. Let $W$ be a nondegenerate 2-space such that $W^\perp$ is either hyperbolic or parabolic. Then $W^\perp$ contains a totally singular subspace of dimension $\lfloor (n-2)/2 \rfloor$ and thus the element $z$ constructed in the previous paragraph fixes a totally singular subspace of dimension $m$ for all $m \leqslant \lfloor (n-2)/2 \rfloor$. To complete the proof, we may assume that $n$ is odd, $W$ is hyperbolic and $m = (n-1)/2$. By Proposition 2.5.4, $\mathrm{SO}(W)$ and $\Omega(W)$ have the same orbits on the set of totally singular 1-dimensional subspaces of $W$, so in defining $z$ above we may assume that $y \in \mathrm{SO}(W) \setminus \Omega(W)$ fixes a totally singular 1-space. Then $z$ fixes a totally singular $m$-space and the result follows. □

Let $Z$ be the group of all scalar matrices in $\mathrm{GL}_n(q)$. We define $\mathrm{PGO}_n^\varepsilon(q) = \mathrm{GO}_n^\varepsilon(q)/Z$ and for $\mathrm{X} \in \{\Omega, \mathrm{SO}, \mathrm{O}\}$ we set

$$\mathrm{PX}_n^\varepsilon(q) = \mathrm{X}_n^\varepsilon(q)Z/Z \cong \mathrm{X}_n^\varepsilon(q)/(\mathrm{X}_n^\varepsilon(q) \cap Z)$$

Note that $\mathrm{O}_n^\varepsilon(q) \cap Z = \{\pm I_n\}$. For a proof of the next result, see [122, Sections 3.7 and 3.8].

**Theorem 2.5.11** *If $nq$ is odd and $n \geqslant 5$, then $\mathrm{P}\Omega_n(q) = \Omega_n(q)$ is a simple group of order*

$$\frac{1}{2}q^{\frac{1}{4}(n-1)^2} \prod_{i=1}^{(n-1)/2} (q^{2i} - 1)$$

*If $n \geqslant 6$ is even then $\mathrm{P}\Omega_n^\varepsilon(q)$ is a simple group of order*

$$\frac{1}{d}q^{\frac{1}{4}n(n-2)}(q^{n/2} - \varepsilon) \prod_{i=1}^{n/2-1} (q^{2i} - 1)$$

*where $d = (4, q^{n/2} - \varepsilon)$.*

The low-dimensional orthogonal groups admit the following exceptional isomorphisms:

$$\begin{aligned} \mathrm{P}\Omega_6^\varepsilon(q) &\cong \mathrm{PSL}_4^\varepsilon(q) & \mathrm{P}\Omega_4^-(q) &\cong \mathrm{PSL}_2(q^2) \\ \Omega_5(q) &\cong \mathrm{PSp}_4(q) & \Omega_3(q) &\cong \mathrm{PSL}_2(q) \end{aligned} \qquad (2.5.2)$$

and

$$\mathrm{P}\Omega_4^+(q) \cong \mathrm{PSL}_2(q) \times \mathrm{PSL}_2(q), \quad \mathrm{O}_2^\varepsilon(q) \cong D_{2(q-\varepsilon)}, \quad \Omega_2^\varepsilon(q) \cong C_{(q-\varepsilon)/(2,q-1)}$$

(see [86, Proposition 2.9.1]).

If $T = \mathrm{P}\Omega_n(q) = \Omega_n(q)$, where $n$ is odd, then we set

$$\mathrm{Inndiag}(T) = \mathrm{PSO}_n(q) = \mathrm{PGO}_n(q)$$

Now assume $n$ is even. For $T = \mathrm{P}\Omega_n^+(q)$ we define $\mathrm{Inndiag}(T)$ to be the index-two subgroup of $\mathrm{PGO}_n^+(q)$ that fixes setwise each of the two $\mathrm{P}\Omega_n^+(q)$-orbits on maximal totally singular subspaces of $V$ (see Proposition 2.5.4). For $T = \mathrm{P}\Omega_n^-(q)$ we need another model in order to define $\mathrm{Inndiag}(T)$. This is given in the following construction, which is based on [86, p. 40].

**Construction 2.5.12** *The embedding $\mathrm{O}_n^-(q) \leqslant \mathrm{O}_n^+(q^2)$.*

Let $V$ be an $n$-dimensional vector space over $\mathbb{F} = \mathbb{F}_q$, equipped with an elliptic quadratic form $Q$ and associated bilinear form $B$. Let $\beta = \{v_1, \ldots, v_n\}$ be a basis for $V$ and let $J_\beta$ be the Gram matrix of $B$ with respect to $\beta$. Set $\mathbb{F}_\# = \mathbb{F}_{q^2}$ and $V_\# = V \otimes_\mathbb{F} \mathbb{F}_\#$, which is an $n$-dimensional vector space over $\mathbb{F}_\#$ (note that scalar multiplication is defined by $\lambda(v \otimes \xi) = v \otimes \lambda\xi$, extended additively). We may regard $V$ as a subset of $V_\#$ by identifying each $v \in V$ with $v \otimes 1$. Note that $\beta_\# = \{v_1 \otimes 1, \ldots, v_n \otimes 1\}$ is a basis for $V_\#$ over $\mathbb{F}_\#$.

Let $B_\#$ be the symmetric bilinear form on $V_\#$ with Gram matrix $(J_\#)_{\beta_\#} = J_\beta$. For $q$ odd, let $Q_\#$ be the quadratic form on $V_\#$ defined by $Q_\#(v) = \frac{1}{2}B_\#(v, v)$ for all $v \in V_\#$, and for $q$ even let $Q_\#$ be the unique quadratic form on $V_\#$ polarising to $B_\#$ such that $Q_\#(v_i \otimes 1) = Q(v_i)$ for each $i$ (see Lemma 2.2.5). In both cases, $Q_\#$ is nondegenerate and $Q_\#(v \otimes 1) = Q(v)$ for all $v \in V$. In particular, if $W$ is a totally singular subspace of $V$ with respect to $Q$ then $W_\# = W \otimes_\mathbb{F} \mathbb{F}_\#$ is totally singular with respect to $Q_\#$.

Write $V = (U \oplus U^*) \perp W$ where $\{U, U^*\}$ is a dual pair of complementary maximal totally singular subspaces and $W$ is an elliptic 2-space. By Proposition 2.2.7, we may assume that $W = \langle x, y \rangle$ where $Q(x) = 1$, $Q(y) = \zeta$, $B(x, y) = 1$ and $\mathbf{x}^2 + \mathbf{x} + \zeta \in \mathbb{F}[\mathbf{x}]$ is irreducible. Fix $\lambda \in \mathbb{F}_\#$ such that $\lambda^2 + \lambda + \zeta = 0$. Then

$$z = x \otimes \lambda + y \otimes 1 \in W_\#$$

is a nonzero singular vector for $Q_\#$ and thus $\langle U \otimes_\mathbb{F} \mathbb{F}_\#, z \rangle$ is a totally singular subspace of dimension $n/2$. Therefore $Q_\#$ is a hyperbolic quadratic form.

If $g \in \mathrm{GL}_n(q)$ is a similarity of $Q$ then $g$ extends to a unique element $g_\# \in \mathrm{GL}_n(q^2)$ such that $(v \otimes \xi)^{g_\#} = v^g \otimes \xi$ for all $v \in V$ and $\xi \in \mathbb{F}_\#$. Moreover, $g_\#$ is a similarity of $Q_\#$. Therefore $\mathrm{GO}_n^-(q) \leqslant \mathrm{GO}_n^+(q^2)$, and we obtain an embedding of $\mathrm{PGO}_n^-(q)$ in $\mathrm{PGO}_n^+(q^2)$.

We now define $\mathrm{Inndiag}(\mathrm{P}\Omega_n^-(q))$ to be the stabiliser in $\mathrm{PGO}_n^-(q)$ of the two classes of maximal totally singular subspaces of $V_\#$ (see Proposition 2.5.4). In particular, observe that $\mathrm{Inndiag}(\mathrm{P}\Omega_n^-(q))$ is a subgroup of $\mathrm{Inndiag}(\mathrm{P}\Omega_n^+(q^2))$. $\qquad \square$

Next we turn to the automorphisms of the simple orthogonal groups. First assume $T = \Omega_n(q)$, where $nq$ is odd and $q = p^f$. In view of (2.5.2), we may assume that $n \geqslant 7$. Recall that

$$\text{Inndiag}(T) = \text{PSO}_n(q) = \text{PGO}_n(q)$$

and $T$ is an index-two subgroup. By [86, Proposition 2.6.1], there exists a basis $\{v_1, \ldots, v_n\}$ and a scalar $\lambda \in \mathbb{F}_q^\times$ such that

$$B_Q(v_i, v_j) = \lambda \delta_{i,j} \qquad (2.5.3)$$

for all $i, j$.

Define $\phi : V \to V$ by

$$\phi : \sum_i \lambda_i v_i \mapsto \lambda^{(p-1)/2} \sum_i \lambda_i^p v_i$$

Then $\phi$ is a semisimilarity of the underlying parabolic quadratic form $Q$, and $\Gamma O_n(q) = \langle GO_n(q), \phi \rangle$. In particular, $P\Gamma O_n(q) = \langle PGO_n(q), \phi \rangle$. Then

$$\text{Aut}(\Omega_n(q)) = P\Gamma O_n(q), \quad \text{Out}(\Omega_n(q)) = C_2 \times C_f$$

(see [46]). The $\text{Aut}(T)$-conjugates of $\phi^i$ for $1 \leqslant i < f$ are the *field automorphisms* of $T$. There are no graph or graph-field automorphisms.

The choice of $\lambda$ in (2.5.3) does not affect the isomorphism type of $T$ and so for convenience we will always assume that $\lambda = 1$, unless stated otherwise.

Now assume $n \geqslant 8$ is even. If $q$ is even then

$$P\Omega_n^\varepsilon(q) = \text{Inndiag}(P\Omega_n^\varepsilon(q)) < \text{PSO}_n^\varepsilon(q) = \text{PO}_n^\varepsilon(q) = \text{PGO}_n^\varepsilon(q)$$

and the proper inclusion has index two. If $q$ is odd then the relevant inclusions are described in Figure 2.5.1. All inclusions are index two unless $D(Q) = \boxtimes$, in which case $P\Omega_n^\varepsilon(q) = \text{PSO}_n^\varepsilon(q)$ (see [86, Table 2.1.D] and Lemma 2.2.9).

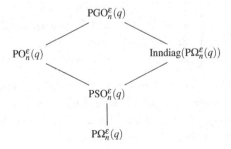

Figure 2.5.1  Some overgroups of $P\Omega_n^\varepsilon(q)$, $n$ even, $q$ odd

Let $T = \mathrm{P\Omega}_n^+(q)$ with $q = p^f$, and let $\{e_1, \ldots, e_{n/2}, f_1, \ldots, f_{n/2}\}$ be a standard basis for $V$ with respect to the defining hyperbolic quadratic form $Q$. Define $\phi : V \to V$ by setting

$$\phi : \sum_i (\lambda_i e_i + \mu_i f_i) \mapsto \sum_i (\lambda_i^p e_i + \mu_i^p f_i)$$

Then $\phi$ is a semisimilarity of $Q$, and we have $\mathrm{\Gamma O}_n^+(q) = \langle \mathrm{GO}_n^+(q), \phi \rangle$ and $\mathrm{P\Gamma O}_n^+(q) = \langle \mathrm{PGO}_n^+(q), \phi \rangle$. The $\mathrm{Aut}(T)$-conjugates of $\phi^i$ (with $1 \leqslant i < f$) are the *field automorphisms* of $T$.

If $n \neq 8$ then $\mathrm{Aut}(T) = \mathrm{P\Gamma O}_n^+(q)$ (see [46]). However, if $n = 8$ then there is a *triality* $\tau$ of the corresponding polar space, which cyclically permutes the three sets $\mathscr{P}_1$, $\mathscr{U}_1$ and $\mathscr{U}_2$, where $\mathscr{P}_1$ is the set of totally singular 1-spaces, and $\mathscr{U}_1$, $\mathscr{U}_2$ are the two $T$-orbits of maximal totally singular subspaces defined in Proposition 2.5.4. Then $\tau$ induces an automorphism of $T$, which we also denote by $\tau$, and we have $\mathrm{Aut}(T) = \langle \mathrm{P\Gamma O}_8^+(q), \tau \rangle$.

If $n \neq 8$ then the elements of $\mathrm{PGO}_n^+(q)$ that interchange the two $T$-orbits of maximal totally singular subspaces are the nontrivial *graph automorphisms* of $T$. These are simply the elements in $\mathrm{PGO}_n^+(q) \setminus \mathrm{Inndiag}(T)$. However, if $n = 8$ then there are additional graph automorphisms; these are the elements of $\langle \mathrm{PGO}_8^+(q), \tau \rangle$ that induce a nontrivial permutation on the set of subspace collections $\{\mathscr{P}_1, \mathscr{U}_1, \mathscr{U}_2\}$.

Fix an involutory graph automorphism $\gamma$ of $T$ (we can take $\gamma$ to be a reflection if $q$ is odd, and a transvection if $q$ is even). The elements that are $\mathrm{Aut}(T)$-conjugate to $\gamma \phi^i$, for some $1 \leqslant i < f$, are the *graph-field automorphisms* of $T$, unless $n = 8$. In the special case $n = 8$, the elements of $\mathrm{Aut}(T)$ that cyclically permute the sets $\mathscr{P}_1$, $\mathscr{U}_1$ and $\mathscr{U}_2$ are called *triality automorphisms*, and they are either graph or graph-field automorphisms (see Section 3.5.7 for more details).

By [86, Proposition 2.7.3], $\mathrm{Out}(\mathrm{P\Omega}_n^+(q))$ has the following structure:

$$\mathrm{Out}(\mathrm{P\Omega}_n^+(q)) = \begin{cases} C_2 \times C_f & q \text{ even}, n \neq 8 \\ S_3 \times C_f & q \text{ even}, n = 8 \\ C_2 \times C_2 \times C_f & q \equiv 3 \ (\mathrm{mod}\ 4), n \equiv 2 \ (\mathrm{mod}\ 4) \\ S_4 \times C_f & q \text{ odd}, n = 8 \\ D_8 \times C_f & \text{otherwise} \end{cases}$$

Now assume $T = \mathrm{P\Omega}_n^-(q)$ with $q = p^f$. Here the definition of a standard field automorphism $\phi$ is more complicated, depending on both $q$ and the discriminant of the underlying elliptic quadratic form $Q$. We follow the description in [86, p. 39]. If $q$ is even, then take a standard basis

$$\{e_1, \ldots, e_{n/2-1}, f_1, \ldots, f_{n/2-1}, x, y\}$$

(with $Q(x) = 1$, $Q(y) = \zeta$ and $B_Q(x,y) = 1$ as in Proposition 2.2.7(ii)) and define $\phi$ by

$$\phi : \alpha x + \beta y + \sum_i (\lambda_i e_i + \mu_i f_i) \mapsto \alpha^2 x + \beta^2 (\zeta x + y) + \sum_i (\lambda_i^2 e_i + \mu_i^2 f_i)$$

Now suppose $q$ is odd and let $\{v_1, \ldots, v_n\}$ be the orthogonal basis for $V$ given in Lemma 2.2.14 (so $B_Q(v_i, v_i) = 1$ for $i \geqslant 2$, and $B_Q(v_1, v_1) = \mu$, where $\mu = 1$ if $D(Q) = \square$, otherwise $\mu$ is a primitive element of $\mathbb{F}_q$). Consequently, if $D(Q) = \square$ then we define $\phi$ by

$$\phi : \sum_i \lambda_i v_i \mapsto \sum_i \lambda_i^p v_i$$

whereas if $D(Q) = \boxtimes$ we define $\phi$ by

$$\phi : \sum_i \lambda_i v_i \mapsto \lambda_1^p \mu^{(p-1)/2} v_1 + \sum_{i>1} \lambda_i^p v_i$$

In both cases $\phi$ is a semisimilarity of $Q$ and we have $\Gamma \mathrm{O}_n^-(q) = \langle \mathrm{GO}_n^-(q), \phi \rangle$, hence $\mathrm{P\Gamma O}_n^-(q) = \langle \mathrm{PGO}_n^-(q), \phi \rangle$. Note that $\mathrm{GO}_n^-(q) \cap \langle \phi \rangle$ may be nontrivial. In particular, $|\phi| = 2f$ if $D(Q) = \boxtimes$ or $q$ is even, otherwise $|\phi| = f$ is odd. By [46], we have $\mathrm{Aut}(T) = \mathrm{P\Gamma O}_n^-(q)$.

Following [67, Definition 2.5.13], the conjugates of $\phi^i$, where $\phi^i \neq 1$ has odd order, are the *field automorphisms* of $T$. Similarly, the *graph automorphisms* of $T$ are the elements in $\mathrm{P\Gamma O}_n^-(q) \setminus \mathrm{Inndiag}(T)$ of even order. The conjugacy classes of graph automorphisms of order two will be discussed in Sections 3.5.2 and 3.5.4. We observe that if $q$ is even then $\phi^f$ is a graph automorphism of type $b_1$ (see Section 3.5.4), and if $q$ is odd and $D(Q) = \boxtimes$, then $\phi^f$ is a graph automorphism of type $\gamma_1$ (see Section 3.5.2.14). There are no graph-field automorphisms of $T$.

By [86, Proposition 2.8.2], $\mathrm{Out}(\mathrm{P\Omega}_n^-(q))$ has the following structure:

$$\mathrm{Out}(\mathrm{P\Omega}_n^-(q)) = \begin{cases} C_{2f} & q \text{ even} \\ D_8 \times C_f & q \equiv 3 \pmod 4, n \equiv 2 \pmod 4 \\ C_2 \times C_{2f} & \text{otherwise} \end{cases}$$

To close this section on orthogonal groups we describe two field extension embeddings.

**Construction 2.5.13** *The embedding* $\mathrm{O}_{n/2}^{\varepsilon'}(q^2) \leqslant \mathrm{O}_n^\varepsilon(q)$, $q$ *odd.*

Let $V$ be an $\frac{n}{2}$-dimensional vector space over $\mathbb{F}_{q^2}$, where $q$ is odd, and let $\overline{Q} : V \to \mathbb{F}_{q^2}$ be a nondegenerate quadratic form with associated bilinear form $\overline{B}$. Then $Q = \mathrm{Tr} \circ \overline{Q} : V \to \mathbb{F}_q$ is a nondegenerate quadratic form on $V$ (viewed

Table 2.5.1 *The relationship between $\overline{Q}$ and $Q$*

| $\overline{Q}(v)$ | $q \pmod 4$ | Restriction of $Q$ to $\langle v\rangle_{\mathbb{F}_{q^2}}$ |
|---|---|---|
| □ | 1 | elliptic |
| □ | 3 | hyperbolic |
| ⊠ | 1 | hyperbolic |
| ⊠ | 3 | elliptic |

as an $n$-dimensional $\mathbb{F}_q$-space) with associated bilinear form $B = \mathrm{Tr} \circ \overline{B}$, where $\mathrm{Tr}: \mathbb{F}_{q^2} \to \mathbb{F}_q$ is the usual trace map.

Let $g \in \mathrm{GL}_{n/2}(q^2)$ be a similarity of $\overline{Q}$. If $\tau(g) \in \mathbb{F}_q$ then $Q(v^g) = \tau(g)Q(v)$ for all $v \in V$ and so $g$ also induces a similarity of $Q$. This allows us to embed the isometry group of $\overline{Q}$ in the isometry group of $Q$.

We consider three cases, according to the type of the quadratic form $\overline{Q}$. Let $\xi$ be an element of order $2(q-1)_2$ in $\mathbb{F}_{q^2}$ and note that

$$\{\alpha \in \mathbb{F}_{q^2} \mid \mathrm{Tr}(\alpha) = 0\} = \{\beta\xi \mid \beta \in \mathbb{F}_q\} \tag{2.5.4}$$

*Case 1. $\overline{Q}$ hyperbolic.* Here $Q$ is also hyperbolic since totally singular $\mathbb{F}_{q^2}$-subspaces with respect to $\overline{Q}$ are totally singular $\mathbb{F}_q$-subspaces with respect to $Q$. It is easy to check that if $\{e_1,\ldots,e_{n/4},f_1,\ldots,f_{n/4}\}$ is a hyperbolic basis for $V$ over $\mathbb{F}_{q^2}$ with respect to $\overline{Q}$, then

$$\left\{e_1,\xi e_1,\ldots,e_{n/4},\xi e_{n/4},\tfrac{1}{2}f_1,\tfrac{1}{2}\xi^{-1}f_1,\ldots,\tfrac{1}{2}f_{n/4},\tfrac{1}{2}\xi^{-1}f_{n/4}\right\}$$

is a hyperbolic basis for $V$ over $\mathbb{F}_q$ with respect to $Q$.

*Case 2. $\overline{Q}$ parabolic.* Here $n \equiv 2 \pmod 4$. Let $v \in V$ be a nonsingular vector with respect to $\overline{Q}$, say $\overline{Q}(v) = a \neq 0$, so $Q(\lambda v) = \mathrm{Tr}(\lambda^2 a)$ for all $\lambda \in \mathbb{F}_{q^2}$. If $a$ is a square in $\mathbb{F}_{q^2}$, then $\{\lambda^2 a \mid \lambda \in \mathbb{F}_{q^2}^\times\}$ is the set of all squares in $\mathbb{F}_{q^2}^\times$ and thus $\langle v\rangle_{\mathbb{F}_{q^2}}$ contains a nonzero singular vector with respect to $Q$ if and only if $\xi$ is a square (see (2.5.4)). In other words, if $a$ is a square then the restriction of $Q$ to $\langle v\rangle_{\mathbb{F}_{q^2}}$ is hyperbolic if and only if $q \equiv 3 \pmod 4$. Similarly, if $a$ is a nonsquare then the restriction is hyperbolic if and only if $q \equiv 1 \pmod 4$. We record this information in Table 2.5.1. Note that if $\langle v\rangle_{\mathbb{F}_{q^2}}$ is a hyperbolic 2-space with respect to $Q$, then

$$\left\{\lambda v, \tfrac{1}{4}\lambda\xi^{-1}v\right\}$$

is a standard basis, where $\lambda^2 = \xi/a$.

If $n \geqslant 6$ then $V = U \perp \langle x \rangle_{\mathbb{F}_{q^2}}$, where $x$ is nonsingular and the restriction of $\overline{Q}$ to $U$ is hyperbolic. The restriction of $Q$ to $U$ (viewed as an $(n-2)$-dimensional space over $\mathbb{F}_q$) is also hyperbolic and so the type of $Q$ on $V$ is determined by $\overline{Q}(x)$. In particular, since $\overline{Q}(x)$ can either be a square or nonsquare (in $\mathbb{F}_{q^2}$), $Q$ can be either hyperbolic or elliptic. This is determined as in Table 2.5.1. For instance, if $\overline{Q}(x)$ is a square and $q \equiv 1 \pmod 4$, then $Q$ is elliptic.

*Case 3. $\overline{Q}$ elliptic.* Let $\{e_1, \ldots, e_{n/4-1}, f_1, \ldots, f_{n/4-1}, x, y\}$ be an elliptic basis for $V$ over $\mathbb{F}_{q^2}$ with respect to $\overline{Q}$, where $\overline{Q}(x) = 1$, $\overline{B}(x,y) = 0$ and $\overline{Q}(y) = \alpha$ with $-\alpha \in \mathbb{F}_{q^2}$ a nonsquare (see Remark 2.2.8). Set $U = \langle x, y \rangle_{\mathbb{F}_{q^2}}$, so $U$ is a 2-dimensional anisotropic space over $\mathbb{F}_{q^2}$. Since $q^2 \equiv 1 \pmod 4$, it follows that $\alpha$ is an $\mathbb{F}_{q^2}$-nonsquare. Hence Table 2.5.1 indicates that one of the $\mathbb{F}_q$-spaces $\langle x, \xi x \rangle$ and $\langle y, \xi y \rangle$ is hyperbolic and the other is elliptic. For example, since $\overline{Q}(y) = \alpha$ is a nonsquare, we deduce that $\langle y, \xi y \rangle$ is hyperbolic if and only if $q \equiv 1 \pmod 4$. Therefore $U$ is a 4-dimensional elliptic $\mathbb{F}_q$-space with respect to $Q$, and thus the quadratic form $Q$ on $V$ is elliptic. $\qquad \square$

**Construction 2.5.14** *The embedding* $\mathrm{GU}_{n/2}(q) \leqslant \mathrm{O}_n^\varepsilon(q)$, $\varepsilon = (-)^{n/2}$.

Let $V$ be an $\frac{n}{2}$-dimensional vector space over $\mathbb{F}_{q^2}$ and let $\overline{B} : V \times V \to \mathbb{F}_{q^2}$ be a nondegenerate hermitian form on $V$. We can view $V$ as an $n$-dimensional space over $\mathbb{F}_q$, in which case the map $Q : V \to \mathbb{F}_q$ defined by $Q(v) = \overline{B}(v,v)$ is a nondegenerate quadratic form on this $n$-space, with associated symmetric bilinear form $B = \mathrm{Tr} \circ \overline{B}$.

Suppose $n \equiv 0 \pmod 4$. Here $Q$ is hyperbolic since totally isotropic $\mathbb{F}_{q^2}$-spaces with respect to $\overline{B}$ are totally singular $\mathbb{F}_q$-spaces with respect to $Q$. Moreover, if $U$ and $W$ are totally singular $\frac{n}{2}$-spaces over $\mathbb{F}_q$ with respect to $Q$, both arising from totally isotropic subspaces for $\overline{B}$, then $\dim U - \dim(U \cap W)$ is even. Hence they lie in the same $\Omega_n^+(q)$-orbit on maximal totally singular subspaces (see Proposition 2.5.4).

Now assume $n \equiv 2 \pmod 4$. By applying Proposition 2.2.2 we may write $V = U \perp \langle x \rangle_{\mathbb{F}_{q^2}}$, where $\overline{B}(x,x) = 1$ and the $\mathbb{F}_{q^2}$-space $U$ contains a totally isotropic subspace of dimension $(n-2)/4$. For all nonzero $\lambda \in \mathbb{F}_{q^2}$ we have $Q(\lambda x) = \lambda^{q+1} \overline{B}(x,x) \neq 0$. Thus the restriction of $Q$ to the 2-dimensional $\mathbb{F}_q$-space $\langle x \rangle_{\mathbb{F}_{q^2}}$ is anisotropic, and the restriction to $U$ (as an $\mathbb{F}_q$-space) is hyperbolic. Hence the quadratic form $Q$ on the $\mathbb{F}_q$-space $V$ is elliptic.

Let $g \in \mathrm{GL}_{n/2}(q^2)$ be a similarity of $\overline{B}$; say $g$ preserves $\overline{B}$ up to a scalar $\tau(g) \in \mathbb{F}_{q^2}$. Since $\overline{B}(v,v) \in \mathbb{F}_q$ it follows that $\tau(g) \in \mathbb{F}_q$ and thus $g$ is a similarity of $Q$. Hence $\Delta \mathrm{U}_{n/2}(q) \leqslant \mathrm{GO}_n^\varepsilon(q)$, where $\varepsilon = (-)^{n/2}$. We also observe that $\mathrm{GU}_{n/2}(q) \leqslant \mathrm{SO}_n^\varepsilon(q)$: if $g \in \mathrm{GU}_{n/2}(q)$ has determinant $\lambda$ as an $\mathbb{F}_{q^2}$-linear map, then $g$ has determinant $\lambda^{q+1} = 1$ when viewed as an $\mathbb{F}_q$-linear map. Moreover, we note that if $n \equiv 0 \pmod 4$ then $\Delta U_{n/2}(q)$ is contained in the stabiliser in $\mathrm{GO}_n^+(q)$ of each $\Omega_n^+(q)$-orbit on maximal totally singular subspaces. In particular, $\mathrm{GU}_{n/2}(q) \leqslant \Omega_n^+(q)$ if $q$ is even.

Let $\{v_1, \ldots, v_{n/2}\}$ be an orthonormal basis for $V$ over $\mathbb{F}_{q^2}$ with respect to $\overline{B}$, and recall the map $\varphi : V \to V$ given by

$$\varphi : \sum_i \lambda_i v_i \mapsto \sum_i \lambda_i^q v_i \tag{2.5.5}$$

Then $\varphi$ is $\mathbb{F}_q$-linear and $Q(v^\varphi) = Q(v)$ for all $v \in V$, so

$$\langle \mathrm{GU}_{n/2}(q), \varphi \rangle \leqslant \mathrm{O}_n^\varepsilon(q)$$

Finally, let us consider standard bases when $q$ is odd. Let $\xi \in \mathbb{F}_{q^2}$ be an element of order $2(q-1)_2$ and note that $\mathrm{Tr}(\xi) = 0$. If $n \equiv 0 \pmod 4$, let $\{e_1, \ldots, e_{n/4}, f_1, \ldots, f_{n/4}\}$ be an hermitian basis for $V$ over $\mathbb{F}_{q^2}$ with respect to $\overline{B}$. Then

$$\left\{ e_1, \xi e_1, \ldots, e_{n/4}, \xi e_{n/4}, \frac{1}{2}f_1, -\frac{1}{2}\xi^{-1}f_1, \ldots, \frac{1}{2}f_{n/4}, -\frac{1}{2}\xi^{-1}f_{n/4} \right\}$$

is a hyperbolic basis for $Q$. If $n \equiv 2 \pmod 4$, let

$$\left\{ e_1, \ldots, e_{(n-2)/4}, f_1, \ldots, f_{(n-2)/4}, x \right\}$$

be an hermitian basis for $V$ over $\mathbb{F}_{q^2}$ with respect to $\overline{B}$, with $\overline{B}(x,x) = 1$. Then

$$\left\{ e_1, \xi e_1, \ldots, e_{(n-2)/4}, \xi e_{(n-2)/4}, \frac{1}{2}f_1, -\frac{1}{2}\xi^{-1}f_1, \ldots \right.$$

$$\left. \ldots, \frac{1}{2}f_{(n-2)/4}, -\frac{1}{2}\xi^{-1}f_{(n-2)/4}, x, \xi x \right\}$$

is a basis for $V$ that satisfies the conditions in Remark 2.2.8 (indeed, observe that $Q(\xi x) = \overline{B}(\xi x, \xi x) = \xi^{q+1} = -\xi^2$ and $\xi^2 \in \mathbb{F}_q$ is a nonsquare). $\qquad\square$

**Remark 2.5.15** Let us consider the embedding $\mathrm{GU}_{n/2}(q) \leqslant \mathrm{O}_n^\varepsilon(q)$ described in Construction 2.5.14. Assume $q$ is odd and let $\xi \in \mathbb{F}_{q^2}$ be an element of order $2(q-1)_2$. Let $\{v_1, \ldots, v_{n/2}\}$ be an orthonormal basis for $V$ (as an $\mathbb{F}_{q^2}$-space) and define the map $\varphi$ as in (2.5.5). Let $U$ be the $\mathbb{F}_q$-span of this basis. Then $U$ is a nondegenerate $\frac{n}{2}$-space over $\mathbb{F}_q$ centralised by $\varphi$, and $U^\perp$ is the $\mathbb{F}_q$-span of $\{\xi v_1, \ldots, \xi v_{n/2}\}$, which coincides with the $(-1)$-eigenspace of $\varphi$. If $\varepsilon = -$ then $n \equiv 2 \pmod 4$ and thus $\varphi \notin \mathrm{SO}_n^-(q)$. Now assume that $\varepsilon = +$, so $n \equiv 0 \pmod 4$ and $U, U^\perp$ are isometric. Since $B(v_i, v_i) = 2$, the determinant of the Gram matrix of $B|_U$ is $2^{n/2}$, which is a square, so Lemma 2.2.9 implies that $U$ is elliptic if $n \equiv 4 \pmod 8$ and $q \equiv 3 \pmod 4$, otherwise $U$ is hyperbolic.

Let $\varepsilon'$ denote the type of $U$, and hence also the type of $U^\perp$. Since $D(Q|_{U^\perp})$ is a square, Lemma 2.5.6 implies that $-I_{n/2} \in \Omega_{n/2}^{\varepsilon'}(q)$ and thus $\varphi \in \Omega_n^+(q)$ by Lemma 2.5.9(ii).

Similarly, if $q$ is even then $\varphi \in \Omega_n^\varepsilon(q)$ if and only if $\varepsilon = +$. Indeed, we note that $\varphi$ has Jordan form $[J_2^{n/2}]$ on the natural $O_n^\varepsilon(q)$-module, where $J_2$ denotes a standard unipotent Jordan block of size 2. Hence, the result follows from Proposition 3.5.16 in Section 3.5.4. Alternatively, note that if $\varepsilon = +$ then $\varphi$ fixes the set of maximal totally singular subspaces of $V$ that arise from appropriate totally isotropic $\mathbb{F}_{q^2}$-spaces with respect to $\overline{B}$, hence $\varphi \in \Omega_n^+(q)$.

## 2.6 Subgroup structure

In the final section of this chapter we briefly discuss the subgroup structure of the finite classical groups, with a particular focus on their maximal subgroups. This information will be essential in our later work on primitive actions of classical groups.

The problem of determining the maximal subgroups of the finite simple groups and their automorphism groups has a long history. For example, in the famous letter that Galois wrote to Chevalier in 1832 [59], he states that if $p \geqslant 13$ is a prime then $p + 1$ is the smallest index of a proper subgroup of $\mathrm{PSL}_2(p)$. By 1901, the maximal subgroups of $\mathrm{PSL}_2(q)$ had been classified by Dickson [45], and analogous results for some of the other low-dimensional classical groups were obtained in the decades that followed. For example, Mitchell [104, 105] and Hartley [74] determined the maximal subgroups of $\mathrm{PSL}_3(q)$, $\mathrm{PSU}_3(q)$, and also $\mathrm{PSp}_4(q)$ for odd $q$. Indeed, most of the classical groups of dimension $n \leqslant 7$ had been dealt with prior to the Classification of Finite Simple Groups.

### 2.6.1 Aschbacher's theorem

Let $G$ be an almost simple classical group over $\mathbb{F}_q$ with socle $T$ and natural module $V$, where $\dim V = n$ and $q$ is a power of a prime $p$.

The main theorem on the subgroup structure of $G$ is a result of Aschbacher [3] from 1984, which provides a powerful analogue of the O'Nan–Scott Theorem for classical groups. Indeed, the theorem roughly states that if $H$ is a maximal subgroup of $G$, not containing $T$, then one of the following holds:

(i) $H$ is a known group with a clearly defined projective action on $V$, or

(ii) $H$ is an almost simple group with socle $S$, whose projective action on $V$ corresponds to an absolutely irreducible representation of the full covering group $\hat{S}$ of $S$ in $\mathrm{GL}(V)$.

(Recall that an $\mathbb{F}\hat{S}$-module $V$ is *absolutely irreducible* if the $\mathbb{E}\hat{S}$-module $V \otimes \mathbb{E}$ is irreducible for all extension fields $\mathbb{E}$ of $\mathbb{F}$.) Therefore, in some sense, Aschbacher's theorem 'reduces' the problem of determining the maximal subgroups of $G$ to that of finding its absolutely irreducible simple subgroups.

In order to state a more detailed version of Aschbacher's theorem, we need to introduce some new terminology. In [3], eight *geometric* collections of subgroups of $G$ are defined. These collections comprise the 'known' subgroups referred to in (i) above, and their members are defined in terms of the underlying geometry of $V$. This includes the stabilisers of appropriate subspaces of $V$, the stabilisers of suitable direct sum and tensor product decompositions of $V$, and the stabilisers of nondegenerate forms on $V$. We use the notation $\mathscr{C}_1, \ldots, \mathscr{C}_8$ to denote these subgroup collections, and we will briefly describe the subgroups in each of these collections in Section 2.6.2 below.

Two additional subgroup collections arise in the statement of Aschbacher's theorem. Firstly, there are the *non-geometric* subgroups that correspond to the irreducibly embedded almost simple groups referred to in (ii) above; we use $\mathscr{S}$ to denote this collection of subgroups. It turns out that an extra collection of so-called *novelty* subgroups arises if $T = \mathrm{PSp}_4(q)'$ ($p = 2$) or $\mathrm{P\Omega}_8^+(q)$; this collection is denoted by $\mathscr{N}$. We will say more about these two collections in Section 2.6.3.

We are now ready to state a version of Aschbacher's theorem. It is worth noting that Aschbacher's original paper [3, Section 15] only gives partial information in the case where $T = \mathrm{P\Omega}_8^+(q)$ and $G$ contains triality automorphisms (see p. 58 for the definition of a triality automorphism). In this special case, the maximal subgroups of $G$ were determined later by Kleidman [85].

**Theorem 2.6.1** *Let $G$ be an almost simple classical group with socle $T$, and let $H$ be a maximal subgroup of $G$ not containing $T$. Then $H$ belongs to one of the subgroup collections $\mathscr{C}_1, \ldots, \mathscr{C}_8, \mathscr{S}$ or $\mathscr{N}$.*

## 2.6.2 Geometric subgroups

A brief description of the ten subgroup collections arising in Aschbacher's theorem is presented in Table 1.4.2. In this section we give more details on the geometric subgroups comprising the eight $\mathscr{C}_i$ collections, $1 \leqslant i \leqslant 8$.

In the following discussion, and throughout this book, we adopt the precise definition of the $\mathscr{C}_i$ collections used by Kleidman and Liebeck [86] (note that this differs slightly from Aschbacher's original set-up; see [86, p. 58]). Their book is the definitive reference for detailed information on the existence, structure, maximality and conjugacy of geometric subgroups (more precisely, see

[86, Section 4.$i$] for specific details on the subgroups in the collection $\mathscr{C}_i$, $1 \leqslant i \leqslant 8$). The recent book [13] by Bray, Holt and Roney-Dougal provides a comprehensive treatment of all the maximal subgroups in the low-dimensional classical groups with $n \leqslant 12$, and Section 3.10 in [122] is another good source of information. The reader is also referred to Section 4.1 for more details on the subgroups comprising the $\mathscr{C}_1$ collection (the *subspace subgroups*), and Section 5.$i$ for those in $\mathscr{C}_i$ with $2 \leqslant i \leqslant 8$ (the geometric *non-subspace subgroups*).

We will write $f$ for the sesquilinear or quadratic form on $V$ associated to $T$, and accordingly we will refer to subspaces of $V$ as being nondegenerate, totally singular (see Remark 2.2.6), etc. Note that if $T = \mathrm{PSL}_n(q)$ then we take $f$ to be the zero bilinear form, so in this situation all subspaces of $V$ are totally singular. For the purposes of the following discussion we set

$$\mathbb{F} = \begin{cases} \mathbb{F}_{q^2} & \text{if } T = \mathrm{PSU}_n(q) \\ \mathbb{F}_q & \text{otherwise} \end{cases} \qquad (2.6.1)$$

We now consider the eight subgroup collections $\mathscr{C}_i$, $1 \leqslant i \leqslant 8$, in turn.

### 2.6.2.1 $\mathscr{C}_1$: Reducible subgroups

A subgroup $H$ of $G$ is in the $\mathscr{C}_1$ collection if one of the following holds:

(i) $H = G_U$ is the stabiliser in $G$ of a nontrivial $m$-dimensional subspace $U$ of $V$ such that one of the following holds:
  (a) $U$ is totally singular;
  (b) $U$ is nondegenerate and $m < n/2$;
  (c) $T = \mathrm{P}\Omega_n^{\pm}(q)$, $q$ is even and $U$ is a nonsingular 1-space.

(ii) $T = \mathrm{PSL}_n(q)$, $G$ contains a graph automorphism and $H$ is the stabiliser in $G$ of a pair of nontrivial subspaces $\{U, W\}$ of $V$, where $\dim U + \dim W = n$, $\dim U < n/2$ and either $U < W$ or $U \cap W = \{0\}$.

Note that if $H$ is one of the subgroups arising in (ii), then $H \cap T$ is non-maximal in $T$ (a maximal subgroup $H$ of $G$ with this property is said to be a *novelty* subgroup). In [3, Section 13], Aschbacher uses the notation $\mathscr{C}_1'$ for the reducible maximal subgroups that arise in this situation; in [86] they are included in the $\mathscr{C}_1$ collection, and we do the same. We refer the reader to [86, Section 4.1] and Section 4.1 (Table 4.1.1 in particular) for further details on the particular subgroups comprising this collection.

### 2.6.2.2 $\mathscr{C}_2$: Imprimitive subgroups

The subgroups $H \in \mathscr{C}_2$ are the stabilisers of appropriate direct sum decompositions

$$V = V_1 \oplus V_2 \oplus \cdots \oplus V_t$$

where $t \geqslant 2$ and $\dim V_i = a$ for all $i$ (so $n = at$). In particular, $H$ fixes $\{V_1, \ldots, V_t\}$ setwise. In addition, one of the following holds:

(i) The $V_i$ are totally singular and either $T = \mathrm{PSL}_n(q)$, or $t = 2$ and $T = \mathrm{PSU}_n(q)$, $\mathrm{PSp}_n(q)$ or $\mathrm{P\Omega}_n^+(q)$.

(ii) The $V_i$ are nondegenerate and pairwise orthogonal, and either they are all isometric, or $t = 2$, $T = \mathrm{P\Omega}_n^{\pm}(q)$ and $nq/2$ is odd.

See [86, Section 4.2] and Section 5.2 for further details.

### 2.6.2.3 $\mathscr{C}_3$: Extension field subgroups

These subgroups arise from prime degree field extensions of $\mathbb{F}$ (where $\mathbb{F}$ is given in (2.6.1)). Let $k$ be a prime divisor of $n$. As observed in Construction 2.1.4, we can embed $\mathrm{GL}_{n/k}(q^k)$ in $\mathrm{GL}_n(q)$, and this yields an embedding of $\mathrm{PSL}_{n/k}(q^k)$ in $\mathrm{PSL}_n(q)$. Then $H = N_G(\mathrm{PSL}_{n/k}(q^k))$ is the corresponding $\mathscr{C}_3$-subgroup of $G$ in the linear case $T = \mathrm{PSL}_n(q)$.

The extension field subgroups of the other classical groups can be defined in a similar way. For example, the embedding $\mathrm{GU}_{n/k}(q^k) < \mathrm{GU}_n(q)$ (with $k$ odd) is highlighted in Construction 2.3.3 (also see Constructions 2.4.2, 2.4.3, 2.5.13 and 2.5.14). We refer the reader to [86, Section 4.3] and Section 5.3 for further information on these subgroups. Some additional properties of extension field subgroups are discussed in [60], and in several cases we determine the precise structure of $H \cap T$ in Section 5.3.2.

### 2.6.2.4 $\mathscr{C}_4$: Tensor product subgroups, I

The subgroups in $\mathscr{C}_4$ are the stabilisers of appropriate tensor product decompositions $V = V_1 \otimes V_2$. As explained in Section 2.2.8, if $f_i$ is a sesquilinear or quadratic form on $V_i$, then we can construct a form $f = f_1 \otimes f_2$ on $V$ (see (2.2.9)). Furthermore, if $x_i$ is an isometry of $V_i$ then the element $x_1 \otimes x_2$ defined as in (2.2.11) is an isometry of the form $f$ on $V$. In this way we can embed a central product of the isometry groups of $f_1$ and $f_2$ in the isometry group of $f$ on $V$. All the $\mathscr{C}_4$-subgroups arise in this way, with the additional condition that $(V_1, f_1)$ is not similar to $(V_2, f_2)$. The specific subgroups in the $\mathscr{C}_4$ collection are listed in Table 5.4.1; see [86, Section 4.4] and Section 5.4 for further details.

### 2.6.2.5 $\mathscr{C}_5$: Subfield subgroups

The $\mathscr{C}_5$-subgroups arise naturally from subfields of $\mathbb{F}$ of prime index, where we define $\mathbb{F}$ as in (2.6.1). Let $\mathbb{F}_{\#}$ be a subfield of $\mathbb{F}$ of prime index $k$, and let $V_{\#}$ be the $\mathbb{F}_{\#}$-span of an $\mathbb{F}$-basis for $V$. Clearly, any invertible $\mathbb{F}_{\#}$-linear transformation of $V_{\#}$ can be extended uniquely to an invertible $\mathbb{F}$-linear transformation of $V$.

This allows us to embed $\mathrm{GL}_n(q^{1/k})$ in $\mathrm{GL}_n(q)$, and similarly $\mathrm{PSL}_n(q^{1/k})$ in $\mathrm{PSL}_n(q)$. Then $H = N_G(\mathrm{PSL}_n(q^{1/k}))$ is the corresponding $\mathscr{C}_5$-subgroup of $G$ when $T = \mathrm{PSL}_n(q)$. The other $\mathscr{C}_5$-subgroups can be constructed in a similar fashion. For example, the subfield subgroup embedding $\mathrm{O}_n^-(q^{1/2}) < \mathrm{O}_n^+(q)$ is described in Construction 2.5.12. In Section 5.5.1, we determine the precise structure of $H \cap T$ for some specific $\mathscr{C}_5$-subgroups $H$.

### 2.6.2.6 $\mathscr{C}_6$: Symplectic-type normalisers

The members of the $\mathscr{C}_6$ collection arise as local subgroups of $G$. Let $k \ne p$ be a prime and let $R$ be a *symplectic-type* $k$-group of (minimal) exponent $k(k,2)$. Here 'symplectic-type' means that every characteristic abelian subgroup of $R$ is cyclic; the structure of such a group is closely related to the extraspecial $k$-groups. More precisely, $R$ is the central product of an extraspecial $k$-group and a group that is either cyclic, dihedral, semidihedral or quaternion (see [4, (23.9)]).

It turns out that $R$ has $|Z(R)| - 1$ inequivalent faithful absolutely irreducible representations over an algebraically closed field of characteristic $p$, and each of these representations has degree $k^m$ for some fixed positive integer $m$ (see [86, Section 4.6]). This allows us to embed $R$ in an appropriate classical group $J$ of dimension $n = k^m$ over a finite field $\mathbb{F}$, where $L = J^\infty$ is quasisimple. If $G$ is an almost simple group with socle $T = L/Z(L)$, then the collection $\mathscr{C}_6$ comprises the subgroups of the form $H = N_G(H_0)$, where $H_0 < T$ is the image (modulo scalars) of $N_L(R)$.

To ensure that $H$ is not contained in a $\mathscr{C}_5$-subgroup of $G$ we also require the condition $\mathbb{F} = \mathbb{F}_{p^e}$, where $e$ is minimal such that $p^e \equiv 1 \pmod{|Z(R)|}$. These are rather restrictive conditions, and the $\mathscr{C}_6$ collection is empty for 'most' classical groups $G$. The various possibilities that do arise are listed in Table 5.6.1, and as usual we refer the reader to [86, Section 4.6] and Section 5.6 for further details.

### 2.6.2.7 $\mathscr{C}_7$: Tensor product subgroups, II

A $\mathscr{C}_7$-subgroup is the stabiliser in $G$ of a tensor product decomposition

$$V = V_1 \otimes V_2 \otimes \cdots \otimes V_t$$

where $t \geqslant 2$ and $\dim V_i = a$ for all $i$ (so $n = a^t$). The above description of the subgroups in $\mathscr{C}_4$ extends naturally to a tensor product of $t$ spaces. Indeed, if $f_i$ is a sesquilinear or quadratic form on $V_i$, then we can construct a form $f = f_1 \otimes \cdots \otimes f_t$ on $V$ so that a central product of the isometry groups of the $f_i$ embeds in the isometry group of $f$ on $V$. All the $\mathscr{C}_7$-subgroups $H$ arise in this way, with the added condition that $(V_1, f_1)$ is similar to $(V_i, f_i)$ for all $i$

(in particular, this means that elements of $H$ may permute the factors $V_i$ in the tensor product decomposition). See Table 5.7.1 for a list of the various tensor product subgroups that comprise the $\mathcal{C}_7$ collection, and see [86, Section 4.7] and Section 5.7 for further details.

### 2.6.2.8 $\mathcal{C}_8$: Classical subgroups

The subgroups in the final geometric collection are defined to be the stabilisers in $G$ of appropriate nondegenerate forms on $V$. More precisely, one of the following holds:

(i) $T = \mathrm{PSL}_n(q)$ and $H$ is the stabiliser in $G$ of a nondegenerate hermitian, alternating or quadratic form on $V$.
(ii) $T = \mathrm{PSp}_n(q)$, $q$ is even and $H$ is the stabiliser in $G$ of a nondegenerate quadratic form on $V$.

Note that in (ii), if $f$ is the associated bilinear form of the relevant quadratic form $Q$, then $f$ is nondegenerate and alternating (since $q$ is even), and so we obtain an embedding of isometry group of $Q$ in the isometry group of $f$. See Table 5.8.1 for a specific list of the subgroups that arise in the $\mathcal{C}_8$ collection, and see [86, Section 4.8] and Section 5.8 for further details.

### 2.6.3 The collections $\mathcal{S}$ and $\mathcal{N}$

According to Aschbacher's theorem, if $H$ is a *non-geometric* maximal subgroup of $G$ then $H$ is almost simple, with socle $S$ say, and the projective action of $S$ on $V$ corresponds to an absolutely irreducible representation of the covering group of $S$ in $\mathrm{GL}(V)$. Following [86], we write $\mathcal{S}$ to denote this family of subgroups. Several additional conditions are imposed to ensure that no subgroup in $\mathcal{S}$ is contained in one of the geometric collections $\mathcal{C}_i$, $1 \leqslant i \leqslant 8$ (see [86, p. 3] for the specific details).

In studying the $\mathcal{S}$-subgroups of $G$ it is natural to make a distinction between the subgroups whose socle is a simple group of Lie type over a field of characteristic $p$ (the so-called *defining characteristic* case) and the remaining subgroups (the *non-defining characteristic* case). Of course, in general it is impossible to write down a complete list of the subgroups in $\mathcal{S}$ of a given finite classical group; we do not even know the degrees of the irreducible representations of quasisimple groups (in defining or non-defining characteristic). However, complete lists are presented in [13] for the classical groups of dimension $n \leqslant 12$ (see Table 6.3.1 for the special case $n \leqslant 5$). In addition, work of Lübeck [98] (defining characteristic), and Hiss and Malle [76] (non-defining characteristic) provides information that can be used to restrict severely the possible $\mathscr{S}$-subgroups of higher-dimensional classical groups. For example, [76] gives a classification of the absolutely irreducible representations of quasisimple groups up to degree 250 in non-defining characteristic, including information on the corresponding character fields and Frobenius–Schur indicators. In any case, it should be clear that the subgroups in $\mathscr{S}$ require special attention, often involving different tools and techniques to those used in the analysis of geometric subgroups, where the projective action on $V$ is transparent.

As previously noted, a small additional subgroup collection, denoted by $\mathscr{N}$, arises when $T = \mathrm{PSp}_4(q)'$ (with $p = 2$) or $\mathrm{P\Omega}_8^+(q)$, and $G$ contains graph-field or triality automorphisms, respectively. The relevant subgroups are novelties. For $T = \mathrm{PSp}_4(q)'$, the subgroups in $\mathscr{N}$ were determined in Aschbacher's original paper (see [3, Section 14]). A few years later, the more difficult case $T = \mathrm{P\Omega}_8^+(q)$ was handled by Kleidman [85], building on earlier work of Aschbacher [3, Section 15].

We refer the reader to Table 5.9.1 in Section 5.9 for a complete list of the subgroups comprising the collection $\mathscr{N}$.

# 3

# Conjugacy classes

Let $G$ be an almost simple classical group over $\mathbb{F}_q$ with socle $T$ and natural module $V$, where $q = p^f$ for a prime $p$ and positive integer $f$. Then

$$T \leqslant G \leqslant \mathrm{Aut}(T)$$

where we identify the nonabelian finite simple group $T$ with its group of inner automorphisms. As described in Chapter 2, the possibilities for $T$ (up to isomorphism) are listed in Table 3.0.1.

In this chapter we will investigate the conjugacy classes of elements of prime order in $\mathrm{Aut}(T)$, providing information on representatives, centralisers and the splitting and fusing of classes. This will be essential for our study of derangements of prime order in Chapters 4, 5 and 6, and more generally we anticipate that our treatment will be useful in a wide range of related problems concerning the finite classical groups.

In various guises, most of the material in this chapter can be found in a range of sources in the literature. We are particularly indebted to Wall's work in [119], and we make extensive use of the analysis of involutions given by Aschbacher and Seitz [5] (in even characteristic), and by Gorenstein, Lyons

Table 3.0.1 *The finite simple classical groups*

| $T$ | Conditions |
|---|---|
| $\mathrm{PSL}_n(q)$ | $n \geqslant 2$, $(n,q) \neq (2,2), (2,3)$ |
| $\mathrm{PSU}_n(q)$ | $n \geqslant 3$, $(n,q) \neq (3,2)$ |
| $\mathrm{PSp}_n(q)'$ | $n \geqslant 4$ |
| $\Omega_n(q)$ | $nq$ odd, $n \geqslant 7$ |
| $\mathrm{P\Omega}_n^{\pm}(q)$ | $n \geqslant 8$ even |

and Solomon [67] (in odd characteristic). For unipotent elements, our main reference is the recent monograph [96] by Liebeck and Seitz. We continue to adopt the notation introduced in Chapter 2.

## 3.1 Preliminaries

We begin by describing the conjugacy classes in the general linear group. Recall that if $\phi(t) = \sum_{i=0}^{m} a_i t^i \in \mathbb{F}_q[t]$ is a monic polynomial of degree $m$, then the *companion matrix* of $\phi$ is the $m \times m$ matrix given by

$$C(\phi) = \left( \begin{array}{c|ccc} 0 & & & \\ \vdots & & I_{m-1} & \\ 0 & & & \\ \hline -a_0 & -a_1 & \cdots & -a_{m-1} \end{array} \right)$$

**Lemma 3.1.1** *Each element* $x \in \mathrm{GL}_n(q)$ *is* $\mathrm{GL}_n(q)$*-conjugate to a unique block-diagonal matrix of the form*

$$[C(\phi_1^{e_{11}}), \ldots, C(\phi_1^{e_{1r_1}}), \ldots, C(\phi_k^{e_{k1}}), \ldots, C(\phi_k^{e_{kr_k}})] \tag{3.1.1}$$

*where the* $\phi_i(t) \in \mathbb{F}_q[t]$ *are distinct monic irreducible polynomials, and the* $e_{ij}$ *are positive integers such that* $e_{i1} \geqslant \cdots \geqslant e_{ir_i}$ *for all* $i$.

For a proof of this basic lemma, see [75, Section 6.7]. The block-diagonal matrix in (3.1.1) is called the *rational canonical form* of $x$. The polynomials $\prod_{i,j} \phi_i(t)^{e_{ij}}$ and $\prod_i \phi_i(t)^{e_{i1}}$ are the characteristic and minimal polynomials of $x$, respectively, and the $\phi_i(t)^{e_{ij}}$ are the elementary divisors of $x$. The following familiar result states that the conjugacy class of $x$ is uniquely determined by its rational canonical form (again, see [75, Section 6.7]).

**Lemma 3.1.2** *Two elements* $x, y \in \mathrm{GL}_n(q)$ *are conjugate if and only if they have the same rational canonical form. In particular,* $x$ *and its transpose* $x^{\mathsf{T}}$ *are* $\mathrm{GL}_n(q)$*-conjugate.*

The next result, which follows from [18, Lemma 3.11], concerns the lifting of elements in a projective classical group to the corresponding matrix group; this will be useful for the analysis of conjugacy classes in the projective groups we are interested in.

**Lemma 3.1.3** *Let* $G, \hat{G}$ *be as follows:*

| $G$ | $\mathrm{PGL}_n^\varepsilon(q)$ | $\mathrm{PGSp}_n(q)$ | $\mathrm{PGO}_n^\varepsilon(q)$ |
|---|---|---|---|
| $\hat{G}$ | $\mathrm{GL}_n^\varepsilon(q)$ | $\mathrm{GSp}_n(q)$ | $\mathrm{GO}_n^\varepsilon(q)$ |

*If* $x \in G$ *has prime order* $r$, *then one of the following holds:*

(i)   $x$ *lifts to a unique element* $\hat{x} \in \hat{G}$ *of order* $r$;
(ii)  $G = \mathrm{PGL}_n^\varepsilon(q)$ *and* $r$ *divides* $q - \varepsilon$;
(iii) $G = \mathrm{PGSp}_n(q)$ *or* $\mathrm{PGO}_n^\varepsilon(q)$, $r = 2$ *and* $q$ *is odd*.

Let $T$ be a finite simple classical group over $\mathbb{F}_q$ with natural module $V$, where $q = p^f$ for a prime $p$. Let $G$ be an almost simple group with socle $T$.

**Definition 3.1.4** Let $x \in G \cap \mathrm{PGL}(V)$ and let $\hat{x} \in \mathrm{GL}(\overline{V})$ be a preimage of $x$, where $\overline{V} = V \otimes \mathbb{F}$ and $\mathbb{F}$ is the algebraic closure of $\mathbb{F}_q$. Following [97, p. 509] we define

$$\nu(x) = \min\{\dim[\overline{V}, \lambda \hat{x}] \mid \lambda \in \mathbb{F}^\times\}$$

where $[\overline{V}, \lambda \hat{x}]$ is the subspace $\langle v - v^{\lambda \hat{x}} \mid v \in \overline{V}\rangle$.

Note that $\nu(x) > 0$ if $x$ is nontrivial. More generally, observe that $\nu(x)$ is equal to the codimension of the largest eigenspace of $\hat{x}$ on $\overline{V}$. This parameter will play an important role in our analysis of derangements in classical groups.

Let $\mathrm{Inndiag}(T)$ be the group of inner-diagonal automorphisms of $T$ and suppose $x \in \mathrm{Inndiag}(T)$ has order $r$. We say that $x$ is *semisimple* if $(r, p) = 1$, in which case an appropriate lift $\hat{x} \in \mathrm{GL}(V)$ of $x$ is diagonalisable over a suitable extension of $\mathbb{F}_q$. The multiset of eigenvalues of $\hat{x}$ (in a splitting field) plays a central role in the analysis of the conjugacy classes of such elements. If $r$ is a $p$-power then $x$ is *unipotent*, which means that 1 is the only eigenvalue of $\hat{x}$ on $V$ (in any field extension of $\mathbb{F}_q$). We will see that the behaviour of unipotent conjugacy classes is essentially determined by the partition of $\dim V$ that encodes the Jordan canonical form of $x$ in its action on $V$.

Note that every element of prime order in $\mathrm{Inndiag}(T)$ is either semisimple or unipotent. More generally, every element $x \in \mathrm{Inndiag}(T)$ has a unique *Jordan decomposition* $x = su = us$, where $s, u \in \mathrm{Inndiag}(T)$ are semisimple and unipotent, respectively.

If $x \in \mathrm{Aut}(T)$ has prime order then either $x$ is an inner-diagonal automorphism (and therefore semisimple or unipotent), or $x$ is a field, graph or graph-field automorphism (see Theorem 2.1.3 and [67, Section 4.9]). For each type of classical group (linear, unitary, symplectic and orthogonal) we will consider each of these five possibilities in turn. In this preliminary section we discuss some of their basic properties, starting with semisimple elements in Section 3.1.1.

To end this preliminary discussion we record a well known number-theoretic result of Zsigmondy [123], which generalises earlier work of Bang [7] in the special case $q = 2$.

**Theorem 3.1.5** *Let $n$ and $q$ be integers with $n, q \geqslant 2$. Assume $(n, q) \neq (6, 2)$, and also assume that $q + 1$ is divisible by an odd prime if $n = 2$. Then there exists a prime $r$ that divides $q^n - 1$, but does not divide $q^i - 1$ if $i < n$.*

Such a prime $r$ is called a *primitive prime divisor* of $q^n - 1$. These prime numbers play an important role in the study of finite groups of Lie type.

Note that if $m$ is a positive integer, then $r$ divides $q^m - 1$ if and only if $n$ divides $m$ (see Lemma A.1(i) in Appendix A). By Fermat's Little Theorem, $r$ divides $q^{r-1} - 1$, so $r \equiv 1 \pmod{n}$.

### 3.1.1 Semisimple elements

Let $r \neq p$ be a prime and set $i = \Phi(r, q)$, where

$$\Phi(r, q) = \min\{i \in \mathbb{N} \mid r \text{ divides } q^i - 1\} \tag{3.1.2}$$

(so $r$ is a primitive prime divisor of $q^i - 1$). Since $r$ divides $q^{r-1} - 1$, it follows that $i < r$.

If $x \in \mathrm{GL}_n(q)$ has order $r$, then $x$ is diagonalisable over $\mathbb{F}_{q^i}$, but not over any proper subfield. Thus $x$ fixes a direct sum decomposition

$$V = U_1 \oplus \cdots \oplus U_s \oplus C_V(x) \tag{3.1.3}$$

of the natural $\mathrm{GL}_n(q)$-module $V$, where

$$C_V(x) = \{v \in V \mid vx = v\}$$

is the 1-eigenspace of $x$, and each $U_j$ is an $i$-dimensional subspace on which $x$ acts irreducibly. Here the $U_j$ are called the *irreducible blocks* of $x$ corresponding to this decomposition. We will sometimes refer to the restriction of $x$ to $U_j$ (which we denote by $x|_{U_j}$) as an irreducible block, and the eigenvalues of $x|_{U_j}$ (in $\mathbb{F}_{q^i}$) as the eigenvalues of the irreducible block $U_j$.

In the statement of the next lemma, we use the term *totally singular* as described in Remark 2.2.6.

**Lemma 3.1.6** *Let $x \in \mathrm{GL}_n(q)$ be a semisimple element that preserves a nondegenerate reflexive sesquilinear or quadratic form $f$ on $V$. Let $U$ be an* *x-invariant subspace of V such that $x|_U$ is irreducible. Then U is either non-degenerate or totally singular, or f is a quadratic form, q is even and U is a nonsingular 1-space. In the latter situation, x acts trivially on U.*

*Proof* First observe that $x$ fixes $U^\perp$, so it also fixes $U \cap U^\perp$. Therefore, the irreducibility of $x|_U$ implies that either $U \cap U^\perp = \{0\}$ and $U$ is nondegenerate, or $U \subseteq U^\perp$ and $U$ is totally isotropic. In the latter case, if $f$ is a quadratic form then $W = \{u \in U \mid f(u) = 0\}$ is an $x$-invariant subspace of $U$. By the irreducibility of $x|_U$, either $W = U$ and thus $U$ is totally singular, or $W = \{0\}$, $q$ is even and $U$ is a nonsingular 1-space. In the latter case, $f|_U : U \to \mathbb{F}_q$ is a bijection since $q$ is even, so $x$ acts trivially on $U$. $\qquad\square$

Let $\mathscr{S}_r$ be the set of nontrivial $r$th roots of unity in $\mathbb{F}_{q^i}$. The Frobenius automorphism $\sigma : \mathbb{F}_{q^i} \to \mathbb{F}_{q^i}$, $\lambda \mapsto \lambda^q$, induces a permutation of $\mathscr{S}_r$. The orbits of $\sigma$ on $\mathscr{S}_r$ have length $i$, so there are $t = (r-1)/i$ distinct $\sigma$-orbits in total, which we can label $\Lambda_1, \ldots, \Lambda_t$, where

$$\Lambda_j = \{\lambda_j, \lambda_j^q, \ldots, \lambda_j^{q^{i-1}}\}$$

for some $\lambda_j \in \mathscr{S}_r$.

The set of eigenvalues of $x$ on $U_j \otimes \mathbb{F}_{q^i}$ in (3.1.3) coincides with some $\sigma$-orbit $\Lambda_k$ and we will often abuse notation by writing

$$x = [\Lambda_1^{a_1}, \ldots, \Lambda_t^{a_t}, I_e]$$

where

$$e = \dim C_V(x) = n - i \sum_j a_j$$

and $a_j$ denotes the multiplicity of $\Lambda_j$ in the multiset of eigenvalues of $x$ on $V \otimes \mathbb{F}_{q^i}$. The next lemma, which follows immediately from Lemma 3.1.2, shows that the corresponding $t$-tuple $(a_1, \ldots, a_t)$ encodes the $\mathrm{GL}_n(q)$-class of $x$.

**Lemma 3.1.7** *There is a bijection*

$$\theta : (a_1, \ldots, a_t) \longmapsto [\Lambda_1^{a_1}, \ldots, \Lambda_t^{a_t}, I_e]^{\mathrm{GL}_n(q)}$$

*from the set of nonzero t-tuples $(a_1, \ldots, a_t) \in \mathbb{N}_0^t$ with $i\sum_j a_j \leqslant n$, to the set of $\mathrm{GL}_n(q)$-classes of semisimple elements of order r in $\mathrm{GL}_n(q)$.*

Let $x = [\Lambda_j] \in \mathrm{GL}_i(q)$, where $\Lambda_j = \{\lambda_j, \lambda_j^q, \ldots, \lambda_j^{q^{i-1}}\}$ is a $\sigma$-orbit. Then the characteristic polynomial

$$(t - \lambda_j)(t - \lambda_j^q) \cdots (t - \lambda_j^{q^{i-1}}) = b_0 + b_1 t + \cdots + b_{i-1} t^{i-1} + t^i \in \mathbb{F}_q[t]$$

of $x$ is irreducible over $\mathbb{F}_q$, so

$$\begin{pmatrix} 0 & & & \\ \vdots & & I_{i-1} & \\ 0 & & & \\ \hline -b_0 & -b_1 & \cdots & -b_{i-1} \end{pmatrix}$$

is the rational canonical form of $x$.

For example, if $(r,q) = (7,2)$ then $i = 3$ and the two $\sigma$-orbits are

$$\Lambda_1 = \{\lambda, \lambda^2, \lambda^4\}, \quad \Lambda_2 = \{\lambda^3, \lambda^6, \lambda^5\}$$

where $\lambda$ is a fixed primitive element of $\mathbb{F}_8$ such that $\lambda^3 + \lambda + 1 = 0$. Now

$$(t - \lambda)(t - \lambda^2)(t - \lambda^4) = t^3 + t + 1$$

and $t^3 + t^2 + 1$ are the respective characteristic polynomials of $x_1, x_2 \in \mathrm{GL}_3(2)$, where $x_j = [\Lambda_j]$. Therefore,

$$x_1 = \begin{pmatrix} 0 & 1 & 0 \\ 0 & 0 & 1 \\ 1 & 1 & 0 \end{pmatrix}, \quad x_2 = \begin{pmatrix} 0 & 1 & 0 \\ 0 & 0 & 1 \\ 1 & 0 & 1 \end{pmatrix}$$

represent the two conjugacy classes of elements of order 7 in $\mathrm{GL}_3(2)$.

Field extensions provide an alternative way to view semisimple elements. Recall that an $a$-dimensional vector space $U$ over $\mathbb{F}_{q^i}$ can be viewed as an $(ai)$-dimensional space over $\mathbb{F}_q$. Moreover, any $\mathbb{F}_{q^i}$-linear map on $U$ is also $\mathbb{F}_q$-linear, so we obtain an embedding $\mathrm{GL}_a(q^i) \leqslant \mathrm{GL}_{ai}(q)$ as in Construction 2.1.4. In particular, $\mathrm{GL}_1(q^i) \leqslant \mathrm{GL}_i(q)$. In terms of this embedding, if $x = [\Lambda] \in \mathrm{GL}_i(q)$ is a semisimple element of order $r$, where $r$ is a primitive prime divisor of $q^i - 1$, then $x$ is in a cyclic subgroup $\mathrm{GL}_1(q^i) \leqslant \mathrm{GL}_i(q)$ and Schur's Lemma asserts that $C_{\mathrm{GL}_i(q)}(x) = \mathrm{GL}_1(q^i)$. More generally, an element $x = [\Lambda^a] \in \mathrm{GL}_{ai}(q)$ of order $r$ can be viewed as a central element $\lambda I_a$ in a subgroup $\mathrm{GL}_a(q^i)$, so $x$ is clearly centralised by $\mathrm{GL}_a(q^i)$.

**Remark 3.1.8** In the statement of Lemma 3.1.9 below, we define $\mathrm{GL}_0(q) = 1$. More generally, for the remainder of this book it will be convenient to adopt the convention that a 0-dimensional classical group is trivial.

**Lemma 3.1.9** Let $x = [\Lambda_1^{a_1}, \ldots, \Lambda_t^{a_t}, I_e] \in \mathrm{GL}_n(q)$ be a semisimple element of prime order $r$ and set $i = \Phi(r,q)$. Then

$$C_{\mathrm{GL}_n(q)}(x) = \mathrm{GL}_e(q) \times \prod_{j=1}^{t} \mathrm{GL}_{a_j}(q^i)$$

*Proof* Let $V$ be the natural $\mathrm{GL}_n(q)$-module. An element $g \in C_{\mathrm{GL}_n(q)}(x)$ fixes each eigenspace of $x$ (on $V \otimes \mathbb{F}_{q^i}$) and so it fixes each subspace in the decomposition

$$V = W_1 \oplus W_2 \oplus \cdots \oplus W_t \oplus C_V(x)$$

where $x|_{W_j} = [\Lambda_j^{a_j}]$. Therefore

$$C_{\mathrm{GL}_n(q)}(x) \leqslant \mathrm{GL}_e(q) \times \prod_{j=1}^{t} \mathrm{GL}_{ia_j}(q)$$

To complete the proof, we may assume that $x = [\Lambda_1^{a_1}]$ and thus $V = W_1$. More precisely, it suffices to show that $C_{\mathrm{GL}_{ia_1}(q)}(x) \leqslant \mathrm{GL}_{a_1}(q^i)$, that is, if $g \in \mathrm{GL}_{ia_1}(q)$ centralises $x$ then it preserves the $\mathbb{F}_{q^i}$-vector space structure of $V$. Recall that $x$ acts on $V$ as multiplication by a scalar $\lambda \in \mathbb{F}_{q^i}$ that lies in no proper subfield of $\mathbb{F}_{q^i}$. The minimal polynomial of $\lambda$ over $\mathbb{F}_q$ has degree $i$, so $\{1, \lambda, \lambda^2, \ldots, \lambda^{i-1}\}$ is an $\mathbb{F}_q$-basis for $\mathbb{F}_{q^i}$ and thus each $\mu \in \mathbb{F}_{q^i}$ has a unique expression of the form $\mu = \sum_{j=0}^{i-1} b_j \lambda^j$ with $b_j \in \mathbb{F}_q$. If $g \in \mathrm{GL}_{ia_1}(q)$ centralises $x$ then

$$(\mu v)^g = \left(\sum_j b_j \lambda^j v\right)^g = \left(\sum_j b_j v^{x^j}\right)^g = \sum_j b_j (v^g)^{x^j} = \sum_j b_j \lambda^j v^g = \mu v^g$$

for all $v \in V$. Hence $g$ preserves the $\mathbb{F}_{q^i}$-vector space structure of $V$, so $g \in \mathrm{GL}_{a_1}(q^i)$ and the result follows. $\qquad\square$

**Corollary 3.1.10** *Let $x \in \mathrm{GL}_n(q)$ be a semisimple element of prime order. Then $x^{\mathrm{SL}_n(q)} = x^{\mathrm{GL}_n(q)}$.*

*Proof* By Lemma 3.1.9 we have $|\det(C_{\mathrm{GL}_n(q)}(x))| = q - 1$ (that is, for any $\lambda \in \mathbb{F}_q^\times$ there exists an element $y \in C_{\mathrm{GL}_n(q)}(x)$ with $\det(y) = \lambda$), so

$$|C_{\mathrm{SL}_n(q)}(x)| = \frac{1}{q-1}|C_{\mathrm{GL}_n(q)}(x)|$$

The result follows. $\qquad\square$

**Remark 3.1.11** In fact, similar reasoning shows that the conclusion to Corollary 3.1.10 holds for any semisimple element $x \in \mathrm{GL}_n(q)$.

From Corollary 3.1.10, we deduce that $x^{\mathrm{PSL}_n(q)} = x^{\mathrm{PGL}_n(q)}$ for any semisimple element $x \in \mathrm{PGL}_n(q)$ of prime order. Much more generally, we have the following result.

**Theorem 3.1.12** *Let $T$ be a finite simple group of Lie type and suppose $x \in$* Inndiag$(T)$ *is a semisimple element of prime order. Then $x^T = x^{\mathrm{Inndiag}(T)}$.*

*Proof*  This is [67, Theorem 4.2.2(j)].  □

For a $\sigma$-orbit $\Lambda$ we define

$$\Lambda^{-1} = \{\lambda^{-1} \mid \lambda \in \Lambda\}, \quad \Lambda^{-q} = \{\lambda^{-q} \mid \lambda \in \Lambda\}$$

Note that both $\Lambda^{-1}$ and $\Lambda^{-q}$ are also $\sigma$-orbits. We conclude this section with a technical result that will be useful when we consider semisimple elements in unitary, symplectic and orthogonal groups.

**Lemma 3.1.13** *Let $r \neq p$ be an odd prime and let $i = \Phi(r,q)$. Let $\lambda$ be a nontrivial $r$th root of unity in $\mathbb{F}_{q^i}$.*

(i) *Let $\Lambda = \{\lambda, \lambda^q, \ldots, \lambda^{q^{i-1}}\}$ be a $\sigma$-orbit. Then $\Lambda = \Lambda^{-1}$ if and only if $i$ is even. Moreover, if $i > 1$ then $\prod_{\mu \in \Lambda} \mu = 1$.*

(ii) *Let $\Lambda = \{\lambda, \lambda^{q^2}, \ldots, \lambda^{q^{2(j-1)}}\}$ be a $\sigma^2$-orbit on $\mathscr{S}_r$, so $j = i$ if $i$ is odd, and $j = i/2$ if $i$ is even. Then $\Lambda = \Lambda^{-q}$ if and only if $i \equiv 2 \pmod 4$. Moreover, if $i > 2$ then $\prod_{\mu \in \Lambda} \mu = 1$.*

*Proof*  Consider (i). The first part follows from [18, Lemma 3.26(i)] and the second from the fact that $r$ divides $(q^i - 1)/(q - 1) = 1 + q + \cdots + q^{i-1}$.

Now let us turn to (ii). Since $\Lambda$ and $\Lambda^{-q}$ are closed under the map $\mu \mapsto \mu^{q^2}$, either $\Lambda = \Lambda^{-q}$ or $\Lambda \cap \Lambda^{-q}$ is empty. Hence $\Lambda = \Lambda^{-q}$ if and only if $\lambda^{-q} = \lambda^{q^{2k}}$ for some integer $k$ in the range $0 \leqslant k < j$. Suppose $\lambda^{-q} = \lambda^{q^{2k}}$. Then $\lambda^{q(q^{2k-1}+1)} = 1$ and so $r$ divides $q^{2k-1} + 1$. Hence $i$ divides $4k - 2$, but not $2k - 1$. It follows that $i$ is even and $i/2$ divides $2k - 1$, so $i \equiv 2 \pmod 4$. Conversely, if $i \equiv 2 \pmod 4$ then $\lambda^{-q} = \lambda^{q^{i/2+1}}$ and $i/2 + 1$ is even, so $\Lambda = \Lambda^{-q}$. Finally, the second part of (ii) holds since $r$ divides $(q^{2j} - 1)/(q^2 - 1)$.  □

### 3.1.2  Unipotent elements

By Lemma 3.1.3, elements of order $p$ in $G$ lift to elements of order $p$ in $\hat{G}$ and so we can work in the matrix group $\hat{G}$. Observe that there is a bijection from the set of distinct rational canonical forms of elements of order $p$ in $\mathrm{GL}_n(q)$, to the set of nontrivial partitions of $n$ with parts of size at most $p$ (here $(1^n)$ is the trivial partition of $n$). Indeed, if $x \in \mathrm{GL}_n(q)$ has order $p$ then its characteristic polynomial is $(t - 1)^n$, so its elementary divisors are of the form $(t - 1)^i$ with $1 \leqslant i \leqslant p$. Moreover, if $(t - 1)^i$ has multiplicity $a_i$ as an elementary divisor of

$x$, then $(p^{a_p}, \ldots, 1^{a_1})$ is a partition of $n$, and the rational canonical form of $x$ is uniquely determined by this partition. Therefore, we obtain the following result as a special case of Lemma 3.1.2 (here $J_i$ denotes a standard (lower-triangular) unipotent Jordan block of size $i$, and $J_i^\ell$ denotes $\ell$ copies of $J_i$).

**Lemma 3.1.14** *There is a bijection*

$$\theta : (p^{a_p}, \ldots, 1^{a_1}) \longmapsto [J_p^{a_p}, \ldots, J_1^{a_1}]^{\mathrm{GL}_n(q)}$$

*from the set of nontrivial partitions of $n$ with parts of size at most $p$, to the set of $\mathrm{GL}_n(q)$-classes of elements of order $p$ in $\mathrm{GL}_n(q)$.*

Note that $x = [J_p^{a_p}, \ldots, J_1^{a_1}]$ fixes a direct sum decomposition

$$V = U_{p,1} \oplus \cdots \oplus U_{p,a_p} \oplus \cdots \oplus U_{1,1} \oplus \cdots \oplus U_{1,a_1}$$

of the natural $\mathrm{GL}_n(q)$-module, where $x$ acts indecomposably on each $i$-space $U_{i,k}$, so $x|_{U_{i,k}} = J_i$. Set $U_j = U_{j,1} \oplus \cdots \oplus U_{j,a_j}$ for $1 \leqslant j \leqslant p$. Then $x$ has Jordan form $[J_j^{a_j}]$ on $U_j$, which is conjugate to the element $J_j \otimes I_{a_j}$ in a tensor product subgroup $\mathrm{GL}_j(q) \otimes \mathrm{GL}_{a_j}(q)$ of $\mathrm{GL}(U_j) = \mathrm{GL}_{ja_j}(q)$. Clearly, $J_j \otimes I_{a_j}$ is centralised by the subgroup $1 \otimes \mathrm{GL}_{a_j}(q)$ and thus $\prod_i \mathrm{GL}_{a_i}(q)$ is a subgroup of $C_{\mathrm{GL}_n(q)}(x)$.

The next lemma describes the precise structure of the centraliser of a unipotent element (for a proof see [96, Theorem 7.1]).

**Lemma 3.1.15** *Let $x = [J_n^{a_n}, \ldots, J_1^{a_1}] \in \mathrm{GL}_n(q)$ be a unipotent element. Then $C_{\mathrm{GL}_n(q)}(x) = QR$ where*

$$|Q| = q^\gamma \quad \text{with} \quad \gamma = 2\sum_{i<j} i a_i a_j + \sum_i (i-1) a_i^2$$

*and $R = \prod_i \mathrm{GL}_{a_i}(q)$.*

In order to state the next lemma, recall that $x \in \mathrm{GL}_n(q)$ is a *regular* unipotent element if it has Jordan form $[J_n]$.

**Lemma 3.1.16** *Let $x \in \mathrm{GL}(V)$ be a regular unipotent element and fix a basis $\{v_1, \ldots, v_n\}$ of $V$ so that $x$ is represented by a standard (lower-triangular) Jordan block with respect to this basis. Let $U$ be a nontrivial $m$-dimensional subspace of $V$ fixed by $x$. The following hold:*

(i) $U = \langle v_1, \ldots, v_m \rangle$.

(ii) *If $x$ preserves a nondegenerate reflexive sesquilinear or quadratic form on $V$, then $U^\perp = \langle v_1, \ldots, v_{n-m} \rangle$. In particular, if $m \leqslant n/2$ then $U$ is totally isotropic.*

*Proof* The order of $x$ is a power of $p$, so $V$ must contain a nonzero vector fixed by $x$. Therefore, part (i) holds when $m = 1$ since $\langle v_1 \rangle$ is the 1-eigenspace of $x$. Now assume $m \geqslant 2$. Again, $U$ must contain a nonzero vector fixed by $x$, so $\langle v_1 \rangle \subset U$. Then $x$ induces a regular unipotent element on $V/\langle v_1 \rangle$, fixing the $(m-1)$-dimensional subspace $U/\langle v_1 \rangle$, so the result follows by induction.

Now consider (ii). Since $x$ fixes $U^\perp$, which has dimension $n - m$, part (i) implies that $U^\perp = \langle v_1, \ldots, v_{n-m} \rangle$. Therefore, $U \subseteq U^\perp$ if $m \leqslant n/2$. $\square$

### 3.1.3 Outer automorphisms

Let $T$ be a simple group of Lie type. As noted in Theorem 2.1.3, every automorphism of $T$ is the product of an inner, a diagonal, a field and a graph automorphism of $T$ (see Chapter 2 for the precise definitions of these automorphisms when $T$ is classical). As the terminology indicates, field automorphisms arise from automorphisms of the underlying field, and graph automorphisms correspond to nontrivial symmetries of the corresponding Dynkin diagram for $T$. Note that if $x \in \mathrm{Aut}(T) \setminus \mathrm{Inndiag}(T)$ has prime order, then $x$ is either a field, graph, or graph-field automorphism.

The following lemma is a useful observation concerning field and graph-field automorphisms of prime order (see [67, Proposition 4.9.1(d)]).

**Lemma 3.1.17** *Let $T$ be a simple group of Lie type and let $x \in \mathrm{Aut}(T)$ be a field or graph-field automorphism of prime order $r$. Then $y \in \mathrm{Inndiag}(T)x$ has order $r$ if and only if $x$ and $y$ are $\mathrm{Inndiag}(T)$-conjugate.*

## 3.2 Linear groups

Throughout this section, let $T = \mathrm{PSL}_n(q)$ and set

$$G = \mathrm{Inndiag}(T) = \mathrm{PGL}_n(q) = \mathrm{GL}_n(q)/Z$$

where $q = p^f$ for a prime $p$, and $Z$ denotes the centre of $\mathrm{GL}_n(q)$, comprising the scalar matrices. Let $V$ be the natural module for $\mathrm{GL}_n(q)$ and set $\delta = [\mu, I_{n-1}]Z \in G$, where $\mathbb{F}_q^\times = \langle \mu \rangle$. Recall that $G = \langle T, \delta \rangle$.

### 3.2.1 Semisimple elements

Let $r \neq p$ be a prime divisor of $|G|$ and set $i = \Phi(r, q)$ (see (3.1.2)). Suppose $x \in G$ has order $r$ and fix $\hat{x} \in \mathrm{GL}_n(q)$ such that $x = \hat{x}Z$ and $\hat{x}$ has the smallest possible order. By Theorem 3.1.12, $x^G = x^T$.

Recall that $\mathscr{S}_r$ denotes the set of nontrivial $r$th roots of unity in $\mathbb{F}_{q^i}$, which is partitioned into $t = (r-1)/i$ distinct $\sigma$-orbits, labelled $\Lambda_1, \ldots, \Lambda_t$, under the permutation $\sigma : \mathscr{S}_r \to \mathscr{S}_r$ defined by $\sigma : \lambda \mapsto \lambda^q$.

**Proposition 3.2.1** *If $i > 1$ then $x \in T$ and the map*

$$\theta : (a_1, \ldots, a_t) \longmapsto ([\Lambda_1^{a_1}, \ldots, \Lambda_t^{a_t}, I_e]Z)^G$$

*is a bijection from the set of nonzero $t$-tuples of the form $(a_1, \ldots, a_t) \in \mathbb{N}_0^t$ with $i \sum_j a_j \leqslant n$, to the set of $G$-classes of elements of order $r$ in $G$.*

*Proof* By Lemma 3.1.3 we may assume $\hat{x}$ has order $r$. Now apply Lemma 3.1.7, noting that $x \in T$ since $|G : T| = (n, q-1)$ is indivisible by $r$.  $\square$

The situation when $i = 1$ is slightly more complicated. In order to state the result, let $\mathscr{T}(n, r)$ be a set of representatives of the $C_r$-orbits of $r$-tuples $(a_0, \ldots, a_{r-1})$ in $\mathbb{N}_0^r$ such that $\sum_j a_j = n$ and $a_j < n$ for all $j$, with respect to the natural cyclic action of $C_r$ on $r$-tuples. Also recall that if $m$ is an integer then $(m)_r$ denotes the largest power of $r$ dividing $m$.

**Proposition 3.2.2** *Suppose $i = 1$. Then up to $G$-conjugacy, either*

(i) $\hat{x} = [I_{a_0}, \lambda I_{a_1}, \ldots, \lambda^{r-1} I_{a_{r-1}}]$ *for some $r$-tuple $(a_0, \ldots, a_{r-1}) \in \mathbb{N}_0^r$, where $\lambda \in \mathbb{F}_q$ is a nontrivial $r$th root of unity, and $a_j < n$ for all $j$, or*

(ii) *$r$ divides $n$ and $\hat{x} = [\Lambda^{n/r}]$ with $\Lambda = \{\mu, \mu^q, \ldots, \mu^{q^{r-1}}\}$ for some $\mu \in \mathbb{F}_{q^r}$ of order $r(q-1)_r$.*

*If $n$ is indivisible by $r$, then there is a bijection*

$$\theta : (a_0, \ldots, a_{r-1}) \longmapsto ([I_{a_0}, \lambda I_{a_1}, \ldots, \lambda^{r-1} I_{a_{r-1}}]Z)^G$$

*from $\mathscr{T}(n, r)$ to the set of $G$-classes of elements of order $r$ in $G$. If $r$ divides $n$ then there are an additional $r - 1$ distinct $G$-classes, represented by elements of type (ii). In (i) we have $x \in T$ if and only if $\lambda^{\sum_j j a_j} = \xi^n$ for some $\xi \in \mathbb{F}_q$, and $x \in T$ in (ii) if and only if $r = 2$ and $q \equiv 3 \pmod 4$, or $(n)_r > (q-1)_r$.*

*Proof* Let $\lambda \in \mathbb{F}_q$ be a nontrivial $r$th root of unity and suppose $x$ lifts to an element $\hat{x} \in \mathrm{GL}_n(q)$ that is diagonalisable over $\mathbb{F}_q$, with eigenvalues $\lambda_1, \ldots, \lambda_n$, say. Now $\hat{x}^r \in Z$, so $\lambda_1^r = \lambda_j^r$ for all $j$. If $\lambda_1^r \neq 1$ then the eigenvalues of $\lambda_1^{-1}\hat{x} \in \hat{x}Z$ are $r$th roots of unity, so without loss we may assume $\hat{x}$ has this property. In particular, if $a_j$ denotes the multiplicity of the eigenvalue $\lambda^j$ then $\hat{x}$ has the diagonal form described in (i), and the given condition for membership in $T$ follows from Lemma 2.1.1. Also notice that

$$[I_{a_0}, \lambda I_{a_1}, \ldots, \lambda^{r-1} I_{a_{r-1}}]Z = [I_{a_{r-1}}, \lambda I_{a_0}, \ldots, \lambda^{r-1} I_{a_{r-2}}]Z \qquad (3.2.1)$$

etc., so the $G$-class of $x$ corresponds to $(a_0, \ldots, a_{r-1}) \in \mathcal{T}(n,r)$.

For the remainder, let us assume $\hat{x}$ is not diagonalisable over $\mathbb{F}_q$. Now $\hat{x}^r \in Z$ and hence $\hat{x}$ is diagonalisable over $\mathbb{F}_{q^r}$. Let $\mu \in \mathbb{F}_{q^r}$ be an eigenvalue of $\hat{x}$ and note that $\mu^q, \ldots, \mu^{q^{r-1}}$ are also eigenvalues. Since

$$\Lambda = \{\mu, \mu^q, \ldots, \mu^{q^{r-1}}\} = \{\gamma \in \mathbb{F}_{q^r} \mid \gamma^r = \mu^r\} \qquad (3.2.2)$$

it follows that $r$ divides $n$ and $\hat{x} = [\Lambda^{n/r}]$, as in part (ii). Now $\mu$ has order $r(q-1)_r \ell$ for some divisor $\ell$ of $q-1$ with $(\ell, r) = 1$, so there exists an integer $a$ such that $a\ell \equiv 1 \pmod{r}$ and $1 \leqslant a < r$. Thus $\mu^{a\ell-1} \in \mathbb{F}_q$ and $\mu(\mu^{a\ell-1}) = \mu^{a\ell}$ has order $r(q-1)_r$, as $(a,r) = 1$. Hence, by replacing $\hat{x}$ with the scalar multiple $\mu^{a\ell-1}\hat{x}$ if necessary, we may assume that $\mu$ has order $r(q-1)_r$, as stated in (ii). Furthermore, we note that there are $(r-1)(q-1)_r/r$ distinct possibilities for $\Lambda$, so this is the number of distinct $\mathrm{GL}_n(q)$-classes of elements of type (ii).

For $\alpha \in \mathbb{F}_q^\times$, observe that the elements in $\alpha\Lambda$ have order $r(q-1)_r$ if and only if the order of $\alpha$ divides $(q-1)_r$. Thus the coset $\hat{x}Z$ contains $(q-1)_r$ elements of order $r(q-1)_r$. Now (3.2.2) implies that $\alpha\Lambda = \Lambda$ if and only if $\alpha^r = 1$, so $\hat{x}Z$ meets precisely $(q-1)_r/r$ distinct $\mathrm{GL}_n(q)$-classes of elements of type (ii) and we deduce that there are $r-1$ distinct $G$-classes of this form (see [18, Lemma 3.35(ii)] for an alternative argument).

Finally, let us consider membership in $T$. By Lemma 2.1.1, $x \in T$ if and only if there exists $\xi \in \mathbb{F}_q$ such that $\xi^n = \det(\hat{x})$, where $\det(\hat{x}) = (\mu^{(q^r-1)/(q-1)})^{n/r}$. If $r = 2$ and $q \equiv 3 \pmod 4$ then $(q^r-1)/(q-1)r = (q+1)/2$ is even and thus $\xi = \mu^{(q^r-1)/(q-1)r} \in \mathbb{F}_q$ has the property that $\xi^n = \det(\hat{x})$, so $x \in T$.

In all other cases, we note that $(q^r-1)/(q-1)r$ is indivisible by $r$ (see Lemma A.4 in Appendix A), so $\omega = \mu^{(q^r-1)/(q-1)r}$ has order $r(q-1)_r$ and $\det(\hat{x}) = \omega^n$. Set $m = n/r$ and write $\omega^r = \gamma^s$, where $q-1 = s(q-1)_r$ and $\mathbb{F}_q^\times = \langle\gamma\rangle$. We claim that $x \in T$ if and only if $(n)_r > (q-1)_r$.

Suppose $x \in T$. By Lemma 2.1.1, $\det(\hat{x}) = \gamma^{sm} = \gamma^{an}$ for some positive integer $a$, so $\gamma^{m(ar-s)} = 1$ and thus $q-1$ divides $m(ar-s)$. Since $ar-s$ is indivisible by $r$, it follows that $(q-1)_r \leqslant (m)_r < (n)_r$ as required. Conversely, if $(n)_r > (q-1)_r$ then $\det(\hat{x}) = \gamma^{sm} = \gamma^{(q-1)b}$ for some positive integer $b$, so $\det(\hat{x}) = 1$ and thus $x \in T$. $\qquad\square$

**Remark 3.2.3** Note that elements $x \in G$ of type (ii) in Proposition 3.2.2 arise from elements $\xi I_{n/r}$ of order $r(q-1)_r$ in the centre of a field extension subgroup $\mathrm{GL}_{n/r}(q^r) < \mathrm{GL}_n(q)$. In particular, if $\{v_1, \ldots, v_{n/r}\}$ is a basis for $V$ as an $\mathbb{F}_{q^r}$-space, then

$$\{v_1, \xi v_1, \ldots, \xi^{r-1}v_1, v_2, \xi v_2, \ldots, \xi^{r-1}v_2, \ldots, v_{n/r}, \xi v_{n/r}, \ldots, \xi^{r-1}v_{n/r}\}$$

is an $\mathbb{F}_q$-basis for $V$. In terms of this basis, $\hat{x}$ is the block-diagonal matrix $[A_1^{n/r}]$, where

$$
A_1 = \left(
\begin{array}{c|ccc}
0 & & & \\
\vdots & & I_{r-1} & \\
0 & & & \\
\hline
\xi^r & 0 & \cdots & 0
\end{array}
\right) \in \mathrm{GL}_r(q)
$$

This coincides with the rational canonical form of $\hat{x}$. Similarly, the block-diagonal matrices $[A_j^{n/r}]$ with $2 \leqslant j < r$, represent the remaining $r - 2$ distinct $G$-classes of this type, where $A_j$ is the same as $A_1$ above, except that the $(r, 1)$-entry is $\xi^{jr}$, rather than $\xi^r$.

**Remark 3.2.4** Let $x \in G$ be a semisimple element of prime order $r$. The order of the centraliser $C_G(x)$ can be computed from Lemma 3.1.9 and it is given in Table B.3 (see Appendix B). Extra care is required if $r$ divides $q - 1$. In particular, if we are in case (ii) of Proposition 3.2.2, or if all $r$th roots of unity occur as eigenvalues of $\hat{x}$ with the same multiplicity, then $\lambda \hat{x}$ and $\hat{x}$ have the same eigenvalues (in an appropriate extension of $\mathbb{F}_q$) for all $\lambda \in \mathbb{F}_q$ of order $r$, so $\lambda \hat{x}$ and $\hat{x}$ are $\mathrm{GL}_n(q)$-conjugate. This explains the additional factor of $r$ in the order of $C_G(x)$.

### 3.2.2 Semisimple involutions

In this section we highlight the special case $r = 2$ in Proposition 3.2.2. This provides an easy reference for later use, and consistency with our treatment of involutions in the other classical groups. Throughout this section we adopt the labelling of conjugacy class representatives given in [67, Table 4.5.1].

#### 3.2.2.1 Involutions of type $t_i$, $1 \leqslant i \leqslant n/2$

Write $t_i = \hat{t}Z \in G$, where $\hat{t} \in \mathrm{GL}_n(q)$ is an involution whose $(-1)$-eigenspace has dimension $i$, so $-\hat{t}$ is an involution with a $(-1)$-eigenspace of dimension $n - i$. Note that $\hat{t}$ and $-\hat{t}$ are $\mathrm{SL}_n(q)$-conjugate if and only if $i = n/2$. Now $\hat{t}$ has determinant $(-1)^i$, so Lemma 2.1.1 implies that $t_i \in T$ if and only if there exists $\xi \in \mathbb{F}_q$ such that $\xi^n = (-1)^i$. Clearly, if $i$ is even then $t_i \in T$, otherwise $t_i \in T$ if and only if $-1$ is an $n$th root of unity. Now $\xi^n = -1$ if and only if $\xi^{(n)_2} = -1$, so if $i$ is odd then $t_i \in T$ if and only if $(q-1)_2 > (n)_2$. In particular, if $n$ is even, then $t_{n/2} \in T$ if and only if $n \equiv 0 \pmod 4$ or $q \equiv 1 \pmod 4$.

Finally, note that if $\hat{x} \in \mathrm{GL}_n(q)$ is diagonalisable over $\mathbb{F}_q$, $|\hat{x}| > 2$ and $\hat{x}Z \in G$ is an involution, then the eigenvalues of $\hat{x}$, say $\lambda_1, \ldots, \lambda_n$, satisfy $\lambda_1^2 = \lambda_j^2$ for all $j$. Therefore $\hat{x} = [-\lambda I_i, \lambda I_{n-i}]$ for some $\lambda \in \mathbb{F}_q^\times$ and integer $i \leqslant n/2$, and thus $\lambda^{-1}\hat{x} = \hat{t}$ as above.

### 3.2.2.2 Involutions of type $t'_{n/2}$

Here $n$ is even and involutions of this type arise from the quadratic field extension $\mathbb{F}_{q^2}/\mathbb{F}_q$. Fix an $\mathbb{F}_{q^2}$-basis $\{v_1, \ldots, v_{n/2}\}$ for $V$ and suppose $\xi \in \mathbb{F}_{q^2}$ has order $2(q-1)_2$. Let $x$ be the central element $\xi I_{n/2}$ of $\mathrm{GL}_{n/2}(q^2)$ and write $t'_{n/2} = \hat{t}Z$, where $\hat{t}$ is the image of $x$ in $\mathrm{GL}_n(q)$ under the natural embedding of $\mathrm{GL}_{n/2}(q^2)$ in $\mathrm{GL}_n(q)$ (see Construction 2.1.4). In terms of the $\mathbb{F}_q$-basis $\{v_1, \xi v_1, \ldots, v_{n/2}, \xi v_{n/2}\}$ for $V$, we have $\hat{t} = [A^{n/2}]$ where

$$A = \begin{pmatrix} 0 & 1 \\ \xi^2 & 0 \end{pmatrix} \tag{3.2.3}$$

Note that $\xi^2 \in \mathbb{F}_q$ and $\hat{t}^2 \in Z$. Also note that every element in the coset $\hat{t}Z \subset \mathrm{GL}_n(q)$ has eigenvalues in $\mathbb{F}_{q^2} \setminus \mathbb{F}_q$, so there are no involutions in this coset. Since $A$ and $-A$ are $\mathrm{SL}_2(q)$-conjugate (see Remark 3.1.11), it follows that $\hat{t}$ and $-\hat{t}$ are $\mathrm{SL}_n(q)$-conjugate. Finally, observe that these involutions arise in part (ii) of Proposition 3.2.2, hence $t'_{n/2} \in T$ if and only if $q \equiv 3 \pmod 4$ or $(n)_2 > (q-1)_2$.

### 3.2.2.3 Involutions: a summary

The next proposition summarises the above discussion. Note that Table B.1 is located in Appendix B; the listed centraliser orders can be read off from [67, Table 4.5.1]. Some additional properties of these involutions are recorded in Table B.2.

**Proposition 3.2.5** *For $q$ odd, Table B.1 (with $\varepsilon = +$) provides a complete set of representatives of the G-classes of involutions $x$ in $G$, together with the order of the corresponding centraliser $C_G(x)$.*

### 3.2.3 Unipotent elements

Let $x \in G$ be a unipotent element, so $x$ has $p$-power order and $x \in T$ since $|G : T| = (n, q-1)$. By Lemma 3.1.3, $x$ lifts to a unique element $\hat{x} \in \mathrm{GL}_n(q)$ of the same order with Jordan form $[J_n^{a_n}, \ldots, J_1^{a_1}]$ for some partition $(n^{a_n}, \ldots, 1^{a_1})$ of $n$. Recall that we refer to $(1^n)$ as the trivial partition of $n$.

**Proposition 3.2.6** *The map*

$$\theta : (p^{a_p}, \ldots, 1^{a_1}) \longmapsto ([J_p^{a_p}, \ldots, J_1^{a_1}]Z)^G$$

*is a bijection from the set of nontrivial partitions of n with parts of size at most p, to the set of G-classes of elements of order p in G.*

*Proof*  This follows immediately from Lemma 3.1.14.                              □

In general, if $x \in G$ is unipotent then the $G$-class of $x$ is a union of several $T$-classes (this is in contrast to the situation for semisimple elements; see Theorem 3.1.12). The next result describes the splitting of $x^G$ into $H$-classes, for any subgroup $H$ of $G$ containing $T$.

**Proposition 3.2.7**  *Let $x \in G$ be an element of order $p$ with Jordan form $[J_p^{a_p}, \ldots, J_1^{a_1}]$ and let $H = T.b$ be a subgroup of $G$. Then $x^G$ splits into precisely*

$$\left( (v, q-1), \frac{(n, q-1)}{b} \right)$$

*H-classes, where $v = \gcd\{j \mid a_j > 0\}$. In particular, $x^G$ splits into $(v, q-1)$ distinct T-classes.*

*Proof*  Recall that $C_{\mathrm{GL}_n(q)}(\hat{x})$ is an extension of a $p$-group by a group $R$ isomorphic to $\prod_j \mathrm{GL}_{a_j}(q)$ (see Lemma 3.1.15). Note that $x$ fixes a decomposition $V = U_1 \oplus \cdots \oplus U_p$, where $\dim U_j = ja_j$. Set $v = \gcd\{j \mid a_j > 0\}$ and observe that $v$ divides $n$.

Now $\hat{x}$ has Jordan form $[J_j^{a_j}]$ on $U_j$. As noted in Section 3.1.2, the restriction $\hat{x}|_{U_j}$ is $\mathrm{GL}(U_j)$-conjugate to the element $J_j \otimes I_{a_j}$ in a tensor product subgroup $\mathrm{GL}_j(q) \otimes \mathrm{GL}_{a_j}(q)$ of $\mathrm{GL}(U_j) = \mathrm{GL}_{ja_j}(q)$. Therefore, if $g = (g_1, \ldots, g_p) \in R$ then $g$ acts on $U_j$ as $I_j \otimes g_j$, hence $g$ has determinant $\prod_j \det(g_j)^j$ as an element of $\mathrm{GL}_n(q)$, which is a $v$th power. There are exactly $(v, q-1)$ elements $\mu \in \mathbb{F}_q$ such that $\mu^v = 1$, so

$$|C_{\mathrm{GL}_n(q)}(\hat{x}) : C_{\mathrm{SL}_n(q)}(\hat{x})| = \frac{q-1}{(v, q-1)}$$

and thus $x^G$ splits into $(v, q-1)$ distinct $T$-classes.

More generally, $G$ cyclically permutes the $T$-classes in $x^G$, with kernel $T.m$, where $m = (n, q-1)/(v, q-1)$. Therefore, if $H = T.b$ then $T.(b, m)$ is the stabiliser in $H$ of a $T$-class in $x^G$, so $x^H$ is a union of $b/(b, m)$ distinct $T$-classes. In particular, $x^G$ splits into

$$\frac{(v, q-1)(b, m)}{b} = \left( (v, q-1), \frac{(n, q-1)}{b} \right)$$

distinct $H$-classes as claimed.                                               □

Representatives of the $T$-classes of elements with Jordan form $[J_p^{a_p}, \ldots, J_1^{a_1}]$ can be described as follows. Let $x = [J_p^{a_p}, \ldots, J_1^{a_1}]Z$ and recall that $G = \langle T, \delta \rangle$, where $\delta = \hat{\delta}Z$, $\hat{\delta} = [\mu, I_{n-1}]$ and $\mathbb{F}_q^\times = \langle \mu \rangle$. The subgroup $\langle \delta \rangle$ cyclically permutes the $(v, q-1)$ distinct $T$-classes in $x^G$. Note that conjugating the matrix $\hat{x}$ by $\hat{\delta}^i$ has the effect of replacing the field element 1 in the $(2,1)$-entry by $\mu^i$. For future reference, we record this observation in the next lemma.

**Lemma 3.2.8** *Let $x \in G$ be an element with Jordan form $[J_p^{a_p}, \ldots, J_1^{a_1}]$. Then $x^{\delta^i}$ and $x^{\delta^j}$ are $T$-conjugate if and only if $j - i$ is divisible by $(v, q-1)$. In particular,*

$$x^{\delta^i}, \quad 0 \leqslant i < (v, q-1)$$

*is a complete set of representatives of the distinct $T$-classes in $x^G$.*

### 3.2.4 Field automorphisms

Fix a basis $\beta = \{v_1, \ldots, v_n\}$ for $V$ and let $\phi : V \to V$ be the map defined in (2.1.1). Recall that $\phi$ induces an automorphism of $T$, and the nontrivial *field automorphisms* of $T$ are the $G$-conjugates of $\phi^i$ with $1 \leqslant i < f$. The $\phi^i$ are called the *standard* field automorphisms of $T$ (with respect to the basis $\beta$).

**Proposition 3.2.9** *Let $x \in \mathrm{Aut}(T)$ be a field automorphism of prime order $r$. Then $r$ divides $f$, say $f = mr$, and $x$ is $G$-conjugate to exactly one of the standard field automorphisms $\phi^{jm}$, where $1 \leqslant j < r$. Moreover, we have $C_G(x) \cong \mathrm{PGL}_n(q^{1/r})$ and $x^G$ splits into*

$$\left( n, \frac{q-1}{q^{1/r}-1} \right)$$

*distinct $T$-classes.*

*Proof* Since $x$ has order $r$ it is $G$-conjugate to $\phi^{jm}$ for some $j$ in the range $1 \leqslant j < r$. The structure of $C_G(x)$ follows from [67, Proposition 4.9.1(b)]. There are precisely $|x^G|/|x^T|$ distinct $T$-classes in $x^G$, and this can be computed via [86, Proposition 4.5.3], which gives the order of $C_T(x)$. $\qquad \square$

### 3.2.5 Graph automorphisms

Recall that a *graph automorphism* of $T$ is an element of the form $\gamma = x\iota$, where $x \in G$ and $\iota$ is the involutory automorphism of $T$ induced from the inverse-transpose map on $\mathrm{SL}_n(q)$. Throughout this section we may assume $n \geqslant 3$ since the inverse-transpose map is an inner automorphism when $n = 2$ (so in this situation there are no graph automorphisms of $T$).

Let $x = J\iota \in G$ and set $\gamma_J = x\iota$. Then

$$\gamma_J : AZ \mapsto J^\mathsf{T} A^{-\mathsf{T}} J^{-\mathsf{T}} Z$$

and

$$
\begin{aligned}
C_G(\gamma_J) &= \{AZ \in G \mid J^\mathsf{T} A^{-\mathsf{T}} J^{-\mathsf{T}} = \lambda A \text{ for some } \lambda \in \mathbb{F}_q^\times\} \\
&= \{AZ \in G \mid J^\mathsf{T} = \lambda A J^\mathsf{T} A^\mathsf{T} \text{ for some } \lambda \in \mathbb{F}_q^\times\} \\
&= \{AZ \in G \mid J = \lambda A J A^\mathsf{T} \text{ for some } \lambda \in \mathbb{F}_q^\times\}
\end{aligned}
\tag{3.2.4}
$$

Therefore, if $B_J : V \times V \to \mathbb{F}_q$ is the nondegenerate bilinear form defined by

$$B_J(v, w) = v J w^\mathsf{T}$$

then (3.2.4) indicates that $C_G(\gamma_J)$ is the group of all similarities of $B_J$ (modulo scalars).

Recall that $\iota$ induces a map on the set of subspaces of $V$, sending a subspace $U$ to

$$\{v \in V \mid B_{I_n}(u, v) = 0 \text{ for all } u \in U\}$$

It follows that $\gamma_J$ maps $U$ to the subspace

$$\{v \in V \mid B_J(u, v) = 0 \text{ for all } u \in U\}$$

Note that this is the orthogonal complement $U^\perp$ with respect to $B_J$; see Section 2.2.4.

We are interested in determining the graph automorphisms of order two. Now $\gamma_J^2 = 1$ if and only if $J\iota J\iota = \lambda I_n$ for some $\lambda \in \mathbb{F}_q^\times$, that is $JJ^{-\mathsf{T}} = \lambda I_n$. Hence $J = \lambda J^\mathsf{T}$. It follows that $J^\mathsf{T} = \lambda J$, so $\lambda^2 = 1$ and thus $\lambda = \pm 1$. Therefore, $J$ is a symmetric or skew-symmetric matrix, and the form $B_J$ is symmetric or skew-symmetric, respectively. Note that if $J$ is skew-symmetric then $n$ is even since $\det(J) = (-1)^n \det(J)$. Here we are essentially deducing part of the Birkhoff–von Neumann Theorem (see Theorem 2.2.4), which states that every polarity of $\mathscr{P}(V)$ arises from a nondegenerate reflexive sesquilinear form that is either symmetric, alternating or hermitian.

Recall that two bilinear forms $B_1, B_2$ on $V$ are said to be *similar* if there exists $A \in \mathrm{GL}_n(q)$ and $\lambda \in \mathbb{F}_q^\times$ such that $B_1(v, w) = \lambda B_2(vA, wA)$ for all $v, w \in V$. In particular, if $J_1, J_2 \in \mathrm{GL}_n(q)$ then $B_{J_1}$ and $B_{J_2}$ are similar if there exists $A \in \mathrm{GL}_n(q)$ and $\lambda \in \mathbb{F}_q^\times$ such that

$$v J_1 w^\mathsf{T} = \lambda v A J_2 (wA)^\mathsf{T}$$

for all $v, w \in V$, that is, if $J_1 = \lambda A J_2 A^\mathsf{T}$. Also observe that $J_1$ and $\lambda^{-1} J_1$ give rise to the same element of $G$.

Now, if $g = AZ \in G$ then

$$(\gamma_J)^g = (A^{-1}Z)JZ\iota(AZ) = A^{-1}J\iota AZ = A^{-1}JA^{-\mathsf{T}}Z\iota = \gamma_{A^{-1}JA^{-\mathsf{T}}}$$

Note that if $J$ is symmetric or skew-symmetric then so is $A^{-1}JA^{-\mathsf{T}}$. In addition, the bilinear forms $B_J$ and $B_{A^{-1}JA^{-\mathsf{T}}}$ are similar. We have now established the following result.

**Proposition 3.2.10** *There is a bijection from the set of G-classes of involutory graph automorphisms of T, to the set of similarity classes of nondegenerate symmetric and skew-symmetric bilinear forms on V.*

Let us now describe the *standard* graph automorphisms. If $n$ is odd, let $\gamma_1$ be the graph automorphism given by $J_1 = I_n$ (that is, $\gamma_1 = \iota$). Now assume $n$ is even, so $n \geqslant 4$. Let $\gamma_1$ be the graph automorphism induced by

$$J_1 = \begin{pmatrix} 0 & 1 & & & \\ -1 & 0 & & & \\ & & \ddots & & \\ & & & 0 & 1 \\ & & & -1 & 0 \end{pmatrix}$$

If $q$ is odd, let $\gamma_2$ and $\gamma_2'$ be the graph automorphisms given by

$$J_2 = \begin{pmatrix} 0 & 1 & & & \\ 1 & 0 & & & \\ & & \ddots & & \\ & & & 0 & 1 \\ & & & 1 & 0 \end{pmatrix} \qquad J_2' = \begin{pmatrix} 0 & 1 & & & & \\ 1 & 0 & & & & \\ & & \ddots & & & \\ & & & 0 & 1 & \\ & & & 1 & 0 & \\ & & & & & 1 \\ & & & & & & \alpha \end{pmatrix}$$

respectively, where $-\alpha \in \mathbb{F}_q$ is a nonsquare. In particular, note that $\det(J_2)$ is an $\mathbb{F}_q$-square if and only if $n(q-1)/4$ is even, and $\det(J_2')$ is an $\mathbb{F}_q$-square if and only if $n(q-1)/4$ is odd. Finally, if both $n$ and $q$ are even then let $\gamma_3$ be the graph automorphism given by $J_3 = I_n$, so $\gamma_3 = \iota$.

We are now ready to describe the $G$-classes of involutory graph automorphisms of $T$. For alternative proofs, see [67, Theorem 4.5.1] when $q$ is odd, and [5, (19.8), (19.9)] for $q$ even. Note that in part (iii), a *transvection* is an element with Jordan form $[J_2, J_1^{n-2}]$.

**Proposition 3.2.11** *Let* $\gamma = x\iota$ *be an involutory graph automorphism of T with* $x = JZ \in G$.

(i) *If n is odd then* $\gamma \in \gamma_1^G$ *and* $C_G(\gamma) \cong \mathrm{PGO}_n(q)$ *(so* $C_G(\gamma) \cong \mathrm{Sp}_{n-1}(q)$ *if q is even).*

(ii) *Suppose n is even and q is odd.*
   (a) *If* $J^\mathsf{T} = -J$ *then* $\gamma \in \gamma_1^G$ *and* $C_G(\gamma) \cong \mathrm{PGSp}_n(q)$.
   (b) *If* $J^\mathsf{T} = J$ *then* $\gamma \in \gamma_2^G$ *if* $\det(J) \equiv \det(J_2) \pmod{(\mathbb{F}_q^\times)^2}$, *otherwise* $\gamma \in (\gamma_2')^G$. *Here* $C_G(\gamma_2) \cong \mathrm{PGO}_n^+(q)$ *and* $C_G(\gamma_2') \cong \mathrm{PGO}_n^-(q)$.

(iii) *Suppose n and q are even. If* $B_J$ *is alternating then* $\gamma \in \gamma_1^G$, *otherwise* $\gamma \in \gamma_3^G$. *Here* $C_G(\gamma_1) \cong \mathrm{Sp}_n(q)$ *and* $C_G(\gamma_3) \cong C_{\mathrm{Sp}_n(q)}(t)$, *where* $t \in \mathrm{Sp}_n(q)$ *is a transvection.*

*Proof* First recall that each $G$-class of involutory graph automorphisms corresponds to a similarity class of nondegenerate symmetric or skew-symmetric bilinear forms (see Proposition 3.2.10). The given centralisers can be read off from (3.2.4); alternatively, see [67, Table 4.5.1] and [5, (19.9)] in the cases $q$ odd and even, respectively.

Suppose $q$ is odd. Nondegenerate skew-symmetric bilinear forms exist only if $n$ is even, and Proposition 2.2.1 indicates that there is a unique similarity class of such forms. This gives (ii)(a). As noted in Section 2.2.6, there is precisely one similarity class of symmetric forms if $n$ is odd, and two if $n$ is even; in the latter case, the two similarity classes can be distinguished by considering $\det(J)$, which is either an $\mathbb{F}_q$-square or nonsquare. This gives (i) for $q$ odd, and also (ii)(b).

Finally, suppose $q$ is even. Here every skew-symmetric form is symmetric. By Proposition 2.2.1 and [108, Corollary 1], there are only two isometry classes of nondegenerate symmetric forms on $V$: alternating and nonalternating. In particular, if $B_J$ is alternating then $B_J$ and $B_{J_1}$ are isometric and thus $\gamma \in \gamma_1^G$. On the other hand, if $B_J$ is nonalternating then [108, Theorem 1] states that $V$ has an orthonormal basis, so in this case $B_J$ and $B_{J_3}$ are isometric and $\gamma \in \gamma_3^G$. $\qquad\qquad\square$

**Remark 3.2.12** Suppose $n$ is even and $\gamma$ is an involutory graph automorphism of $T$. If $C_G(\gamma) \cong \mathrm{PGSp}_n(q)$ then we say that $\gamma$ is a *symplectic-type* graph automorphism, otherwise $\gamma$ is *non-symplectic*. This terminology will be useful later.

**Corollary 3.2.13** *The G-class containing the inverse-transpose graph automorphism* $\iota$ *is recorded in Table 3.2.1.*

Table 3.2.1 $\mathrm{PGL}_n(q)$-conjugacy of the inverse-transpose automorphism $\iota$

| Conditions on $n$ and $q$ | $\mathrm{PGL}_n(q)$-class containing $\iota$ |
| --- | --- |
| $n$ odd | $\gamma_1$ |
| $n$ even and $q$ even | $\gamma_3$ |
| $n$ even, $q$ odd and $n(q-1)/4$ even | $\gamma_2$ |
| $n$ even, $q$ odd and $n(q-1)/4$ odd | $\gamma_2'$ |

Finally, let us consider the splitting of $G$-classes of involutory graph automorphisms into $T$-classes.

**Proposition 3.2.14** *Let $\gamma \in \mathrm{Aut}(T)$ be an involutory graph automorphism. Then the $G$-class of $\gamma$ splits into $c$ distinct $T$-classes, where*

$$c = \begin{cases} (n,q-1)/2 & n \text{ even, } q \text{ odd, } \gamma \in \{\gamma_2^G, \gamma_2'^G\} \\ (n/2,q-1) & n \text{ even, } q \text{ odd, } \gamma \in \gamma_1^G \\ (n,q-1) & \text{otherwise} \end{cases}$$

*Proof* This is a routine calculation, using the information in [86, Section 4.8] to compute the order of $C_T(\gamma)$ (note that if $n$ and $q$ are even and $\gamma \in \gamma_3^G$, then $\mathrm{Sp}_n(q) < T$ and thus $C_T(\gamma) = C_G(\gamma)$). $\qquad\square$

If we write $G = \langle T, \delta \rangle$, where $\delta = [\mu, I_{n-1}]Z \in G$ and $\mathbb{F}_q^\times = \langle \mu \rangle$, then

$$\gamma^{\delta^i}, \ 0 \leqslant i < c$$

is a set of representatives for the $T$-classes in $\gamma^G$, where $c$ is the integer defined in Proposition 3.2.14.

### 3.2.6 Graph-field automorphisms

Recall that the graph-field automorphisms of $T$ are the $G$-conjugates of elements of the form $\psi = \phi^i \iota$, where $1 \leqslant i < f$ and $\phi$ is a standard field automorphism of order $f$ (with respect to a fixed basis of $V$). If $\psi$ has prime order then $\psi$ and $\phi^i$ are both involutions, so in particular $f$ is even. It follows that $T$ has at most one $G$-class of involutory graph-field automorphisms.

Assume $f$ is even. Let $\varphi = \phi^{f/2}$ be the standard involutory field automorphism of $T$ and write $x = JZ \in G$ with $J \in \mathrm{GL}_n(q)$. Set $\psi_J = x\varphi \iota$ and let $AZ \in G$. Then

$$\psi_J : AZ \mapsto J^{\varphi\mathsf{T}} A^{-\varphi\mathsf{T}} J^{-\varphi\mathsf{T}} Z$$

and

$$C_G(\psi_J) = \{AZ \in G \mid J^{\varphi\mathsf{T}}A^{-\varphi\mathsf{T}}J^{-\varphi\mathsf{T}} = \lambda A \text{ for some } \lambda \in \mathbb{F}_q^\times\}$$
$$= \{AZ \in G \mid J^{\varphi\mathsf{T}} = \lambda A J^{\varphi\mathsf{T}}A^{\varphi\mathsf{T}} \text{ for some } \lambda \in \mathbb{F}_q^\times\}$$
$$= \{AZ \in G \mid J = \lambda A J A^{\varphi\mathsf{T}} \text{ for some } \lambda \in \mathbb{F}_q^\times\} \qquad (3.2.5)$$

Therefore, if $B_J : V \times V \to \mathbb{F}_q$ is the nondegenerate sesquilinear form with

$$B_J(v, w) = vJw^{\varphi\mathsf{T}}$$

for all $v, w \in V$, then $C_G(\psi_J)$ is the group of all similarities of $B_J$ (modulo scalars). In addition, if $U$ is a subspace of $V$ then it is easy to see that $\psi_J$ maps $U$ to the subspace

$$\{v \in V \mid B_J(u, v) = 0 \text{ for all } u \in U\}$$

which is the orthogonal complement $U^\perp$ with respect to $B_J$.

Let us describe the involutory graph-field automorphisms. Set $q_0 = q^{1/2}$. Now $\psi_J^2 = 1$ if and only if $J\varphi\iota J\varphi\iota = \lambda I_n$ for some $\lambda \in \mathbb{F}_q^\times$. In particular, $JJ^{-\varphi\mathsf{T}} = \lambda I_n$ and so $J = \lambda J^{\varphi\mathsf{T}}$. It follows that $J^{\varphi\mathsf{T}} = \lambda^\varphi J$ and so $J = \lambda^{1+q_0}J$. Hence $\lambda^{1+q_0} = 1$ and thus $\lambda \in \langle \mu^{q_0-1} \rangle$ where $\mu$ is a primitive element of $\mathbb{F}_q$. Therefore, $\lambda = \xi^{q_0-1}$ for some $\xi \in \mathbb{F}_q$. Let $K = \xi J$. Then $\psi_K = \psi_J$ and

$$K^{\varphi\mathsf{T}} = (\xi J)^{\varphi\mathsf{T}} = \xi^{q_0}J^{\varphi\mathsf{T}} = \xi^{q_0}\lambda^{-1}J = \xi^{q_0}\xi^{1-q_0}J = \xi J = K$$

so involutory graph-field automorphisms of $T$ correspond to invertible hermitian matrices, and hence nondegenerate hermitian forms. Therefore, if $\psi_J$ is an involutory graph-field automorphism, then without loss of generality we may assume that $J$ is hermitian. (Note that this gives us the remaining case appearing in the Birkhoff–von Neumann Theorem.)

If $J_1, J_2 \in \mathrm{GL}_n(q)$ then the sesquilinear forms $B_{J_1}$ and $B_{J_2}$ are similar if there exists $A \in \mathrm{GL}_n(q)$ and $\lambda \in \mathbb{F}_q^\times$ such that

$$vJ_1w^{\varphi\mathsf{T}} = \lambda v A J_2 (wA)^{\varphi\mathsf{T}}$$

for all $v, w \in V$, that is, if $J_1 = \lambda A J_2 A^{\varphi\mathsf{T}}$. Note also that $J_1$ and $\lambda^{-1}J_1$ give rise to the same element of $G$.

Now, if $g = AZ \in G$ then

$$(\psi_J)^g = (A^{-1}Z)JZ\varphi\iota(AZ) = A^{-1}J\varphi\iota AZ = A^{-1}JA^{-\varphi\mathsf{T}}Z\varphi\iota = \psi_{A^{-1}JA^{-\varphi\mathsf{T}}}$$

Note that if $J$ is hermitian then so is $A^{-1}JA^{-\varphi\mathsf{T}}$, and the two hermitian forms $B_J$ and $B_{A^{-1}JA^{-\varphi\mathsf{T}}}$ are similar. By Proposition 2.2.2, all nondegenerate hermitian forms on $V$ are isometric, so this provides another way to see that there is a unique $G$-class of involutory graph-field automorphisms.

**Proposition 3.2.15** *Let $x \in \mathrm{Aut}(T)$ be an involutory graph-field automorphism. Then $f$ is even, $x$ is $G$-conjugate to $\varphi\iota$ and $C_G(x) \cong \mathrm{PGU}_n(q^{1/2})$. Moreover, the $G$-class of $x$ splits into $(n, q^{1/2} - 1)$ distinct $T$-classes.*

*Proof*  By definition, $x$ is $G$-conjugate to $\varphi\iota$. The structure of $C_G(x)$ follows from (3.2.5). Finally, the order of $C_T(x)$ is given in [86, Proposition 4.8.5] and we quickly deduce that the $G$-class of $x$ splits into

$$\frac{(n, q-1)\left(q^{1/2}+1, \frac{q-1}{(n,q-1)}\right)}{q^{1/2}+1} = (n, q^{1/2} - 1)$$

distinct $T$-classes.                                                    □

Observe that

$$x^{\delta^i}, \ 0 \leqslant i < (n, q^{1/2} - 1)$$

is a set of representatives for the $T$-classes in $x^G$, where $\delta = [\mu, I_{n-1}]Z \in G$ and $\mathbb{F}_q^\times = \langle \mu \rangle$.

### 3.2.7 Linear groups: a summary

Here we summarise some of the above results on the conjugacy classes of elements of prime order in $\mathrm{Aut}(T)$, where $T = \mathrm{PSL}_n(q)$ and $G = \mathrm{Inndiag}(T) = \mathrm{PGL}_n(q)$. (Note that Table B.3 is located in Appendix B.)

**Proposition 3.2.16** *Let $x \in \mathrm{Aut}(T)$ be an element of prime order $r$. Then the possibilities for $x$ up to $G$-conjugacy are listed in Table B.3, together with the order of the centraliser $C_G(x)$.*

*Proof*  By Theorem 2.1.3, the possibilities for $x$ are recorded in Table 3.2.2 (in Table B.3, the *type* of $x$ is recorded in the second column). If $x$ is a field, graph or graph-field automorphism, then the result follows from Propositions 3.2.9, 3.2.11 and 3.2.15, respectively. Similarly, if $x$ is an inner-diagonal automorphism, then $x$ is semisimple or unipotent, and the result follows by applying Propositions 3.2.1, 3.2.2 and 3.2.6 (the relevant centraliser orders can be read off from Lemmas 3.1.9 and 3.1.15).                                    □

## 3.3  Unitary groups

Set $T = \mathrm{PSU}_n(q)$ and $G = \mathrm{Inndiag}(T) = \mathrm{PGU}_n(q) = \mathrm{GU}_n(q)/Z$, where

$$Z = \{\mu I_n \mid \mu \in \mathbb{F}_{q^2}, \mu^{q+1} = 1\}$$

Table 3.2.2 *Types of elements of prime order r in* $\mathrm{Aut}(\mathrm{PSL}_n(q))$

| Type | Description | Conditions |
|------|-------------|------------|
| s  | semisimple | $r \neq p$ |
| u  | unipotent | $r = p$ |
| f  | field automorphism | $q = q_0^r$ |
| gf | graph-field automorphism | $n \geqslant 3, q = q_0^2, r = 2$ |
| g  | graph automorphism | $n \geqslant 3, r = 2$ |

denotes the centre of $\mathrm{GU}_n(q)$. Recall that $\mathrm{GU}_n(q) < \mathrm{GL}_n(q^2)$ and $G = \langle T, \delta \rangle$, where $\delta = [\mu, I_{n-1}]Z \in G$ and $\mu \in \mathbb{F}_{q^2}$ has order $q+1$. Also recall that $\mathrm{PSU}_2(q)$ and $\mathrm{PSL}_2(q)$ are isomorphic, so throughout this section we will assume $n \geqslant 3$. Let $V$ be the natural module for $T$, which is an $n$-dimensional vector space over $\mathbb{F}_{q^2}$. Define the map $\varphi : \mathrm{GL}_n(q^2) \to \mathrm{GL}_n(q^2)$ as in (2.2.6), so $\varphi : (a_{ij}) \mapsto (a_{ij}^q)$ with respect to a fixed orthonormal basis of $V$.

Membership and conjugacy in $\mathrm{GU}_n(q)$ is given by the following proposition (see [119, p. 34]).

**Proposition 3.3.1** *Let $A \in \mathrm{GL}_n(q^2)$. Then $A$ is similar to an element of $\mathrm{GU}_n(q)$ if and only if $A$ and $A^{-\varphi\mathsf{T}}$ are conjugate in $\mathrm{GL}_n(q^2)$. Moreover, $A, B \in \mathrm{GU}_n(q)$ are conjugate in $\mathrm{GU}_n(q)$ if and only if they are conjugate in $\mathrm{GL}_n(q^2)$.*

### 3.3.1 Semisimple elements

We begin our analysis of the conjugacy classes in $\mathrm{Aut}(T)$ by considering the semisimple elements of odd prime order (semisimple involutions will be discussed separately in Section 3.3.2). Let $r \neq p$ be an odd prime divisor of $|G|$ and define $i = \Phi(r, q)$ as before (see (3.1.2)). Let $\mathscr{S}_r$ denote the set of nontrivial $r$th roots of unity in $\mathbb{F}_{q^i}$, and define a permutation $\sigma$ of $\mathscr{S}_r$ by $\sigma : \lambda \mapsto \lambda^q$. It will be convenient to set

$$b = b(i) = \begin{cases} i & i \text{ odd} \\ i/2 & i \text{ even} \end{cases} \tag{3.3.1}$$

Let $\Lambda_1, \ldots, \Lambda_s$ denote the distinct $\sigma^2$-orbits in $\mathscr{S}_r$. By Lemma 3.1.13(ii), we have $|\Lambda_j| = b$ for all $j$, so $s = (r-1)/b$. Recall that

$$\Lambda_j^{-q} = \{\lambda^{-q} \mid \lambda \in \Lambda_j\}$$

If $i \not\equiv 2 \pmod 4$ then $s$ is even and Lemma 3.1.13(ii) implies that $\Lambda_j \neq \Lambda_j^{-q}$ for all $j$, so in this situation we will label the $\Lambda_j$ so that $\Lambda_j^{-q} = \Lambda_{s/2+j}$ for all $1 \leqslant j \leqslant s/2$.

Suppose $x \in G$ has order $r$ and fix $\hat{x} \in \mathrm{GU}_n(q)$ such that $x = \hat{x}Z$ and $\hat{x}$ has the smallest possible order. By Theorem 3.1.12 we have $x^G = x^T$. Recall that $\{U, U^*\}$ is a *dual pair* of totally isotropic subspaces of $V$ if $U \oplus U^*$ is nondegenerate.

**Proposition 3.3.2** *If $i \neq 2$ then $x \in T$ and one of the following holds:*

(i) *If $i \not\equiv 2 \pmod 4$, then $\hat{x}$ fixes an orthogonal decomposition of the form*

$$V = (U_{1,1} \oplus U_{1,1}^*) \perp \ldots \perp (U_{1,a_1} \oplus U_{1,a_1}^*) \perp \ldots$$
$$\perp (U_{s/2,1} \oplus U_{s/2,1}^*) \perp \ldots \perp (U_{s/2,a_{s/2}} \oplus U_{s/2,a_{s/2}}^*) \perp C_V(\hat{x})$$

*where each $\{U_{j,k}, U_{j,k}^*\}$ is a dual pair of totally isotropic b-spaces on which $\hat{x}$ acts irreducibly, and $C_V(\hat{x})$ is nondegenerate (or trivial). Moreover, the map*

$$\theta_1 : (a_1, \ldots, a_{s/2}) \longmapsto ([(\Lambda_1, \Lambda_1^{-q})^{a_1}, \ldots, (\Lambda_{s/2}, \Lambda_{s/2}^{-q})^{a_{s/2}}, I_e]Z)^G$$

*is a bijection from the set of nonzero $\frac{s}{2}$-tuples $(a_1, \ldots, a_{s/2}) \in \mathbb{N}_0^{s/2}$ such that $2b\sum_j a_j \leqslant n$, to the set of G-classes of elements of order $r$ in G.*

(ii) *If $i \equiv 2 \pmod 4$, then $\hat{x}$ fixes an orthogonal decomposition*

$$V = U_{1,1} \perp \ldots \perp U_{1,a_1} \perp \ldots \perp U_{s,1} \perp \ldots \perp U_{s,a_s} \perp C_V(\hat{x})$$

*where each $U_{j,k}$ is a nondegenerate $\frac{i}{2}$-space on which $\hat{x}$ acts irreducibly, and $C_V(\hat{x})$ is nondegenerate (or trivial). Moreover, the map*

$$\theta_2 : (a_1, \ldots, a_s) \longmapsto ([\Lambda_1^{a_1}, \ldots, \Lambda_s^{a_s}, I_e]Z)^G$$

*is a bijection from the set of nonzero s-tuples $(a_1, \ldots, a_s) \in \mathbb{N}_0^s$ such that $i\sum_j a_j \leqslant 2n$, to the set of G-classes of elements of order $r$ in G.*

*Proof* Notice that $|G : T| = (n, q+1)$ is indivisible by $r$, so $x \in T$ and we may assume that $\hat{x}$ has order $r$ (see Lemma 3.1.3).

First assume $i \not\equiv 2 \pmod 4$. If $i$ is odd then the smallest even integer $k$ such that $r$ divides $q^k - 1$ is $k = 2i$, whence the irreducible blocks of $\hat{x}$ have dimension $b = i$ (recall that $V$ is a vector space over $\mathbb{F}_{q^2}$). Similarly, if $i \equiv 0 \pmod 4$ then the irreducible blocks have dimension $b = i/2$. Let $\Lambda_j$ be a $\sigma^2$-orbit on $\mathscr{S}_r$ and note that $|\Lambda_j| = b$ and $\Lambda_j \neq \Lambda_j^{-q}$ (see Lemma 3.1.13). By applying Lemma 3.1.6 and Proposition 3.3.1, we deduce that the irreducible blocks of $\hat{x}$ are totally isotropic and arise in dual pairs $U_j \oplus U_j^*$ with eigenvalues $\Lambda_j \cup \Lambda_j^{-q}$ (in $\mathbb{F}_{q^i}$). In particular, if $a_j$ denotes the multiplicity of $\Lambda_j$ in the multiset of eigenvalues of $\hat{x}$, then $\hat{x}$ fixes an orthogonal decomposition of $V$ as described in part (i). Moreover, by applying Lemma 3.1.7 and Proposition 3.3.1 we deduce that $\theta_1(a_1,\ldots,a_{s/2}) = x^G$ and thus $\theta_1$ is surjective. The injectivity of $\theta_1$ follows from Lemma 3.1.7, whence $\theta_1$ is a bijection.

Now assume $i \equiv 2 \pmod 4$ and $i > 2$. Here the irreducible blocks of $\hat{x}$ have dimension $i/2$, with eigenvalues

$$\Lambda_j = \{\lambda, \lambda^{q^2}, \ldots, \lambda^{q^{2(i/2-1)}}\}$$

for some $r$th root of unity $\lambda \in \mathbb{F}_{q^i}$. Now $\Lambda_j = \Lambda_j^{-q}$ by Lemma 3.1.13, so Proposition 3.3.1 implies that $[\Lambda_j] \in \mathrm{SU}_{i/2}(q)$. Therefore, there exists an element $y = \hat{y}Z \in G$ of order $r$ such that $\hat{y}$ and $\hat{x}$ have the same eigenvalues, and $\hat{y}$ fixes an orthogonal decomposition of $V$ as given in part (ii). In particular, if $a_j$ denotes the multiplicity of $\Lambda_j$ in the multiset of eigenvalues of $\hat{x}$, then a combination of Lemma 3.1.7 and Proposition 3.3.1 implies that

$$\theta_2(a_1,\ldots,a_s) = x^G = y^G$$

Therefore, $\theta_2$ is surjective and $\hat{x}$ fixes an orthogonal decomposition as in (ii). Finally, by applying Lemma 3.1.7 again we deduce that $\theta_2$ is a bijection. $\quad\square$

The case $i = 2$ requires special attention. In the statement of the next result, the set $\mathscr{T}(n,r)$ is defined in the paragraph preceding Proposition 3.2.2.

**Proposition 3.3.3** *Suppose $i = 2$. Then up to G-conjugacy one of the following holds:*

(i) $\hat{x} = [I_{a_0}, \lambda I_{a_1}, \ldots, \lambda^{r-1} I_{a_{r-1}}]$ *for some r-tuple* $(a_0,\ldots,a_{r-1}) \in \mathbb{N}_0^r$, *where* $\lambda \in \mathbb{F}_{q^2}$ *is a nontrivial rth root of unity, and $a_j < n$ for all $j$. Moreover, $\hat{x}$ fixes an orthogonal decomposition of the form*

$$V = U_1 \perp \ldots \perp U_n$$

*where each $U_j$ is a nondegenerate 1-space.*

(ii) *r divides n and $\hat{x} = [\Lambda^{n/r}]$ with $\Lambda = \{\mu, \mu^{q^2}, \ldots, \mu^{q^{2(r-1)}}\}$ for some $\mu \in \mathbb{F}_{q^{2r}}$ of order $r(q+1)_r$. Further, $\hat{x}$ fixes a decomposition*

$$V = U_1 \perp \ldots \perp U_{n/r}$$

*where each $U_j$ is a nondegenerate r-space on which $\hat{x}$ acts irreducibly.*

*If n is indivisible by r, then there is a bijection*

$$\theta : (a_0,\ldots,a_{r-1}) \longmapsto ([I_{a_0}, \lambda I_{a_1}, \ldots, \lambda^{r-1} I_{a_{r-1}}]Z)^G$$

*from $\mathscr{T}(n,r)$ to the set of G-classes of elements of order r in G. If r divides n then there are an additional $r - 1$ distinct G-classes, represented by elements of type (ii). In (i) we have $x \in T$ if and only if $\lambda^{\sum_j j a_j} = \xi^n$ for some $\xi \in \mathbb{F}_{q^2}$ with $\xi^{q+1} = 1$, and $x \in T$ in (ii) if and only if $(n)_r > (q+1)_r$.*

*Proof* This is very similar to the proof of Proposition 3.2.2. First assume $\hat{x}$ is diagonalisable over $\mathbb{F}_{q^2}$, with eigenvalues $\lambda_1,\ldots,\lambda_n$. Now $\hat{x}^r \in Z$ so $\lambda_1^r = \lambda_j^r$ for all $j$, and $\lambda_1^{r(q+1)} = 1$ (recall that $Z$ is a cyclic group of order $q+1$). Since $r(q+1)$ does not divide $q^2 - 1$ it follows that $\lambda_1^{q+1} = 1$, so by replacing $\hat{x}$ with $\lambda_1^{-1}\hat{x} \in \hat{x}Z$ if necessary, we may assume that the eigenvalues of $\hat{x}$ are $r$th roots of unity. Since $\lambda_j = \lambda_j^{-q}$ for all $j$, Proposition 3.3.1 implies that there exists an element $\hat{y} \in \mathrm{GU}_n(q)$ of order $r$ such that $\hat{y}$ and $\hat{x}$ have the same eigenvalues, and $\hat{y}$ fixes an orthogonal decomposition of $V$ as in (i). By Lemma 3.1.7 and Proposition 3.3.1, $\hat{x}$ and $\hat{y}$ are $\mathrm{GU}_n(q)$-conjugate, so $\hat{x}$ also fixes such a decomposition. In addition, (3.2.1) holds (where $\lambda \in \mathbb{F}_{q^2}$ has order $r$) and thus the $G$-class of $x$ corresponds to $(a_0,\ldots,a_{r-1}) \in \mathscr{T}(n,r)$.

Now suppose $\hat{x}$ is not diagonalisable over $\mathbb{F}_{q^2}$. Here $\hat{x}^r \in Z$ and thus $\hat{x}$ is diagonalisable over $\mathbb{F}_{q^{2r}}$. Let $\mu$ be an eigenvalue of $\hat{x}$, so the order of $\mu^r \in \mathbb{F}_{q^2}$ divides $q+1$. Moreover, $\mu^{q^2},\ldots,\mu^{q^{2(r-1)}}$ are also eigenvalues of $\hat{x}$, and since

$$\Lambda = \{\mu, \mu^{q^2},\ldots,\mu^{q^{2(r-1)}}\} = \{\gamma \in \mathbb{F}_{q^{2r}} \mid \gamma^r = \mu^r\}$$

it follows that $r$ divides $n$ and $\hat{x} = [\Lambda^{n/r}]$. Now $(\mu^{-q})^r = \mu^{r(q+1)-rq} = \mu^r$, so $\gamma^r = \mu^r$ for all $\gamma \in \Lambda^{-q}$ and thus $\Lambda^{-q} = \Lambda$. By arguing as in the previous case, applying Lemma 3.1.7 and Proposition 3.3.1, we deduce that $\hat{x}$ fixes a decomposition of $V$ into nondegenerate $r$-spaces as in part (ii).

Now $\mu$ has order $r(q+1)_r\ell$ for some divisor $\ell$ of $q+1$ with $(\ell,r)=1$. Let $a$ be an integer such that $a\ell \equiv 1 \pmod r$ and $1 \leqslant a < r$. Then $\mu^{a\ell-1} \in \mathbb{F}_{q^2}$ has order dividing $q+1$, and $\mu(\mu^{a\ell-1}) = \mu^{a\ell}$ has order $r(q+1)_r$ since $(a,r)=1$. Therefore, replacing $\hat{x}$ by the scalar multiple $\mu^{a\ell-1}\hat{x}$ if necessary, we may assume that $\mu$ has order $r(q+1)_r$, so $\hat{x}$ has the form given in part (ii). In addition, there are $(r-1)(q+1)_r/r$ distinct possibilities for $\Lambda$, and by arguing as in the proof of Proposition 3.2.2 we deduce that there are precisely $r-1$ distinct $G$-classes of elements of this type.

Finally, let us consider the membership of these elements in $T$. In (i), the stated criterion follows at once from Lemma 2.3.1, so let us turn to (ii). First observe that $\det(\hat{x}) = \omega^n$, where $\omega = \mu^{(q^{2r}-1)/(q^2-1)r}$ and $(q^{2r}-1)/(q^2-1)r$ is indivisible by $r$ (by Lemma A.4). In particular, $\omega$ has order $r(q+1)_r$. Write $q+1 = s(q+1)_r$ and $\omega^r = \gamma^s$, where $\gamma \in \mathbb{F}_{q^2}$ has order $q+1$. Then $\det(\hat{x}) = \gamma^{sm}$, where $m = n/r$. By applying Lemma 2.3.1, and by arguing as in the proof of Proposition 3.2.2, we deduce that $x \in T$ if and only if $(n)_r > (q+1)_r$. $\square$

**Remark 3.3.4** Explicit representatives for the semisimple elements arising in part (ii) of Proposition 3.3.3 can be given as in Remark 3.2.3. Also note that if $x \in G$ is a semisimple element of prime order, then

$$C_{\mathrm{GU}_n(q)}(\hat{x}) = \mathrm{GU}_n(q) \cap C_{\mathrm{GL}_n(q^2)}(\hat{x})$$

and the structure of $C_{\mathrm{GL}_n(q^2)}(\hat{x})$ is given in Lemma 3.1.9. More precisely, if $\hat{x}$ is given as in Proposition 3.3.2(ii) then

$$C_{\mathrm{GU}_n(q)}(\hat{x}) = \mathrm{GU}_{a_1}(q^{i/2}) \times \cdots \times \mathrm{GU}_{a_s}(q^{i/2}) \times \mathrm{GU}_e(q)$$

Similarly, if $\hat{x}$ is as in Proposition 3.3.2(i) then

$$C_{\mathrm{GU}_n(q)}(\hat{x}) = \mathrm{GL}_{a_1}(q^{2b}) \times \cdots \times \mathrm{GL}_{a_{s/2}}(q^{2b}) \times \mathrm{GU}_e(q)$$

(with $b$ defined in (3.3.1)). Again, extra care is needed if $r$ divides $q+1$. The order of $C_G(x)$ is recorded in Table B.4, which is located in Appendix B.

**Remark 3.3.5** Note that if $i \equiv 2 \pmod 4$ and $\Lambda$ is a $\sigma^2$-orbit on $\mathscr{S}_r$ of length $i/2$ then $\mathrm{GU}_i(q)$ contains an element $x = [\Lambda^2]$ that fixes an orthogonal decomposition of the natural $\mathrm{GU}_i(q)$-module $V$ into a pair of nondegenerate $\frac{i}{2}$-spaces. Since $\Lambda = \Lambda^{-q}$, Lemma 2.2.17 implies that $\mathrm{GU}_i(q)$ also contains an element $y = [\Lambda^2]$ that fixes a dual pair of totally isotropic $\frac{i}{2}$-spaces. By Proposition 3.3.1, $x$ and $y$ are conjugate in $\mathrm{GU}_i(q)$ and so $x$ also fixes such a dual pair.

An alternative way to see that $x$ fixes a pair of orthogonal nondegenerate $\frac{i}{2}$-spaces and also a dual pair of totally isotropic $\frac{i}{2}$-spaces is as follows. Recall from Section 2.2.8 that if $U$ and $W$ are equipped with nondegenerate hermitian forms $B_1$ and $B_2$ respectively, then we can construct an hermitian form $B$ on the tensor product $V = U \otimes W$ (see (2.2.9)). If $\dim U = 2$ and $\dim W = i/2$ then $x = I_2 \otimes [\Lambda]$ preserves $B$ and has the form $[\Lambda^2]$ on $V$. Let $\{u_1, u_2\}$ be an orthonormal basis for $U$ with respect to $B_1$. Then $x$ fixes the orthogonal decomposition

$$V = (\langle u_1 \rangle \otimes W) \perp (\langle u_2 \rangle \otimes W)$$

into nondegenerate $\frac{i}{2}$-spaces. If we now take an hermitian basis $\{e_1, f_1\}$ for $U$, then $x$ also fixes the decomposition

$$V = (\langle e_1 \rangle \otimes W) \oplus (\langle f_1 \rangle \otimes W)$$

into complementary totally isotropic $\frac{i}{2}$-spaces.

Similarly, if $i = 2$, $r$ divides $n$, and $x \in G$ is one of the elements described in part (ii) of Proposition 3.3.3, then $x$ fixes an orthogonal decomposition

$$V = (W_1 \oplus W_1^*) \perp \ldots \perp (W_k \oplus W_k^*) \perp W$$

where each $\{W_j, W_j^*\}$ is a dual pair of totally isotropic $r$-spaces, and either $n/r$ is even and $W$ is trivial, or $n/r$ is odd and $W$ is a nondegenerate $r$-space.

### 3.3.2 Semisimple involutions

Next we describe the conjugacy classes of involutions in $G$ when $q$ is odd, using the notation given in [67, Table 4.5.1].

#### 3.3.2.1 Involutions of type $t_i$, $1 \leqslant i \leqslant n/2$

Take $t_i = \hat{t}Z$, where $\hat{t} \in \mathrm{GU}_n(q)$ is an involution with an $i$-dimensional nondegenerate $(-1)$-eigenspace. Note that the coset $\hat{t}Z$ also contains the involution $-\hat{t}$, which is $\mathrm{SU}_n(q)$-conjugate to $\hat{t}$ if and only if $i = n/2$. Now $\det(\hat{t}) = (-1)^i$, so $t_i \in T$ if $i$ is even. On the other hand, if $i$ is odd then $\det(\hat{t}) = -1$ and thus Lemma 2.3.1 indicates that $t_i \in T$ if and only if there exists $\lambda \in \mathbb{F}_{q^2}$ such that $\lambda^{q+1} = 1$ and $\lambda^n = -1$. Now $\lambda^n = -1$ if and only if $\lambda^{(n)_2} = -1$, so $t_i \in T$ if and only if $(q+1)_2 > (n)_2$. In particular, if $n$ is even then $t_{n/2} \in T$ if and only if $n \equiv 0 \pmod 4$ or $q \equiv 3 \pmod 4$.

Fix an hermitian basis for $V$, as described in Proposition 2.2.2(ii). In order to give an explicit representative of an involution of type $t_i$, it is sufficient to describe its $(-1)$-eigenspace, which we denote by $U$. By applying Lemma 2.2.3, we can take

$$U = \begin{cases} \langle e_1, \ldots, e_{i/2}, f_1, \ldots, f_{i/2} \rangle & i \text{ even} \\ \langle e_1, \ldots, e_{(i-1)/2}, f_1, \ldots, f_{(i-1)/2}, e_{(i+1)/2} + f_{(i+1)/2} \rangle & i \text{ odd} \end{cases}$$

#### 3.3.2.2 Involutions of type $t'_{n/2}$

Here $n$ is even and these elements fix a pair of maximal totally isotropic subspaces. Let $t'_{n/2} = \hat{t}Z$, where $\hat{t} = [\xi I_{n/2}, \xi^{-q} I_{n/2}] \in \mathrm{GU}_n(q)$ (with respect to an hermitian basis $\{e_1, \ldots, e_{n/2}, f_1, \ldots, f_{n/2}\}$) and $\xi \in \mathbb{F}_{q^2}$ has order $2(q+1)_2$. Note that $\xi^q \hat{t} = [\xi^{q+1} I_{n/2}, I_{n/2}] = [-I_{n/2}, I_{n/2}]$. Now $\xi^q I_n$ has order $2(q+1)_2$, so it does not lie in $\mathrm{GU}_n(q)$. In particular, the involutions $\pm \xi^q \hat{t}$ lie in $\mathrm{GU}_n(q)Z_0 \setminus \mathrm{GU}_n(q)$, where $Z_0$ denotes the centre of $\Delta \mathrm{U}_n(q)$, and there are no involutions in the coset $\hat{t}Z \subset \mathrm{GU}_n(q)$. Also observe that $\xi^{-q} = -\xi$, so $\hat{t}$ and $-\hat{t}$ are conjugate under the element of $\mathrm{SU}_n(q)$ that maps each $e_i$ to $\alpha f_i$, and $f_i$ to $-\alpha^{-1} e_i$, where $\alpha \in \mathbb{F}_{q^2}$ has order $2(q-1)$.

Now $\det(\hat{t}) = (\xi^{1-q})^{n/2}$ and so Lemma 2.3.1 implies that $t'_{n/2} \in T$ if and only if $\xi^{(1-q)n/2} = \lambda^n$ for some $\lambda \in \mathbb{F}_{q^2}$ with $\lambda^{q+1} = 1$. Let $\omega = \xi^{(1-q)/2}$. If $q \equiv 1 \pmod 4$ then $(1-q)/2$ is even, so the order of $\omega$ divides $(q+1)_2$ and thus $t'_{n/2} \in T$. Now assume $q \equiv 3 \pmod 4$. Here $(1-q)/2$ is odd, so $\omega$ has order $2(q+1)_2$. Write $q+1 = s(q+1)_2$ and $\omega^2 = \gamma^s$ for some $\gamma \in \mathbb{F}_{q^2}$ of order $q+1$, so $\det(\hat{t}) = \gamma^{sn/2}$. We claim that $t'_{n/2} \in T$ if and only if $(n)_2 > (q+1)_2$. First assume that $t'_{n/2} \in T$. Then Lemma 2.3.1 implies that $\det(\hat{t}) = \gamma^{sn/2} = \gamma^{an}$ for some positive integer $a$, so $\gamma^{(s-2a)n/2} = 1$ and thus $q+1$ divides $(s-2a)n/2$.

Since $s - 2a$ is odd, it follows that $(n)_2 > (q+1)_2$. Conversely, if $(n)_2 > (q+1)_2$ then $\det(\hat{t}) = \gamma^{sn/2} = \gamma^{(q+1)b}$ for some positive integer $b$, so $\det(\hat{t}) = 1$ and thus $t'_{n/2} \in T$.

### 3.3.2.3 Involutions: a summary
The above information is summarised in the following proposition. The centraliser orders recorded in Table B.1 (see Appendix B) can be read off from [67, Table 4.5.1], and some additional information is given in Table B.2.

**Proposition 3.3.6** *For $q$ odd, Table B.1 (with $\varepsilon = -$) provides a complete set of representatives of the G-classes of involutions $x$ in G, together with the order of the corresponding centraliser $C_G(x)$.*

## 3.3.3 Unipotent elements

Next let us turn to the elements of order $p$ in $G$. Our main result is the following.

**Proposition 3.3.7** *Let $x = \hat{x}Z \in G$ be an element of order p. Then $x \in T$ and $\hat{x}$ fixes an orthogonal decomposition of the form*

$$V = U_{p,1} \perp \ldots \perp U_{p,a_p} \perp \ldots \perp U_{1,1} \perp \ldots \perp U_{1,a_1}$$

*where each $U_{i,j}$ is a nondegenerate i-space on which $\hat{x}$ acts indecomposably. Moreover, the map*

$$\theta : (p^{a_p}, \ldots, 1^{a_1}) \longmapsto ([J_p^{a_p}, \ldots, J_1^{a_1}]Z)^G$$

*is a bijection from the set of nontrivial partitions of n with parts of size at most p, to the set of G-classes of elements of order p in G.*

*Proof* First observe that $x \in T$ since $p$ does not divide $|G:T| = (n, q+1)$. Let $a_i$ denote the multiplicity of $J_i$ in the Jordan form of $\hat{x}$ on $V$. By Lemma 3.1.14, $J_i$ is $\mathrm{GL}_i(q^2)$-conjugate to $J_i^{-\varphi^T}$ (where $\varphi$ is the map in (2.2.6)). Therefore, Proposition 3.3.1 implies that there exists an element $y = \hat{y}Z \in G$ of order $p$ such that $\hat{y}$ and $\hat{x}$ have the same Jordan form on $V$, and $\hat{y}$ fixes an orthogonal decomposition of $V$ as in the statement of the proposition. By applying Lemma 3.1.14 and Proposition 3.3.1, we deduce that $\theta(p^{a_p}, \ldots, 1^{a_1}) = x^G = y^G$. Therefore, $\theta$ is surjective and $\hat{x}$ also fixes such a decomposition. Finally, by applying Lemma 3.1.14 again, we deduce that $\theta$ is a bijection. $\square$

The structure of the centraliser of a unipotent element is described in the next lemma (see [96, Theorem 7.1]).

**Lemma 3.3.8** *Let* $x = [J_n^{a_n}, \ldots, J_1^{a_1}] \in \mathrm{GU}_n(q)$ *be a unipotent element. Then* $C_{\mathrm{GU}_n(q)}(x) = QR$, *where*

$$|Q| = q^\gamma \quad \text{with} \quad \gamma = 2 \sum_{i<j} i a_i a_j + \sum_i (i-1) a_i^2$$

*and* $R = \prod_i \mathrm{GU}_{a_i}(q)$.

**Remark 3.3.9** By arguing as in Remark 3.3.5, we observe that an element $[J_i^2] \in \mathrm{GU}_{2i}(q)$ fixes a pair of orthogonal nondegenerate $i$-spaces, and also a dual pair of totally isotropic $i$-spaces of the natural $\mathrm{GU}_{2i}(q)$-module.

**Proposition 3.3.10** *Let* $x \in G$ *be an element of order* $p$ *with Jordan form* $[J_p^{a_p}, \ldots, J_1^{a_1}]$ *and let* $H = T.b$ *be a subgroup of* $G$. *Then* $x^G$ *splits into precisely*

$$\left( (v, q+1), \frac{(n, q+1)}{b} \right)$$

*$H$-classes, where* $v = \gcd\{ j \mid a_j > 0 \}$. *In particular,* $x^G$ *splits into* $(v, q+1)$ *distinct $T$-classes.*

*Proof* This is entirely similar to the proof of Proposition 3.2.7. □

We can describe representatives of the $T$-classes of unipotent elements as we did for linear groups in Section 3.2.3. Let $x = [J_p^{a_p}, \ldots, J_1^{a_1}]Z$ and recall that $G = \langle T, \delta \rangle$, where $\delta = [\mu, I_{n-1}]Z$ and $\mu \in \mathbb{F}_{q^2}$ has order $q+1$. The subgroup $\langle \delta \rangle$ cyclically permutes the $(v, q+1)$ distinct $T$-classes in $x^G$. The next result is the unitary group analogue of Lemma 3.2.8.

**Lemma 3.3.11** *Let* $x \in G$ *be an element with Jordan form* $[J_p^{a_p}, \ldots, J_1^{a_1}]$. *Then* $x^{\delta^i}$ *and* $x^{\delta^j}$ *are $T$-conjugate if and only if* $j - i$ *is divisible by* $(v, q+1)$. *In particular,*

$$x^{\delta^i}, \quad 0 \leqslant i < (v, q+1)$$

*is a complete set of representatives of the distinct $T$-classes in* $x^G$.

### 3.3.4 Field automorphisms

As before, let $\phi : \mathrm{GL}_n(q^2) \to \mathrm{GL}_n(q^2)$ be a standard field automorphism of order $2f$ with respect to a fixed orthonormal $\mathbb{F}_{q^2}$-basis of $V$ (see Section 2.3). Recall that $x \in \mathrm{Aut}(T)$ is a field automorphism if $x$ is $G$-conjugate to $\phi^i$, where $1 \leqslant i < 2f$ and $\phi^i$ has odd order. The next proposition is the unitary group version of Proposition 3.2.9, and its proof is entirely similar.

**Proposition 3.3.12** *Let $x \in \mathrm{Aut}(T)$ be a field automorphism of odd prime order $r$. Then $r$ divides $f$, say $f = mr$, and $x$ is $G$-conjugate to exactly one of the standard field automorphisms $\phi^{2jm}$, where $1 \leqslant j < r$. Moreover, $C_G(x) \cong \mathrm{PGU}_n(q^{1/r})$ and $x^G$ splits into*

$$\left( n, \frac{q+1}{q^{1/r}+1} \right)$$

*distinct $T$-classes.*

### 3.3.5 Graph automorphisms

Fix an orthonormal basis for $V$ and recall that $\varphi : \mathrm{GL}_n(q^2) \to \mathrm{GL}_n(q^2)$ is the map $\phi^f : (a_{ij}) \mapsto (a_{ij}^q)$ induced by the involutory automorphism of $\mathbb{F}_{q^2}$. The restriction of $\varphi$ to $\mathrm{GU}_n(q)$ induces the inverse-transpose graph automorphism on $\mathrm{GU}_n(q)$, and subsequently on $T$. More generally, a graph automorphism of $T$ is an element of the form $\gamma = x\varphi$ for some $x \in G$.

Let us define

$$L = \{ \lambda \in \mathbb{F}_{q^2} \mid \lambda^{q+1} = 1 \}$$

and

$$K = \{ \lambda \in L \mid \lambda = \mu^2 \text{ for some } \mu \in L \}$$

We call $K$ the set of *unitary-squares* in $\mathbb{F}_{q^2}$. Note that $K = L$ if $q$ is even. However, if $q$ is odd then exactly half of the elements in $L$ are unitary-squares (since $L$ is a cyclic group of order $q + 1$), and the remaining elements in $L$ are the *unitary-nonsquares*. Note that $-1$ is a unitary-square if and only if $q \equiv 3 \pmod 4$.

For $x = JZ \in G$ we set $\gamma_J = x\varphi$. As in Section 3.2.5, $\gamma_J$ is an involution if and only if $J$ is a symmetric or skew-symmetric matrix. Now $\gamma_J$ maps $AZ \in G$ to $J^\mathsf{T} A^{-\mathsf{T}} J^{-\mathsf{T}} Z$ and thus

$$
\begin{aligned}
C_G(\gamma_J) &= \{ AZ \in G \mid J^\mathsf{T} A^{-\mathsf{T}} J^{-\mathsf{T}} = \lambda A \text{ for some } \lambda \in L \} \\
&= \{ AZ \in G \mid J^\mathsf{T} = \lambda A J^\mathsf{T} A^\mathsf{T} \text{ for some } \lambda \in L \} \\
&= \{ AZ \in G \mid J = \lambda A J A^\mathsf{T} \text{ for some } \lambda \in L \} \quad (3.3.2)
\end{aligned}
$$

In particular, $C_G(\gamma_J)$ is the intersection of $G$ with the subgroup of $\mathrm{GL}_n(q^2)$ that preserves the bilinear form $B_J$ corresponding to the matrix $J$, up to scalars.

We say that two bilinear forms $B_{J_1}, B_{J_2}$ with $J_1, J_2 \in \mathrm{GU}_n(q)$ are *unitarily similar* if there exists $A \in \mathrm{GU}_n(q)$ and $\lambda \in L$ such that $J_1 = \lambda A J_2 A^\mathsf{T}$. As noted in Section 3.2.5, if $g = AZ \in G$ then $(\gamma_J)^g = \gamma_{A^{-1} J A^{-\mathsf{T}}}$. This gives the following result.

**Proposition 3.3.13** *There is a bijection from the set of G-classes of involutory graph automorphisms of T, to the set of unitary similarity classes of nondegenerate symmetric and skew-symmetric bilinear forms on V.*

**Remark 3.3.14** Note that if $J \in \mathrm{GU}_n(q)$ then $\det(J)^{q+1} = 1$, so $\det(J)$ is an $\mathbb{F}_{q^2}$-square. For $A \in \mathrm{GU}_n(q)$ and $\lambda \in L$, we have

$$\det(\lambda A J A^{\mathsf{T}}) = \lambda^n \det(A)^2 \det(J)$$

Therefore, if $n$ is even and $B_{J_1}$ and $B_{J_2}$ are unitarily similar, then $\det(J_1)$ and $\det(J_2)$ are either both unitary-squares or both unitary-nonsquares.

We now describe the *standard* graph automorphisms. If $n$ is odd then let $\gamma_1$ be the graph automorphism given by $J_1 = I_n$ (so $\gamma_1 = \varphi$). Now assume $n \geqslant 4$ is even. Define the matrices $J_1, J_2, J_2'$ and $J_3$ as in Section 3.2.5, in terms of an orthonormal basis of $V$ (the only difference being that in the definition of $J_2'$ we take $-\alpha \in \mathbb{F}_{q^2}$ to be a unitary-nonsquare). Note that all of these matrices are elements of $\mathrm{GU}_n(q)$. Let $\gamma_1$ be the graph automorphism induced by $J_1$. Similarly, if $q$ is odd then let $\gamma_2$ and $\gamma_2'$ be the graph automorphisms induced by $J_2$ and $J_2'$, respectively. Note that $\det(J_2)$ is a unitary-square if and only if $n(q+1)/4$ is even, and $\det(J_2')$ is a unitary-square if and only if $n(q+1)/4$ is odd.

Finally, if both $n$ and $q$ are even then let $\gamma_3$ be the graph automorphism induced by $J_3 = I_n$ (that is, $\gamma_3 = \varphi$). Our main result on the involutory graph automorphisms of $T$ is the following.

**Proposition 3.3.15** *Let $\gamma = x\varphi$ be an involutory graph automorphism of $T$ with $x = JZ \in G$.*

(i) *If $n$ is odd then $\gamma \in \gamma_1^G$ and $C_G(\gamma) \cong \mathrm{PGO}_n(q)$ (so $C_G(\gamma) \cong \mathrm{Sp}_{n-1}(q)$ if $q$ is even).*

(ii) *Suppose $n$ is even and $q$ is odd.*
   (a) *If $J^{\mathsf{T}} = -J$ then $\gamma \in \gamma_1^G$ and $C_G(\gamma) \cong \mathrm{PGSp}_n(q)$.*
   (b) *If $J^{\mathsf{T}} = J$ then $\gamma \in \gamma_2^G$ if $\det(J) \equiv \det(J_2) \pmod{K}$, otherwise $\gamma \in (\gamma_2')^G$. Here $C_G(\gamma_2) \cong \mathrm{PGO}_n^+(q)$ and $C_G(\gamma_2') \cong \mathrm{PGO}_n^-(q)$.*

(iii) *Suppose $n$ and $q$ are even. If $B_J$ is alternating then $\gamma \in \gamma_1^G$, otherwise $\gamma \in \gamma_3^G$. Here $C_G(\gamma_1) \cong \mathrm{Sp}_n(q)$ and $C_G(\gamma_3) \cong C_{\mathrm{Sp}_n(q)}(t)$, where $t \in \mathrm{Sp}_n(q)$ is a transvection.*

*Proof* The number of $G$-classes of involutory graph automorphisms is given in [67, Table 4.5.1] when $q$ is odd, and in [5, (19.8)] when $q$ is even. If $n$ is

Table 3.3.1 $\mathrm{PGU}_n(q)$-*conjugacy of the inverse-transpose automorphism* $\varphi$

| Conditions on $n$ and $q$ | $\mathrm{PGU}_n(q)$-class containing $\varphi$ |
|---|---|
| $n$ odd | $\gamma_1$ |
| $n$ even and $q$ even | $\gamma_3$ |
| $n$ even, $q$ odd and $n(q+1)/4$ even | $\gamma_2$ |
| $n$ even, $q$ odd and $n(q+1)/4$ odd | $\gamma_2'$ |

odd then there is a unique class and so (i) follows. There are precisely three classes if $n$ is even and $q$ is odd, so part (ii) follows from Proposition 3.3.13 and Remark 3.3.14. Finally, if $n$ and $q$ are both even then there are precisely two $G$-classes. Since a nonalternating form is not similar to an alternating form, part (iii) follows from Proposition 3.3.13. In all cases, the given centralisers can be deduced from (3.3.2); alternatively, see [67, Table 4.5.1] (if $q$ is odd) and [5, (19.9)] ($q$ even). $\qquad\square$

As in Remark 3.2.12, if $n$ is even then we say that $\gamma$ is a *symplectic-type* graph automorphism if $C_G(\gamma) \cong \mathrm{PGSp}_n(q)$, otherwise $\gamma$ is *non-symplectic*.

**Corollary 3.3.16** *The $G$-class containing the inverse-transpose graph automorphism $\varphi$ is recorded in Table 3.3.1.*

Finally, let us record the unitary group analogue of Proposition 3.2.14.

**Proposition 3.3.17** *Let $\gamma \in \mathrm{Aut}(T)$ be an involutory graph automorphism. Then the $G$-class of $\gamma$ splits into $c$ distinct $T$-classes, where*

$$c = \begin{cases} (n, q+1)/2 & n \text{ even, } q \text{ odd, } \gamma \in \{\gamma_2^G, \gamma_2'^G\} \\ (n/2, q+1) & n \text{ even, } q \text{ odd, } \gamma \in \gamma_1^G \\ (n, q+1) & \text{otherwise} \end{cases}$$

*Proof* This is entirely similar to the proof of Proposition 3.2.14, using the relevant results in [86, Section 4.5] to compute the order of $C_T(\gamma)$. $\qquad\square$

Let $c$ be the integer defined in the statement of Proposition 3.3.17. Then

$$\gamma^{\delta^i}, \ 0 \leqslant i < c$$

is a set of representatives for the $T$-classes in $\gamma^G$, where $\delta = [\mu, I_{n-1}]Z \in G$ and $\mu \in \mathbb{F}_{q^2}$ has order $q+1$.

### 3.3.6 Unitary groups: a summary

The final proposition of this section summarises some of the above results on unitary groups. Note that in Table B.4 (see Appendix B) we use the notation defined in Table 3.2.2 to describe the types of elements of prime order in $\text{Aut}(T)$.

**Proposition 3.3.18** *Let* $x \in \text{Aut}(T)$ *be an element of prime order r. Then the possibilities for x up to G-conjugacy are listed in Table B.4, together with the order of the centraliser* $C_G(x)$.

## 3.4 Symplectic groups

Let $Z \cong C_{q-1}$ be the centre of $\text{GSp}_n(q)$ (which coincides with the centre of $\text{GL}_n(q)$) and set

$$T = \text{PSp}_n(q) = \text{Sp}_n(q)Z/Z \cong \text{Sp}_n(q)/(Z \cap \text{Sp}_n(q))$$

and

$$G = \text{Inndiag}(T) = \text{PGSp}_n(q) = \text{GSp}_n(q)/Z$$

Note that $Z \cap \text{Sp}_n(q) \cong C_{(2,q-1)}$. As before, write $q = p^f$ where $p$ is prime, and let $V$ be the natural $\text{Sp}_n(q)$-module. Membership in $\text{Sp}_n(q)$ is given by the following lemma (see [119, p. 36] if $q$ is odd, and [119, Section 3.7] and [96, Section 7.2] if $q$ is even).

**Lemma 3.4.1** *Let* $A \in \text{GL}_n(q)$. *Then A is similar to an element of* $\text{Sp}_n(q)$ *if and only if*

(i) *A and* $A^{-1}$ *are conjugate in* $\text{GL}_n(q)$, *and*
(ii) *each Jordan block of odd size and eigenvalue* $\pm 1$ *of A occurs with even multiplicity.*

*In particular, if A is unipotent with Jordan form* $[J_n^{a_n}, \ldots, J_1^{a_1}]$, *then A is similar to an element of* $\text{Sp}_n(q)$ *if and only if* $a_i$ *is even for all odd i.*

### 3.4.1 Semisimple elements

Our first result follows from [119, p. 36] (see [14, Proposition 2.9(2)] for an alternative proof using Lang's Theorem).

**Lemma 3.4.2** *Two semisimple elements in* $\text{Sp}_n(q)$ *are conjugate in* $\text{Sp}_n(q)$ *if and only if they are conjugate in* $\text{GL}_n(q)$.

Let $r \neq p$ be an odd prime divisor of $|G|$ and define $i = \Phi(r,q)$ as in (3.1.2). As before, let $\mathscr{S}_r$ be the set of nontrivial $r$th roots of unity in $\mathbb{F}_{q^i}$ and let $\Lambda_1,\ldots,\Lambda_t$ be the $\sigma$-orbits on $\mathscr{S}_r$ corresponding to the permutation $\sigma : \mathscr{S}_r \to \mathscr{S}_r$ defined by $\sigma : \lambda \mapsto \lambda^q$. Recall that $\Lambda_j^{-1} = \{\lambda^{-1} \mid \lambda \in \Lambda_j\}$. Also note that $|\Lambda_j| = i$ for all $j$, so $t = (r-1)/i$. If $i$ is odd then $\Lambda_j \neq \Lambda_j^{-1}$ by Lemma 3.1.13, and in this situation we label the $\sigma$-orbits so that $\Lambda_j^{-1} = \Lambda_{t/2+j}$ for all $1 \leqslant j \leqslant t/2$.

**Proposition 3.4.3** *Let $x = \hat{x}Z \in G$ be an element of order $r$ and set $i = \Phi(r,q)$. Then $x \in T$, $x^G = x^T$ and one of the following holds:*

(i) *If $i$ is odd, then $\hat{x}$ fixes an orthogonal decomposition of the form*

$$V = (U_{1,1} \oplus U_{1,1}^*) \perp \ldots \perp (U_{1,a_1} \oplus U_{1,a_1}^*) \perp \ldots$$
$$\perp (U_{t/2,1} \oplus U_{t/2,1}^*) \perp \ldots \perp (U_{t/2,a_{t/2}} \oplus U_{t/2,a_{t/2}}^*) \perp C_V(\hat{x})$$

*where each $\{U_{j,k}, U_{j,k}^*\}$ is a dual pair of totally isotropic $i$-spaces on which $\hat{x}$ acts irreducibly, and $C_V(\hat{x})$ is nondegenerate (or trivial). Moreover, the map*

$$\theta_1 : (a_1,\ldots,a_{t/2}) \longmapsto ([(\Lambda_1,\Lambda_1^{-1})^{a_1},\ldots,(\Lambda_{t/2},\Lambda_{t/2}^{-1})^{a_{t/2}}, I_e]Z)^G$$

*is a bijection from the set of nonzero $\frac{t}{2}$-tuples $(a_1,\ldots,a_{t/2}) \in \mathbb{N}_0^{t/2}$ such that $2i\sum_j a_j \leqslant n$, to the set of $G$-classes of elements of order $r$ in $G$.*

(ii) *If $i$ is even, then $\hat{x}$ fixes an orthogonal decomposition*

$$V = U_{1,1} \perp \ldots \perp U_{1,a_1} \perp \ldots \perp U_{t,1} \perp \ldots \perp U_{t,a_t} \perp C_V(\hat{x})$$

*where each $U_{j,k}$ is a nondegenerate $i$-space on which $\hat{x}$ acts irreducibly, and $C_V(\hat{x})$ is nondegenerate (or trivial). Moreover, the map*

$$\theta_2 : (a_1,\ldots,a_t) \longmapsto ([\Lambda_1^{a_1},\ldots,\Lambda_t^{a_t}, I_e]Z)^G$$

*is a bijection from the set of nonzero $t$-tuples $(a_1,\ldots,a_t) \in \mathbb{N}_0^t$ such that $i\sum_j a_j \leqslant n$, to the set of $G$-classes of elements of order $r$ in $G$.*

*Proof* First observe that $x \in T$ since $|G:T| = (2,q-1)$ is indivisible by $r$. In addition, $x^G = x^T$ by Theorem 3.1.12. Now $\hat{x}$ is diagonalisable over $\mathbb{F}_{q^i}$ but no proper subfield, so the irreducible blocks of $\hat{x}$ have dimension $i$. If $i$ is odd, then Lemma 3.1.13 implies that $\Lambda_j \neq \Lambda_j^{-1}$ and so Lemmas 3.1.6 and 3.4.1 imply that the irreducible blocks are totally isotropic and arise in dual pairs $U_j \oplus U_j^*$ with eigenvalues $\Lambda_j \cup \Lambda_j^{-1}$. In particular, if $a_j$ denotes the multiplicity of $\Lambda_j$ in the multiset of eigenvalues of $\hat{x}$, then $\hat{x}$ fixes an orthogonal decomposition of

$V$ as described in part (i). Moreover, by applying Lemmas 3.1.7 and 3.4.2 we deduce that $\theta_1(a_1,\ldots,a_{t/2}) = x^G$ and $\theta_1$ is a bijection.

Now assume $i$ is even. Here $\Lambda_j = \Lambda_j^{-1}$ (see Lemma 3.1.13), so Lemma 3.4.1 implies that $[\Lambda_j] \in \mathrm{Sp}_i(q)$. Therefore, there exists an element $y = \hat{y}Z \in G$ of order $r$ such that $\hat{y}$ and $\hat{x}$ have the same eigenvalues, and $\hat{y}$ fixes an orthogonal decomposition of $V$ as given in part (ii). If $a_j$ denotes the multiplicity of $\Lambda_j$ in the multiset of eigenvalues of $\hat{x}$, then Lemmas 3.1.7 and 3.4.2 imply that $\theta_2(a_1,\ldots,a_t) = x^G = y^G$. In particular, $\theta_2$ is surjective and $\hat{x}$ fixes an orthogonal decomposition as in (ii). Finally, a further application of Lemma 3.1.7 shows that $\theta_2$ is a bijection. □

**Remark 3.4.4** Let $x = \hat{x}Z \in G$ be a semisimple element of odd prime order $r$. Set $i = \Phi(r,q)$ as before. Then

$$C_{\mathrm{GSp}_n(q)}(\hat{x}) = \mathrm{GSp}_n(q) \cap C_{\mathrm{GL}_n(q)}(\hat{x})$$

and the structure of $C_{\mathrm{GL}_n(q)}(\hat{x})$ is given in Lemma 3.1.9. Suppose that $i$ is odd and the $G$-class of $x$ corresponds to the $\frac{t}{2}$-tuple $(a_1,\ldots,a_{t/2})$ as in Proposition 3.4.3(i). For $1 \leqslant j \leqslant t/2$, set

$$U_j = (U_{j,1} \oplus U_{j,1}^*) \perp \ldots \perp (U_{j,a_j} \oplus U_{j,a_j}^*)$$

which is a nondegenerate $2ia_j$-space fixed by $\hat{x}$. Then

$$\begin{aligned}
C_{\mathrm{GSp}_n(q)}(\hat{x}) &= C_{\mathrm{Sp}_n(q)}(\hat{x})\langle \delta' \rangle \\
&= (\mathrm{GL}_{a_1}(q^i) \times \cdots \times \mathrm{GL}_{a_{t/2}}(q^i) \times \mathrm{Sp}_e(q)).(q-1)
\end{aligned}$$

where $\delta'$ acts as a similarity on $C_V(\hat{x})$ and each $U_j$.

Similarly, if $i$ is even and $(a_1,\ldots,a_t)$ is the corresponding $t$-tuple in Proposition 3.4.3(ii), then

$$C_{\mathrm{GSp}_n(q)}(\hat{x}) = (\mathrm{GU}_{a_1}(q^{i/2}) \times \cdots \times \mathrm{GU}_{a_t}(q^{i/2}) \times \mathrm{Sp}_e(q)).(q-1)$$

For example, suppose $n = ai$ and $\hat{x} = [\Lambda^a] \in \mathrm{Sp}_n(q)$. Then

$$\mathrm{GU}_a(q^{i/2}) \leqslant \mathrm{Sp}_{2a}(q^{i/2}) \leqslant \mathrm{Sp}_n(q)$$

and the centre of $\mathrm{GU}_a(q^{i/2})$ contains an element of order $r$ (the embedding $\mathrm{GU}_a(q^{i/2}) \leqslant \mathrm{Sp}_{2a}(q^{i/2})$ is discussed in Construction 2.4.3). Therefore

$$\mathrm{GU}_a(q^{i/2}) \leqslant C_{\mathrm{Sp}_n(q)}(\hat{x}) \leqslant C_{\mathrm{GL}_n(q)}(\hat{x}) = \mathrm{GL}_a(q^i)$$

and one can show that $C_{\mathrm{Sp}_n(q)}(\hat{x}) = \mathrm{GU}_a(q^{i/2})$ (see [119, p. 36, p. 60]). The order of the centraliser $C_G(x)$ is recorded in Table B.7 (see Appendix B).

**Remark 3.4.5** Note that if $i$ is even and $\Lambda$ is a $\sigma$-orbit on $\mathscr{S}_r$, then $\mathrm{Sp}_{2i}(q)$ contains an element $x = [\Lambda^2]$ that fixes an orthogonal decomposition of the natural $\mathrm{Sp}_{2i}(q)$-module into two nondegenerate $i$-spaces. Since $\Lambda = \Lambda^{-1}$, $\mathrm{Sp}_{2i}(q)$ also contains an element $y = [\Lambda^2]$ fixing a dual pair of totally isotropic $i$-spaces (see Lemma 2.2.17). By Lemma 3.4.2, $x$ and $y$ are conjugate in $\mathrm{Sp}_{2i}(q)$, so $x$ also fixes such a dual pair.

As in Remark 3.3.5, we can also see this via a tensor product construction. Indeed, recall from Section 2.2.8 that if $U$ is equipped with a hyperbolic quadratic form with associated bilinear form $B_1$, and $W$ is equipped with a nondegenerate alternating form $B_2$, then we can construct a nondegenerate alternating form $B$ on the tensor product $V = U \otimes W$, as defined in (2.2.9). If $\dim U = 2$ and $\dim W = i$ then the element $x = I_2 \otimes [\Lambda]$ preserves $B$ and has the form $[\Lambda^2]$ on $V$. Let $\{u_1, u_2\}$ be an orthonormal basis for $U$ with respect to $B_1$. Then $x$ fixes the orthogonal decomposition

$$V = (\langle u_1 \rangle \otimes W) \perp (\langle u_2 \rangle \otimes W)$$

into nondegenerate $i$-spaces. Similarly, if $\{e_1, f_1\}$ is a hyperbolic basis for $U$ then $x$ also fixes the decomposition

$$V = (\langle e_1 \rangle \otimes W) \oplus (\langle f_1 \rangle \otimes W)$$

of $V$ into complementary totally isotropic $i$-spaces.

### 3.4.2 Semisimple involutions

Now let us turn our attention to semisimple involutions. Here $q$ is odd and representatives for the $G$-classes of involutions in $G$ are listed in [67, Table 4.5.1], and we adopt the notation therein. Note that if $x \in G$ is an involution then $x^G = x^T$ by Theorem 3.1.12. Recall that if $\hat{x} \in \mathrm{GSp}_n(q)$ then there exists a scalar $\tau(\hat{x}) \in \mathbb{F}_q^\times$ such that

$$B(u\hat{x}, v\hat{x}) = \tau(\hat{x})B(u, v)$$

for all $u, v \in V$, where $B$ is the underlying nondegenerate alternating form on $V$. Note that if $\hat{x} \in \mathrm{Sp}_n(q)Z$, then $\tau(\hat{x})$ is a square. Let

$$\{e_1, \ldots, e_{n/2}, f_1, \ldots, f_{n/2}\}$$

be a symplectic basis of $V$ as in Proposition 2.2.1.

### 3.4.2.1 Involutions of type $t_i$, $1 \leqslant i < n/4$

Here $t_i = \hat{t}Z$, where $\hat{t} \in \mathrm{Sp}_n(q)$ is an involution whose $(-1)$-eigenspace is a nondegenerate $2i$-space. The $(-1)$-eigenspace of $-\hat{t} \in \mathrm{Sp}_n(q)$ is $(n-2i)$-dimensional, so $\hat{t}$ and $-\hat{t}$ are not conjugate. We can take $\hat{t}$ to be the matrix $[-I_{2i}, I_{n-2i}]$, with respect to the specific ordered basis $\{e_1, f_1, \ldots, e_{n/2}, f_{n/2}\}$.

### 3.4.2.2 Involutions of type $t_{n/4}$, $n \equiv 0 \pmod 4$

Here $t_{n/4} = \hat{t}Z$, where $\hat{t} \in \mathrm{Sp}_n(q)$ is an involution whose $(-1)$-eigenspace is a nondegenerate $\frac{n}{2}$-space. Note that $-\hat{t}$ is $\mathrm{Sp}_n(q)$-conjugate to $\hat{t}$. We can take $\hat{t} = [-I_{n/2}, I_{n/2}]$ with respect to the basis $\{e_1, f_1, \ldots, e_{n/2}, f_{n/2}\}$.

### 3.4.2.3 Involutions of type $t'_{n/4}$, $n \equiv 0 \pmod 4$

These involutions arise from elements $\xi I_{n/2}$ of order $2(q-1)_2$ in the centre of $\mathrm{GSp}_{n/2}(q^2)$, which embed in $\mathrm{GSp}_n(q)$. We will adopt the set-up given in Construction 2.4.2, where this embedding is described.

Let $t'_{n/4} = \hat{t}Z$, where $\hat{t}$ is the image of $\xi I_{n/2}$ in $\mathrm{GSp}_n(q)$. Then with respect to the symplectic basis

$$\left\{ e_1, \xi e_1, \ldots, e_{n/4}, \xi e_{n/4}, \frac{1}{2}f_1, \frac{1}{2}\xi^{-1}f_1, \ldots, \frac{1}{2}f_{n/4}, \frac{1}{2}\xi^{-1}f_{n/4} \right\}$$

of $V$ (see Construction 2.4.2), $\hat{t}$ is the block-diagonal matrix $[A^{n/4}, (A^\mathsf{T})^{n/4}]$ where $A$ is the $2 \times 2$ matrix defined in (3.2.3). Now $\tau(\hat{t}) = \xi^2 \in \mathbb{F}_q$ has order $(q-1)_2$, so $\tau(\hat{t})$ is not an $\mathbb{F}_q$-square and thus $\hat{t} \in \mathrm{GSp}_n(q) \setminus \mathrm{Sp}_n(q)Z$. In particular, $t'_{n/4} \in G \setminus T$. Note that every element in the coset $\hat{t}Z \subset \mathrm{GSp}_n(q)$ has eigenvalues in $\mathbb{F}_{q^2} \setminus \mathbb{F}_q$, so this coset does not contain any involutions.

Finally, observe that $A$ and $-A$ have the same eigenvalues (in $\mathbb{F}_{q^2}$), so there exists $X \in \mathrm{GL}_2(q)$ such that $X^{-1}AX = -A$ (see Lemma 3.1.7). Therefore, the element $[X, X^{-\mathsf{T}}] \in \mathrm{Sp}_4(q)$ conjugates $[A, A^\mathsf{T}]$ to $[-A, -A^\mathsf{T}]$, and we deduce that $\hat{t}$ and $-\hat{t}$ are $\mathrm{Sp}_n(q)$-conjugate.

### 3.4.2.4 Involutions of type $t_{n/2}$

Define $t_{n/2} = \hat{t}Z$ in terms of a symplectic basis $\{e_1, \ldots, e_{n/2}, f_1, \ldots, f_{n/2}\}$, where $\hat{t} = [\lambda I_{n/2}, \lambda^{-1} I_{n/2}] \in \mathrm{Sp}_n(q)$ if $q \equiv 1 \pmod 4$ (with $\lambda \in \mathbb{F}_q$ of order 4) and $\hat{t} = [-I_{n/2}, I_{n/2}] \in \mathrm{GSp}_n(q)$ if $q \equiv 3 \pmod 4$. In the latter case, $\hat{t} \notin \mathrm{Sp}_n(q)Z$ since $Z$ does not contain any elements of order 4, so $t_{n/2} \in T$ if and only if $q \equiv 1 \pmod 4$. Note that the eigenspaces of $\hat{t}$ are maximal totally isotropic subspaces. Also note that the coset $\hat{t}Z$ in $\mathrm{GSp}_n(q)$ contains the involutions $[-I_{n/2}, I_{n/2}]$ and $[I_{n/2}, -I_{n/2}]$, neither of which is contained in $\mathrm{Sp}_n(q)$. It is clear that $\hat{t}$ and $-\hat{t}$ are $\mathrm{Sp}_n(q)$-conjugate.

### 3.4.2.5 Involutions of type $t'_{n/2}$

Write $t'_{n/2} = \hat{t}Z$, where $\hat{t} \in \mathrm{GSp}_n(q)$ is the image of the central element $\xi I_{n/2} \in \Delta U_{n/2}(q)$ embedded in $\mathrm{GSp}_n(q)$, where $\xi \in \mathbb{F}_{q^2}$ has order $2(q-1)_2$ (see Construction 2.4.3). Since $\hat{t}$ fixes totally isotropic and nondegenerate 1-spaces over $\mathbb{F}_{q^2}$, it also fixes totally isotropic and nondegenerate 2-spaces over $\mathbb{F}_q$. Note that $\tau(\hat{t}) = \xi^{q+1}$.

If $q \equiv 3 \pmod 4$ then $\tau(\hat{t}) = 1$, so $\hat{t} \in \mathrm{Sp}_n(q)$ and thus $t'_{n/2} \in T$. On the other hand, if $q \equiv 1 \pmod 4$ then $\xi^{q+1} \in \mathbb{F}_q \setminus \{0,1\}$ and thus $\hat{t} \in \mathrm{GSp}_n(q) \setminus \mathrm{Sp}_n(q)$. Moreover, $\xi^{q+1}$ is an $\mathbb{F}_q$-nonsquare and so $\hat{t} \notin \mathrm{Sp}_n(q)Z$. It follows that $t'_{n/2} \in G \setminus T$ if $q \equiv 1 \pmod 4$. For any $q$, every element in $\hat{t}Z \subset \mathrm{GSp}_n(q)$ has eigenvalues in $\mathbb{F}_{q^2} \setminus \mathbb{F}_q$, so there are no involutions in this coset.

Notice that if $n \equiv 0 \pmod 4$ then these involutions are distinct from those of type $t'_{n/4}$ described above. Indeed, we have observed that $t'_{n/2} \in T$ if and only if $q \equiv 3 \pmod 4$, whereas $t'_{n/4} \in T$ if and only if $q \equiv 1 \pmod 4$. An alternative way to see this is as follows. Let

$$\left\{ e_1, \xi e_1, \ldots, e_{n/4}, \xi e_{n/4}, -\frac{1}{2}\xi^{-1}f_1, \frac{1}{2}\xi^{-2}f_1, \ldots, -\frac{1}{2}\xi^{-1}f_{n/4}, \frac{1}{2}\xi^{-2}f_{n/4} \right\}$$

be the symplectic basis for $V$ over $\mathbb{F}_q$ given in Construction 2.4.3. With respect to this basis, $\hat{t} = [A^{n/4}, (-A^{\mathrm{T}})^{n/4}]$, where $A$ is defined as in (3.2.3), so the distinction between these involutions and those of type $t'_{n/4}$ is transparent. In addition, by arguing as in Section 3.4.2.3, we deduce that $\hat{t}$ and $-\hat{t}$ are $\mathrm{Sp}_n(q)$-conjugate.

Finally, note that if $n \equiv 2 \pmod 4$ then we can take

$$\hat{t} = \begin{pmatrix} A & & & & & & \\ & \ddots & & & & & \\ & & A & & & & -2\xi^2 \\ & & & 0 & -A^{\mathrm{T}} & & \\ & & & & \ddots & & \\ & & & & & -A^{\mathrm{T}} & \\ & -\frac{1}{2} & & & & & 0 \end{pmatrix}$$

with respect to the appropriate symplectic basis given in Construction 2.4.3 obtained from an hermitian $\mathbb{F}_{q^2}$-basis. By Remark 3.1.11, the matrix

$$\begin{pmatrix} 0 & -2\xi^2 \\ -\frac{1}{2} & 0 \end{pmatrix}$$

is conjugate in $\mathrm{SL}_2(q) = \mathrm{Sp}_2(q)$ to its negative (since both matrices have the same eigenvalues). By arguing as before, we deduce that $\hat{t}$ is $\mathrm{Sp}_n(q)$-conjugate to $-\hat{t}$.

### 3.4.2.6 Involutions: a summary

Let us summarise some of the above information in the following proposition. We refer the reader to [67, Table 4.5.1] for the various centraliser orders listed in Table B.5. The fact that each involution is centralised by a group of the given order is clear from the above constructions. Some additional properties of these involutions are recorded in Table B.6.

**Proposition 3.4.6** *For $q$ odd, Table B.5 provides a complete set of representatives of the $G$-classes of involutions $x$ in $G$, together with the order of the corresponding centraliser $C_G(x)$.*

## 3.4.3 Unipotent elements, $p > 2$

First observe that all unipotent elements in $G$ belong to $T$ since $|G : T| = (2, q-1)$. The behaviour of unipotent classes varies according to the parity of $p$, and in this section we will assume $p$ is odd (the case $p = 2$ is handled in Section 3.4.4).

We will use a tensor product construction introduced in Section 2.2.8 to study the conjugacy classes of unipotent elements. Let $a > 0$ be an even integer and let $U$ be an $a$-dimensional $\mathbb{F}_q$-space equipped with a nondegenerate alternating form $B_1$. Further, let $W$ be a $b$-dimensional $\mathbb{F}_q$-space equipped with a nondegenerate quadratic form of type $\varepsilon$ and associated bilinear form $B_2$. Then $U \otimes W$ is an $(ab)$-dimensional vector space with a nondegenerate alternating form $B$ defined by

$$B(u_1 \otimes w_1, u_2 \otimes w_2) = B_1(u_1, u_2)B_2(w_1, w_2)$$

Note that if $u \in U$ then the subspace $\langle u \rangle \otimes W$ is totally isotropic with respect to $B$. Similarly, $U \otimes \langle w \rangle$ is totally isotropic if $w \in W$ is singular, and nondegenerate if $w$ is nonsingular.

There are three cases to consider. Note that $[J_a] \in \mathrm{Sp}_a(q)$ by Lemma 3.4.1. In fact, [96, Theorem 7.1] implies that there are two conjugacy classes of such unipotent elements in $\mathrm{Sp}_a(q)$.

*Case 1. $b$ even.* By [96, Theorem 7.1], there are precisely two $\mathrm{Sp}_{ab}(q)$-classes of unipotent elements $x \in \mathrm{Sp}_{ab}(q)$ with Jordan form $[J_a^b]$. Here $C_{\mathrm{Sp}_{ab}(q)}(x) = QR$, where $Q$ is a $p$-group and $R = \mathrm{O}_b^\varepsilon(q)$ with $\varepsilon = \pm$. We can describe representatives of these two classes as follows. First observe that $\mathrm{Sp}_{ab}(q)$ contains a tensor product subgroup of type $\mathrm{Sp}_a(q) \otimes \mathrm{O}_b^\varepsilon(q)$. This subgroup contains an element $J_a \otimes I_b$ with Jordan form $[J_a^b]$, which fixes an orthogonal decomposition of $U \otimes W$ into $b$ nondegenerate $a$-spaces (since $W$ has an orthogonal basis). Such an element is clearly centralised in $\mathrm{Sp}_{ab}(q)$ by $I_a \otimes O_b^\varepsilon(q)$, so we see that the two conjugacy classes of elements with Jordan form $[J_a^b]$ arise from the two isometry types of nondegenerate quadratic forms on the $b$-space $W$. We denote these two classes by $[J_a^{b,\varepsilon}]$, with $\varepsilon = \pm$.

The following lemma allows us to distinguish the two classes.

**Lemma 3.4.7** *A unipotent element of type $[J_a^{b,\varepsilon}]$ fixes a pair of complementary maximal totally isotropic subspaces of the natural $\mathrm{Sp}_{ab}(q)$-module if and only if $\varepsilon = +$.*

*Proof* Suppose that $\varepsilon = +$. Then $W$ has a pair $\{W_1, W_2\}$ of complementary maximal totally isotropic subspaces and $x = J_a \otimes I_b = [J_a^{b,+}]$ fixes $U \otimes W_1$ and $U \otimes W_2$, which are complementary maximal totally isotropic subspaces of the natural $\mathrm{Sp}_{ab}(q)$-module. Moreover, $x$ has Jordan form $[J_a^{b/2}]$ on each subspace.

Seeking a contradiction, suppose that $y = [J_a^{b,-}]$ fixes a pair of complementary maximal totally isotropic subspaces of the natural $\mathrm{Sp}_{ab}(q)$-module. Since $\mathrm{Sp}_{ab}(q)$ is transitive on the set of all such pairs (see Lemma 2.2.18), we may assume that $y$ also fixes $U \otimes W_1$ and $U \otimes W_2$, and has Jordan form $[J_a^{b/2}]$ on each subspace. By Lemma 2.2.17, the stabiliser in $\mathrm{Sp}_{ab}(q)$ of a pair of maximal totally isotropic subspaces is isomorphic to $\mathrm{GL}_{ab/2}(q)$, which has a unique conjugacy class of unipotent elements of the form $[J_a^{b/2}]$ by Lemma 3.1.14. This implies that $x$ and $y$ are $\mathrm{Sp}_{ab}(q)$-conjugate, which is a contradiction. The result follows. □

*Case 2. b odd, orthogonal group centralisers.* First we require a lemma (recall that $x \in \mathrm{GL}_n(q)$ is a *regular* unipotent element if it has Jordan form $[J_n]$).

**Lemma 3.4.8** *Let $x \in \mathrm{GSp}_n(q)$ be a regular unipotent element. Then $x^{\mathrm{GSp}_n(q)}$ splits into two $\mathrm{Sp}_n(q)$-classes.*

*Proof* Consider the algebraic group $\overline{G} = \mathrm{Sp}_n(\mathbb{F})$, where $\mathbb{F}$ is the algebraic closure of $\mathbb{F}_q$, and set $\overline{Z} = Z(\mathrm{GL}_n(\mathbb{F})) \cong \mathbb{F}^\times$. By [96, Proposition 3.1] we have $C_{\overline{G}}(x) = U_{n/2} \times \langle -I_n \rangle$ and thus $C_{\overline{Z}\overline{G}}(x) = U_{n/2} \times \overline{Z}$, where $U_{n/2}$ is a unipotent group of dimension $n/2$. By taking fixed points under an appropriate Frobenius morphism of $\overline{G}$, we deduce that

$$|C_{\mathrm{Sp}_n(q)}(x)| = 2q^{\frac{1}{2}n}, \quad |C_{\mathrm{GSp}_n(q)}(x)| = (q-1)q^{\frac{1}{2}n}$$

The result follows. □

Let $J_{a,+}$ and $J_{a,-}$ represent the two classes of regular unipotent elements in $\mathrm{Sp}_a(q)$ and fix $z \in \mathrm{GSp}_a(q)$ such that $(J_{a,+})^z = J_{a,-}$. Consider the elements $J_{a,+} \otimes I_b$ and $J_{a,-} \otimes I_b$ in a tensor product subgroup $\mathrm{Sp}_a(q) \otimes O_b(q) < \mathrm{Sp}_{ab}(q)$. Both elements have Jordan form $[J_a^b]$ and they are centralised by the subgroup $I_a \otimes O_b(q)$. Moreover, they are $\mathrm{GSp}_{ab}(q)$-conjugate since

$$(J_{a,+} \otimes I_b)^{z \otimes I_b} = J_{a,-} \otimes I_b$$

but we claim that they represent distinct $\mathrm{Sp}_{ab}(q)$-classes. To justify the claim, first observe that the $\mathrm{GSp}_{ab}(q)$-class of $x = J_{a,+} \otimes I_b$ splits into two $\mathrm{Sp}_{ab}(q)$-classes (this can be established by arguing as in the proof of Lemma 3.4.8). Therefore,

$$C_{\mathrm{GSp}_{ab}(q)}(x) = C_{Z\mathrm{Sp}_{ab}(q)}(x)$$

with $Z = Z(\mathrm{GSp}_{ab}(q))$ and the claim follows since $z \otimes I_b \notin Z\mathrm{Sp}_{ab}(q)$. We will write $[J_a^{b,\varepsilon}]$ with $\varepsilon = \pm$ to denote the two $\mathrm{Sp}_{ab}(q)$-class representatives that arise in this way.

**Remark 3.4.9** It is worth noting that if $b$ is even then the elements $J_{a,+} \otimes I_b$ and $J_{a,-} \otimes I_b$ in $\mathrm{Sp}_a(q) \otimes O_b^\varepsilon(q)$ are conjugate in $\mathrm{Sp}_{ab}(q)$. Indeed, by Lemma 2.5.2 there exists an element $z' \in \mathrm{GO}_b^\varepsilon(q)$ such that $z \otimes z' \in \mathrm{Sp}_{ab}(q)$, so

$$(J_{a,+} \otimes I_b)^{z \otimes z'} = J_{a,-} \otimes I_b$$

as required.

*Case 3. b odd, symplectic group centralisers.* By [96, Theorem 7.1], if $b$ is odd then there is a unique $\mathrm{Sp}_{ab}(q)$-class of unipotent elements in $\mathrm{Sp}_{ab}(q)$ with Jordan form $[J_b^a]$, which can be represented by an element $x = I_a \otimes J_b$ in a tensor product subgroup $\mathrm{Sp}_a(q) \otimes O_b(q)$. In terms of a symplectic basis $\{e_1, \ldots, e_{a/2}, f_1, \ldots, f_{a/2}\}$ for $U$, we note that $x$ fixes the orthogonal decomposition

$$U \otimes W = ((\langle e_1 \rangle \otimes W) \oplus (\langle f_1 \rangle \otimes W)) \perp \ldots \perp ((\langle e_{a/2} \rangle \otimes W) \oplus (\langle f_{a/2} \rangle \otimes W))$$

where each $\{\langle e_i \rangle \otimes W, \langle f_i \rangle \otimes W\}$ is a dual pair of totally isotropic $b$-spaces. Moreover, $x$ is centralised by $\mathrm{Sp}_a(q) \otimes I_b$.

Our main result on unipotent classes in odd characteristic is Proposition 3.4.10 below. To state this result, let $\mathscr{P}(n, p)$ denote the set of nontrivial *signed partitions* of $n$ of the form

$$(p^{a_p}, \varepsilon_{p-1}(p-1)^{a_{p-1}}, (p-2)^{a_{p-2}}, \ldots, \varepsilon_2 2^{a_2}, 1^{a_1})$$

where $\sum_i i a_i = n$, $\varepsilon_{2i} = \pm$, and $a_i$ is even when $i$ is odd (as before, *nontrivial* means that $a_1 < n$).

**Proposition 3.4.10** *Let $x = \hat{x}Z \in G$ be an element of order $p$, where $p$ is odd. Then $x \in T$ and $\hat{x}$ fixes an orthogonal decomposition of the form*

$$V = \bigoplus_{i=1}^{(p-1)/2} (U_{2i,1} \perp \ldots \perp U_{2i,a_{2i}}) \perp$$

$$\bigoplus_{i=1}^{(p+1)/2} \left( (U_{2i-1,1} \oplus U_{2i-1,1}^*) \perp \ldots \perp (U_{2i-1,a_{2i-1}/2} \oplus U_{2i-1,a_{2i-1}/2}^*) \right)$$

*where each $U_{2i,k}$ is a nondegenerate $2i$-space on which $\hat{x}$ acts indecomposably, and $\{U_{2i-1,k}, U_{2i-1,k}^*\}$ is a dual pair of totally isotropic subspaces of dimension $2i - 1$ on which $\hat{x}$ acts indecomposably. Moreover, the map*

$$\theta : (p^{a_p}, \varepsilon_{p-1}(p-1)^{a_{p-1}}, \ldots, \varepsilon_2 2^{a_2}, 1^{a_1}) \longmapsto ([J_p^{a_p}, J_{p-1}^{a_{p-1},\varepsilon_{p-1}}, \ldots, J_2^{a_2,\varepsilon_2}, J_1^{a_1}]Z)^T$$

*is a bijection from $\mathscr{P}(n,p)$ to the set of $T$-classes of elements of order $p$ in $T$.*

*Proof* As previously noted, $x \in T$ since $|G : T| = (2, q-1)$. By Lemma 3.4.1, $\hat{x}$ has Jordan form $[J_p^{a_p}, \ldots, J_1^{a_1}]$, where $\lambda = (p^{a_p}, \ldots, 1^{a_1})$ is a partition of $n$ such that $a_i$ is even for all odd $i$.

If $i$ is even, then Lemma 3.4.1 implies that $\mathrm{Sp}_i(q)$ contains an element with Jordan form $[J_i]$. Moreover, as discussed above in Cases 1 and 2, $\mathrm{Sp}_{ia_i}(q)$ has two classes of elements with Jordan form $[J_i^{a_i}]$, with representatives $[J_i^{a_i,+}]$ and $[J_i^{a_i,-}]$. Similarly, if $i$ is odd then Lemma 2.2.17 implies that $\mathrm{Sp}_{2i}(q)$ contains an element with Jordan form $[J_i^2]$, which fixes a dual pair of $i$-dimensional totally isotropic subspaces of the natural $\mathrm{Sp}_{2i}(q)$-module. In this way, we see that each signed partition

$$(p^{a_p}, \varepsilon_{p-1}(p-1)^{a_{p-1}}, \ldots, \varepsilon_2 2^{a_2}, 1^{a_1}) \in \mathscr{P}(n,p)$$

corresponds to an element $[J_p^{a_p}, J_{p-1}^{a_{p-1},\varepsilon_{p-1}}, \ldots, J_2^{a_2,\varepsilon_2}, J_1^{a_1}] \in \mathrm{Sp}_n(q)$ of order $p$, and the map $\theta$ is injective.

Set $k = |\{i \mid i \text{ even}, a_i > 0\}|$. The injectivity of $\theta$ implies that there are at least $2^k$ distinct $T$-classes of unipotent elements with Jordan form $[J_p^{a_p}, \ldots, J_1^{a_1}]$. By [96, Theorem 7.1], there are precisely $2^k$ such classes, so $\theta$ is surjective and $\hat{x}$ is $\mathrm{Sp}_n(q)$-conjugate to one of the unipotent elements constructed above. In particular, $\hat{x}$ fixes an appropriate orthogonal decomposition of $V$. $\qquad\square$

**Lemma 3.4.11** *Suppose $p$ is odd. Let $x \in \mathrm{Sp}_n(q)$ be a unipotent element corresponding to the signed partition $(p^{a_p}, \varepsilon_{p-1}(p-1)^{a_{p-1}}, \ldots, \varepsilon_2 2^{a_2}, 1^{a_1})$. Then $C_{\mathrm{Sp}_n(q)}(x) = QR$ where*

$$|Q| = q^{\gamma} \quad \text{with} \quad \gamma = \sum_{i<j} i a_i a_j + \frac{1}{2} \sum_i (i-1) a_i^2 + \frac{1}{2} \sum_{i \text{ even}} a_i$$

$$R = \prod_{i \text{ even}} O_{a_i}^{\varepsilon_i}(q) \times \prod_{i \text{ odd}} Sp_{a_i}(q)$$

and $O_{a_i}^{\pm}(q) = O_{a_i}(q)$ if $a_i$ is odd.

*Proof* This follows from [96, Theorem 7.1], combined with the discussion in Cases 1–3 above. $\qquad\qquad\qquad\qquad\qquad\qquad\qquad\qquad\qquad\qquad\qquad\qquad$ □

**Proposition 3.4.12** *Suppose $p$ is odd. Let $x \in T$ be a unipotent element corresponding to the signed partition $(p^{a_p}, \varepsilon_{p-1}(p-1)^{a_{p-1}}, \ldots, \varepsilon_2 2^{a_2}, 1^{a_1})$. Set $\alpha = 1$ if $a_i$ is odd for some $i$, otherwise $\alpha = 0$.*

  (i) *$x^G$ splits into $2^{\alpha}$ distinct $T$-classes;*
 (ii) *$|C_G(x)| = 2^{-\alpha}|Q||R|$, where $Q$ and $R$ are defined in Lemma 3.4.11.*

*Proof* First consider the centraliser in $GSp_{ab}(q)$ of a unipotent element $y$ with Jordan form $[J_a^b]$, where $a$ is even. Recall that such elements are of the form $J_a \otimes I_b$ in a tensor product subgroup $Sp_a(q) \otimes O_b^{\varepsilon}(q)$. By Lemma 3.4.11, the centraliser of $y$ in $Sp_{ab}(q)$ is a $p$-group $P$ extended by $I_a \otimes O_b^{\varepsilon}(q)$. It follows that the centraliser of $y$ in $GSp_{ab}(q)$ is an extension of $P$ by $\langle \mu I_a \rangle \otimes GO_b^{\varepsilon}(q) = I_a \otimes GO_b^{\varepsilon}(q)$, where $\mu$ is a primitive element of $\mathbb{F}_q$. Now

$$|GO_b^{\varepsilon}(q) : O_b^{\varepsilon}(q)| = (q-1)/(b+1,2), \quad |GSp_{ab}(q) : Sp_{ab}(q)| = q-1$$

so the two $Sp_{ab}(q)$-classes of elements with Jordan form $[J_a^b]$ fuse in $GSp_{ab}(q)$ if and only if $b$ is odd.

As noted in Lemma 3.4.11, the centraliser of $\hat{x}$ in $Sp_n(q)$ is an extension of a $p$-group $Q$ by a subgroup $R = \prod_{i \text{ even}} O_{a_i}^{\varepsilon_i}(q) \times \prod_{i \text{ odd}} Sp_{a_i}(q)$, where we define $O_{a_i}^{\pm}(q) = O_{a_i}(q)$ if $a_i$ is odd. Clearly, two classes in $T$ are fused in $G$ only if the corresponding signed partitions are compatible in the sense that they encode elements with isomorphic centralisers. This happens only if $a_i$ is odd for some (even) $i$. Moreover, in this situation the argument in the first paragraph implies that the $T$-class of $x$ is fused in $G$ with the $T$-class corresponding to the dual signed partition

$$(p^{a_p}, \varepsilon'_{p-1}(p-1)^{a_{p-1}}, \ldots, \varepsilon'_2 2^{a_2}, 1^{a_1})$$

where $\varepsilon'_i = (-)^{a_i} \varepsilon_i$. This proves part (i), and now (ii) follows immediately from Lemma 3.4.11. $\qquad\qquad\qquad\qquad\qquad\qquad\qquad\qquad\qquad\qquad\qquad\qquad\qquad\qquad$ □

### 3.4.4 Unipotent involutions

Now assume $p = 2$, so $G = T$. As before, let $B$ be the underlying nondegenerate alternating form on $V$. The conjugacy classes of unipotent involutions are described by Aschbacher and Seitz in [5, Section 7], and we will provide a brief summary of the main results.

First some notation. Given a nondegenerate 4-space $V_i$ with symplectic basis $\{e_{i1}, e_{i2}, f_{i1}, f_{i2}\}$, let $x_i \in \mathrm{Sp}(V_i)$ be the involution interchanging $e_{i1}$ and $e_{i2}$, and $f_{i1}$ and $f_{i2}$. Similarly, if $U_i$ is a nondegenerate 2-space with symplectic basis $\{e_{i1}, f_{i1}\}$, then let $y_i \in \mathrm{Sp}(U_i)$ be the involution that swaps $e_{i1}$ and $f_{i1}$.

For each even integer $s = 2\ell \leqslant n/2$ we can consider a decomposition

$$V = V_1 \perp V_2 \perp \ldots \perp V_\ell \perp W$$

where each $V_i$ is a nondegenerate 4-space and $W$ is a nondegenerate subspace of dimension $n - 2s$. Set

$$a_s = [x_1, \ldots, x_\ell, I_{n-2s}] \in \mathrm{Sp}_n(q)$$

with respect to the above decomposition. Then $a_s$ has Jordan form $[J_2^s, J_1^{n-2s}]$ and $B(v, va_s) = 0$ for all $v \in V$. We refer to involutions that are $\mathrm{Sp}_n(q)$-conjugate to $a_s$ as being of *type $a_s$*.

Similarly, for each even integer $s = 2(\ell + 1) \leqslant n/2$ we have a decomposition

$$V = U_1 \perp U_2 \perp V_1 \perp V_2 \perp \ldots \perp V_\ell \perp W$$

where each $U_i$, $V_j$ and $W$ is a nondegenerate subspace of dimension 2, 4 and $n - 2s$, respectively. Set

$$c_s = [y_1, y_2, x_1, \ldots, x_\ell, I_{n-2s}] \in \mathrm{Sp}_n(q)$$

Then $c_s$ has Jordan form $[J_2^s, J_1^{n-2s}]$ and $B(e_{11}, e_{11}c_s) = 1$ for $e_{11} \in U_1$. Thus $c_s$ is not of type $a_s$. Involutions that are $\mathrm{Sp}_n(q)$-conjugate to $c_s$ are said to be of *type $c_s$*.

Finally, for each odd integer $s = 2\ell + 1 \leqslant n/2$, let

$$V = U_1 \perp V_1 \perp V_2 \perp \ldots \perp V_\ell \perp W$$

where the $U_i$, $V_j$ and $W$ are as above. Set

$$b_s = [y_1, x_1, \ldots, x_\ell, I_{n-2s}] \in \mathrm{Sp}_n(q)$$

and note that $B(e_{11}, e_{11}b_s) = 1$ for $e_{11} \in U_1$. Evidently $b_s$ is neither of type $a_s$ nor of type $c_s$ since $b_s$ has Jordan form $[J_2^s, J_1^{n-2s}]$ and $s$ is odd. Any $\mathrm{Sp}_n(q)$-conjugate of $b_s$ is said to be an involution of *type $b_s$*.

The following result is proved in [5, Section 7].

Table 3.4.1 *Involutions in* $G = \mathrm{PGSp}_n(q)$, $p = 2$

| $x$ | Conditions | $|C_G(x)|$ |
|---|---|---|
| $a_s$ | $s = 2\ell,\ 1 \leqslant \ell \leqslant n/4$ | $q^{ns-3s^2/2+s/2}|\mathrm{Sp}_s(q)||\mathrm{Sp}_{n-2s}(q)|$ |
| $b_s$ | $s = 2\ell-1,\ 1 \leqslant \ell \leqslant (n+2)/4$ | $q^{ns-3s^2/2+s/2}|\mathrm{Sp}_{s-1}(q)||\mathrm{Sp}_{n-2s}(q)|$ |
| $c_s$ | $s = 2\ell,\ 1 \leqslant \ell \leqslant n/4$ | $q^{ns-3s^2/2+3s/2-1}|\mathrm{Sp}_{s-2}(q)||\mathrm{Sp}_{n-2s}(q)|$ |

**Proposition 3.4.13** *Suppose $p = 2$ and $x \in G$ is an involution with Jordan form $[J_2^s, J_1^{n-2s}]$. If $s$ is even then $x$ is of type $a_s$ or $c_s$, otherwise $x$ is of type $b_s$. Furthermore, $|C_G(x)|$ is given in Table 3.4.1.*

We will also require the following lemma for combining involutions of different types. If $x$ and $y$ are labels for two conjugacy classes of involutions, we write $[x, y]$ to denote a block-diagonal matrix with an involution of type $x$ in the first block and an involution of type $y$ in the second.

**Lemma 3.4.14** *Suppose $p = 2$. Let $x \in \mathrm{Sp}_n(q)$ and $y \in \mathrm{Sp}_m(q)$ be involutions. Set $z = [x, y] \in \mathrm{Sp}_{n+m}(q)$. Then the conjugacy class of $z$ is recorded in the following table:*

|  | $y = a_t$ | $b_t$ | $c_t$ |
|---|---|---|---|
| $x = a_s$ | $a_{s+t}$ | $b_{s+t}$ | $c_{s+t}$ |
| $b_s$ | $b_{s+t}$ | $c_{s+t}$ | $b_{s+t}$ |
| $c_s$ | $c_{s+t}$ | $b_{s+t}$ | $c_{s+t}$ |

*Proof* Let $V = U \perp W$ be the natural $\mathrm{Sp}_{n+m}(q)$-module, where $U$ and $W$ are nondegenerate subspaces corresponding to the natural modules for $\mathrm{Sp}_n(q)$ and $\mathrm{Sp}_m(q)$, respectively. In particular, our notation indicates that $z|_U = x$ and $z|_W = y$. If $x$ is of type $b_s$ or $c_s$ then there exists $v \in V$ such that $B(v, vx) \neq 0$, so $z$ must be of type $b_{s+t}$ or $c_{s+t}$, according to the parity of $s+t$. The same conclusion holds if $y$ is of type $b_t$ or $c_t$. Clearly, $z$ is of type $a_{s+t}$ if and only if both $x$ and $y$ are $a$-type involutions. $\qquad\square$

### 3.4.5 Field automorphisms

The next proposition describes the conjugacy classes of prime order field automorphisms in $\mathrm{Aut}(T)$. As described in Section 2.4, fix a standard field automorphism $\phi \in \mathrm{Aut}(T)$ of order $f$ (with respect to a fixed basis of $V$), where $q = p^f$.

**Proposition 3.4.15** *Let $x \in \mathrm{Aut}(T)$ be a field automorphism of prime order $r$. Then $r$ divides $f$, say $f = mr$, and $x$ is $G$-conjugate to exactly one of the standard field automorphisms $\phi^{jm}$, where $1 \leqslant j < r$. Moreover, $C_G(x) \cong \mathrm{PGSp}_n(q^{1/r})$ and $x^G$ splits into $(2, q-1, r)$ distinct $T$-classes.*

*Proof* First note that $x \in \mathrm{P\Gamma Sp}_n(q) = \langle G, \phi \rangle$. By definition, $x$ is $G$-conjugate to $\phi^i$ for some $i$. Therefore, $r$ divides $f$, say $f = mr$, and it follows that $x$ is $G$-conjugate to $\phi^{jm}$ for some $j$ in the range $1 \leqslant j < r$. The structure of $C_G(x)$ follows from [67, Proposition 4.9.1(b)], and the stated $G$-class splitting is computed in the usual way, via [86, Proposition 4.5.4]. $\qquad\qquad\square$

### 3.4.6 Graph-field automorphisms

Recall that the *polar space* associated to $V$ is the set of totally isotropic subspaces of $V$. If $(n, p) = (4, 2)$ then there is a duality of the polar space that interchanges the sets of totally isotropic 1-spaces and totally isotropic 2-spaces (see Section 2.4). This duality induces an exceptional automorphism $\rho$ of $T$ such that $\rho^2 = \phi$, where $\phi$ is a standard field automorphism of order $f$. The automorphism $\rho$ can be obtained by composing the isomorphisms

$$\psi_1 : \mathrm{Sp}_4(q) \to \mathrm{O}_5(q), \quad \psi_2 : \mathrm{O}_5(q) \to \mathrm{Sp}_4(q)$$

where $\psi_1$ arises from the *Klein correspondence* (see [118, Corollary 12.32]), and $\psi_2$ is the isomorphism discussed in Remark 2.5.1.

As in Section 2.4, we define the graph-field automorphisms of $T$ to be the $\mathrm{Aut}(T)$-conjugates of $\phi^i \rho$, with $1 \leqslant i \leqslant f$. Note that if $f$ is odd then $\mathrm{Aut}(T)$ contains involutory graph-field automorphisms, such as $\tau = \rho^f$, for example. The next result follows from [5, (19.5)].

**Proposition 3.4.16** *Suppose $(n, p) = (4, 2)$ and $f$ is odd. If $x \in \mathrm{Aut}(T)$ is an involutory graph-field automorphism, then $x$ is $T$-conjugate to $\tau$. Moreover, $C_G(x) \cong \mathrm{Sz}(q)$.*

We refer the reader to [18, Proposition 3.52] for information on the action of $\tau$ on the set of $G$-classes of elements of prime order in $G$. In particular, we note that $\tau$ fuses the $G$-classes of involutions of type $b_1$ and $a_2$.

### 3.4.7 Symplectic groups: a summary

The following proposition summarises some of the above results on the conjugacy classes in symplectic groups.

**Proposition 3.4.17** *Let $x \in \mathrm{Aut}(T)$ be an element of prime order $r$. Then the possibilities for $x$ up to $G$-conjugacy are listed in Table B.7, together with the order of the centraliser $C_G(x)$.*

*Proof* If $x$ is a field or graph-field automorphism then we appeal to Propositions 3.4.15 and 3.4.16, so let us assume $x \in G$. If $x$ is semisimple and $r$ is odd, then Proposition 3.4.3 applies (the structure of $C_G(x)$ is given in Remark 3.4.4), and we use Proposition 3.4.6 if $x$ is a semisimple involution. Similarly, if $x$ is unipotent of odd order then the result follows by combining Propositions 3.4.10 and 3.4.12, and for unipotent involutions we use Proposition 3.4.13. $\square$

## 3.5 Orthogonal groups

Let $V$ be an $n$-dimensional vector space over $\mathbb{F}_q$ equipped with a nondegenerate quadratic form $Q$ and associated bilinear form $B_Q$. Recall from Section 2.2.5 that if $n$ is odd (in which case $q$ is also odd) then we say that $Q$ is *parabolic*, while $Q$ is either *hyperbolic* or *elliptic* if $n$ is even. We use $\mathrm{O}_n^+(q)$ to denote the isometry group of a hyperbolic quadratic form on $V$, and the corresponding groups in the elliptic and parabolic cases are denoted by $\mathrm{O}_n^-(q)$ and $\mathrm{O}_n(q)$ (or occasionally $\mathrm{O}_n^\circ(q)$), respectively. Also recall that if $q$ is odd then the determinant of the Gram matrix of $B_Q$ is either a square or nonsquare in $\mathbb{F}_q$, and we define the *discriminant* of $Q$, denoted $D(Q) \in \{\square, \boxtimes\}$, accordingly. We write $\mathrm{GO}_n^\varepsilon(q)$ for the group of similarities of a nondegenerate quadratic form on $V$ of type $\varepsilon \in \{\pm, \circ\}$. By definition, if $\hat{x} \in \mathrm{GO}_n^\varepsilon(q)$ then there exists a scalar $\tau(\hat{x}) \in \mathbb{F}_q^\times$ such that

$$Q(v\hat{x}) = \tau(\hat{x})Q(v) \tag{3.5.1}$$

for all $v \in V$.

Our first result follows from a theorem of Wall [119, p. 38].

**Lemma 3.5.1** *Suppose $A \in \mathrm{GL}_n(q)$ and $q$ is odd. Then $A$ is similar to an element of $\mathrm{O}_n^\varepsilon(q)$, for some $\varepsilon$, if and only if*

(i) *$A$ and $A^{-1}$ are conjugate in $\mathrm{GL}_n(q)$, and*
(ii) *each Jordan block of even size and eigenvalue $\pm 1$ occurs with even multiplicity.*

*In particular, if $A$ is unipotent with Jordan form $[J_n^{a_n}, \ldots, J_1^{a_1}]$, then $A$ is similar to an element of $\mathrm{O}_n^\varepsilon(q)$, for some $\varepsilon$, if and only if $a_i$ is even for all even $i$.*

**Lemma 3.5.2** *Suppose* $A \in \mathrm{GL}_n(q)$ *and both n and q are even.*

(i) *A is similar to an element of* $\mathrm{O}_n^\varepsilon(q)$, *for some* $\varepsilon$, *if and only if A is similar to an element of* $\mathrm{Sp}_n(q)$.

(ii) *If A is semisimple then A is similar to an element of* $\mathrm{O}_n^\varepsilon(q)$, *for some* $\varepsilon$, *if and only if A and* $A^{-1}$ *are conjugate in* $\mathrm{GL}_n(q)$.

(iii) *If A is unipotent with Jordan form* $[J_n^{a_n}, \ldots, J_1^{a_1}]$, *then A is similar to an element of* $\mathrm{O}_n^\varepsilon(q)$, *for some* $\varepsilon$, *if and only if* $a_i$ *is even for all odd i.*

*Proof*  First observe that $\mathrm{O}_n^\pm(q) < \mathrm{Sp}_n(q)$ since $q$ is even. Part (i) is equivalent to the following well known fact (see [111, Lemma 4.1], for example):

$$\mathrm{Sp}_n(q) = \bigcup_{i=1}^{2} \bigcup_{g \in \mathrm{Sp}_n(q)} H_i^g$$

where $H_1 = \mathrm{O}_n^+(q)$ and $H_2 = \mathrm{O}_n^-(q)$ (so in the notation of Remark 1.1.3(i), we have $\gamma(\mathrm{Sp}_n(q)) = 2$). Parts (ii) and (iii) now follow immediately from Lemma 3.4.1. $\qquad\square$

Let $Z$ be the centre of $\mathrm{GL}_n(q)$ and recall that $\mathrm{PGO}_n^\varepsilon(q) = \mathrm{GO}_n^\varepsilon(q)/Z$. For $\mathrm{X} \in \{\mathrm{O}, \mathrm{SO}, \Omega\}$ we set

$$\mathrm{PX}_n^\varepsilon(q) = \mathrm{X}_n^\varepsilon(q)Z/Z \cong \mathrm{X}_n^\varepsilon(q)/(\mathrm{X}_n^\varepsilon(q) \cap Z)$$

Note that $\mathrm{O}_n^\varepsilon(q) \cap Z \cong C_{(2,q-1)}$. Throughout this section we set

$$T = \mathrm{P}\Omega_n^\varepsilon(q), \quad G = \mathrm{Inndiag}(T), \quad \widetilde{G} = \mathrm{PGO}_n^\varepsilon(q)$$

so $T \leqslant G \leqslant \widetilde{G}$ and $|\widetilde{G} : T|$ divides 8.

### 3.5.1 Semisimple elements

The next lemma follows from [119, p. 38] (see also [14, Proposition 2.11]).

**Lemma 3.5.3** *Two semisimple elements of odd order in* $\mathrm{O}_n^\varepsilon(q)$ *are conjugate in* $\mathrm{O}_n^\varepsilon(q)$ *if and only if they are conjugate in* $\mathrm{GL}_n(q)$.

Let $r \neq p$ be an odd prime divisor of $|G|$ and set $i = \Phi(r, q)$ as in (3.1.2). As usual, let $\mathscr{S}_r$ be the set of nontrivial $r$th roots of unity in $\mathbb{F}_{q^i}$ and let $\Lambda_1, \ldots, \Lambda_t$ be the $\sigma$-orbits on $\mathscr{S}_r$ corresponding to the permutation $\sigma : \mathscr{S}_r \to \mathscr{S}_r$ defined by $\sigma : \lambda \mapsto \lambda^q$. Note that $|\Lambda_j| = i$ for all $j$, so $t = (r-1)/i$. If $i$ is odd then $t$ is even and $\Lambda_j \neq \Lambda_j^{-1}$ by Lemma 3.1.13; in this situation, we label the $\sigma$-orbits so that $\Lambda_j^{-1} = \Lambda_{t/2+j}$ for all $1 \leqslant j \leqslant t/2$.

Our main result on the conjugacy of semisimple elements of order $r$ is Proposition 3.5.4 below. In order to state this result, we need to introduce some additional notation:

(i) For $t$ even, let $\mathscr{T}_1(n,i,t,\varepsilon)$ be the set of nonzero $\frac{t}{2}$-tuples $(a_1,\dots,a_{t/2}) \in \mathbb{N}_0^{t/2}$ such that $2i\sum_j a_j \leqslant n$, with equality only if $\varepsilon = +$.

(ii) Similarly, let $\mathscr{T}_2(n,i,t,\varepsilon)$ be the set of nonzero $t$-tuples $(a_1,\dots,a_t) \in \mathbb{N}_0^t$ such that $i\sum_j a_j \leqslant n$, with equality only if $(-)^{\sum_j a_j} = \varepsilon$.

**Proposition 3.5.4** *Let $x = \hat{x}Z \in \widetilde{G}$ be an element of order $r$ and set $i = \Phi(r,q)$. Then $x \in T$ and one of the following holds:*

(i) *If $i$ is odd, then $\hat{x}$ fixes an orthogonal decomposition of the form*

$$V = \bigoplus_{j=1}^{t/2} \left( (U_{j,1} \oplus U_{j,1}^*) \perp \dots \perp (U_{j,a_j} \oplus U_{j,a_j}^*) \right) \perp C_V(\hat{x})$$

*where each $\{U_{j,k}, U_{j,k}^*\}$ is a dual pair of totally singular $i$-spaces on which $\hat{x}$ acts irreducibly, and $C_V(\hat{x})$ is nondegenerate (or trivial). Moreover, the map*

$$\theta_1 : (a_1,\dots,a_{t/2}) \longmapsto ([(\Lambda_1,\Lambda_1^{-1})^{a_1},\dots,(\Lambda_{t/2},\Lambda_{t/2}^{-1})^{a_{t/2}}, I_e]Z)^{\mathrm{PO}_n^\varepsilon(q)}$$

*is a bijection from $\mathscr{T}_1(n,i,t,\varepsilon)$ to the set of $\mathrm{PO}_n^\varepsilon(q)$-classes of elements of order $r$ in $\widetilde{G}$.*

(ii) *If $i$ is even, then $\hat{x}$ fixes an orthogonal decomposition*

$$V = U_{1,1} \perp \dots \perp U_{1,a_1} \perp \dots \perp U_{t,1} \perp \dots \perp U_{t,a_t} \perp C_V(\hat{x})$$

*where each $U_{j,k}$ is an elliptic $i$-space on which $\hat{x}$ acts irreducibly, and $C_V(\hat{x})$ is nondegenerate (or trivial). Moreover, the map*

$$\theta_2 : (a_1,\dots,a_t) \longmapsto ([\Lambda_1^{a_1},\dots,\Lambda_t^{a_t}, I_e]Z)^{\mathrm{PO}_n^\varepsilon(q)}$$

*is a bijection from $\mathscr{T}_2(n,i,t,\varepsilon)$ to the set of $\mathrm{PO}_n^\varepsilon(q)$-classes of elements of order $r$ in $\widetilde{G}$.*

*Proof* Clearly, $x \in T$ since $|\widetilde{G}:T|$ divides 8, so we may assume that $\hat{x} \in \Omega_n^\varepsilon(q)$ has order $r$ (see Lemma 3.1.3). Since $\hat{x}$ is diagonalisable over $\mathbb{F}_{q^i}$ (but no proper subfield), the irreducible blocks of $\hat{x}$ are $i$-dimensional.

If $i$ is odd, then Lemma 3.1.13 implies that $\Lambda_j \neq \Lambda_j^{-1}$ and so Lemmas 3.1.6, 3.5.1 and 3.5.2 imply that the irreducible blocks of $\hat{x}$ are totally singular and arise in dual pairs $U_j \oplus U_j^*$ with eigenvalues $\Lambda_j \cup \Lambda_j^{-1}$. In particular, if $a_j$ denotes the multiplicity of $\Lambda_j$ in the multiset of eigenvalues of $\hat{x}$, then $\hat{x}$ fixes an orthogonal decomposition of $V$ as described in part (i). Moreover, by applying Lemmas 3.1.7 and 3.5.3, we deduce that $\theta_1(a_1,\ldots,a_{t/2}) = x^{\mathrm{PO}_n^{\varepsilon}(q)}$ and $\theta_1$ is a bijection.

If $i$ is even, then $\Lambda_j = \Lambda_j^{-1}$ (see Lemma 3.1.13), so Lemmas 3.5.1 and 3.5.2 imply that $[\Lambda_j] \in O_i^{\varepsilon'}(q)$ for some $\varepsilon'$. However, $r$ does not divide the order of $O_i^+(q)$ and thus $\varepsilon' = -$. Therefore, there exists an element $y = \hat{y}Z \in \widetilde{G}$ of order $r$ such that $\hat{y}$ and $\hat{x}$ have the same eigenvalues, and $\hat{y}$ fixes an orthogonal decomposition of $V$ as in part (ii). If $a_j$ denotes the multiplicity of $\Lambda_j$ in the multiset of eigenvalues of $\hat{x}$, then Lemmas 3.1.7 and 3.5.3 imply that

$$\theta_2(a_1,\ldots,a_t) = x^{\mathrm{PO}_n^{\varepsilon}(q)} = y^{\mathrm{PO}_n^{\varepsilon}(q)}$$

In particular, $\theta_2$ is surjective and $\hat{x}$ also fixes an orthogonal decomposition as in (ii). A further application of Lemma 3.1.7 shows that $\theta_2$ is a bijection. $\square$

**Remark 3.5.5** Let $x \in \widetilde{G}$ be a semisimple element of odd prime order $r$. Define $i = \Phi(r,q)$ as before and let $\hat{x} \in \Omega_n^{\varepsilon}(q)$ be the unique lift of $x$ of order $r$. As stated in Proposition 3.5.4, the 1-eigenspace of $\hat{x}$, denoted by $C_V(\hat{x})$ above, is nondegenerate (or trivial). The specific type of $C_V(\hat{x})$ is determined as follows:

(i) If $n$ is odd then $C_V(\hat{x})$ is parabolic.
(ii) Suppose $\varepsilon = +$. If $i$ is odd then $C_V(\hat{x})$ is hyperbolic (or trivial), while if $i$ is even then either $\sum_j a_j$ is even and $C_V(\hat{x})$ is hyperbolic (or trivial), or $\sum_j a_j$ is odd and $C_V(\hat{x}) \neq \{0\}$ is elliptic.
(iii) Suppose $\varepsilon = -$. If $i$ is odd then $C_V(\hat{x}) \neq \{0\}$ is elliptic, while if $i$ is even then either $\sum_j a_j$ is odd and $C_V(\hat{x})$ is hyperbolic (possibly trivial), or $\sum_j a_j$ is even and $C_V(\hat{x}) \neq \{0\}$ is elliptic.

**Remark 3.5.6** Let $x = \hat{x}Z \in \widetilde{G}$ be a semisimple element of odd prime order $r$. Set $i = \Phi(r,q)$. Then

$$C_{\mathrm{GO}_n^{\varepsilon}(q)}(\hat{x}) = \mathrm{GO}_n^{\varepsilon}(q) \cap C_{\mathrm{GL}_n(q)}(\hat{x})$$

and the structure of $C_{\mathrm{GL}_n(q)}(\hat{x})$ is given in Lemma 3.1.9. More precisely, if $i$ is odd and the $\mathrm{PO}_n^{\varepsilon}(q)$-class of $x$ corresponds to the tuple $(a_1,\ldots,a_{t/2})$ as in Proposition 3.5.4(i), then

$$C_{\mathrm{GO}_n^{\varepsilon}(q)}(\hat{x}) = (\mathrm{GL}_{a_1}(q^i) \times \cdots \times \mathrm{GL}_{a_{t/2}}(q^i) \times O_e^{\varepsilon}(q)).(q-1)/(2,n-1)$$

Here the cyclic group $(q-1)/(2,n-1)$ arises from the fact that if $e \neq 0$ then

$$|\mathrm{GO}_e^{\varepsilon}(q) : O_e^{\varepsilon}(q)| = (q-1)/(2,e-1)$$

and $e$ is odd if and only if $n$ is odd. Similarly, if $i$ is even and the $\mathrm{PO}_n^{\varepsilon}(q)$-class of $x$ corresponds to $(a_1, \ldots, a_t)$ as in Proposition 3.5.4(ii), then

$$C_{\mathrm{GO}_n^{\varepsilon}(q)}(\hat{x}) = (\mathrm{GU}_{a_1}(q^{i/2}) \times \cdots \times \mathrm{GU}_{a_t}(q^{i/2}) \times \mathrm{O}_e^{\varepsilon'}(q)).(q-1)/(2, n-1)$$

for some $\varepsilon'$ (see Remark 3.5.5). To see this, observe that

$$\mathrm{GU}_a(q^{i/2}) \leqslant \mathrm{SO}_{2a}^{\varepsilon''}(q^{i/2}) \leqslant \mathrm{SO}_{ai}^{\varepsilon''}(q)$$

where $\varepsilon'' = (-)^a$ (see Construction 2.5.14), and the centre of $\mathrm{GU}_a(q^{i/2})$ contains an element of order $r$. The precise structure of $C_{\mathrm{O}_n^{\varepsilon}(q)}(\hat{x})$ is given in [119, p. 39, p. 60], and we record the order of $C_G(x)$ in Table B.12.

**Remark 3.5.7** Suppose $i$ is even and $\Lambda$ is a $\sigma$-orbit on $\mathscr{S}_r$. Then $\mathrm{O}_{2i}^+(q)$ contains an element $x = [\Lambda^2]$ that fixes an orthogonal decomposition of the natural $\mathrm{O}_{2i}^+(q)$-module into two elliptic $i$-spaces. Since $\Lambda = \Lambda^{-1}$, $\mathrm{O}_{2i}^+(q)$ also contains an element $y = [\Lambda^2]$ fixing a dual pair of totally singular $i$-spaces. By Lemma 3.5.3, $x$ and $y$ are conjugate in $\mathrm{O}_{2i}^+(q)$, so $x$ also fixes such a dual pair.

We can also see this by considering a suitable tensor product decomposition of the natural $\mathrm{O}_{2i}^+(q)$-module $V$. We may view $x = I_2 \otimes [\Lambda]$ as an element of a tensor product subgroup $\mathrm{O}_2^+(q) \otimes \mathrm{O}_i^-(q) < \mathrm{O}_{2i}^+(q)$. Then $x$ fixes the decomposition

$$V = (\langle e_1 \rangle \otimes W) \oplus (\langle f_1 \rangle \otimes W)$$

of $V$ into a dual pair of totally singular $i$-spaces, where $\{e_1, f_1\}$ is a hyperbolic basis for the natural $\mathrm{O}_2^+(q)$-module $U$, and $W$ is the natural $\mathrm{O}_i^-(q)$-module. Similarly, $x$ fixes the orthogonal decomposition

$$V = (\langle u \rangle \otimes W) \perp (\langle v \rangle \otimes W)$$

into elliptic $i$-spaces, where $\{u, v\}$ is a basis for $U$ consisting of nonsingular orthogonal vectors.

Let $x \in \widetilde{G}$ be a semisimple element of odd prime order $r$. Note that Proposition 3.5.4 is stated in terms of $\mathrm{PO}_n^{\varepsilon}(q)$-classes. By Lemma 3.5.3, two semisimple elements in $\widetilde{G}$ of order $r$ are $\mathrm{PO}_n^{\varepsilon}(q)$-conjugate if and only if they are $\widetilde{G}$-conjugate, whence $x^{\mathrm{PO}_n^{\varepsilon}(q)} = x^{\widetilde{G}}$. In fact, if $n$ is odd then

$$G = \mathrm{PSO}_n(q) = \mathrm{PO}_n(q) = \mathrm{PGO}_n(q) = \widetilde{G}$$

and Theorem 3.1.12 implies that $x^T = x^{\widetilde{G}}$.

Now assume $n$ is even. If $q$ is even then

$$\mathrm{P\Omega}_n^{\varepsilon}(q) = \mathrm{Inndiag}(\mathrm{P\Omega}_n^{\varepsilon}(q)) < \mathrm{PSO}_n^{\varepsilon}(q) = \mathrm{PO}_n^{\varepsilon}(q) = \mathrm{PGO}_n^{\varepsilon}(q)$$

and the proper inclusion has index two. If $q$ is odd, then these five groups are related as in Figure 2.5.1 (see Section 2.5).

**Proposition 3.5.8** *Suppose $n$ is even and let $x \in \widetilde{G}$ be a semisimple element of odd prime order $r$. The following hold:*

(i) $x^T = x^{\mathrm{PSO}_n^\varepsilon(q)} = x^G$ *and* $x^{\mathrm{PO}_n^\varepsilon(q)} = x^{\widetilde{G}}$;

(ii) $x^G = x^{\widetilde{G}}$, *unless* $C_V(\hat{x}) = \{0\}$ *in which case* $x^{\widetilde{G}}$ *splits into two $G$-classes.*

*Proof* Part (i) follows immediately from Theorem 3.1.12 and Lemma 3.5.3. Part (ii) can be deduced by computing class lengths. Set $i = \Phi(r, q)$ and for convenience let us assume $i$ is odd (an entirely similar argument applies if $i$ is even). In terms of Proposition 3.5.4(i), let $(a_1, \ldots, a_{t/2}) \in \mathscr{T}_1(n, i, t, \varepsilon)$ be the corresponding $\frac{t}{2}$-tuple and set $e = \dim C_V(\hat{x})$, where $x = \hat{x}Z$. Now, if $q$ is even then

$$|x^G| = \frac{|\Omega_n^\varepsilon(q)|}{|\Omega_e^\varepsilon(q)| \prod_j |\mathrm{GL}_{a_j}(q^i)|}, \quad |x^{\widetilde{G}}| = \frac{|O_n^\varepsilon(q)|}{|O_e^\varepsilon(q)| \prod_j |\mathrm{GL}_{a_j}(q^i)|}$$

(see Remark 3.5.6) and the conclusion in part (ii) follows immediately since $|O_n^\varepsilon(q) : \Omega_n^\varepsilon(q)| = 2$. Similarly, observe that if $q$ is odd then

$$|x^{\mathrm{PSO}_n^\varepsilon(q)}| = \frac{|\mathrm{SO}_n^\varepsilon(q)|}{|\mathrm{SO}_e^\varepsilon(q)| \prod_j |\mathrm{GL}_{a_j}(q^i)|}$$

and

$$|x^{\mathrm{PO}_n^\varepsilon(q)}| = \frac{|O_n^\varepsilon(q)|}{|O_e^\varepsilon(q)| \prod_j |\mathrm{GL}_{a_j}(q^i)|}$$

Therefore $|x^{\widetilde{G}}| = (1 + \delta_{0,e})|x^G|$ and part (ii) follows. $\qquad\square$

**Remark 3.5.9** We can give an alternative, more geometric, explanation for the splitting in part (ii) of the previous proposition. We will assume $i$ is odd.

If $e = 0$ then $\varepsilon = +$ (see Remark 3.5.5) and $\hat{x}$ fixes a maximal totally singular subspace $W$ of $V$ so that $\hat{x}|_W = [\Lambda_1^{a_1}, \ldots, \Lambda_{t/2}^{a_{t/2}}]$. Now $T$ also contains an element $y = \hat{y}Z$ that is $\widetilde{G}$-conjugate to $x$ with the property that $\hat{y}$ fixes a maximal totally singular subspace $U$, where $\hat{x}|_U = [\Lambda_1^{a_1}, \ldots, \Lambda_{t/2}^{a_{t/2}}]$ and $\dim U - \dim(U \cap W) = i$. Then $U$ and $W$ are in distinct $T$-orbits (see Proposition 2.5.4), so $x$ and $y$ are not $T$-conjugate and thus the $\widetilde{G}$-class of $x$ splits into two $G$-classes.

Now assume $e \neq 0$. Here $\hat{x}$ fixes a totally singular subspace $W$ such that $\hat{x}|_W = [\Lambda_1^{a_1}, \ldots, \Lambda_{t/2}^{a_{t/2}}]$ and either $\varepsilon \neq +$ or $W$ is non-maximal. Then $W^G = W^{\widetilde{G}}$ and thus $x^G = x^{\widetilde{G}}$.

### 3.5.2 Semisimple involutions

Now assume $p$ is odd and let us turn to the inner-diagonal and graph automorphisms of order two in $\mathrm{Aut}(T)$. We will adopt the notation for class representatives given in [67, Table 4.5.1]. As before, Theorem 3.1.12 implies that $x^G = x^T$ for all involutions $x \in G$. Also note that $Z \cap \mathrm{O}_n^\varepsilon(q) = \{\pm I_n\}$, so if $\hat{t} \in \mathrm{O}_n^\varepsilon(q)$ and $\hat{t}Z \in \mathrm{P\Omega}_n^\varepsilon(q)$, then either $\hat{t} \in \Omega_n^\varepsilon(q)$ or $-\hat{t} \in \Omega_n^\varepsilon(q)$.

We begin by assuming $n \geqslant 7$ is odd, so

$$G = \mathrm{SO}_n(q) = \mathrm{PSO}_n(q) = \mathrm{PO}_n(q) = \mathrm{PGO}_n(q) = \widetilde{G}$$

and $T = \Omega_n(q)$ is an index-two subgroup of $G$. Let

$$\{e_1, \ldots, e_{(n-1)/2}, f_1, \ldots, f_{(n-1)/2}, x\}$$

be a standard basis for $V$, as described in part (iii) of Proposition 2.2.7.

#### 3.5.2.1 Involutions of type $t_i$, $1 \leqslant i \leqslant (n-1)/2$, $n$ odd

Write $t_i = \hat{t}Z$, where $\hat{t} \in \mathrm{SO}_n(q)$ is an involution with a hyperbolic $(-1)$-eigenspace of dimension $2i$ (we can take $\hat{t} = [-I_{2i}, I_{n-2i}]$ with respect to the specific basis ordering $\{e_1, f_1, \ldots, e_{(n-1)/2}, f_{(n-1)/2}, x\}$). Note that the coset $\hat{t}Z$ also contains the involution $-\hat{t} \in \mathrm{O}_n(q) \setminus \mathrm{SO}_n(q)$, whose $(-1)$-eigenspace is a parabolic space of dimension $n - 2i$. Clearly, $\hat{t}$ is not conjugate to $-\hat{t}$.

By Lemma 2.5.9(ii), we have $\hat{t} \in \Omega_n(q)$ if and only if $-I_{2i} \in \Omega_{2i}^+(q)$. Combining Lemmas 2.2.9 and 2.5.6, it follows that $\hat{t} \in \Omega_n(q)$ if and only if $i(q-1)/2$ is even, which holds if and only if $q^i \equiv 1 \pmod 4$. Thus $t_i \in T$ if and only if $q^i \equiv 1 \pmod 4$.

#### 3.5.2.2 Involutions of type $t_i'$, $1 \leqslant i \leqslant (n-1)/2$, $n$ odd

Here $t_i' = \hat{t}Z$, where $\hat{t} \in \mathrm{SO}_n(q)$ is an involution with an elliptic $(-1)$-eigenspace of dimension $2i$. (Such a subspace is easily constructed from a standard basis of $V$, using Lemma 2.2.13(i).) The coset $\hat{t}Z$ also contains the involution $-\hat{t} \in \mathrm{O}_n(q) \setminus \mathrm{SO}_n(q)$, whose $(-1)$-eigenspace is a parabolic $(n - 2i)$-space. Clearly, $\hat{t}$ is not conjugate to $-\hat{t}$. It is also clear that involutions of type $t_i'$ are not conjugate to those of type $t_i$; indeed, the isometry type of the restriction of the underlying quadratic form $Q$ to the respective $(-1)$-eigenspaces is different in each case.

Arguing as above, we deduce that $\hat{t} \in \Omega_n(q)$ if and only if $-I_{2i} \in \Omega_{2i}^-(q)$. Thus $\hat{t} \in \Omega_n(q)$ if and only if $i(q-1)/2$ is odd, which holds if and only if $q^i \equiv 3 \pmod 4$. Therefore $t_i' \in T$ if and only if $q^i \equiv 3 \pmod 4$.

For the remainder of this discussion of semisimple involutions, we may assume that $T = \mathrm{P\Omega}_n^\varepsilon(q)$ and $n \geqslant 8$ is even. Fix a standard basis for $V$ as in Proposition

2.2.7. Note that if $\hat{x} \in O_n^\varepsilon(q)Z$, then $\tau(\hat{x}) \in \mathbb{F}_q$ is a square (where $\tau$ is the map in (3.5.1)).

### 3.5.2.3 Involutions of type $t_i$, $1 \leqslant i < n/4$, $n$ even

Set $t_i = \hat{t}Z$, where $\hat{t} \in SO_n^\varepsilon(q)$ is an involution with a hyperbolic $2i$-dimensional $(-1)$-eigenspace (in terms of a standard basis, we can take $\langle e_1, f_1, \ldots, e_i, f_i \rangle$). Note that the $(-1)$-eigenspace of $-\hat{t} \in SO_n^\varepsilon(q)$ is a nondegenerate subspace of dimension $n - 2i$ and type $\varepsilon$. Clearly, $\hat{t}$ and $-\hat{t}$ are not conjugate.

By Lemma 2.5.9(ii), $\hat{t} \in \Omega_n^\varepsilon(q)$ if and only if $-I_{2i} \in \Omega_{2i}^+(q)$, so Lemmas 2.2.9 and 2.5.6 imply that $\hat{t} \in \Omega_n^\varepsilon(q)$ if and only if $i(q-1)/2$ is even. Similarly, $-\hat{t} \in \Omega_n^\varepsilon(q)$ if and only if $(n - 2i)(q-1)/4$ is even (respectively, odd) when $\varepsilon = +$ (respectively, $\varepsilon = -$). Therefore, $t_i \in T$ if and only if either $q^i \equiv 1$ (mod 4) or $\varepsilon = (-)^{n/2-i}$.

### 3.5.2.4 Involutions of type $t_i'$, $1 \leqslant i < n/4$, $n$ even

Here $t_i' = \hat{t}Z$, where $\hat{t}$ is an involution in $SO_n^\varepsilon(q)$ whose $(-1)$-eigenspace is an elliptic $2i$-space. (Such a subspace is easily identified from a standard basis for $V$ when $\varepsilon = -$, and it can be constructed using Lemma 2.2.13(ii) when $\varepsilon = +$.) Note that the $(-1)$-eigenspace of $-\hat{t} \in SO_n^\varepsilon(q)$ is a nondegenerate subspace of dimension $n - 2i$ and type $-\varepsilon$. The elements $\hat{t}$ and $-\hat{t}$ are not conjugate.

By Lemma 2.5.9(ii), $\hat{t} \in \Omega_n^\varepsilon(q)$ if and only if $-I_{2i} \in \Omega_{2i}^-(q)$. As before we deduce that $\hat{t} \in \Omega_n^\varepsilon(q)$ if and only if $i(q-1)/2$ is odd, and $-\hat{t} \in \Omega_n^\varepsilon(q)$ if and only if $(n - 2i)(q-1)/4$ is odd or even as $\varepsilon = +$ or $-$. Thus $t_i' \in T$ if and only if $q^i \equiv 3$ (mod 4), or $q \equiv 1$ (mod 4) and $\varepsilon = -$, or $q \equiv 3$ (mod 4) and $\varepsilon = (-)^{n/2-i+1}$.

### 3.5.2.5 Involutions of type $t_{n/4}$, $n \equiv 0$ (mod 4), $\varepsilon = +$

Here $t_{n/4} = \hat{t}Z$, where $\hat{t} \in SO_n^+(q)$ is an involution with a hyperbolic $(-1)$-eigenspace of dimension $n/2$ (we can take $\langle e_1, f_1, \ldots, e_{n/4}, f_{n/4} \rangle$). Similarly, the $(-1)$-eigenspace of $-\hat{t}$ is also a hyperbolic $\frac{n}{2}$-space, whence $\hat{t}$ and $-\hat{t}$ are $\Omega_n^+(q)$-conjugate since $\Omega_n^+(q)$ acts transitively on the set of hyperbolic $\frac{n}{2}$-spaces (see Lemma 2.5.10).

By Lemma 2.5.9(ii), $\hat{t} \in \Omega_n^+(q)$ if and only if $-I_{n/2} \in \Omega_{n/2}^+(q)$. By applying Lemmas 2.2.9 and 2.5.6, we see that the latter condition holds if and only if $n(q-1)/8$ is even, whence $t_{n/4} \in T$ if and only if $q^{n/4} \equiv 1$ (mod 4).

### 3.5.2.6 Involutions of type $t_{n/4}'$, $n \equiv 0$ (mod 4), $\varepsilon = +$

Here we set $t_{n/4}' = \hat{t}Z$, where $\hat{t} \in SO_n^+(q)$ is an involution with an elliptic $(-1)$-eigenspace of dimension $n/2$ (such a subspace can be constructed using Lemma 2.2.13(ii)). As in the previous case, we deduce that $\hat{t}$ and $-\hat{t}$ are $\Omega_n^+(q)$-conjugate since $\Omega_n^+(q)$ is transitive on elliptic $\frac{n}{2}$-spaces.

By Lemma 2.5.9(ii), $\hat{t} \in \Omega_n^+(q)$ if and only if $-I_{n/2} \in \Omega_{n/2}^-(q)$, which is equivalent to the condition that $n(q-1)/8$ is odd (see Lemmas 2.2.9 and 2.5.6). Therefore, $t'_{n/4} \in T$ if and only if $q^{n/4} \equiv 3 \pmod 4$.

### 3.5.2.7 Involutions of type $t''_{n/4}, t'''_{n/4}$, $n \equiv 0 \pmod 4$, $\varepsilon = +$

These involutions arise from elements $\hat{t}$ of order $2(q-1)_2$ in the centre of $\mathrm{GO}_{n/2}^+(q^2)$, which can be naturally embedded in $\mathrm{GO}_n^+(q)$ as described in Construction 2.5.13. Define the quadratic forms $\overline{Q}$ and $Q$ as in Construction 2.5.13, where $\overline{Q}$ and $Q = \mathrm{Tr} \circ \overline{Q}$ are hyperbolic and $\mathrm{Tr} : \mathbb{F}_{q^2} \to \mathbb{F}_q$ is the usual trace map. Fix $\xi \in \mathbb{F}_{q^2}$ of order $2(q-1)_2$ and note that $\mathrm{Tr}(\xi) = 0$. With respect to the hyperbolic basis

$$\left\{ e_1, \xi e_1, \ldots, e_{n/4}, \xi e_{n/4}, \frac{1}{2}f_1, \frac{1}{2}\xi^{-1}f_1, \ldots, \frac{1}{2}f_{n/4}, \frac{1}{2}\xi^{-1}f_{n/4} \right\}$$

of $V$ given in Construction 2.5.13, we have $\hat{t} = [A^{n/4}, (A^\mathsf{T})^{n/4}]$ where

$$A = \begin{pmatrix} 0 & 1 \\ \xi^2 & 0 \end{pmatrix} \tag{3.5.2}$$

Now $\tau(\hat{t}) = \xi^2 \in \mathbb{F}_q$ (see (3.5.1)) is an $\mathbb{F}_q$-nonsquare (it has order $(q-1)_2$), so $\hat{t} \notin \mathrm{O}_n^+(q)Z$ and thus the involutions we obtain lie in $\widetilde{G} \setminus \mathrm{PO}_n^+(q)$.

Let $X \in \mathrm{SL}_2(q)$ be an element such that $X^{-1}AX = -A$ (see Remark 3.1.11). Then $[X, X^{-\mathsf{T}}] \in \Omega_4^+(q)$ (see Lemma 2.5.7), and this element conjugates $[A, A^\mathsf{T}]$ to $[-A, -A^\mathsf{T}]$. It follows that $\hat{t}$ is $\Omega_n^+(q)$-conjugate to $-\hat{t}$. We also note that the coset $\hat{t}Z$ contains no involutions since every element in this coset has eigenvalues in $\mathbb{F}_{q^2} \setminus \mathbb{F}_q$.

By construction, any involution $x$ of this type fixes a maximal totally singular subspace of $V$, so $x \in G \setminus \mathrm{PSO}_n^+(q)$. Now any two totally singular $\frac{n}{2}$-dimensional $\mathbb{F}_q$-subspaces $U, W$ of $V$ arising from appropriate $\mathbb{F}_{q^2}$-spaces have the property that $\dim U - \dim(U \cap W)$ is even, so they belong to the same $\Omega_n^+(q)$-orbit on such subspaces. In particular, all maximal totally singular subspaces fixed by $x$ lie in the same $\Omega_n^+(q)$-orbit. Moreover, the two $G$-orbits of maximal totally singular subspaces give rise to two $G$-classes of embeddings of $\mathrm{O}_{n/2}^+(q^2)$ in $\mathrm{O}_n^+(q)$, and consequently two $G$-classes of involutions in $G \setminus \mathrm{PSO}_n^+(q)$ of the above form. Every element in $\widetilde{G} \setminus G$ interchanges the two types of maximal totally singular subspaces, and therefore fuses the two $G$-classes.

Finally, let us explain how to distinguish the two $G$-classes. Let $x, y \in G$ be involutions of this form, fixing maximal totally singular subspaces $U$ and $W$, respectively. Then $x$ and $y$ are $G$-conjugate if and only if $\dim U - \dim(U \cap W)$ is even.

### 3.5.2.8 Involutions of type $t_{n/4}$, $n \equiv 0 \pmod 4$, $\varepsilon = -$

Here $t_{n/4} = \hat{t}Z$, where $\hat{t} \in \mathrm{SO}_n^-(q)$ is an involution whose $(-1)$-eigenspace is a hyperbolic subspace of dimension $n/2$ (in terms of a standard basis, we can take $\langle e_1, f_1, \ldots, e_{n/4}, f_{n/4} \rangle$). Note that the $(-1)$-eigenspace of $-\hat{t}$ is an elliptic $\frac{n}{2}$-space, whence $\hat{t}$ and $-\hat{t}$ are not $\mathrm{GO}_n^-(q)$-conjugate. By Lemma 2.5.9, $t_{n/4} \in T$ if and only if $-I_{n/2} \in \Omega_{n/2}^+(q)$ or $-I_{n/2} \in \Omega_{n/2}^-(q)$. In view of Lemmas 2.2.9 and 2.5.6, we conclude that $t_{n/4} \in T$.

### 3.5.2.9 Involutions of type $t'_{n/4}$, $n \equiv 0 \pmod 4$, $\varepsilon = -$

These involutions arise from elements in the centre of $\mathrm{GO}_{n/2}^-(q^2)$ of order $2(q-1)_2$, which can be embedded in $\mathrm{GO}_n^-(q)$ via Construction 2.5.13. Define $\overline{Q}$, $Q$, $\overline{B}$ and $B$ as in Construction 2.5.13, with $\overline{Q}$ elliptic. Note that $Q = \mathrm{Tr} \circ \overline{Q}$ and $B = \mathrm{Tr} \circ \overline{B}$, where $\mathrm{Tr} : \mathbb{F}_{q^2} \to \mathbb{F}_q$ is the trace map. Let $\xi \in \mathbb{F}_{q^2}$ be an element of order $2(q-1)_2$ and set $\hat{t} = \xi I_{n/2} \in \mathrm{GO}_{n/2}^-(q^2)$. Since $\tau(\hat{t}) = \xi^2 \in \mathbb{F}_q$ is an $\mathbb{F}_q$-nonsquare, it follows that $\hat{t}$ embeds as an element of $\mathrm{GO}_n^-(q) \setminus \mathrm{O}_n^-(q)Z$. We also note that $\hat{t}$ fixes singular 1-spaces and nonsingular 1-spaces of both discriminants over $\mathbb{F}_{q^2}$, and so it fixes totally singular, hyperbolic and elliptic 2-spaces over $\mathbb{F}_q$ (see Table 2.5.1). Set $t'_{n/4} = \hat{t}Z$.

We claim that $t'_{n/4} \in G \setminus \mathrm{PSO}_n^-(q)$. Let $\{e_1, \ldots, e_{n/4-1}, f_1, \ldots, f_{n/4-1}, x, y\}$ be an elliptic basis for $V$ over $\mathbb{F}_{q^2}$ such that $\langle e_1, \ldots, e_{n/4-1}, f_1, \ldots, f_{n/4-1} \rangle$ is hyperbolic and $\overline{Q}(x) = 1$, $\overline{B}(x,y) = 0$ and $\overline{Q}(y) = \alpha$, where $-\alpha \in \mathbb{F}_{q^2}$ is a nonsquare (see Remark 2.2.8). For each $v \in V$, denote $\xi v$ by $\overline{v}$. Note that

$$Q(x) = 2, \quad Q(\overline{x}) = 2\xi^2, \quad Q(y) = \mathsf{Tr}(\alpha), \quad Q(\overline{y}) = \xi^2 \mathsf{Tr}(\alpha)$$

$$B(x,y) = B(x,\overline{y}) = B(\overline{x},y) = B(\overline{x},\overline{y}) = 0$$

$$B(x,\overline{x}) = \mathsf{Tr}(2\xi) = 0, \quad B(y,\overline{y}) = 2\mathsf{Tr}(\alpha\xi)$$

Let $V_\# = V \otimes_{\mathbb{F}_q} \mathbb{F}_{q^2}$ (so $V_\#$ is an $n$-dimensional vector space over $\mathbb{F}_{q^2}$) and recall the hyperbolic quadratic form $Q_\#$ on $V_\#$ defined in Construction 2.5.12, with associated bilinear form $B_\#$. Note that

$$Q_\#(v \otimes 1) = Q(v), \quad B_\#(u \otimes 1, v \otimes 1) = B(u,v)$$

for all $u, v \in V$.

Let $\hat{t}_\# \in \mathrm{GL}(V_\#)$ be the linear map that sends each $v \otimes \lambda$ to $(v\hat{t}) \otimes \lambda$. Then $\hat{t}_\#$ is a similarity of $Q_\#$, which fixes the totally singular subspace

$$U_\# = \langle e_1 \otimes 1, \overline{e}_1 \otimes 1, \ldots, e_{n/4-1} \otimes 1, \overline{e}_{n/4-1} \otimes 1 \rangle$$

setwise. Fix $i \in \mathbb{F}_{q^2}$ such that $i^2 = -1$. Then

$$
\begin{aligned}
Q_\#(x \otimes (i\xi) + \bar{x} \otimes 1) &= Q_\#(x \otimes (i\xi)) + Q_\#(\bar{x} \otimes 1) + B_\#(x \otimes (i\xi), \bar{x} \otimes 1) \\
&= -\xi^2 Q(x) + Q(\bar{x}) + i\xi B(x, \bar{x}) \\
&= 0
\end{aligned}
\tag{3.5.3}
$$

For any $\lambda \in \mathbb{F}_{q^2}$ we have

$$
Q_\#(y \otimes \lambda + \bar{y} \otimes 1) = \lambda^2 \operatorname{Tr}(\alpha) + 2\lambda \operatorname{Tr}(\alpha\xi) + \xi^2 \operatorname{Tr}(\alpha)
$$

The expression on the right is a quadratic polynomial in $\lambda$ with coefficients in $\mathbb{F}_q$, so there exists $\lambda \in \mathbb{F}_{q^2}$ such that $Q_\#(y \otimes \lambda + \bar{y} \otimes 1) = 0$. It is straightforward to check that $\lambda \neq \pm\xi$. Since

$$
B_\#(x \otimes (i\xi) + \bar{x} \otimes 1, y \otimes \lambda + \bar{y} \otimes 1) = 0
$$

it follows that

$$
W = \langle x \otimes (i\xi) + \bar{x} \otimes 1, y \otimes \lambda + \bar{y} \otimes 1 \rangle
$$

is totally singular and thus $X = \langle U_\#, W \rangle$ is a maximal totally singular subspace of $V_\#$. Now

$$
(x \otimes (i\xi) + \bar{x} \otimes 1)\hat{t}_\# = \bar{x} \otimes (i\xi) + (\xi^2 x) \otimes 1 = i\xi(\bar{x} \otimes 1 + x \otimes (-i\xi)) \notin W
$$

and similarly

$$
(y \otimes \lambda + \bar{y} \otimes 1)\hat{t}_\# = \bar{y} \otimes \lambda + (\xi^2 y) \otimes 1 = \begin{cases} \lambda(\bar{y} \otimes 1 + y \otimes (\xi^2 \lambda^{-1})) & \lambda \neq 0 \\ (\xi^2 y) \otimes 1 & \lambda = 0 \end{cases}
$$

Clearly, $(\xi^2 y) \otimes 1 \notin W$. Similarly, if $\lambda \neq 0$ then $\xi^2 \lambda^{-1} \neq \lambda$ and we deduce that $\lambda(\bar{y} \otimes 1 + y \otimes (\xi^2 \lambda^{-1})) \notin W$. Therefore $W \cap W\hat{t}_\# = \{0\}$ and thus $X$ and $X\hat{t}_\#$ belong to the same $\Omega_n^+(q^2)$-orbit on maximal totally singular subspaces (see Proposition 2.5.4). In view of the definition of $G = \operatorname{Inndiag}(\mathrm{P}\Omega_n^-(q))$ (see Construction 2.5.12), we conclude that $t'_{n/4} \in G \setminus \mathrm{PSO}_n^-(q)$.

**3.5.2.10 Involutions of type** $t_{n/2}, t_{n/2-1}, n \equiv 0 \pmod 4, \varepsilon = +$
These involutions fix a pair of complementary maximal totally singular subspaces $U, W$ of $V$, so they are contained in $G$. Moreover, since $n/2$ is even, $U$ and $W$ lie in the same $\Omega_n^+(q)$-orbit, so we obtain two $G$-classes of involutions of this type, which are fused in $\tilde{G}$.

Define maximal totally singular subspaces

$$
U = \langle e_1, \ldots, e_{n/2} \rangle, \quad W = \langle f_1, \ldots, f_{n/2} \rangle
$$

and

$$U' = \langle f_1, e_2, \ldots, e_{n/2} \rangle, \quad W' = \langle e_1, f_2, \ldots, f_{n/2} \rangle$$

with respect to a hyperbolic basis $\{e_1, \ldots, e_{n/2}, f_1, \ldots, f_{n/2}\}$ of $V$. Note that $\dim U - \dim(U \cap U') = 1$, so $U$ and $U'$ are in different $\Omega_n^+(q)$-orbits (and similarly, $W$ and $W'$ are in different orbits).

Suppose $q \equiv 1 \pmod 4$ and let $\hat{t} = [\lambda I_{n/2}, \lambda^{-1} I_{n/2}]$ with respect to the decomposition $V = U \oplus W$, where $\lambda \in \mathbb{F}_q$ has order 4. Since $n/2$ is even, Lemma 2.5.7 implies that $\hat{t} \in \Omega_n^+(q)$ and thus $\hat{t}Z \in T$. As $\lambda^2 = -1$, it follows that $\lambda \hat{t} = [-I_{n/2}, I_{n/2}] \in \hat{t}Z$. Clearly, neither $\lambda \hat{t}$ nor $-\lambda \hat{t}$ lies in $O_n^+(q)$, so $\lambda \hat{t}, -\lambda \hat{t} \in \Omega_n^+(q)Z \setminus \Omega_n^+(q)$. Similarly, if we set $\hat{t} = [\lambda I_{n/2}, \lambda^{-1} I_{n/2}]$ with respect to the decomposition $V = U' \oplus W'$, then we obtain a representative in the other $G$-class.

Now assume $q \equiv 3 \pmod 4$ and set $\hat{t} = [-I_{n/2}, I_{n/2}]$ and $t = \hat{t}Z$, again with respect to the decomposition $V = U \oplus W$ (or alternatively, $V = U' \oplus W'$ to obtain an element in the second $G$-class). Neither $\hat{t}$ nor $-\hat{t}$ lies in $O_n^+(q)$. Moreover, there is no scalar $\lambda \in \mathbb{F}_q$ such that $\lambda^{-1} = -\lambda$, so Lemma 2.5.7 implies that no element of $\hat{t}Z$ lies in $O_n^+(q)$. We conclude that $t \in G \setminus \mathrm{PSO}_n^+(q)$.

In both cases, Lemma 2.5.8 implies that $\Omega_n^+(q)$ contains an element interchanging the two eigenspaces of $\hat{t}$, and so $\hat{t}$ and $-\hat{t}$ are $\Omega_n^+(q)$-conjugate.

### 3.5.2.11 Involutions of type $t'_{n/2}, t'_{n/2-1}$, $n \equiv 0 \pmod 4$, $\varepsilon = +$

These involutions arise from the embedding of the centre of the unitary group $\Delta U_{n/2}(q)$ in $\mathrm{GO}_n^+(q)$, as described in Construction 2.5.14. Define the forms $\overline{B}$, $B$ and $Q$ as in Construction 2.5.14 and note that $Q$ is hyperbolic since $n/2$ is even. Let $\xi \in \mathbb{F}_{q^2}$ be an element of order $2(q-1)_2$ and set $\hat{t} = \xi I_{n/2} \in \Delta U_{n/2}(q)$. In terms of the hyperbolic basis

$$\left\{ e_1, \xi e_1, \ldots, e_{n/4}, \xi e_{n/4}, \frac{1}{2} f_1, -\frac{1}{2} \xi^{-1} f_1, \ldots, \frac{1}{2} f_{n/4}, -\frac{1}{2} \xi^{-1} f_{n/4} \right\}$$

for $V$ given in Construction 2.5.14, we have $\hat{t} = [A^{n/4}, (-A^{\mathsf{T}})^{n/4}]$, where $A$ is given in (3.5.2). In particular, $\hat{t}$ fixes a maximal totally singular subspace of $V$, hence the involution $t = \hat{t}Z$ is contained in $G$. In addition, since $A$ and $-A$ are $\mathrm{SL}_2(q)$-conjugate, the usual argument (see Section 3.5.2.7) shows that $\hat{t}$ is $\Omega_n^+(q)$-conjugate to $-\hat{t}$. Since $B = \mathrm{Tr} \circ \overline{B}$, we also observe that

$$B(v, \xi v) = \overline{B}(v, \xi v) + \overline{B}(v, \xi v)^q = \xi^q \overline{B}(v, v) + \xi \overline{B}(v, v) = 0$$

for all $v \in V$, so $\hat{t}$ maps each vector $v$ to an element of $\langle v \rangle^{\perp}$.

Now, for any $v \in V$ we have

$$Q(v\hat{t}) = Q(\xi v) = \overline{B}(\xi v, \xi v) = \xi^{q+1} \overline{B}(v, v) = \xi^{q+1} Q(v)$$

and thus $\tau(\hat{t}) = \xi^{q+1} \in \mathbb{F}_q$, so $\hat{t} \in \mathrm{GO}_n^+(q)$. If $q \equiv 3 \pmod 4$ then $\det(A^{n/4}) = (-\xi^2)^{n/4} = 1$, so Lemma 2.5.7 implies that $\hat{t} \in \Omega_n^+(q)$ and thus $t \in T$. Now, if $q \equiv 1 \pmod 4$ then $\xi^{q+1} \neq 1$ and so $\hat{t} \notin \mathrm{O}_n^+(q)$. Moreover, the order of $\xi^{q+1}$ is divisible by $(q-1)_2$ and thus $\xi^{q+1}$ is an $\mathbb{F}_q$-nonsquare. Therefore, there is no scalar $\lambda \in \mathbb{F}_q^\times$ such that $\lambda \hat{t} \in \mathrm{O}_n^+(q)Z$ and so $t \in G \setminus \mathrm{PSO}_n^+(q)$.

Notice that if $U, W$ are maximal totally singular $\mathbb{F}_q$-subspaces of $V$ arising from suitable $\mathbb{F}_{q^2}$-spaces, then $\dim U - \dim(U \cap W)$ is even and thus $U$ and $W$ belong to the same $\Omega_n^+(q)$-orbit. Therefore, we obtain two distinct $G$-classes of involutions of this type, which are fused in $\widetilde{G}$. We fix a labelling of the two classes so that an involution of type $t'_{n/2}$ is the product of an involution of type $t''_{n/4}$ and an involution of type $t_{n/2}$. Similarly, an involution of type $t'_{n/2-1}$ is the product of an involution of type $t'''_{n/4}$ and one of type $t_{n/2-1}$.

As in Section 3.5.2.7, all maximal totally singular subspaces fixed by an involution $x \in G$ of this form lie in the same $\Omega_n^+(q)$-orbit. In particular, if $x$ fixes a maximal totally singular subspace $U$ and $y \in G$ is another involution of this form that fixes a totally singular subspace $W$, then $x$ and $y$ are $G$-conjugate if and only if $\dim U - \dim(U \cap W)$ is even.

### 3.5.2.12 Involutions of type $t_{n/2}$, $n \equiv 2 \pmod 4$, $\varepsilon = +$

Here $t_{n/2} = \hat{t}Z$, where $\hat{t}$ fixes a pair $U, W$ of maximal totally singular subspaces, so $t_{n/2} \in G$. Since $n/2$ is odd, it follows that $U$ and $W$ belong to distinct $T$-orbits on maximal totally singular subspaces (see Proposition 2.5.4), and this explains why there is a unique $G$-class of elements of this type (in contrast to the two classes that arise when $n/2$ is even, as described in Section 3.5.2.10).

First assume $q \equiv 1 \pmod 4$ and fix $\lambda \in \mathbb{F}_q$ of order 4. Set $\hat{t} = [\lambda I_{n/2}, \lambda^{-1} I_{n/2}]$, with respect to a hyperbolic basis. Then $\hat{t} \in \mathrm{SO}_n^+(q)$, and by applying Lemma 2.5.7 we deduce that $\hat{t} \in \Omega_n^+(q)$ if and only if $\lambda^{n/2}$ is an $\mathbb{F}_q$-square. Since $n/2$ is odd, this holds if and only if $\lambda$ is an $\mathbb{F}_q$-square, that is, if and only if $q \equiv 1 \pmod 8$. The same condition is also necessary for the membership of $-\hat{t}$ in $\Omega_n^+(q)$, whence $t_{n/2} \in T$ if and only if $q \equiv 1 \pmod 8$. Note that $\lambda \hat{t} = [-I_{n/2}, I_{n/2}] \in \hat{t}Z$ does not preserve $Q$, so $\lambda \hat{t} \in \mathrm{SO}_n^+(q)Z \setminus \mathrm{SO}_n^+(q)$.

Now suppose $q \equiv 3 \pmod 4$. Here we set $\hat{t} = [-I_{n/2}, I_{n/2}]$, again with respect to a hyperbolic basis. Since $n/2$ is odd, $\hat{t}Z$ and $-\hat{t}Z$ are not $G$-conjugate, but they are $\widetilde{G}$-conjugate. Clearly, $\hat{t} \notin \mathrm{O}_n^+(q)$ and there is no scalar $\lambda \in \mathbb{F}_q$ such that $\lambda^2 = -1$, so Lemma 2.5.7 implies that no element of $\hat{t}Z$ lies in $\mathrm{O}_n^+(q)$ and thus $t_{n/2} \in G \setminus \mathrm{PSO}_n^+(q)$.

Finally, observe that the eigenspaces of $\hat{t}$ represent the two $\mathrm{SO}_n^+(q)$-orbits on maximal totally singular subspaces. Therefore, $\hat{t}$ and $-\hat{t}$ are $\mathrm{O}_n^+(q)$-conjugate, but they are not $\mathrm{SO}_n^+(q)$-conjugate.

**3.5.2.13 Involutions of type** $t_{n/2}$, $n \equiv 2 \pmod 4$, $\varepsilon = -$
These involutions are obtained by embedding an element $\hat{t}$ of order $2(q-1)_2$
from the centre of $\Delta U_{n/2}(q)$ in $GO_n^-(q)$, as described in Construction 2.5.14.
Define the forms $\overline{B}$ and $Q$ as in Construction 2.5.14, and note that $Q$ is elliptic
since $n/2$ is odd. Set $\hat{t} = \xi I_{n/2} \in \Delta U_{n/2}(q)$ where $\xi \in \mathbb{F}_{q^2}$ has order $2(q-1)_2$.
Note that

$$Q(v\hat{t}) = \overline{B}(\xi v, \xi v) = \xi^{q+1}\overline{B}(v,v) = \xi^{q+1}Q(v)$$

for all $v \in V$, so $\hat{t} \in GO_n^-(q)$ since $\xi^{q+1} \in \mathbb{F}_q$. As in Section 3.5.2.11, we also
observe that $\hat{t}$ maps each vector $v$ to an element of $\langle v \rangle^\perp$.

Let $\{e_1, \ldots, e_{(n-2)/4}, f_1, \ldots, f_{(n-2)/4}, x\}$ be a hermitian basis for $V$ over $\mathbb{F}_{q^2}$
with respect to $\overline{B}$, and recall from Construction 2.5.14 that

$$\left\{ e_1, \xi e_1, \ldots, e_{(n-2)/4}, \xi e_{(n-2)/4}, \frac{1}{2}f_1, -\frac{1}{2}\xi^{-1}f_1, \ldots \right.$$

$$\left. \ldots, \frac{1}{2}f_{(n-2)/4}, -\frac{1}{2}\xi^{-1}f_{(n-2)/4}, x, \xi x \right\}$$

is an elliptic $\mathbb{F}_q$-basis for $V$. We have $\hat{t} = [A^{(n-2)/4}, (-A^\top)^{(n-2)/4}, A]$ with
respect to this basis, where $A$ is defined in (3.5.2). Therefore $\hat{t}$ has determinant
$(-\xi^2)^{n/2}$. Set $t_{n/2} = \hat{t}Z$. We claim that

$$t_{n/2} \in \begin{cases} T & q \equiv 7 \pmod 8 \\ PSO_n^-(q) \setminus T & q \equiv 3 \pmod 8 \\ G \setminus PSO_n^-(q) & q \equiv 1 \pmod 4 \end{cases}$$

First assume $q \equiv 3 \pmod 4$. Here $\xi^2 = -1$ and thus $\hat{t} \in SO_n^-(q)$. Moreover,
$[A^{(n-2)/4}, (-A^\top)^{(n-2)/4}] \in \Omega_{n-2}^+(q)$ by Lemma 2.5.7. Now $SO_2^-(q) \cong C_{q+1}$
and $\Omega_2^-(q) \cong C_{(q+1)/2}$. In particular, if $q \equiv 7 \pmod 8$ then $A \in \Omega_2^-(q)$ and so
Lemma 2.5.9(i) implies that $\hat{t} \in \Omega_n^-(q)$ and thus $t_{n/2} \in T$. Now, if $q \equiv 3 \pmod 8$
then $A \in SO_2^-(q) \setminus \Omega_2^-(q)$ and so Lemma 2.5.9(ii) implies that $\hat{t} \in SO_n^-(q) \setminus$
$\Omega_n^-(q)$. Moreover, $D(Q) = \square$ since $n(q-1)/4$ is odd (see Lemma 2.2.9),
so Lemma 2.5.6 implies that $-I_n \in \Omega_n^-(q)$ and thus $-\hat{t} \in SO_n^-(q) \setminus \Omega_n^-(q)$.
Therefore $\hat{t} \in SO_n^-(q) \setminus \Omega_n^-(q)Z$ and we deduce that $t_{n/2} \in PSO_n^-(q) \setminus T$.

Now assume $q \equiv 1 \pmod 4$. Here $Q(v\hat{t}) = -\xi^2 Q(v)$ and $-\xi^2 \neq 1$, so $\hat{t} \in$
$GO_n^-(q) \setminus O_n^-(q)$. Moreover, $-\xi^2$ is an $\mathbb{F}_q$-nonsquare and thus $\hat{t} \in GO_n^-(q) \setminus$
$O_n^-(q)Z$. Let $V_\# = V \otimes_{\mathbb{F}_q} \mathbb{F}_{q^2}$ and recall the hyperbolic quadratic form $Q_\#$ on $V_\#$
defined in Construction 2.5.12, with associated bilinear form $B_\#$. To avoid any
possible confusion, for $v \in V$ we denote $\xi v$ by $\overline{v}$. Note that

$$Q(x) = \overline{B}(x,x) = 1, \quad Q(\overline{x}) = \overline{B}(\overline{x}, \overline{x}) = \xi^{q+1}$$

Let $\hat{t}_\# \in \mathrm{GL}(V_\#)$ be the linear map that sends each $v \otimes \lambda$ to $(v\hat{t}) \otimes \lambda$, and note that $\hat{t}_\#$ is a similarity of $Q_\#$. Now $\hat{t}_\#$ fixes the totally singular subspace

$$U_\# = \langle e_1 \otimes 1, \bar{e}_1 \otimes 1, \ldots, e_{(n-2)/4} \otimes 1, \bar{e}_{(n-2)/4} \otimes 1 \rangle$$

of dimension $n/2 - 1$. Moreover,

$$
\begin{aligned}
Q_\#(x \otimes \xi + \bar{x} \otimes 1) &= Q_\#(x \otimes \xi) + Q_\#(\bar{x} \otimes 1) + B_\#(x \otimes \xi, \bar{x} \otimes 1) \\
&= \xi^2 Q_\#(x \otimes 1) + Q_\#(\bar{x} \otimes 1) + \xi B_\#(x \otimes 1, \bar{x} \otimes 1) \\
&= \xi^2 Q(x) + Q(\bar{x}) + \xi B(x, \bar{x}) \\
&= \xi^2 + \xi^{q+1} + \xi (\overline{B}(x, \xi x) + \overline{B}(x, \xi x)^q) \\
&= \xi^2 - \xi^2 + \xi(\xi^q + \xi) \\
&= 0
\end{aligned}
$$

Now

$$(x \otimes \xi + \bar{x} \otimes 1)\hat{t}_\# = \bar{x} \otimes \xi + (\xi^2 x) \otimes 1 = \xi(\bar{x} \otimes 1 + x \otimes \xi)$$

and thus $\hat{t}_\#$ fixes the maximal totally singular subspace $\langle U_\#, x \otimes \xi + \bar{x} \otimes 1 \rangle$ of $V_\#$. Therefore $t_{n/2} \in G \setminus \mathrm{PSO}_n^-(q)$ as claimed.

### 3.5.2.14 Involutions of type $\gamma_i$, $1 \leqslant i \leqslant (n+2)/4$, $n$ even

Set $\gamma_i = \hat{t}Z$, where $\hat{t} \in \mathrm{O}_n^\varepsilon(q) \setminus \mathrm{SO}_n^\varepsilon(q)$ is an involution with a parabolic $(-1)$-eigenspace of dimension $2i - 1$. (Note that $Q(e_1 + \lambda f_1) = \lambda$ for all $\lambda \in \mathbb{F}_q$, so it is easy to construct such a subspace from a standard basis of $V$.) Since the only scalars in $\mathrm{O}_n^\varepsilon(q)$ are $\pm I_n$, it follows that no element of $\hat{t}Z$ lies in $\mathrm{SO}_n^\varepsilon(q)$ and thus $\gamma_i \in \mathrm{PO}_n^\varepsilon(q) \setminus \mathrm{PSO}_n^\varepsilon(q)$.

Suppose $i < (n+2)/4$. The $(-1)$-eigenspace of $-\hat{t}$ has dimension $n - 2i + 1 \neq 2i - 1$, so $\hat{t}$ and $-\hat{t}$ are not conjugate. Moreover, the $\mathrm{GO}_n^\varepsilon(q)$-class of $\hat{t}$ splits into two $\mathrm{O}_n^\varepsilon(q)$-classes, which correspond to the two isometry types of parabolic $(2i - 1)$-spaces. Since $T$ acts transitively on the set of parabolic subspaces of a given isometry type (see Lemma 2.5.10) it follows that the two $\mathrm{PO}_n^\varepsilon(q)$-classes are $T$-classes. Therefore, $\gamma_i^G$ splits into two $T$-classes.

Now assume $i = (n+2)/4$, so $n/2$ is odd. Let $U$ be an $\frac{n}{2}$-dimensional parabolic subspace of $V$. If $D(Q) = \square$ (equivalently, if $q \equiv \varepsilon \pmod 4$) then $U$ and $U^\perp$ are isometric, so $\hat{t}$ is $\mathrm{O}_n^\varepsilon(q)$-conjugate to $-\hat{t}$. In this case, the $G$-class of $\gamma_{(n+2)/4}$ splits into two $\mathrm{PO}_n^\varepsilon(q)$-classes, corresponding to the two $\mathrm{O}_n^\varepsilon(q)$-orbits on the set of parabolic $\frac{n}{2}$-spaces. Moreover, Lemma 2.5.10 implies that each $\mathrm{PO}_n^\varepsilon(q)$-class is also a $T$-class. However, if $D(Q) = \boxtimes$ then $U$ and $U^\perp$ are similar, but not isometric, and so $\hat{t}$ and $-\hat{t}$ are $\mathrm{GO}_n^\varepsilon(q)$-conjugate but not $\mathrm{O}_n^\varepsilon(q)$-conjugate. Thus $\hat{t}Z$ contains representatives from both $\mathrm{O}_n^\varepsilon(q)$-classes and so the $G$-class and $\mathrm{PO}_n^\varepsilon(q)$-class of $\gamma_{(n+2)/4}$ coincide. Moreover, since

$T$ is transitive on the set of nondegenerate subspaces of a given isometry type, it follows that $\gamma_{(n+2)/4}^G = \gamma_{(n+2)/4}^T$. To summarise, we have shown that $\gamma_{(n+2)/4}^G = \gamma_{(n+2)/4}^T$ if $q \equiv -\varepsilon \pmod 4$ (see Lemma 2.2.9), otherwise $\gamma_{(n+2)/4}^G$ splits into two $T$-classes.

Note that a parabolic subspace of dimension $2i - 1 > 1$ is the orthogonal sum of a nondegenerate $(2i-2)$-space and a parabolic 1-space. In particular, by fixing a parabolic 1-space and taking either a hyperbolic or elliptic complementary $(2i - 2)$-space, we obtain a space in each of the two $T$-orbits on parabolic $(2i - 1)$-spaces. Therefore, if the $G$-class of $\gamma_i$ splits into two $T$-classes, then we can find involutions $t$ and $t'$ in $G$ of type $t_{i-1}$ and $t'_{i-1}$, respectively, and an involution $\gamma$ of type $\gamma_1$ such that $t\gamma$ and $t'\gamma$ represent the two $T$-classes of involutions of type $\gamma_i$.

### 3.5.2.15 Involutions of type $\gamma_{(n+2)/4}$, $n \equiv 2 \pmod 4$

These involutions arise from the embedding of an element $\hat{t}$ of order $2(q-1)_2$ from the centre of $\mathrm{GO}_{n/2}(q^2)$ in $\mathrm{GO}_n^\varepsilon(q)$, as described in Construction 2.5.13. Define the quadratic forms $\overline{Q}$ and $Q = \mathrm{Tr} \circ \overline{Q}$ as in Construction 2.5.13, where $\overline{Q}$ is parabolic. By varying the discriminant of $\overline{Q}$, the quadratic form $Q$ is either hyperbolic or elliptic. Let $\xi \in \mathbb{F}_{q^2}$ be an element of order $2(q-1)_2$ and set $\hat{t} = \xi I_{n/2} \in \mathrm{GO}_{n/2}(q^2)$. Since $\tau(\hat{t}) = \xi^2 \in \mathbb{F}_q$ is a nonsquare, it follows that $\hat{t}$ embeds in $\mathrm{GO}_n^\varepsilon(q) \setminus \mathrm{O}_n^\varepsilon(q)Z$. Observe that $\hat{t}$ fixes totally singular, hyperbolic and elliptic 2-dimensional subspaces of $V$ over $\mathbb{F}_q$.

Let $\{e_1, \ldots, e_{(n-2)/4}, f_1, \ldots, f_{(n-2)/4}, x\}$ be a parabolic $\mathbb{F}_{q^2}$-basis for $V$ with respect to $\overline{Q}$. In terms of the $\mathbb{F}_q$-basis

$$\{e_1, \xi e_1, \ldots, e_{(n-2)/4}, \xi e_{(n-2)/4}, f_1, \xi f_1 \ldots, f_{(n-2)/4}, \xi f_{(n-2)/4}, x, \xi x\}$$

we have $\hat{t} = [A^{n/2}]$, where $A$ is the matrix in (3.5.2). In particular, $\hat{t}$ fixes the $\mathbb{F}_q$-subspace

$$U = \langle e_1, \xi e_1, \ldots, e_{(n-2)/4}, \xi e_{(n-2)/4} \rangle$$

which is totally singular and has dimension $n/2 - 1$. Since the eigenvalues of $\hat{t}$ lie in $\mathbb{F}_{q^2} \setminus \mathbb{F}_q$, it follows that $\hat{t}$ does not fix any 1-spaces over $\mathbb{F}_q$. Set $t = \hat{t}Z$, which is an involution of type $\gamma_{(n+2)/4}$.

Suppose $Q$ is hyperbolic and let $W$ be a maximal totally singular subspace of $V$ containing $U$. Then $U = W \cap W\hat{t}$ and thus $\hat{t}$ interchanges the two $T$-orbits on maximal totally singular subspaces. Therefore $t \in \widetilde{G} \setminus G$.

Now suppose $Q$ is elliptic. Consider the vector space $V_\# = V \otimes_{\mathbb{F}_q} \mathbb{F}_{q^2}$, equipped with the hyperbolic quadratic form $Q_\#$ given in Construction 2.5.12. For each $v \in V$, denote $\xi v$ by $\overline{v}$. Let $\hat{t}_\# \in \mathrm{GL}(V_\#)$ be the linear map that sends each $v \otimes \lambda$ to $(v\hat{t}) \otimes \lambda$, and note that $\hat{t}_\#$ is a similarity of $Q_\#$. If $i \in \mathbb{F}_{q^2}$ has order 4, then by repeating the calculation in (3.5.3) we deduce that

$$Q_\#(x \otimes (i\xi) + \overline{x} \otimes 1) = 0$$

Moreover, $\hat{t}_\#$ does not fix the $\mathbb{F}_{q^2}$-span of $x \otimes (i\xi) + \overline{x} \otimes 1$. Thus, if we set $W_\# = \langle U_\#, x \otimes (i\xi) + \overline{x} \otimes 1 \rangle$, where $U_\# = U \otimes_{\mathbb{F}_q} \mathbb{F}_{q^2}$, then $W_\#$ is a maximal totally singular subspace of $V_\#$ with respect to $Q_\#$, and we have $U_\# = W_\# \cap W_\# \hat{t}_\#$. Therefore $\hat{t}_\#$ interchanges the two classes of maximal totally singular subspaces with respect to $Q_\#$, so once again we deduce that $t \in \widetilde{G} \setminus G$.

By [86, Proposition 4.3.20], there is a unique $T$-class of extension field subgroups of type $O_{n/2}(q^2)$ when $D(Q) = \boxtimes$, and there are two such classes if $D(Q) = \square$. Therefore, the splitting of the $G$-class of $\gamma'_{(n+2)/4}$ into $T$-classes is the same as for $\gamma_{(n+2)/4}$.

### 3.5.2.16 Involutions of type $\gamma_1^*, \gamma_1^{**}, \gamma_2^*, \gamma_2^{**}$, $(n, \varepsilon) = (8, +)$
These involutions are the images of $\gamma_1$ and $\gamma_2$ under triality graph automorphisms $\tau$ and $\tau^2$ of $T$. Since the $G$-classes of $\gamma_1$ and $\gamma_2$ split into two $T$-classes, so do the $G$-classes of $\gamma_1^*, \gamma_1^{**}, \gamma_2^*$ and $\gamma_2^{**}$. Note that these involutions lie in $\langle \mathrm{PGO}_8^+(q), \tau \rangle \setminus \mathrm{PGO}_8^+(q)$.

### 3.5.2.17 Involutions: a summary
The next proposition summarises some of the above analysis. We refer the reader to [67, Table 4.5.1] for the various centraliser orders listed in Tables B.8, B.10 and B.11 (see Appendix B).

**Proposition 3.5.10** *For $q$ odd, Tables B.8, B.10 and B.11 provide a complete set of representatives of the $G$-classes of inner-diagonal and graph automorphisms $x \in \mathrm{Aut}(\mathrm{P}\Omega_n^\varepsilon(q))$ of order two, together with the order of the corresponding centraliser $C_G(x)$.*

Some additional properties of the involutions in $G$ are recorded in Table B.9. In order to justify the information in this table, it remains to show that $\hat{t}$ and $-\hat{t}$ are $\Omega_n^-(q)$-conjugate when $t$ is of type $t'_{n/4}$ with $n \equiv 0 \pmod 4$, or type $t_{n/2}$ with $n \equiv 2 \pmod 4$. In both cases, the desired result follows from Proposition 3.5.10 as they are the only classes of involutions in $G$ that lift to elements in $\mathrm{GO}_n^\varepsilon(q)$ with eigenvalues in $\mathbb{F}_{q^2} \setminus \mathbb{F}_q$.

From the information in Sections 3.5.2.14–3.5.2.16, we obtain the following result on graph automorphisms of order two.

**Proposition 3.5.11** *Suppose $q$ is odd and $\gamma \in \mathrm{Aut}(T)$ is an involutory graph automorphism. Then one of the following holds:*

(i) $n \equiv 2 \pmod 4$, $q \equiv -\varepsilon \pmod 4$, $\gamma$ *is of type* $\gamma_{(n+2)/4}$ *or* $\gamma'_{(n+2)/4}$, *and* $\gamma^G = \gamma^T$;

(ii) $\gamma^G$ *splits into two T-classes.*

### 3.5.3 Unipotent elements, $p > 2$

Next we turn to the unipotent elements in $\tilde{G} = \mathrm{PGO}_n^\varepsilon(q)$. As observed in the symplectic case, the description of unipotent conjugacy classes differs according to the parity of $p$. In this section we will assume $p$ is odd (see Section 3.5.4 for the case $p = 2$).

As in the symplectic case (see Section 3.4.3), we begin with a tensor product construction. Let $U$ and $W$ be $\mathbb{F}_q$-spaces of dimension $a$ and $b$ respectively, equipped with respective bilinear forms $B_1$ and $B_2$. Then $U \otimes W$ is an $(ab)$-dimensional space endowed with the bilinear form $B$ defined by

$$B(u_1 \otimes w_1, u_2 \otimes w_2) = B_1(u_1, u_2)B_2(w_1, w_2)$$

We consider two cases.

*Case 1. a and b even.* Let $B_1$ and $B_2$ be nondegenerate alternating forms on $U$ and $W$, respectively, so the form $B$ defined above is a nondegenerate symmetric form on $U \otimes W$. As noted in Section 2.2.8, the corresponding quadratic form $Q$ on $U \otimes W$ is hyperbolic and thus $\mathrm{O}_{ab}^+(q)$ contains a tensor product subgroup $\mathrm{Sp}_a(q) \otimes \mathrm{Sp}_b(q)$. This subgroup contains an element $J_a \otimes I_b$ with Jordan form $[J_a^b]$, where each Jordan block of size $a$ fixes a totally singular $a$-space. In particular, such an element fixes a decomposition of $U \otimes W$ into a dual pair of totally singular subspaces of dimension $ab/2$, and it is clearly centralised in $\mathrm{O}_{ab}^+(q)$ by $I_a \otimes \mathrm{Sp}_b(q)$.

*Case 2. a odd.* Now assume $B_1$ and $B_2$ are nondegenerate symmetric bilinear forms, so that $B_1$ corresponds to a parabolic quadratic form $Q_1$ on $U$, and $B_2$ to a quadratic form $Q_2$ on $W$ of type $\varepsilon \in \{\pm, \circ\}$. The above form $B$ on $U \otimes W$ is symmetric, and the corresponding quadratic form $Q$ is of type $\varepsilon$ (see Table 2.2.2). Moreover,

$$D(Q) = D(Q_1)^b D(Q_2)^a$$

(see (2.2.10)).

It follows that $\mathrm{O}_{ab}^\varepsilon(q)$ contains a tensor product subgroup of type $\mathrm{O}_a(q) \otimes \mathrm{O}_b^\varepsilon(q)$. This subgroup contains an element $J_a \otimes I_b$ with Jordan form $[J_a^b]$, which fixes an orthogonal decomposition of $U \otimes W$ into $b$ nondegenerate $a$-spaces since $W$ has an orthogonal basis. An element of this form is clearly centralised in $\mathrm{O}_{ab}^\varepsilon(q)$ by $I_a \otimes \mathrm{O}_b^\varepsilon(q)$.

Now consider a unipotent element $x \in O_n^\varepsilon(q)$ with Jordan form $[J_{a_1}^{b_1}, J_{a_2}^{b_2}]$, where $n = a_1 b_1 + a_2 b_2$ and $a_1, a_2$ are distinct odd integers. Let $V$ be the natural module for $O_n^\varepsilon(q)$ and let $Q$ be the underlying quadratic form on $V$. Then $x$ fixes an orthogonal decomposition $V = (U_1 \otimes W_1) \perp (U_2 \otimes W_2)$ with $\dim U_i = a_i$ and $\dim W_i = b_i$. There are essentially two distinct decompositions of $V$ of this form: the discriminant of the restriction of $Q$ to $U_1 \otimes W_1$ can be chosen arbitrarily, and this choice uniquely determines the discriminant on $U_2 \otimes W_2$. Since no element of $O_n^\varepsilon(q)$ can take one such decomposition to another, there must be two $O_n^\varepsilon(q)$-classes of unipotent elements of the form $[J_{a_1}^{b_1}, J_{a_2}^{b_2}]$. We label the two classes by $[J_{a_1}^{b_1, \delta_1}, J_{a_2}^{b_2, \delta_2}]$, where $\delta_i \in \{\square, \boxtimes\}$ denotes the discriminant of the restriction of $Q$ to $U_i \otimes W_i$, and $D(Q) = \delta_1 \delta_2$. Notice that these classes are fused in $GO_n^\varepsilon(q)$ if and only if $b_1$ and $b_2$ are both odd.

Our main result on unipotent classes in orthogonal groups is Proposition 3.5.12 below. In the statement, $\mathscr{Q}^\delta(n, p)$ denotes the set of nontrivial *labelled partitions* of $n$ of the form

$$(\delta_p p^{a_p}, (p-1)^{a_{p-1}}, \ldots, 2^{a_2}, \delta_1 1^{a_1})$$

where $\sum_i i a_i = n$, $a_i$ is even for all even $i$, $\delta_{2j+1} \in \{\square, \boxtimes\}$ and $\prod_j \delta_{2j+1} = \delta$ (so $(\delta 1^n)$ is the trivial labelled partition of $n$). Here $\delta = D(Q)$ is the discriminant of the quadratic form $Q$ with isometry group $O_n^\varepsilon(q)$ (if $n$ is even, then $\delta$ can be computed via Lemma 2.2.9). If $\varepsilon = -$ then we impose one more condition on the labelled partitions in $\mathscr{Q}^\delta(n, p)$: there exists an odd integer $i$ with $a_i > 0$.

**Proposition 3.5.12** *Let $x = \hat{x}Z \in \widetilde{G}$ be an element of order $p$, where $p$ is odd. Then $x \in T$ and $\hat{x}$ fixes an orthogonal decomposition of the form*

$$V = \bigoplus_{i=1}^{(p+1)/2} \left(U_{2i-1,1} \perp \ldots \perp U_{2i-1,a_{2i-1}}\right) \perp$$
$$\bigoplus_{i=1}^{(p-1)/2} \left((U_{2i,1} \oplus U_{2i,1}^*) \perp \ldots \perp (U_{2i,a_{2i}/2} \oplus U_{2i,a_{2i}/2}^*)\right)$$

*where each $U_{2i-1,k}$ is a nondegenerate $(2i-1)$-space on which $\hat{x}$ acts indecomposably, and $\{U_{2i,k}, U_{2i,k}^*\}$ is a dual pair of totally singular $2i$-spaces on which $\hat{x}$ acts indecomposably. Furthermore, $D(Q|_{U_{2i-1}}) = \delta_{2i-1}$, where*

$$U_{2i-1} = U_{2i-1,1} \perp \ldots \perp U_{2i-1,a_{2i-1}}$$

*In addition, the map*

$$\theta : (\delta_p p^{a_p}, (p-1)^{a_{p-1}}, \ldots, 2^{a_2}, \delta_1 1^{a_1}) \longmapsto ([J_p^{a_p, \delta_p}, J_{p-1}^{a_{p-1}}, \ldots, J_1^{a_1, \delta_1}]Z)^{PO_n^\varepsilon(q)}$$

*is a bijection from $\mathscr{D}^\delta(n,p)$ to the set of $\mathrm{PO}_n^\varepsilon(q)$-classes of elements of order*
*p in T, where $\delta$ is the discriminant of the underlying quadratic form Q.*

*Proof* First observe that $x \in T$ since $|\widetilde{G} : T|$ is indivisible by $p$. By Lemma
3.5.1, $\hat{x}$ has Jordan form $[J_p^{a_p}, \ldots, J_1^{a_1}]$, where $\lambda = (p^{a_p}, \ldots, 1^{a_1})$ is a partition
of $n$ such that $a_i$ is even for all even $i$.

If $i$ is even, then Lemma 2.2.17 implies that $\mathrm{O}_{2i}^+(q)$ contains an element with
Jordan form $[J_i^2]$, which fixes a dual pair of maximal totally singular subspaces
of the natural $\mathrm{O}_{2i}^+(q)$-module. Note that the quadratic form corresponding to
$\mathrm{O}_{2i}^+(q)$ has a square discriminant since $2i(q-1)/4$ is even. From the discussion
in Case 2 above, we see that each labelled partition

$$(\delta_p p^{a_p}, (p-1)^{a_{p-1}}, \ldots, 2^{a_2}, \delta_1 1^{a_1}) \in \mathscr{D}^\delta(n,p)$$

corresponds to an element $[J_p^{a_p, \delta_p}, J_{p-1}^{a_{p-1}}, \ldots, J_1^{a_1, \delta_1}] \in \mathrm{O}_n^\varepsilon(q)$ of order $p$, and the
map $\theta$ is injective.

Set $k = |\{i \mid i \text{ odd}, a_i > 0\}|$. If $k = 0$ then $n$ is even and either $\varepsilon = +$ and there
is a unique $\mathrm{PO}_n^\varepsilon(q)$-class of unipotent elements with Jordan form $[J_p^{a_p}, \ldots, J_1^{a_1}]$,
or $\varepsilon = -$ and there are no such classes (see [96, Theorem 7.1]). If $k > 0$, then
the injectivity of $\theta$ implies that there are at least $2^{k-1}$ distinct $\mathrm{PO}_n^\varepsilon(q)$-classes
of elements with Jordan form $[J_p^{a_p}, \ldots, J_1^{a_1}]$. But [96, Theorem 7.1] states that
there are precisely $2^{k-1}$ such classes. Therefore, $\theta$ is surjective and $\hat{x}$ is $\mathrm{O}_n^\varepsilon(q)$-
conjugate to one of the unipotent elements constructed above. In particular, $\hat{x}$
fixes an orthogonal decomposition of $V$ as described in the statement of the
proposition. □

**Lemma 3.5.13** *Suppose p is odd. Let $x \in \mathrm{O}_n^\varepsilon(q)$ be a unipotent element*
*corresponding to the labelled partition $(\delta_p p^{a_p}, (p-1)^{a_{p-1}}, \ldots, 2^{a_2}, \delta_1 1^{a_1})$.*
*Then $C_{\mathrm{O}_n^\varepsilon(q)}(x) = QR$ and*

$$|Q| = q^\gamma \quad \text{with} \quad \gamma = \sum_{i<j} i a_i a_j + \frac{1}{2} \sum_i (i-1) a_i^2 - \frac{1}{2} \sum_{i \text{ even}} a_i$$

$$R = \prod_{i \text{ odd}} \mathrm{O}_{a_i}^{\varepsilon_i}(q) \times \prod_{i \text{ even}} \mathrm{Sp}_{a_i}(q)$$

*where $\mathrm{O}_{a_i}^\pm(q) = \mathrm{O}_{a_i}(q)$ if $a_i$ is odd, and*

$$\varepsilon_i = \begin{cases} (-)^{i a_i (q-1)/4} & \delta_i = \square \\ (-)^{1 + i a_i (q-1)/4} & \delta_i = \boxtimes \end{cases} \tag{3.5.4}$$

*if $a_i$ is even.*

*Proof* This follows from [96, Theorem 7.1], combined with the comments in Cases 1 and 2 above. The signs $\varepsilon_i$ (with $a_i$ even) given in (3.5.4) can be computed via Lemma 2.2.9. □

Proposition 3.5.12 is stated in terms of $\mathrm{PO}_n^\varepsilon(q)$-classes; the next result deals with conjugacy in the other almost simple orthogonal groups that we are interested in.

**Proposition 3.5.14** *Let $x = \hat{x}Z \in \widetilde{G}$ be an element of order $p$, where $p$ is odd and $x$ corresponds to the labelled partition $\lambda = (\delta_p p^{a_p}, \ldots, 2^{a_2}, \delta_1 1^{a_1})$.*

(i) *$x^{\mathrm{PSO}_n^\varepsilon(q)}$ splits into two $T$-classes if and only if $a_i \leqslant 1$ for all odd $i$, with $\varepsilon = (-)^{n(q-1)/4}$ if $n$ is even.*

(ii) *Furthermore, if $n$ is even then the following hold:*
   (a) *$x^{\mathrm{PO}_n^\varepsilon(q)} = x^{\widetilde{G}}$ if and only if $a_i$ is even for all $i$;*
   (b) *$x^{\mathrm{PSO}_n^\varepsilon(q)} = x^G$ if and only if $a_i$ is even for all $i$;*
   (c) *$x^{\mathrm{PSO}_n^\varepsilon(q)} = x^{\mathrm{PO}_n^\varepsilon(q)}$ if and only if $a_i > 0$ for some odd $i$.*
   *In particular, either $x^{\mathrm{PO}_n^\varepsilon(q)} = x^{\widetilde{G}}$ or one of the following holds:*
   (a)' *$a_i = 0$ for all odd $i$, in which case $x^{\mathrm{PO}_n^\varepsilon(q)}$ is a union of two $G$-classes;*
   (b)' *$a_i$ is odd for some odd $i$, in which case $x^G$ is a union of two $\mathrm{PO}_n^\varepsilon(q)$-classes.*

*Proof* Part (i) is [96, Proposition 7.2]. Note that if $n$ is even, then the additional requirement $\varepsilon = (-)^{n(q-1)/4}$ is equivalent to the condition $\mathrm{PSO}_n^\varepsilon(q) \neq T$.

Now consider (ii), so $n$ is even. Consider the orthogonal decomposition of $V$ in Proposition 3.5.12, which is fixed by $\hat{x}$. If $i$ is odd then set

$$V_i = U_{i,1} \perp \ldots \perp U_{i,a_i}$$

where each $U_{i,j}$ is a nondegenerate $i$-space, and if $i$ is even we set

$$V_i = (U_{i,1} \oplus U_{i,1}^*) \perp \ldots \perp (U_{i,a_i/2} \oplus U_{i,a_i/2}^*)$$

where each $\{U_{i,j}, U_{i,j}^*\}$ is a dual pair of totally singular $i$-spaces. Note that $D(Q|_{V_i}) = \delta_i$ if $i$ is odd. Then

$$V = V_1 \perp \ldots \perp V_p \tag{3.5.5}$$

is an orthogonal decomposition of $V$ that is fixed by $\hat{x}$.

If each $a_i$ is even then the $\mathrm{PO}_n^\varepsilon(q)$-class of $x^g$ has the same labelled partition $\lambda$ for each $g \in \widetilde{G}$, whence $x^{\mathrm{PO}_n^\varepsilon(q)} = x^{\widetilde{G}}$. Now assume $a_i$ is odd for some (odd) $i$ and suppose $g \in \mathrm{GO}_n^\varepsilon(q)$ centralises $\hat{x}$. Then $g$ fixes the decomposition in (3.5.5) and so $g|_{V_i}$ is a similarity of $Q|_{V_i}$. Since $\tau(g|_{V_i}) = \tau(g)$, Lemma 2.5.2

implies that $\tau(g)$ is an $\mathbb{F}_q$-square. By the same lemma, there exists an element $h \in \mathrm{GO}_n^\varepsilon(q)$ such that $\tau(h)$ is a nonsquare in $\mathbb{F}_q$, so $g$ lies in the index-two subgroup $\mathrm{O}_n^\varepsilon(q)Z$ of $\mathrm{GO}_n^\varepsilon(q)$. It follows that $\widetilde{G}$ fuses the $\mathrm{PO}_n^\varepsilon(q)$-classes corresponding to the labelled partitions $\lambda$ (namely, the $\mathrm{PO}_n^\varepsilon(q)$-class of $x$) and $(\delta_p' p^{a_p}, \ldots, 2^{a_2}, \delta_1' 1^{a_1})$, where $\delta_j' = \delta_j$ if $a_j$ is even, and $\delta_j' = \boxtimes \delta_j$ if $a_j$ is odd. This establishes part (ii)(a). Since the elements in $G \setminus \mathrm{PSO}_n^\varepsilon(q)$ are not isometries, and they lift to elements in $\mathrm{GO}_n^\varepsilon(q) \setminus \mathrm{O}_n^\varepsilon(q)Z$, the same argument yields (ii)(b).

Next assume that $a_i = 0$ for all odd $i$. Then $\varepsilon = +$, $n \equiv 0 \pmod 4$ and $\hat{x}$ fixes a decomposition $V = W \oplus W^*$ into totally singular subspaces such that $\hat{x}|_W$ and $\hat{x}|_{W^*}$ have Jordan form $[J_{p-1}^{a_{p-1}/2}, \ldots, J_2^{a_2/2}]$. Note that $W$ and $W^*$ lie in the same $\mathrm{SO}_n^+(q)$-orbit on maximal totally singular subspaces. Moreover, if $\hat{x}$ fixes any other decomposition $V = U \oplus U^*$ into totally singular subspaces, then $\dim W - \dim(W \cap U)$ is even, and so $U$ and $W$ are in the same $\mathrm{SO}_n^+(q)$-orbit. It follows that $x^{\mathrm{PO}_n^+(q)}$ splits into two $\mathrm{PSO}_n^+(q)$-classes.

Finally, if $a_i > 0$ for some odd $i$ then $\hat{x}|_{V_i} = J_i \otimes I_{a_i}$ is centralised by $I_i \otimes t$, where $t \in \mathrm{O}_{a_i}^{\varepsilon'}(q)$ is an involution of type $\gamma_1$ (see Section 3.5.2.14). Now $I_i \otimes t$ has determinant $(-1)^i = -1$, so $\hat{x}$ is centralised by an element of $\mathrm{O}_n^\varepsilon(q) \setminus \mathrm{SO}_n^\varepsilon(q)$. This implies that $x^{\mathrm{PO}_n^\varepsilon(q)} = x^{\mathrm{PSO}_n^\varepsilon(q)}$, so part (ii)(c) holds. $\qquad\square$

**Remark 3.5.15** As observed in the proof of Proposition 3.5.14, if $a_i$ is odd for some odd $i$ then the $\mathrm{PO}_n^\varepsilon(q)$-classes corresponding to the labelled partitions

$$(\delta_p p^{a_p}, \ldots, 2^{a_2}, \delta_1 1^{a_1}), \quad (\delta_p' p^{a_p}, \ldots, 2^{a_2}, \delta_1' 1^{a_1})$$

are fused in $\mathrm{PGO}_n^\varepsilon(q)$, where $\delta_j' = \delta_j$ if $a_j$ is even, and $\delta_j' = \boxtimes \delta_j$ if $a_j$ is odd.

### 3.5.4 Unipotent involutions

Now let us consider involutions in $\widetilde{G}$ when $p = 2$. Here $n$ is even, $G = T = \Omega_n^\varepsilon(q)$, $\widetilde{G} = \mathrm{O}_n^\varepsilon(q)$ and $\widetilde{G}/T \cong C_2$. Since $\mathrm{O}_n^\varepsilon(q) < \mathrm{Sp}_n(q)$, the analysis of the conjugacy classes of involutions in $\mathrm{O}_n^\varepsilon(q)$ is similar to the symplectic case considered in Section 3.4.4. A complete description is given by Aschbacher and Seitz [5], which we briefly recall below. As before, let $Q$ be the underlying nondegenerate quadratic form on $V$, with associated bilinear form $B_Q$.

First let us introduce some notation. Given a hyperbolic 4-space $V_i$ with standard basis $\{e_{i1}, e_{i2}, f_{i1}, f_{i2}\}$, let $g_i \in \mathrm{O}^+(V_i)$ be the involution that interchanges $e_{i1}$ and $e_{i2}$, and $f_{i1}$ and $f_{i2}$. For a hyperbolic 2-space $Y^+$ with standard basis $\{e_1, f_1\}$, let $h^+ \in O^+(Y^+)$ be the involution that interchanges $e_1$ and $f_1$. Similarly, if $Y^-$ is an elliptic 2-space then we can choose a basis $\{x, y\}$ such that $Q(x) = Q(y) = 1$ (since $q$ is even), and then we let $h^- \in O^-(Y^-)$ be the involution that interchanges $x$ and $y$.

For each even integer $s = 2\ell \leqslant n/2$ (with $s < n/2$ if $\varepsilon = -$), we have a decomposition

$$V = V_1 \perp V_2 \perp \ldots \perp V_\ell \perp W$$

where each $V_i$ is a hyperbolic 4-space and $W$ is a nondegenerate subspace of type $\varepsilon$ and dimension $n - 2s$. In terms of this decomposition, set

$$a_s = [g_1, \ldots, g_\ell, I_{n-2s}] \in O_n^\varepsilon(q)$$

Then $a_s$ has Jordan form $[J_2^s, J_1^{n-2s}]$ and $B_Q(v, va_s) = 0$ for all $v \in V$. We refer to the involutions in $a_s^{O_n^\varepsilon(q)}$ as being of *type $a_s$*. Note that $a_{n/2}$ is not defined when $\varepsilon = -$.

Similarly, for each even integer $s = 2(\ell + 1) \leqslant n/2$ we can consider a decomposition

$$V = Y^+ \perp Y^\varepsilon \perp V_1 \perp V_2 \perp \ldots \perp V_\ell \perp W$$

where each $V_i$ is a hyperbolic 4-space and $W$ is a hyperbolic subspace of dimension $n - 2s$. Let

$$c_s = [h^+, h^\varepsilon, g_1, \ldots, g_\ell, I_{n-2s}] \in O_n^\varepsilon(q)$$

Then $c_s$ has Jordan form $[J_2^s, J_1^{n-2s}]$ and we note that $B_Q(e_1, e_1 c_s) = 1$ for $e_1 \in Y^+$. Thus $c_s$ is not of type $a_s$. Involutions that are $O_n^\varepsilon(q)$-conjugate to $c_s$ are said to be of *type $c_s$*.

Finally, for each odd integer $s = 2\ell + 1 \leqslant n/2$ consider the decomposition

$$V = Y^\varepsilon \perp V_1 \perp V_2 \perp \ldots \perp V_\ell \perp W$$

where each $V_i$ is a hyperbolic 4-space and $W$ is a hyperbolic subspace of dimension $n - 2s$. Set

$$b_s = [h^\varepsilon, g_1, \ldots, g_\ell, I_{n-2s}] \in O_n^\varepsilon(q)$$

so $b_s$ has Jordan form $[J_2^s, J_1^{n-2s}]$ and thus it is neither of type $a_s$ nor of type $c_s$ (since $s$ is odd). Note that there exists $v \in Y^\varepsilon$ such that $B_Q(v, vb_s) = 1$. We say that involutions in $b_s^{O_n^\varepsilon(q)}$ are of *type $b_s$*.

The following result is proved in [5, Section 8].

Table 3.5.1 *Involutions in* $O_n^\varepsilon(q)$, $p = 2$

| $x$ | Conditions | $|C_G(x)|$ |
|---|---|---|
| $a_s$ | $s = 2\ell, 1 \leqslant \ell \leqslant m$ | $q^{ns-3s^2/2-s/2}|\mathrm{Sp}_s(q)||\Omega_{n-2s}^\varepsilon(q)|$ |
| $b_s$ | $s = 2\ell - 1, 1 \leqslant \ell \leqslant (n+2)/4$ | $q^{n(s-1)-3s^2/2+3s/2}|\mathrm{Sp}_{s-1}(q)||\mathrm{Sp}_{n-2s}(q)|$ |
| $c_s$ | $s = 2\ell, 1 \leqslant \ell \leqslant n/4$ | $q^{n(s-1)-3s^2/2+5s/2-1}|\mathrm{Sp}_{s-2}(q)||\mathrm{Sp}_{n-2s}(q)|$ |

$G = \mathrm{Inndiag}(\mathrm{P}\Omega_n^\varepsilon(q)) = \Omega_n^\varepsilon(q)$
In the first row, $m = n/4$ if $\varepsilon = +$, otherwise $m = (n-2)/4$

**Proposition 3.5.16** *Suppose* $p = 2$, $n$ *is even and* $x \in \widetilde{G}$ *is an involution with Jordan form* $[J_2^s, J_1^{n-2s}]$. *If* $s$ *is even then* $x \in T$ *and* $x$ *is of type* $a_s$ *or* $c_s$, *otherwise* $x \in \widetilde{G} \setminus T$ *is of type* $b_s$. *Furthermore,* $|C_G(x)|$ *is given in Table 3.5.1.*

**Remark 3.5.17** Recall that $O_n^\varepsilon(q) < \mathrm{Sp}_n(q)$ when $p = 2$. It is clear from the definitions that an involution of type $a_s$ in $O_n^\varepsilon(q)$ is also an $a_s$-type involution when viewed as an element of $\mathrm{Sp}_n(q)$. Similarly, involutions of type $b_s$ and $c_s$ correspond accordingly. In particular, the conclusion of Lemma 3.4.14 holds for involutions in a subgroup $O_n^{\varepsilon'}(q) \times O_m^{\varepsilon''}(q) < O_{n+m}^\varepsilon(q)$.

**Remark 3.5.18** Suppose $p = 2$ and let $x \in \widetilde{G}$ be an involution. Then either $x^T = x^{\widetilde{G}}$, or $\varepsilon = +$, $n \equiv 0 \pmod 4$ and $x$ is of type $a_{n/2}$ (see [5, (8.12)]). In the exceptional case, $x^{\widetilde{G}}$ splits into two $T$-classes, corresponding to the two $T$-orbits on the set of maximal totally singular subspaces of $V$ (see Proposition 2.5.4). If $x$ is an $a_{n/2}$-type involution, then by definition $x$ fixes complementary maximal totally singular subspaces, say

$$U = \langle e_1, \ldots, e_{n/2} \rangle, \quad W = \langle f_1, \ldots, f_{n/2} \rangle$$

Moreover, if $U'$ is any maximal totally singular subspace fixed by $x$, then $\dim U - \dim(U \cap U')$ is even, so $U$ and $U'$ are in the same $T$-orbit. However, it is easy to construct another $a_{n/2}$-type involution $x'$ that fixes

$$U'' = \langle f_1, e_2, \ldots, e_{n/2} \rangle$$

in which case $x$ and $x'$ are not $T$-conjugate since $\dim U - \dim(U \cap U'') = 1$. We write $a_{n/2}$ and $a'_{n/2}$ for representatives of the two $T$-classes that arise in this situation.

In contrast, note that if $\varepsilon = +$ and $x = c_{n/2}$ then $x$ fixes maximal totally singular subspaces in both $T$-orbits. For example, if we fix a standard hyperbolic basis $\{e_1, \ldots, e_{n/2}, f_1, \ldots, f_{n/2}\}$, then we may assume that $x$ acts on $V$ by interchanging basis vectors as follows:

$$e_1 \leftrightarrow f_1, \quad e_2 \leftrightarrow f_2, \quad e_{2i-1} \leftrightarrow e_{2i}, \quad f_{2i-1} \leftrightarrow f_{2i}$$

for $2 \leqslant i \leqslant n/4$. Therefore, $x$ fixes the totally singular subspaces

$$\langle e_1 + e_2, f_1 + f_2, e_3, \dots, e_{n/2} \rangle$$

and

$$\langle e_1 + e_2, f_1 + f_2, e_3, \dots, e_{n/2-2}, e_{n/2-1} + e_{n/2}, f_{n/2-1} + f_{n/2} \rangle$$

We conclude that $x^T = x^{\widetilde{G}}$ in this case.

Let $m$ and $n$ be positive integers divisible by 4. For $x \in O_n^+(q)$ and $y \in O_m^+(q)$, let $[x, y]$ be the corresponding block-diagonal matrix in a subgroup

$$O_n^+(q) \times O_m^+(q) < O_{n+m}^+(q)$$

If $x$ is $\Omega_n^+(q)$-conjugate to $a_{n/2}$, and $y$ is $\Omega_m^+(q)$-conjugate to $a_{m/2}$, then we adopt the convention that $[x, y]$ is $\Omega_{n+m}^+(q)$-conjugate to $a_{(n+m)/2}$. It follows that the $\Omega_{n+m}^+(q)$-class of $[x, y]$, for $x \in \{a_{n/2}, a'_{n/2}\}$ and $y \in \{a_{m/2}, a'_{m/2}\}$, is as follows:

|  | $y = a_{m/2}$ | $a'_{m/2}$ |
|---|---|---|
| $x = a_{n/2}$ | $a_{(n+m)/2}$ | $a'_{(n+m)/2}$ |
| $a'_{n/2}$ | $a'_{(n+m)/2}$ | $a_{(n+m)/2}$ |

**Remark 3.5.19** Suppose $p = 2$. Note that $\widetilde{G} = O_8^+(q)$ is not normalised by a triality automorphism $\tau$ of $T = \Omega_8^+(q)$. In particular, since $b_1, b_3 \in O_8^+(q) \setminus \Omega_8^+(q)$, their images $b_1^*, b_1^{**}$ and $b_3^*, b_3^{**}$ under $\tau$ and $\tau^2$ are involutions in $\mathrm{Aut}(T) \setminus \widetilde{G}$. In particular, there are exactly eleven $T$-classes of involutions in $\mathrm{Aut}(T)$, namely,

$$a_2, a_4, a'_4, b_1, b_1^*, b_1^{**}, b_3, b_3^*, b_3^{**}, c_2, c_4$$

### 3.5.5 Field automorphisms

The next proposition describes the $G$-classes of prime order field automorphisms of $T$ (recall that a *standard* field automorphism $\phi \in \mathrm{Aut}(T)$ is defined in Section 2.5, with respect to an appropriate basis of $V$).

**Proposition 3.5.20** *Let $x \in \mathrm{Aut}(T)$ be a field automorphism of prime order $r$. Then $r$ divides $f$, say $f = mr$, and $x$ is $G$-conjugate to exactly one of the standard field automorphisms $\phi^{jm}$ with $1 \leqslant j < r$. Moreover, $C_G(x) \cong$ Inndiag$(P\Omega_n^\varepsilon(q^{1/r}))$ and $x^G$ splits into $c$ distinct $T$-classes, where*

$$c = \begin{cases} 1 & r \text{ odd or } p = 2 \\ 2 & n \text{ odd, or } n \equiv 2 \ (\mathrm{mod}\ 4) \text{ and } q^{1/2} \equiv 1 \ (\mathrm{mod}\ 4) \\ 4 & \text{otherwise} \end{cases}$$

*Proof* This is similar to the proof of Proposition 3.4.15. First recall that $r$ is odd if $\varepsilon = -$. The given centraliser structure follows from [67, Proposition 4.9.1(b)], and the stated $G$-class splitting is computed via [86, Propositions 4.5.8 and 4.5.10]. □

### 3.5.6 Graph-field automorphisms

As noted in Section 2.5, the only simple orthogonal groups that admit graph-field automorphisms are those of type $\mathrm{P}\Omega_n^+(q)$. Here we focus on graph-field automorphisms of order 2; those of order 3 (which can only arise if $n = 8$) will be discussed in Section 3.5.7.

Suppose that $q = p^f$ for a prime $p$ and even integer $f$. Fix a hyperbolic basis $\{e_1, \ldots, e_{n/2}, f_1, \ldots, f_{n/2}\}$ and let $\gamma \in \mathrm{PO}_n^+(q)$ be the involution that interchanges $e_1$ and $f_1$, and fixes the remaining basis vectors. Note that $\gamma$ is of type $\gamma_1$ if $q$ is odd (see Section 3.5.2.14), and $\gamma$ is of type $b_1$ if $q$ is even (see Section 3.5.4). Also observe that $\gamma$ commutes with the standard field automorphism $\phi^{f/2}$, so $\gamma\phi^{f/2}$ is an involution. In addition, if $n = 8$ then let $\gamma^*$ and $\gamma^{**}$ denote the images of $\gamma$ under triality automorphisms $\tau$ and $\tau^2$ (see Section 3.5.2.16 and Remark 3.5.19).

**Proposition 3.5.21** *Let $T = \mathrm{P}\Omega_n^+(q)$ and let $x \in \mathrm{Aut}(T)$ be an involutory graph-field automorphism. Then $f$ is even and one of the following holds:*

(i) *$n > 8$ and $x$ is $G$-conjugate to $\gamma\phi^{f/2}$;*
(ii) *$n = 8$ and $x$ is $G$-conjugate to one of $\gamma\phi^{f/2}$, $\gamma^*\phi^{f/2}$, or $\gamma^{**}\phi^{f/2}$.*

*Moreover, $C_G(x) \cong \mathrm{Inndiag}(\mathrm{P}\Omega_n^-(q^{1/2}))$ and $x^G$ splits into $c$ distinct $T$-classes, where*

$$c = \begin{cases} 1 & p = 2 \\ 4 & n \equiv 2 \ (\mathrm{mod}\ 4) \text{ and } q^{1/2} \equiv 1 \ (\mathrm{mod}\ 4) \\ 2 & \text{otherwise} \end{cases}$$

*Proof* By definition, $x$ is $\mathrm{Aut}(T)$-conjugate to $\gamma\phi^{f/2}$. The structure of $C_G(x)$ is given in [67, Proposition 4.9.1(b)], and the $G$-class splitting can be computed via [86, Proposition 4.5.10]. For $n > 8$, $\text{Aut}(T) = \langle G, \gamma, \phi \rangle$ with $[\gamma, \phi] = 1$, so it follows that the $\text{Aut}(T)$-class and the $G$-class of $\gamma \phi^{f/2}$ coincide. However, if $n = 8$ then

$$\text{Aut}(T)/G \cong S_3 \times C_f \quad (3.5.6)$$

and the desired conclusion follows since none of the elements $\gamma, \gamma^*$ and $\gamma^{**}$ are $G$-conjugate. $\qquad \square$

### 3.5.7 Triality automorphisms

Let $T = \text{P}\Omega_8^+(q)$ where $q = p^f$. Recall that there is a triality $\tau$ of the polar space associated to $V$ that cyclically permutes the subspace collections $\mathscr{P}_1$, $\mathscr{U}_1$ and $\mathscr{U}_2$, where $\mathscr{P}_1$ is the set of totally singular 1-spaces, and $\mathscr{U}_1$, $\mathscr{U}_2$ are the two $T$-orbits of maximal totally singular subspaces of $V$ (see Proposition 2.5.4). Now $\tau$ induces a *triality automorphism* of $T$, and more generally, any automorphism that cyclically permutes the sets $\mathscr{P}_1$, $\mathscr{U}_1$ and $\mathscr{U}_2$ is called a triality automorphism. The next proposition, which is [66, Theorem 9.1], concerns the triality automorphisms of prime order.

**Proposition 3.5.22** *Let* $T = \text{P}\Omega_8^+(q)$ *and let* $x \in \text{Aut}(T)$ *be a triality automorphism of prime order* $r$. *Then* $r = 3$ *and up to* $G$-*conjugacy, one of the following holds:*

(i) *x is a graph automorphism and one of the following holds:*
   (a) $C_T(x) \cong G_2(q)$;
   (b) $q \equiv \varepsilon \pmod 3$ *and* $C_T(x) \cong \text{PGL}_3^\varepsilon(q)$;
   (c) $p = 3$ *and* $C_T(x) \cong [q^5].\text{SL}_2(q)$.
(ii) $q = q_0^3$, *x is a graph-field automorphism and* $C_T(x) \cong {}^3D_4(q_0)$.

Set $d = (2, q - 1)$ and let $x \in \text{Aut}(T)$ be a triality graph automorphism of order 3. In view of Proposition 3.5.22(i), we say that $x$ is a $G_2$-*type triality* if $C_T(x) \cong G_2(q)$, otherwise $x$ is a *non-$G_2$ triality*. Let $\tau \in \text{Aut}(T)$ be a standard $G_2$-type triality graph automorphism arising from the order 3 rotational symmetry of the corresponding Dynkin diagram of type $D_4$. We may assume that $[\tau, \phi] = 1$, where $\phi$ is a standard field automorphism of order $f$.

By [66, Theorem 9.1], there are precisely four $G$-classes of triality graph automorphisms of order 3 in $\text{Aut}(T)$, two of each type (note that $x$ and $x^{-1}$ have the same type and (3.5.6) implies that they are $\widetilde{G}$-conjugate, but not $G$-conjugate). In addition, $C_G(x) = C_T(x)$ so each $G$-class splits into $d^2$ $T$-classes. Similarly, if $q = q_0^3$ then there are also four $G$-classes of triality graph-field automorphisms of order 3 in $\text{Aut}(T)$ (the elements $\tau\phi^{f/3}$, $\tau^{-1}\phi^{f/3}$, $\tau\phi^{2f/3}$ and $\tau^{-1}\phi^{2f/3}$ belong to distinct cosets of $G$ in $\text{Aut}(T)$, so they represent the four $G$-classes). Again, $C_G(x) = C_T(x)$ and thus each $G$-class splits into $d^2$ $T$-classes.

We will write $\tau_1^{\pm 1}$ to denote representatives of the two $G$-classes of $G_2$-type triality graph automorphisms of $T$. Similarly, $\tau_2^{\pm 1}$ represent the two $G$-classes of non-$G_2$ triality graph automorphisms. The next proposition summarises the above discussion.

**Proposition 3.5.23** *Let $T = \mathrm{P\Omega}_8^+(q)$ and let $x \in \mathrm{Aut}(T)$ be a triality automorphism of order 3.*

(i) *If $x$ is a graph automorphism then either $x$ is $G$-conjugate to $\tau_1^{\pm 1}$ and $C_G(x) \cong G_2(q)$, or $x$ is $G$-conjugate to $\tau_2^{\pm 1}$ and either $q \equiv \varepsilon \pmod 3$ and $C_G(x) \cong \mathrm{PGL}_3^\varepsilon(q)$, or $p = 3$ and $C_G(x) \cong [q^5].\mathrm{SL}_2(q)$.*
(ii) *If $x$ is a graph-field automorphism then $q = q_0^3$, $C_G(x) \cong {}^3D_4(q_0)$ and $x$ is $G$-conjugate to $\tau_1^{\pm 1}\phi^{f/3}$ or $\tau_1^{\pm 1}\phi^{2f/3}$.*

We refer the reader to [18, Proposition 3.55] for information on the action of a triality graph automorphism on the set of $G$-classes of elements of prime order in $G$. For future reference, we record a special case of this result for involutions (here we use the notation for involutions introduced in Sections 3.5.2 and 3.5.4).

**Proposition 3.5.24** *Let $T = \mathrm{P\Omega}_8^+(q)$ and let $x \in \mathrm{Aut}(T)$ be a triality graph automorphism of order 3.*

(i) *If $p = 2$ then $x$ cyclically permutes the $G$-classes represented by the elements $\{c_2, a_4, a_4'\}$, and all other involution classes in $G$ are fixed.*
(ii) *If $p \neq 2$ then $x$ cyclically permutes the $G$-classes represented by $\{t_1, t_3, t_4\}$, $\{t_1', t_3', t_4'\}$ and $\{t_2', t_2'', t_2'''\}$, and all other involution classes in $G$ are fixed.*

### 3.5.8 Orthogonal groups: a summary

The next proposition summarises some of the above results on elements of prime order in $\mathrm{Aut}(T) \setminus \widetilde{G}$. For convenience, let us write $x \sim y$ if $x, y \in \mathrm{Aut}(T)$ are $G$-conjugate. We adopt the notation for elements introduced in the previous sections, and we write $q = p^f$ with $p$ a prime.

**Proposition 3.5.25** *Let $x \in \mathrm{Aut}(T) \setminus \widetilde{G}$ be an element of prime order $r$.*

(i) *If $n$ is odd then $r$ divides $f$ and $x \sim \phi^{fj/r}$ for some integer $1 \leqslant j < r$.*
(ii) *If $\varepsilon = -$ then $r$ is odd, $r$ divides $f$ and $x \sim \phi^{fj/r}$ for some integer $1 \leqslant j < r$.*
(iii) *If $n > 8$ and $\varepsilon = +$, then either $r$ divides $f$ and $x \sim \phi^{fj/r}$ for some integer $1 \leqslant j < r$, or $r = 2$, $f$ is even and $x \sim \gamma\phi^{f/2}$.*

(iv) *If $(n,\varepsilon) = (8,+)$ then one of the following holds:*
  (a) *$r$ divides $f$ and $x \sim \phi^{fj/r}$ for some integer $1 \leqslant j < r$;*
  (b) *$r = 3$ and either $x \sim \tau_1^{\pm 1}, \tau_2^{\pm 1}$, or $3$ divides $f$ and $x \sim \tau_1^{\pm 1} \phi^{fj/3}$ with $j = 1$ or $2$;*
  (c) *$r = 2$, $p > 2$ and $x \sim \gamma_1^*, \gamma_1^{**}, \gamma_2^*$ or $\gamma_2^{**}$;*
  (d) *$r = p = 2$ and $x \sim b_1^*, b_1^{**}, b_3^*$ or $b_3^{**}$;*
  (e) *$r = 2$, $f$ is even and $x \sim \gamma \phi^{f/2}$, $\gamma^* \phi^{f/2}$ or $\gamma^{**} \phi^{f/2}$.*

*Proof* If $n$ is odd, then $x$ is a field automorphism and the result follows from Proposition 3.5.20.

Next suppose $n$ is even and $\varepsilon = +$. Once again, if $x$ is a field automorphism then Proposition 3.5.20 applies, and we appeal to Proposition 3.5.21 if $x$ is an involutory graph-field automorphism. Finally, suppose $n = 8$ and $x$ is a graph or triality automorphism of $T$. If $r = 2$ and $p$ is odd then Proposition 3.5.10 implies that $x$ is $G$-conjugate to one of the graph automorphisms $\gamma_1^*, \gamma_1^{**}, \gamma_2^*$ or $\gamma_2^{**}$. The case $r = p = 2$ is explained in Remark 3.5.19. On the other hand, if $r = 3$ then $x$ is a triality automorphism and Proposition 3.5.23 applies.

Finally, let us assume $\varepsilon = -$, so $\mathrm{Aut}(T) = \langle \widetilde{G}, \phi \rangle$. If $r = 2$ then $f$ is even and so either $q$ is even, or $q \equiv 1 \pmod 4$ and $D(Q) = \boxtimes$. In particular, $\phi$ has order $2f$ and $\phi^f \in \widetilde{G}$, so there are no involutions in $\mathrm{Aut}(T) \setminus \widetilde{G}$. Therefore, $r$ is odd, $x$ is a field automorphism and the result follows from Proposition 3.5.20. $\quad\square$

**Proposition 3.5.26** *Let $x \in \mathrm{Aut}(T)$ be an element of odd prime order $r$. Then the possibilities for $x$ up to $G$-conjugacy are listed in Table B.12, together with the order of the centraliser $C_G(x)$.*

*Proof* This is simply a matter of combining some of the earlier results in this section. If $x \in \mathrm{Aut}(T) \setminus \widetilde{G}$ then Proposition 3.5.25 applies, so we may assume $x \in \widetilde{G}$ is semisimple or unipotent. The $\mathrm{PO}_n^\varepsilon(q)$-classes of semisimple elements of odd prime order are determined in Proposition 3.5.4, and a description of the corresponding $G$-classes is given in Proposition 3.5.8. In addition, the order of $C_G(x)$ can be deduced from Remark 3.5.6. Similarly, Proposition 3.5.12 describes the $\mathrm{PO}_n^\varepsilon(q)$-classes of elements of order $p$ (with $p$ odd); the corresponding $G$-classes are described in Proposition 3.5.14, and centraliser orders can be read off from Lemma 3.5.13. $\quad\square$

### 3.5.9 Subgroups of $\mathrm{PGO}_n^\varepsilon(q)$, $p \neq 2$

We close this chapter with a brief discussion of the subgroups of $\mathrm{PGO}_n^\varepsilon(q)$ that contain $\mathrm{P\Omega}_n^\varepsilon(q)$, where $q$ is odd. As before, we set

$$T = \mathrm{P\Omega}_n^\varepsilon(q), \quad G = \mathrm{Inndiag}(T), \quad \widetilde{G} = \mathrm{PGO}_n^\varepsilon(q)$$

and we adopt the notation introduced in Section 3.5.2 for involution class representatives. Note that if $n$ is odd then

$$T < \mathrm{PSO}_n(q) = G = \mathrm{PO}_n(q) = \widetilde{G}$$

and $\widetilde{G}$ is a split extension of $T$ by $C_2$, so we will assume throughout that $n$ is even and $q$ is odd. Note that $\widetilde{G}/\mathrm{PSO}_n^{\varepsilon}(q) \cong C_2 \times C_2$. In addition (see p. 57),

$$|\mathrm{PSO}_n^{\varepsilon}(q) : T| = \begin{cases} 2 & D(Q) = \square \\ 1 & D(Q) = \boxtimes \end{cases}$$

There are several cases to consider. In Tables B.10 and B.11 we indicate whether or not a particular involution in $\widetilde{G}$ belongs to $\mathrm{PSO}_n^{\varepsilon}(q)$, so here we will focus on the involutions in $\widetilde{G} \setminus \mathrm{PSO}_n^{\varepsilon}(q)$.

**3.5.9.1 The case $\varepsilon = +, n \equiv 2 \pmod 4, q \equiv 3 \pmod 4$**
Here $D(Q) = \boxtimes$, so $T = \mathrm{PSO}_n^+(q)$ and $\widetilde{G}/T \cong C_2 \times C_2$. The $G$-classes of involutions in $\widetilde{G} \setminus \mathrm{PSO}_n^+(q)$ are represented by the following elements:

$$
\begin{array}{rcl}
t_{n/2} & \in & G \setminus \mathrm{PSO}_n^+(q) \\
\gamma_1, \dots, \gamma_{(n+2)/4} & \in & \mathrm{PO}_n^+(q) \setminus \mathrm{PSO}_n^+(q) \\
\gamma'_{(n+2)/4} & \in & \widetilde{G} \setminus (G \cup \mathrm{PO}_n^+(q))
\end{array}
$$

It follows that $\widetilde{G}$ has precisely three subgroups that are split extensions of $T$ by $C_2$, namely

$$\mathrm{PO}_n^+(q) = \langle \mathrm{PSO}_n^+(q), \gamma_1 \rangle, \quad G = \langle \mathrm{PSO}_n^+(q), t_{n/2} \rangle, \quad \langle \mathrm{PSO}_n^+(q), \gamma'_{(n+2)/4} \rangle$$

each of which has index two in $\widetilde{G}$. Since an involution $x$ of type $\gamma_i$ is the product of an involution of type $\gamma_1$ and one of type $t_{i-1}$ or $t'_{i-1}$ (depending on the $T$-class of $x$), $\mathrm{PO}_n^+(q)$ contains involutions in both $T$-classes of type $\gamma_i$.

**3.5.9.2 The case $\varepsilon = +, n \equiv 2 \pmod 4, q \equiv 1 \pmod 4$**
Here $\widetilde{G}/T \cong D_8$ and the $G$-classes of involutions in $\widetilde{G} \setminus \mathrm{PSO}_n^+(q)$ are represented by the following elements:

$$
\begin{array}{rcl}
\gamma_1, \dots, \gamma_{(n+2)/4} & \in & \mathrm{PO}_n^+(q) \setminus \mathrm{PSO}_n^+(q) \\
\gamma'_{(n+2)/4} & \in & \widetilde{G} \setminus (G \cup \mathrm{PO}_n^+(q))
\end{array}
$$

Therefore, there are five split extensions of $T$ by $C_2$: $\mathrm{PSO}_n^+(q) = \langle T, t'_1 \rangle$, two subgroups of type $\langle T, \gamma_1 \rangle$ corresponding to the two $T$-classes of involutions of type $\gamma_1$, and two subgroups of type $\langle T, \gamma'_{(n+2)/4} \rangle$ corresponding to the two

$T$-classes of involutions of type $\gamma'_{(n+2)/4}$. There are also two subgroups of the form $T.2^2$, namely

$$\mathrm{PO}_n^+(q) = \langle \mathrm{PSO}_n^+(q), \gamma_1 \rangle, \quad \langle \mathrm{PSO}_n^+(q), \gamma'_{(n+2)/4} \rangle$$

Finally, we note that $G = T.4$.

### 3.5.9.3 The case $\varepsilon = +$, $n \equiv 0 \pmod 4$, $q \equiv 1 \pmod 4$

In this case $\widetilde{G}/T \cong D_8$ and representatives of the $G$-classes of involutions in $\widetilde{G} \setminus \mathrm{PSO}_n^+(q)$ are as follows:

$$
\begin{aligned}
t'_{n/2}, t'_{n/2-1}, t''_{n/4}, t'''_{n/4} &\in G \setminus \mathrm{PSO}_n^+(q) \\
\gamma_1, \dots, \gamma_{n/4} &\in \mathrm{PO}_n^+(q) \setminus \mathrm{PSO}_n^+(q)
\end{aligned}
$$

Again there are five split extensions of $T$ by $C_2$, namely

$$\mathrm{PSO}_n^+(q), \quad \langle T, t'_{n/2} \rangle = \langle T, t''_{n/4} \rangle, \quad \langle T, t'_{n/2-1} \rangle = \langle T, t'''_{n/4} \rangle$$

and two groups of the form $\langle T, \gamma_1 \rangle$ corresponding to the two $T$-classes of involutions of type $\gamma_1$. In addition, $\widetilde{G}$ has three subgroups of index two: $G$ and $\mathrm{PO}_n^+(q)$ are extensions of $T$ by $C_2 \times C_2$, and there is an additional subgroup $T.4$, which is a nonsplit extension $\mathrm{PSO}_n^+(q).2$.

### 3.5.9.4 The case $\varepsilon = +$, $n \equiv 0 \pmod 4$, $q \equiv 3 \pmod 4$

Here $\widetilde{G}/T \cong D_8$ and representatives of the $G$-classes of involutions in $\widetilde{G} \setminus \mathrm{PSO}_n^+(q)$ are as follows:

$$
\begin{aligned}
t_{n/2}, t_{n/2-1}, t''_{n/4}, t'''_{n/4} &\in G \setminus \mathrm{PSO}_n^+(q) \\
\gamma_1, \dots, \gamma_{n/4} &\in \mathrm{PO}_n^+(q) \setminus \mathrm{PSO}_n^+(q)
\end{aligned}
$$

We see that there are five split extensions of $T$ by $C_2$, namely

$$\mathrm{PSO}_n^+(q), \quad \langle T, t_{n/2} \rangle = \langle T, t''_{n/4} \rangle, \quad \langle T, t_{n/2-1} \rangle = \langle T, t'''_{n/4} \rangle$$

and two groups of the form $\langle T, \gamma_1 \rangle$. In addition, $G$ and $\mathrm{PO}_n^+(q)$ are index-two subgroups of $\widetilde{G}$, which are both extensions of $T$ by $C_2 \times C_2$. As in the previous case, there is one more index-two subgroup of $\widetilde{G}$, which is a nonsplit extension $\mathrm{PSO}_n^+(q).2$ of the form $T.4$.

### 3.5.9.5 The case $\varepsilon = -$, $n \equiv 2 \pmod 4$, $q \equiv 3 \pmod 4$

Here $\widetilde{G}/T \cong D_8$ and representatives of the $G$-classes of involutions in $\widetilde{G} \setminus \mathrm{PSO}_n^-(q)$ are as follows:

$$
\begin{aligned}
\gamma_1, \dots, \gamma_{(n+2)/4} &\in \mathrm{PO}_n^-(q) \setminus \mathrm{PSO}_n^-(q) \\
\gamma'_{(n+2)/4} &\in \widetilde{G} \setminus (G \cup \mathrm{PO}_n^-(q))
\end{aligned}
$$

There are five split extensions of $T$ by $C_2$; we get $\mathrm{PSO}_n^-(q)$ and two groups each of the form $\langle T, \gamma_1 \rangle$ and $\langle T, \gamma'_{(n+2)/4} \rangle$. In addition, we have $G = T.4$, $\mathrm{PO}_n^-(q) = T.2^2$ and $\langle \mathrm{PSO}_n^-(q), \gamma'_{(n+2)/4} \rangle = T.2^2$.

**3.5.9.6 The case $\varepsilon = -$, $n \equiv 2 \pmod 4$, $q \equiv 1 \pmod 4$**

Here $D(Q) = \boxtimes$, so $T = \mathrm{PSO}_n^-(q)$ and $\widetilde{G}/T \cong C_2 \times C_2$. The $G$-classes of involutions in $\widetilde{G} \setminus \mathrm{PSO}_n^-(q)$ are represented by the following elements:

$$
\begin{array}{ll}
t_{n/2} & \in \quad G \setminus \mathrm{PSO}_n^-(q) \\
\gamma_1, \ldots, \gamma_{(n+2)/4} & \in \quad \mathrm{PO}_n^-(q) \setminus \mathrm{PSO}_n^-(q) \\
\gamma'_{(n+2)/4} & \in \quad \widetilde{G} \setminus (G \cup \mathrm{PO}_n^-(q))
\end{array}
$$

There are three groups of the form $T.2$, namely $G$, $\mathrm{PO}_n^-(q)$ and $\langle T, \gamma'_{(n+2)/4} \rangle$.

**3.5.9.7 The case $\varepsilon = -$, $n \equiv 0 \pmod 4$**

Again we have $D(Q) = \boxtimes$, so $T = \mathrm{PSO}_n^-(q)$ and $\widetilde{G}/T \cong C_2 \times C_2$. Representatives of the $G$-classes of involutions in $\widetilde{G} \setminus \mathrm{PSO}_n^-(q)$ are as follows:

$$
\begin{array}{ll}
t'_{n/4} & \in \quad G \setminus \mathrm{PSO}_n^-(q) \\
\gamma_1, \ldots, \gamma_{n/4} & \in \quad \mathrm{PO}_n^-(q) \setminus \mathrm{PSO}_n^-(q)
\end{array}
$$

There are two split extensions of the form $T.2$, namely $G$ and $\mathrm{PO}_n^-(q)$. In addition, there is a nonsplit extension $T.2$.

# 4

# Subspace actions

## 4.1 Introduction

In this chapter we begin our study of derangements of prime order in finite primitive classical groups. Let $G \leqslant \mathrm{Sym}(\Omega)$ be a primitive almost simple classical group over $\mathbb{F}_q$ with point stabiliser $H$. Let $T$ denote the socle of $G$, let $V$ be the natural module for $T$ and write $q = p^f$, where $p$ is a prime. We may assume that $n$ and $q$ satisfy the conditions in Table 3.0.1, where $n = \dim V$. Let $r$ be a prime divisor of $|\Omega|$, and recall that $G$ is $r$-elusive if it does not contain a derangement of order $r$ (see Definition 1.3.6).

Here we focus on the special case where $\Omega$ is a set of isometric subspaces (or appropriate pairs of subspaces) of $V$. In terms of Aschbacher's subgroup structure theorem (see Section 2.6.1), $H$ is a maximal subgroup of $G$ in the collection $\mathscr{C}_1$ (recall that we follow [86] in defining the subgroups in $\mathscr{C}_1$, which differs slightly from Aschbacher's original set-up in [3]). The specific cases we will consider are listed in Table 4.1.1, where we adopt the abbreviations *t.i.*, *n.d.*, *t.s.* and *n.s.* for the terms *totally isotropic*, *nondegenerate*, *totally singular* and *nonsingular*, respectively. (Recall that a nonsingular 1-space is a subspace spanned by a nonsingular vector.) Furthermore, if $T$ is an orthogonal group and $\Omega$ consists of nondegenerate subspaces then we say that $\Omega$ is *hyperbolic*, *elliptic* or *parabolic* according to the particular type of the subspaces comprising $\Omega$ (see Section 2.2.5). The degrees of the primitive groups listed in Table 4.1.1 are easily computed from the information in [86, Section 4.1], and they are recorded in Table 4.1.2.

Throughout this chapter we continue to adopt the notation introduced in Chapters 2 and 3 for classical groups and their elements.

**Remark 4.1.1** In Cases I$'$ and I$''$ of Table 4.1.1, $T = \mathrm{PSL}_n(q)$, $n \geqslant 3$ and $\Omega$ is the set of pairs of proper subspaces $\{U, W\}$ of $V$ with the stated properties

149

Table 4.1.1 *The collection $\mathscr{C}_1$*

| Case | $T$ | Type of $m$-space | Conditions |
|------|-----|-------------------|------------|
| I | $\mathrm{PSL}_n(q)$ | | $m \leqslant n/2$ |
| I$'$ | $\mathrm{PSL}_n(q)$ | $\{U, W\}$ | $V = U \oplus W$, $\dim U = m < n/2$ |
| I$''$ | $\mathrm{PSL}_n(q)$ | $\{U, W\}$ | $U < W$, $\dim U = m$, $\dim W = n - m$ |
| II | $\mathrm{PSU}_n(q)$ | t.i. | $m \leqslant n/2$ |
| III | $\mathrm{PSU}_n(q)$ | n.d. | $m < n/2$ |
| IV | $\mathrm{PSp}_n(q)$ | t.i. | $m \leqslant n/2$ |
| V | $\mathrm{PSp}_n(q)$ | n.d. | $m$ even, $m < n/2$ |
| VI | $\mathrm{P\Omega}_n^+(q)$ | t.s. | $m \leqslant n/2$ |
| VII | $\mathrm{P\Omega}_n^-(q)$ | t.s. | $m < n/2$ |
| VIII | $\Omega_n(q)$ | t.s. | $nq$ odd, $m < n/2$ |
| IX | $\Omega_n^\varepsilon(q)$ | n.s. 1-space | $n, q$ even |
| X | $\mathrm{P\Omega}_n^+(q)$ | n.d. hyperbolic | $m$ even, $m < n/2$ |
| XI | $\mathrm{P\Omega}_n^+(q)$ | n.d. elliptic | $m$ even, $m < n/2$ |
| XII | $\mathrm{P\Omega}_n^+(q)$ | n.d. parabolic | $mq$ odd, $m < n/2$ |
| XIII | $\mathrm{P\Omega}_n^-(q)$ | n.d. elliptic | $m$ even |
| XIV | $\mathrm{P\Omega}_n^-(q)$ | n.d. parabolic | $mq$ odd, $m < n/2$ |
| XV | $\Omega_n(q)$ | n.d. hyperbolic | $nq$ odd, $m$ even |
| XVI | $\Omega_n(q)$ | n.d. elliptic | $nq$ odd, $m$ even |

(note that the given conditions imply that $\dim U < n/2$ in Case I$''$). In these two cases, $T$ is transitive on $\Omega$ and the corresponding stabilisers are maximal subgroups of $G$ if and only if $G \not\leqslant \mathrm{P\Gamma L}_n(q)$.

**Remark 4.1.2** By applying Witt's Lemma (see Lemma 2.2.15) and Lemma 2.5.10, we observe that $T$ acts transitively on the set of all $m$-spaces of a given type, with the exception of the cases labelled VI (with $m = n/2$), XII and XIV. In Case VI, there are two $T$-orbits on totally singular subspaces of dimension $n/2$, and we recall that two such subspaces $U, W$ are in the same $T$-orbit if and only if $\dim U - \dim(U \cap W)$ is even (see Proposition 2.5.4). In Cases XII and XIV, there are two $T$-orbits on parabolic $m$-spaces as there are two isometry classes of such spaces (but only one similarity class). In the latter two cases, the two orbits can be distinguished by considering the discriminant of the restriction of the underlying quadratic form on $V$, and the actions of $T$ on the two orbits are equivalent.

**Remark 4.1.3** Note that an element of $\mathrm{P\Omega}_n^-(q)$ fixes an elliptic $m$-space if and only if it fixes a hyperbolic $(n - m)$-space (see Lemma 2.2.16). Therefore, by suppressing the usual restriction $m < n/2$ in Case XIII, we do not need to deal separately with the action of $\mathrm{P\Omega}_n^-(q)$ on hyperbolic $m$-spaces. Similarly,

Table 4.1.2 *Degrees of the subspace actions in Table 4.1.1*

| Case | Description | $\|\Omega\|$ |
|---|---|---|
| I | $m$-spaces | $\dfrac{\prod_{i=n-m+1}^{n}(q^i-1)}{\prod_{i=1}^{m}(q^i-1)}$ |
| I′ | pairs $\{U,W\}$, $V=U\oplus W$ | $\dfrac{q^{m(n-m)}\prod_{i=n-m+1}^{n}(q^i-1)}{\prod_{i=1}^{m}(q^i-1)}$ |
| I″ | pairs $\{U,W\}$, $U<W$ | $\dfrac{\prod_{i=n-2m+1}^{n}(q^i-1)}{\prod_{i=1}^{m}(q^i-1)^2}$ |
| II | t.i. $m$-spaces | $\dfrac{\prod_{i=n-2m+1}^{n}(q^i-(-1)^i)}{\prod_{i=1}^{m}(q^{2i}-1)}$ |
| III | n.d. $m$-spaces | $\dfrac{q^{m(n-m)}\prod_{i=n-m+1}^{n}(q^i-(-1)^i)}{\prod_{i=1}^{m}(q^i-(-1)^i)}$ |
| IV | t.i. $m$-spaces | $\dfrac{\prod_{i=n/2-m+1}^{n/2}(q^{2i}-1)}{\prod_{i=1}^{m}(q^i-1)}$ |
| V | n.d. $m$-spaces | $\dfrac{q^{m(n-m)/2}\prod_{i=(n-m+2)/2}^{n/2}(q^{2i}-1)}{\prod_{i=1}^{m/2}(q^{2i}-1)}$ |
| VI, VII | t.s. $m$-spaces, $\varepsilon=\pm$, $m<n/2$ | $\dfrac{(q^{n/2}-\varepsilon)(q^{(n-2m)/2}+\varepsilon)\prod_{i=(n-2m)/2+1}^{n/2-1}(q^{2i}-1)}{\prod_{i=1}^{m}(q^i-1)}$ |
| VI | t.s. $\frac{n}{2}$-spaces | $\prod_{i=1}^{n/2-1}(q^i+1)$ |
| VIII | t.s. $m$-spaces | $\dfrac{\prod_{i=(n-2m+1)/2}^{(n-1)/2}(q^{2i}-1)}{\prod_{i=1}^{m}(q^i-1)}$ |
| IX | n.s. 1-spaces, $\varepsilon=\pm$, $p=2$ | $q^{n/2-1}(q^{n/2}-\varepsilon)$ |
| X, XI | n.d. $m$-spaces, type $\varepsilon'=\pm$ | $\dfrac{q^{m(n-m)/2}(q^{n/2}-1)\prod_{i=(n-m)/2}^{n/2-1}(q^{2i}-1)}{2(q^{m/2}-\varepsilon')(q^{(n-m)/2}-\varepsilon')\prod_{i=1}^{m/2-1}(q^{2i}-1)}$ |
| XII, XIV | n.d. $m$-spaces, $\varepsilon=\pm$, $m>1$ odd | $\dfrac{q^{(mn-m^2-1)/2}(q^{n/2}-\varepsilon)\prod_{i=(n-m+1)/2}^{n/2-1}(q^{2i}-1)}{2\prod_{i=1}^{(m-1)/2}(q^{2i}-1)}$ |
| XII, XIV | n.d. 1-spaces, $\varepsilon=\pm$ | $\frac{1}{2}q^{n/2-1}(q^{n/2}-\varepsilon)$ |
| XIII | elliptic $m$-spaces | $\dfrac{q^{m(n-m)/2}(q^{n/2}+1)\prod_{i=(n-m)/2}^{n/2-1}(q^{2i}-1)}{2(q^{m/2}+1)(q^{(n-m)/2}-1)\prod_{i=1}^{m/2-1}(q^{2i}-1)}$ |
| XV, XVI | n.d. $m$-spaces, type $\varepsilon'=\pm$ | $\dfrac{q^{m(n-m)/2}\prod_{i=(n-m+1)/2}^{(n-1)/2}(q^{2i}-1)}{2(q^{m/2}-\varepsilon')\prod_{i=1}^{m/2-1}(q^{2i}-1)}$ |

if $n$ is odd and $x \in \Omega_n(q)$ fixes a nondegenerate $m$-space with $m$ even, then it also fixes a nondegenerate $(n-m)$-space. Therefore, by omitting the condition $m < n/2$ in Cases XV and XVI, we do not have to deal separately with the action of $\Omega_n(q)$ on parabolic subspaces.

Let us now state the main results of this chapter. First we consider derangements of order $p$; in order to state Theorem 4.1.4 below, we need to introduce the following number-theoretic conditions on partitions of $n$. For $m < n$, we say that a partition $\lambda$ of $n$ *contains* a partition $\lambda'$ of $m$ if each part of $\lambda'$ is also a part of $\lambda$. For example, the partition $(3^2, 2, 1^4)$ of 12 contains the partition $(3, 1^2)$ of 5.

P1 Every partition of $n$ with parts of size at most $p$ contains a partition of $m$.
P2 Every partition of $n$ with parts of size at most $p$ and all odd parts with even multiplicity contains a partition of $m$ with all odd parts of even multiplicity.
P3 Every partition of $n$ with parts of size at most $p$ and all even parts with even multiplicity contains a partition of $m$ with all even parts of even multiplicity.

**Theorem 4.1.4** *Let $G \leqslant \mathrm{Sym}(\Omega)$ be a primitive almost simple classical group with socle $T$ and point stabiliser $H \in \mathscr{C}_1$. Suppose $p > 2$ and $p$ divides $|\Omega|$. Then $T$ is $p$-elusive only if one of the following holds:*

(i) *P1 holds in Cases $\mathrm{I}'$ or III of Table 4.1.1.*
(ii) *P2 holds in Case V.*
(iii) *P3 holds in Cases X, XIII and XV, with $n \equiv m \pmod 4$ in Case XIII, and $m \equiv 0 \pmod 4$ in Case XV.*

*Moreover, in (i) and (ii), the given conditions are also sufficient for $p$-elusivity.*

**Remark 4.1.5** Note that $p$ does not divide $|\Omega|$ in Cases I, $\mathrm{I}''$, II, IV, VI, VII and VIII. As indicated in the theorem, in Cases X, XIII and XV we have not determined necessary and sufficient conditions for $p$-elusivity. We refer the reader to Remarks 4.3.5 and 4.5.3 for further comments in Cases $\mathrm{I}'$, III and V.

Next we turn to derangements of odd prime order $r \neq p$.

**Theorem 4.1.6** *Let $G \leqslant \mathrm{Sym}(\Omega)$ be a primitive almost simple classical group with socle $T$ and point stabiliser $H \in \mathscr{C}_1$. Let $r \neq p$ be an odd prime. Then $T$ is $r$-elusive if and only if $(G, H, r)$ is one of the cases recorded in Tables 4.3.1, 4.4.1, 4.5.1 or 4.6.1, and $r$ divides $|\Omega|$.*

Table 4.1.3 2-*elusive actions,* $H \in \mathscr{C}_1$

| Case | $q$ | Conditions |
|------|-----|------------|
| I | odd | $n$ odd; $m$ even; $(q-1)_2 \geqslant (n)_2, q \equiv 1 \pmod 4$ |
| I' | odd | $n$ odd; $m$ even; $(q-1)_2 \geqslant (n)_2, q \equiv 1 \pmod 4$ |
|  | even | $n$ odd; $m$ even |
| I'' | odd | $n$ odd; $m$ even; $(q-1)_2 \geqslant (n)_2, q \equiv 1 \pmod 4$ |
| II | odd | $m < n/2$; $(q+1)_2 \leqslant (n)_2$ |
| III | odd | $n$ odd; $m$ even; $(q+1)_2 \geqslant (n)_2, q \equiv 3 \pmod 4$ |
|  | even | $n$ odd; $m$ even |
| IV | odd | $m$ even; $q \equiv 1 \pmod 4$ |
| V | odd | always |
|  | even | $n \equiv 2 \pmod 4$; $m \equiv 0 \pmod 4$ |
| VI | odd | $q \equiv 1 \pmod 4$; $m \leqslant n/2 - 2$, $\begin{cases} m \text{ even} \\ n \equiv 2 \pmod 4 \end{cases}$ |
| VII | odd | $n \equiv 0 \pmod 4$; $m$ even; $q \not\equiv 7 \pmod 8$ |
| VIII | odd | $m < (n-1)/2$; $q \equiv 1 \pmod 4$ |
| IX | even | $\varepsilon = -$; $n \equiv 2 \pmod 4$ |
| X | odd | $n \equiv 2 \pmod 4$; $q \equiv 1 \pmod 4$; $m \equiv 0 \pmod 4$ |
|  | even | $n \equiv 2 \pmod 4$; $m \equiv 0 \pmod 4$ |
| XI | odd | $\begin{cases} n \equiv 2 \pmod 4, q \not\equiv 1 \pmod 8 \\ n \equiv 0 \pmod 4, q \equiv 3 \pmod 4, m \equiv 2 \pmod 4 \end{cases}$ |
| XII | odd | $n \equiv 2 \pmod 4, q \not\equiv 1 \pmod 8$ |
| XIII | odd | $n \equiv 0 \pmod 4$; $q \not\equiv 7 \pmod 8$; $m \equiv 2 \pmod 4$ |
|  | even | $n \equiv 0 \pmod 4$; $m \equiv 2 \pmod 4$ |
| XIV | odd | $n \equiv 0 \pmod 4$; $q \not\equiv 7 \pmod 8$ |
| XV | odd | always |
| XVI | odd | always |

Finally, our main result for involutions is the following:

**Theorem 4.1.7** *Let* $G \leqslant \mathrm{Sym}(\Omega)$ *be a primitive almost simple classical group with socle* $T$ *and point stabiliser* $H \in \mathscr{C}_1$. *Then* $T$ *is 2-elusive if and only if* $|\Omega|$ *is even and* $(G, H)$ *is one of the cases in Table 4.1.3.*

**Remark 4.1.8** Let us explain how to read the conditions recorded in the final column of Table 4.1.3. For example, in Case VI, $T$ is 2-elusive if and only if $|\Omega|$ is even, $q$ is odd and either

(i) $q \equiv 1 \pmod 4$, or

(ii) $m \leqslant n/2 - 2$ and either $m$ is even, or $n \equiv 2 \pmod 4$.

In particular, a comma in the final column of Table 4.1.3 should be read as 'and', while a semicolon denotes 'or'.

Derangements in subspace actions of finite classical groups are studied by Fulman and Guralnick in [58], which is the third in a series of papers concerning a conjecture of Boston and Shalev on the proportion of derangements in transitive actions of finite simple groups (see Theorem 1.2.2). In the context of the subspace actions of classical groups that we are considering here, they prove that there is an absolute constant $\varepsilon > 0$ such that $\delta(T) \geqslant \varepsilon$ (where $\delta(T)$ denotes the proportion of derangements in $T$), and they show that $\delta(T)$ tends to 1 as $m$ tends to infinity (see [58, Theorem 1.1(2)], and note that $\delta(T)$ does *not* tend to 1 if $m$ is bounded).

We also refer the reader to [107] for results on the proportion of derangements in the context of a finite classical group acting on the set of 1-dimensional subspaces of its natural module.

## 4.2 Preliminaries

In this section, we begin by recording several preliminary results that will be needed in the proofs of Theorems 4.1.4, 4.1.6 and 4.1.7.

**Lemma 4.2.1** *Let $\mathbb{F}$ be a field and let $x \in \mathrm{GL}_n(\mathbb{F})$ be a semisimple element that fixes a decomposition*

$$V = U_1 \oplus \cdots \oplus U_s \oplus C_V(x) \qquad (4.2.1)$$

*of the natural $\mathrm{GL}_n(\mathbb{F})$-module, where each $U_i$ is an a-space on which $x$ acts irreducibly. Let $m < n$ be a positive integer and write $m \equiv \ell \pmod{a}$ with $0 \leqslant \ell < a$.*

*Then $x$ fixes an m-dimensional subspace of $V$ if and only if one of the following conditions holds:*

(i) $\dim C_V(x) \geqslant m$;
(ii) $a \leqslant m$ and $\dim C_V(x) \geqslant \ell$.

*In addition, if (ii) holds and $x$ preserves a nondegenerate reflexive sesquilinear or quadratic form on $V$ such that each $U_i$ is nondegenerate, the decomposition in (4.2.1) is orthogonal, and $V$ contains nondegenerate m-spaces, then $x$ fixes a nondegenerate m-space.*

*Proof* First we show that $x$ fixes an $m$-space if (i) or (ii) holds. Clearly, if (i) holds then $x$ fixes an $m$-dimensional subspace of $C_V(x)$. Now assume (ii) holds. Let $U$ be an $\ell$-dimensional subspace of $C_V(x)$. If $as \geqslant m$ then $x$ fixes the

$m$-space spanned by $U$ and $(m - \ell)/a$ of the $U_i$. On the other hand, if $as < m$ then $x$ fixes the $m$-space spanned by $U_1 \oplus \cdots \oplus U_s$ and a subspace $U'$ of $C_V(x)$ of dimension $m - as$.

Next assume (ii) holds and $x$ preserves an appropriate nondegenerate form on $V$ as in the statement of the lemma. Note that $C_V(x)$ is either trivial or nondegenerate (it is the orthogonal complement of the nondegenerate space $U_1 \perp \ldots \perp U_s$). To see that $x$ fixes a nondegenerate $m$-space we simply repeat the above argument, noting that we may assume that $U$ and $U'$ are nondegenerate (or trivial).

To complete the proof, we need to show that if $x$ fixes an $m$-space $W$ then either (i) or (ii) holds. Clearly, if $x$ acts trivially on $W$ then $W \leqslant C_V(x)$ and (i) holds, so assume otherwise. Then $x$ fixes a decomposition

$$W = W_1 \oplus \cdots \oplus W_d \oplus C_W(x) \tag{4.2.2}$$

where $d \geqslant 1$ and each $W_i$ is an $a$-space on which $x$ acts irreducibly. Therefore $a \leqslant m$ and $\dim C_V(x) \geqslant \dim C_W(x) \geqslant \ell$, so (ii) holds. $\qquad\square$

In Lemmas 4.2.2, 4.2.3 and 4.2.4 to follow,

$$G \in \{\mathrm{GU}_n(q), \mathrm{Sp}_n(q), \mathrm{O}_n^\varepsilon(q)\} \tag{4.2.3}$$

and $V$ is the natural module for $G$. Let $x \in G$ be a semisimple element of odd prime order $r$. As in Section 3.1.1, we set $i = \Phi(r, q)$, which is the smallest positive integer such that $r$ divides $q^i - 1$. In addition, we define

$$a = a(i) = \begin{cases} i/2 & G = \mathrm{GU}_n(q) \text{ and } i \text{ even} \\ i & \text{otherwise} \end{cases} \tag{4.2.4}$$

Recall that if $V$ is equipped with a nondegenerate sesquilinear form then we sometimes use the term *totally singular* to describe a totally isotropic subspace of $V$. This convenient abuse of terminology was introduced in Remark 2.2.6. Also recall that a pair $\{U, U^*\}$ of complementary totally singular subspaces of $V$ is a *dual pair* if $U \oplus U^*$ is nondegenerate.

**Lemma 4.2.2** *Let $x \in G$ be a semisimple element of odd prime order $r$ with $i = \Phi(r, q)$, where $i$ is odd if $G$ is symplectic or orthogonal, and $i \not\equiv 2 \pmod 4$ if $G$ is unitary. Let $m$ be a positive integer such that $V$ contains a totally singular $m$-space, and write $m \equiv \ell \pmod a$ with $0 \leqslant \ell < a$, where $a$ is the integer defined in (4.2.4).*

*Then $x$ fixes a totally singular $m$-space if and only if one of the following holds:*

(i) *$C_V(x)$ contains a totally singular $m$-space;*
(ii) *$a \leqslant m$ and $C_V(x)$ contains a totally singular $\ell$-space.*

*Moreover, if $a \leqslant m$ and $x$ fixes a totally singular $m$-space $W$, then $x$ also fixes a totally singular $m$-space $W'$ such that $W \cap W'$ has codimension $a$ in $W$.*

*Proof*   By applying Propositions 3.3.2(i), 3.4.3(i) and 3.5.4(i), we see that $x$ fixes an orthogonal decomposition of the form

$$V = (U_1 \oplus U_1^*) \perp (U_2 \oplus U_2^*) \perp \ldots \perp (U_t \oplus U_t^*) \perp C_V(x) \qquad (4.2.5)$$

where each $\{U_j, U_j^*\}$ is a dual pair of totally singular $a$-spaces, and $x$ acts irreducibly on each $U_j$ and $U_j^*$.

First assume that (i) holds, say $C_V(x)$ contains a totally singular $m$-space $W$. Clearly, $x$ fixes $W$. Moreover, since each $U_j \oplus U_j^*$ is nondegenerate it follows that $C_V(x)$ is also nondegenerate, so $\dim C_V(x) \geqslant 2m$. In particular, if $a \leqslant m$ then $C_V(x)$ contains a totally singular $m$-space $W'$ such that $W \cap W'$ has codimension $a$ in $W$.

Next suppose (ii) holds, so $a \leqslant m$ and $C_V(x)$ contains a totally singular $\ell$-space $U$. Note that $(m - \ell)/a \geqslant 1$. If $at \geqslant m$ then $x$ fixes the totally singular $m$-space spanned by $U$ and $(m - \ell)/a$ of the $U_j$. On the other hand, if $at < m$ then

$$U' := U_1 \oplus U_2 \oplus \cdots \oplus U_t$$

is a totally singular $at$-space, which is contained in a totally singular $m$-space $W$. Moreover, $W = U' \oplus U''$ where $U'' \leqslant C_V(x)$, and so $W$ is fixed by $x$. In both cases, $x$ also fixes the totally singular $m$-space $W'$ obtained by replacing $U_1$ in $U'$ with $U_1^*$. Then $W \cap W'$ has codimension $a$ in $W$.

To complete the proof, let us assume that $x$ fixes a totally singular $m$-space $W$. Then $x$ fixes a decomposition of $W$ as in (4.2.2), where $d \geqslant 0$ and each $W_j$ is a totally singular $a$-space on which $x$ acts irreducibly. If $a > m$ then $W \leqslant C_V(x)$ and thus (i) holds. On the other hand, if $a \leqslant m$ then $\dim C_W(x) \geqslant \ell$ and we deduce that $C_V(x)$ contains a totally singular $\ell$-space.                                    $\square$

The next result is a version of Lemma 4.2.2 for nondegenerate subspaces. Again, $G$ is one of the classical groups in (4.2.3).

**Lemma 4.2.3**   *Let $x \in G$ be a semisimple element of odd prime order $r$ with $i = \Phi(r, q)$, where $i$ is odd if $G$ is symplectic or orthogonal, and $i \not\equiv 2 \pmod 4$ if $G$ is unitary. Let $m < n$ be a positive integer such that $V$ contains a nondegenerate $m$-space, and write $m \equiv \ell \pmod{2a}$ with $0 \leqslant \ell < 2a$, where $a$ is the integer defined in (4.2.4).*

*Then $x$ fixes a nondegenerate $m$-space if and only if one of the following holds:*

(i) $C_V(x)$ contains a nondegenerate $m$-space;

(ii) $2a \leqslant m$ and $C_V(x)$ contains a nondegenerate $\ell$-space.

*Proof* As in the proof of the previous lemma, $x$ fixes an orthogonal decomposition of $V$ as in (4.2.5), with $x$ acting irreducibly on each $U_j$ and $U_j^*$. In particular, let $t$ be the positive integer in (4.2.5).

Clearly, if (i) holds then $x$ fixes a nondegenerate $m$-space. Now assume that (ii) holds, so $2a \leqslant m$ and $C_V(x)$ contains a nondegenerate $\ell$-space $U$. If $2at \geqslant m$ then $x$ fixes the nondegenerate $m$-space spanned by $U$ and $(m-\ell)/2a$ of the $2a$-spaces $U_j \oplus U_j^*$. Now assume $2at < m$ and set

$$U' := (U_1 \oplus U_1^*) \perp \ldots \perp (U_t \oplus U_t^*)$$

Note that $U'$ is a nondegenerate $2at$-space. Since $V$ contains nondegenerate $m$-spaces, and $m$ and $m - 2at$ have the same parity, it follows that $V$ also contains nondegenerate $(m - 2at)$-spaces. Therefore, since $C_V(x)$ is nondegenerate (it is the orthogonal complement of $U'$), we deduce that $C_V(x)$ contains a nondegenerate $(m - 2at)$-space $U''$. We now conclude that $x$ fixes the nondegenerate $m$-space $U' \perp U''$.

Now assume that $x$ fixes a nondegenerate $m$-space $W$. By applying Propositions 3.3.2, 3.4.3 and 3.5.4, we see that $x$ fixes an orthogonal decomposition of $W$ as a sum of $C_W(x)$ and $s \geqslant 0$ nondegenerate $2a$-spaces of the form $X \oplus X^*$, where $x$ acts irreducibly on the totally singular $a$-spaces $X$ and $X^*$. On the one hand, if $2a > m$ then $s = 0$ and $W \leqslant C_V(x)$, so (i) holds. On the other hand, if $2a \leqslant m$ then $\dim C_W(x) \geqslant \ell$ and thus $C_V(x)$ contains a nondegenerate $\ell$-space. $\square$

In the next lemma, we use the notation for semisimple elements of prime order that was introduced in Section 3.1.1. Once again, $G$ denotes one of the groups in (4.2.3).

**Lemma 4.2.4** *Let $x \in G$ be a semisimple element of odd prime order $r$ with $i = \Phi(r, q)$, where $i$ is even if $G$ is symplectic or orthogonal, and $i \equiv 2 \pmod 4$ if $G$ is unitary. Define the integer $a$ as in (4.2.4). We may write*

$$x = [\Lambda_1^{a_1}, \ldots, \Lambda_t^{a_t}, I_e]$$

*with respect to an orthogonal decomposition*

$$V = (U_{1,1} \perp \ldots \perp U_{1,a_1}) \perp \ldots \perp (U_{t,1} \perp \ldots \perp U_{t,a_t}) \perp C_V(x)$$

*where each $U_{j,k}$ is a nondegenerate $a$-space and each $x|_{U_{j,k}} = \Lambda_j$ is irreducible. Let $m$ be a positive integer such that $V$ contains a totally singular $m$-space, and write $m \equiv \ell \pmod a$ with $0 \leqslant \ell < a$.*

*Then x fixes a totally singular m-space if and only if one of the following holds:*

(i) $C_V(x)$ *contains a totally singular m-space;*
(ii) $a \leqslant m$, $C_V(x)$ *contains a totally singular $(aa_0/2 + \ell)$-space for some even integer $a_0 \geqslant 0$, and*

$$m - \ell \leqslant a \sum_{j=0}^{t} \lfloor a_j/2 \rfloor$$

*Moreover, if $a \leqslant m$ and x fixes a totally singular m-space W, then x also fixes a totally singular m-space $W'$ such that $W \cap W'$ has codimension a in W.*

*Proof* It will be convenient to set $\mathbb{F} = \mathbb{F}_{q^u}$, where $u = 2$ if $G$ is a unitary group, otherwise $u = 1$. Note that if $G = O_n^{\varepsilon}(q)$, then the $U_{j,k}$ are elliptic spaces.

First assume that $C_V(x)$ contains a totally singular $m$-space $W$. Clearly, $x$ fixes $W$. Moreover, since each $U_{j,k}$ is nondegenerate, $C_V(x)$ is also nondegenerate and thus $\dim C_V(x) \geqslant 2m$. Therefore, if $a \leqslant m$ then $C_V(x)$ contains a totally singular subspace $W'$ such that $W \cap W'$ has codimension $a$ in $W$.

Now assume that (ii) holds, but not (i). Here $a \leqslant m$ and there exists an even integer $a_0 \geqslant 0$ such that $C_V(x)$ contains a totally singular $(aa_0/2 + \ell)$-space $U$ and $m - \ell \leqslant au$, where we set

$$u = \sum_{j=0}^{t} \lfloor a_j/2 \rfloor$$

(Note that $u > a_0/2$ since we are assuming that $C_V(x)$ does not contain a totally singular $m$-space.) As noted in Remarks 3.3.5, 3.4.5 and 3.5.7, the element $[\Lambda_j^2] \in \mathrm{GL}_{2a}(\mathbb{F})$ fixes a dual pair of totally singular $a$-spaces. Therefore, $[\Lambda_j^{a_j}] \in \mathrm{GL}_{a_j a}(\mathbb{F})$ fixes a totally singular subspace of dimension $a \lfloor a_j/2 \rfloor$, and thus $x$ fixes a totally singular subspace $U'$ of $V$ of dimension $a(u - a_0/2)$. In fact, $x$ fixes a totally singular subspace of $U'$ of each dimension that is a multiple of $a$, up to and including $a(u - a_0/2)$.

If $a(u - a_0/2) \geqslant m - \ell$ then $x$ fixes the totally singular $m$-space $W$ spanned by an $\ell$-dimensional subspace of $U$ and an $(m - \ell)$-dimensional subspace of $U'$. Now assume that $a(u - a_0/2) < m - \ell$. Here $U'$ is contained in a totally singular $m$-space of the form $W = U' \oplus U''$, where $U'' \leqslant U$. In particular, $x$ fixes $W$. In both cases, by replacing an appropriate totally singular $a$-space $X$ by $X^*$ in one of the relevant dual pairs $\{X, X^*\}$ used in the construction of $U'$, we deduce that $x$ also fixes a totally singular $m$-space $W'$ with the property that $W \cap W'$ has codimension $a$ in $W$.

To complete the proof, suppose $x$ fixes a totally singular $m$-space $W$. Then $x$ fixes a decomposition

$$W = W_1 \oplus \cdots \oplus W_d \oplus C_W(x)$$

where $d \geqslant 0$ and each $W_j$ is a totally singular $a$-space on which $x$ acts irreducibly as some $\Lambda_k$. If $a > m$ then $d = 0$ and $W \leqslant C_V(x)$, so (i) holds.

Now assume $a \leqslant m$. Here $\dim C_W(x) \geqslant \ell$ and thus $C_V(x)$ contains a totally singular $\ell$-space. By Lemma 2.2.17(ii), there exists a totally singular $m$-space $W^*$ such that $\{W, W^*\}$ is a dual pair fixed by $x$. In terms of matrices, if $x|_W = A$ then $x|_{W \oplus W^*} = [A, (A^{\varphi \mathsf{T}})^{-1}]$ is block-diagonal, where $\varphi$ is the map defined in (2.2.6) if $G = \mathrm{GU}_n(q)$, and $\varphi = 1$ in all other cases. Moreover, since the irreducible blocks $U_{j,k}$ are nondegenerate, it follows that $A$ and $(A^{\varphi \mathsf{T}})^{-1}$ are $\mathrm{GL}_m(\mathbb{F})$-conjugate. For instance, if $G$ is a unitary group then Lemma 3.1.13(ii) implies that $\Lambda_j = \Lambda_j^{-q}$ for each $j$.

In particular, if

$$x|_W = [\Lambda_1^{b_1}, \ldots, \Lambda_t^{b_t}, I_{e'}] \quad \text{and} \quad x|_{(W \oplus W^*)^\perp} = [\Lambda_1^{c_1}, \ldots, \Lambda_t^{c_t}, I_{e''}]$$

then $e' = a a_0/2 + \ell$ for some even integer $a_0 \geqslant 0$, and

$$x = [\Lambda_1^{2b_1 + c_1}, \ldots, \Lambda_t^{2b_t + c_t}, I_{2e' + e''}]$$

Now $C_W(x) \leqslant C_V(x)$ is a totally singular subspace of dimension $a a_0/2 + \ell$. In addition, since $\lfloor a_j/2 \rfloor \geqslant b_j$ for all $1 \leqslant j \leqslant t$, we have

$$a \sum_{j=0}^{t} \lfloor a_j/2 \rfloor \geqslant a a_0/2 + a \sum_{j=1}^{t} b_j = m - \ell$$

Therefore, (ii) holds. □

Finally, let us record a useful combinatorial lemma.

**Lemma 4.2.5** *Let $S$ be a set of size $n$ and let $m$ be a positive integer such that $2m \leqslant n$. Given any colouring of the elements of $S$ from a palette of size $k$, there exist $m$ mutually disjoint pairs of equally coloured elements of $S$ if and only if $n - 2(m-1) > k$.*

*Proof* Suppose first that $n - 2(m-1) \leqslant k$. Then we can colour $2(m-1)$ elements of $S$ with the same colour and the remaining $n - 2(m-1)$ elements can be given distinct colours. This colouring does not have $m$ mutually disjoint pairs of equally coloured elements.

Conversely, suppose $n - 2(m-1) > k$ and $\mathscr{C}$ is a colouring of $S$. We proceed by induction on $m$. If $m = 1$ then $n > k$ and so $\mathscr{C}$ must have two elements with the same colour. Assume the statement is true for $m = t$ and let $m = t + 1$. Since $n - 2(m - 1) > k$ we have $n - 2(t - 1) > k$, so the inductive hypothesis implies that $\mathscr{C}$ has $t$ mutually disjoint pairs of equally coloured elements. Consider the remaining $n - 2t$ elements of $S$. Since $n - 2t = n - 2(m - 1) > k$ it follows that at least two of these elements have the same colour. The result follows. $\qquad\square$

## 4.3 Linear groups

Let $G \leqslant \mathrm{Sym}(\Omega)$ be a primitive almost simple classical group with socle $T = \mathrm{PSL}_n(q)$. Here we are interested in the cases labelled I, I' and I'' in Table 4.1.1, which involve $m$-dimensional subspaces of the natural $T$-module $V$. Let $r$ be a prime divisor of $|\Omega|$. If $r \neq p$ then set $i = \Phi(r, q)$, where $\Phi(r, q)$ is defined in (3.1.2) (so $i$ is the smallest integer such that $r$ divides $q^i - 1$), and write

$$n \equiv k \pmod{i}, \quad m \equiv \ell \pmod{i}$$

with $0 \leqslant k, \ell < i$. Let $Z$ denote the centre of $\mathrm{GL}_n(q)$. Recall that if $a$ is an integer then $(a)_r$ is the largest $r$-power dividing $a$.

**Remark 4.3.1** The relevant $r$-elusive actions with $r \neq p$ and $r > 2$ are recorded in Table 4.3.1. Here $i = \Phi(r, q)$ and the required conditions are listed in the final column. For instance, if $i = 1$ then Table 4.3.1 indicates that $T$ is $r$-elusive in Case I if and only if $r$ divides $m$, or $(q - 1)_r \geqslant (n)_r$.

**Proposition 4.3.2** *In Case I of Table 4.1.1, $T$ is $r$-elusive if and only if one of the following holds:*

(i) *$r \neq p$, $i = 1$ and one of the following holds:*
   (a) *$r$ divides $m$;*
   (b) *$(q - 1)_r \geqslant (n)_r$, with $q \equiv 1 \pmod{4}$ if $r = 2$ and $n$ is even.*
(ii) *$r \neq p$, $2 \leqslant i \leqslant m$ and $k \geqslant \ell$.*

*Proof* Here $T = \mathrm{PSL}_n(q)$ and $\Omega$ is the set of $m$-dimensional subspaces of $V$, with $m \leqslant n/2$. The degree $|\Omega|$ is given in Table 4.1.2; it follows that $r \neq p$ and $r$ divides $q^j - 1$ for some integer $j$ in the range $n - m + 1 \leqslant j \leqslant n$. Let $j$ be the largest such integer and note that $n - j = k$ and $i$ divides $j$. In particular, observe that $k < m$. Let $x = \hat{x}Z \in T$ be an element of order $r$.

Suppose $i = 1$. The possibilities for $\hat{x}$ are described in Proposition 3.2.2. If $\hat{x}$ is diagonalisable over $\mathbb{F}_q$ then $x$ visibly fixes an $m$-space, so let us assume $r$ divides $n$ and $\hat{x} = [\Lambda^{n/r}]$, as in Proposition 3.2.2(ii). Then $x$ is a derangement if

Table 4.3.1 *r-elusive subspace actions, $T = \mathrm{PSL}_n(q)$, $r \neq p$, $r > 2$*

| Case | $i$ | Conditions |
|------|-----|------------|
| I, I′ | $i > 1$ | $i \leqslant m$ and $k \geqslant \ell$ |
|      | $i = 1$ | $r$ divides $m$ or $(q-1)_r \geqslant (n)_r$ |
| I″   | $i > 1$ | $i \leqslant m$ and $k \geqslant 2\ell$ |
|      | $i = 1$ | $r$ divides $m$ or $(q-1)_r \geqslant (n)_r$ |

$n \equiv k \pmod{i}$, $m \equiv \ell \pmod{i}$

and only if $(m, r) = 1$, and these elements are in $T$ if and only if $r = 2$ and $q \equiv 3$ (mod 4), or $(n)_r > (q-1)_r$. This explains the conditions recorded in part (i).

Now assume $i \geqslant 2$ and set $\hat{x} = [\Lambda^{j/i}, I_{n-j}]$, where $\Lambda$ is an irreducible block of size $i$ and order $r$ (see Proposition 3.2.1). Since $\dim C_V(\hat{x}) = n - j = k < m$, Lemma 4.2.1 implies that $x$ is a derangement if $i > m$, or if $i \leqslant m$ and $k < \ell$. Finally, suppose $2 \leqslant i \leqslant m$ and $k \geqslant \ell$. By Proposition 3.2.1, every element $y = \hat{y}Z \in T$ of order $r$ is of the form $\hat{y} = [\Lambda_1^{a_1}, \ldots, \Lambda_t^{a_t}, I_{n-iu}]$, where $t = (r-1)/i$ and $u = \sum_v a_v$. Since $\dim C_V(\hat{y}) \geqslant k \geqslant \ell$, Lemma 4.2.1 implies that $y$ fixes an $m$-space, so $T$ is $r$-elusive. $\qquad\qquad\square$

**Remark 4.3.3** Note that it is possible that a prime $r$ satisfies one of the conditions (i) or (ii) in the statement of Proposition 4.3.2, but does not divide $|\Omega|$, in which case $T$ trivially contains no derangements of order $r$. For example, if $q$ is odd then $|\Omega|$ may be odd (for instance, if $(n, m) = (3, 1)$ then $|\Omega| = q^2 + q + 1$). On the other hand, we can also find prime divisors $r$ of $|\Omega|$ that do satisfy the conditions in (i) or (ii). For example, if $q$ is odd and $(n, m, r) = (5, 2, 2)$ then $|\Omega|$ is even and $T$ is 2-elusive.

**Proposition 4.3.4** *In Case I′ of Table 4.1.1, $T$ is $r$-elusive if and only if one of the following holds:*

(i) $r = p > 2$ *and every partition of $n$ with parts of size at most $p$ contains a partition of $m$.*

(ii) $r = p = 2$ *and either $n$ is odd or $m$ is even.*

(iii) $r \neq p$, $i = 1$ *and one of the following holds:*
  (a) $r$ *divides $m$;*
  (b) $(q-1)_r \geqslant (n)_r$, *with $q \equiv 1$ (mod 4) if $r = 2$ and $n$ is even.*

(iv) $r \neq p$, $2 \leqslant i \leqslant m$ *and $k \geqslant \ell$.*

*Proof* Here $T = \mathrm{PSL}_n(q)$ (with $n \geqslant 3$) and $\Omega$ is the set of pairs of subspaces $\{U, W\}$ such that $V = U \oplus W$ and $\dim U = m < n/2$. First assume $r = p$. Recall that the conjugacy classes of elements of order $p$ in $\mathrm{PGL}_n(q)$ are in bijective correspondence with the set of nontrivial partitions of $n$ with parts of size at most $p$ (see Proposition 3.2.6). Let $x = \hat{x}Z \in T$ be an element of order $p$, and assume that $x$ has fixed points on $\Omega$. Then $\hat{x} \in \mathrm{GL}_m(q) \times \mathrm{GL}_{n-m}(q) =: L$ is $L$-conjugate to $[J_p^{a_p}, \ldots, J_1^{a_1}, J_p^{b_p}, \ldots, J_1^{b_1}]$, where $\sum_v v a_v = m$ and $\sum_v v b_v = n-m$, so $\hat{x}$ is $\mathrm{GL}_n(q)$-conjugate to $[J_p^{a_p+b_p}, \ldots, J_1^{a_1+b_1}]$. We conclude that $T$ is $p$-elusive if and only if every partition of $n$ with parts of size at most $p$ contains a partition of $m$ (this is the condition labelled P1 in the statement of Theorem 4.1.4). In particular, if $p = 2$ then every partition $(2^s, 1^{n-2s})$ of $n$ has this property, unless $n$ is even and $m$ is odd (in this situation, $(2^{n/2})$ does not contain a partition of $m$). We refer the reader to Remark 4.3.5 for further comments on this number-theoretic condition when $p > 2$.

Finally, if $r \neq p$ then an element in $T$ of order $r$ fixes an $m$-space if and only if it fixes a complementary $(n-m)$-space. Therefore, in this situation the required conditions for $r$-elusivity are precisely those given in the statement of Proposition 4.3.2.                                    $\square$

**Remark 4.3.5** Consider the number-theoretic condition (denoted by P1 in Section 4.1) appearing in part (i) of Proposition 4.3.4 (and also in Proposition 4.4.2(i)). Clearly, if $d \leqslant p$ is an integer dividing $n$, but not $m$, then $\lambda = (d^{n/d})$ does not contain a partition of $m$. In particular, if $p = 3$ and $n$ is divisible by 6 then either $m$ is also divisible by 6, in which case every partition of $n$ with parts of size at most 3 contains a partition of $m$ (and thus $T$ is 3-elusive), or $m$ is indivisible by 6 and either $(2^{n/2})$ or $(3^{n/3})$ does not contain a partition of $m$.

**Proposition 4.3.6** *In Case $I''$ of Table 4.1.1, $T$ is $r$-elusive if and only if one of the following holds:*

(i) $r \neq p$, $i = 1$ and one of the following holds:
  (a) $r$ divides $m$;
  (b) $(q-1)_r \geqslant (n)_r$, with $q \equiv 1 \pmod 4$ if $r = 2$ and $n$ is even.
(ii) $r \neq p$, $2 \leqslant i \leqslant m$ and $k \geqslant 2\ell$.

*Proof* In this case $T = \mathrm{PSL}_n(q)$ and $\Omega$ is the set of pairs of subspaces $\{U, W\}$ such that $U < W$, $\dim U = m < n/2$ and $\dim W = n - m$. Since $r$ divides $|\Omega|$, it follows that $r \neq p$ (see Table 4.1.2). Clearly, if $x \in T$ does not fix an $m$-space then it does not fix an element of $\Omega$, so we only need to consider the particular primes arising in the statement of Proposition 4.3.2. Set $x = \hat{x}Z \in T$ of order $r$.

Suppose $i = 1$. If $\hat{x}$ is diagonalisable over $\mathbb{F}_q$ then $x$ clearly fixes an element of $\Omega$, so we may assume otherwise. Then $r$ divides $n$ and we observe that such an element has fixed points on $\Omega$ if and only if $r$ divides $m$ (see Proposition 3.2.2(ii)). Moreover, we note that $T$ contains elements of this form if and only if $(n)_r > (q-1)_r$, or $r = 2$ and $q \equiv 3 \pmod 4$.

Now assume $i \geqslant 2$. By Proposition 4.3.2(ii), if every element of order $r$ in $T$ fixes an $m$-space then $i \leqslant m$ and $k \geqslant \ell$, so we may assume that these conditions hold. Set $\hat{x} = [\Lambda^{(n-k)/i}, I_k]$. If $U$ is an $m$-space fixed by $\hat{x}$ then $\hat{x}|_U = [\Lambda^{(m-\ell)/i}, I_\ell]$, and similarly $\hat{x}|_W = [\Lambda^{(n-m-k+\ell)/i}, I_{k-\ell}]$ if $W$ is an $(n-m)$-space fixed by $\hat{x}$. In particular, if $x$ fixes an element of $\Omega$ then $\ell \leqslant k - \ell$, so $x$ is a derangement if $k < 2\ell$.

Finally, let us assume $2 \leqslant i \leqslant m$ and $k \geqslant 2\ell$, so $n - 2m \equiv k - 2\ell \pmod i$. We claim that $T$ is $r$-elusive. To see this, let $y = \hat{y}Z \in T$ be an element of order $r$. As in Proposition 3.2.1 we may write $\hat{y} = [\Lambda_1^{a_1}, \ldots, \Lambda_t^{a_t}, I_{di+k}]$ with $d \geqslant 0$. Let $U$ be an $m$-space fixed by $\hat{y}$, say $\hat{y}|_U = [\Lambda_1^{b_1}, \ldots, \Lambda_t^{b_t}, I_{d'i+\ell}]$ with $d' \geqslant 0$. Note that $d' \leqslant d$, so $(d - d')i + k - \ell \geqslant k - \ell$. To see that $y$ has fixed points it suffices to show that

$$\hat{z} = [\Lambda_1^{a_1-b_1}, \ldots, \Lambda_t^{a_t-b_t}, I_{(d-d')i+k-\ell}] \in \mathrm{GL}_{n-m}(q)$$

fixes an $(n-2m)$-space. Since $n - 2m \equiv k - 2\ell \pmod i$, we may write $n - 2m = ci + k - 2\ell$ for some $c \geqslant 0$. If $c = 0$ then $(d - d')i + k - \ell \geqslant k - \ell \geqslant n - 2m$, so $\hat{z}$ fixes an $(n-2m)$-space by Lemma 4.2.1(i). On the other hand, if $c > 0$ then $n - 2m \geqslant i > \ell$ and thus Lemma 4.2.1(ii) implies that $\hat{z}$ fixes an $(n-2m)$-space. We conclude that $T$ is $r$-elusive, as claimed. $\qquad\square$

## 4.4 Unitary groups

In this section we turn our attention to the cases labelled II and III in Table 4.1.1. Here $G \leqslant \mathrm{Sym}(\Omega)$ is a primitive group with socle $T = \mathrm{PSU}_n(q)$, and $\Omega$ is a set of $m$-dimensional subspaces of the natural $T$-module $V$, which are all either totally isotropic or nondegenerate. Let $Z$ denote the centre of $\mathrm{GU}_n(q)$.

Let $r \neq p$ be a prime divisor of $|\Omega|$ and set $i = \Phi(r,q)$,

$$b = b(i) = \begin{cases} i/2 & i \text{ even} \\ i & i \text{ odd} \end{cases} \tag{4.4.1}$$

and

$$c = c(i) = \begin{cases} i & i \equiv 0 \pmod 4 \\ i/2 & i \equiv 2 \pmod 4 \\ 2i & i \text{ odd} \end{cases} \tag{4.4.2}$$

Table 4.4.1 *r-elusive subspace actions,* $T = \mathrm{PSU}_n(q)$, $r \neq p$, $r > 2$

| Case | $i$ | Conditions |
|------|-----|------------|
| II | $i \not\equiv 2 \ (\mathrm{mod}\ 4)$ | $b \leqslant m$ and $k \geqslant 2\ell$ |
|  | $i \equiv 2 \ (\mathrm{mod}\ 4),\ i > 2$ | $b \leqslant m$ and $k \geqslant 2\ell$ and $n' - 2m' \geqslant s$ |
|  | $i = 2$ | $b \leqslant m$ and $k \geqslant 2\ell$ and $n' - 2m' \geqslant s$ |
|  |  | and either $r$ divides $m$ or $(q+1)_r \geqslant (n)_r$ |
| III | $i \neq 2$ | $c \leqslant m$ and $k \geqslant \ell'$ |
|  | $i = 2$ | $r$ divides $m$ or $(q+1)_r \geqslant (n)_r$ |

$n \equiv k \ (\mathrm{mod}\ c)$, $m \equiv \ell \ (\mathrm{mod}\ b)$, $m \equiv \ell' \ (\mathrm{mod}\ c)$
$n' = \lfloor n/b \rfloor$, $m' = \lfloor m/b \rfloor$, $s = (r-1)/b$

Write

$$n \equiv k \ (\mathrm{mod}\ c), \quad m \equiv \ell \ (\mathrm{mod}\ b), \quad m \equiv \ell' \ (\mathrm{mod}\ c)$$

where $0 \leqslant k, \ell' < c$ and $0 \leqslant \ell < b$. Note that if $i \neq 2$ and $x = \hat{x}Z \in T$ has order $r$, then Proposition 3.3.2 implies that the irreducible blocks of $\hat{x}$ have dimension $b$, and $c$ is the minimal dimension of an $\hat{x}$-invariant nondegenerate subspace on which $\hat{x}$ acts nontrivially.

**Proposition 4.4.1** *In Case II of Table 4.1.1, $T$ is $r$-elusive if and only if one of the following holds:*

(i) $r = 2$, $p \neq 2$ *and either* $m < n/2$ *or* $(q+1)_2 \leqslant (n)_2$.
(ii) $r \neq p$, $r > 2$, $b \leqslant m$, $k \geqslant 2\ell$ *and either* $i \not\equiv 2 \ (\mathrm{mod}\ 4)$, *or* $i \equiv 2 \ (\mathrm{mod}\ 4)$, $n' - 2m' \geqslant s$ *and one of the following holds:*
   (a) $i > 2$;
   (b) $i = 2$ *and either* $r$ *divides* $m$, *or* $(q+1)_r \geqslant (n)_r$,
   *where* $n' = \lfloor n/b \rfloor$, $m' = \lfloor m/b \rfloor$ *and* $s = (r-1)/b$.

*Proof* Here $T = \mathrm{PSU}_n(q)$ and $\Omega$ is the set of totally isotropic $m$-dimensional subspaces of $V$, with $m \leqslant n/2$. The degree $|\Omega|$ is recorded in Table 4.1.2, and it follows that $r \neq p$ and $r$ divides $q^j - (-1)^j$ for some integer $j$ in the range $n - 2m + 1 \leqslant j \leqslant n$. Let $x = \hat{x}Z \in T$ be an element of order $r$.

*Case 1.* $r = 2$. First assume $r = 2$, so $p$ is odd. The conjugacy classes of involutions in $T$ are described in Section 3.3.2, and we adopt the notation therein. Clearly, if $n$ is even then involutions of type $t'_{n/2}$ have fixed points, so we may assume that $x = t_i$ with $1 \leqslant i \leqslant n/2$. Suppose $x$ fixes a totally isotropic $m$-space $U$, so $\hat{x}|_U = [-I_s, I_{m-s}]$ for some integer $s$ in the range $0 \leqslant s \leqslant m$. Then the

$(-1)$-eigenspace of $\hat{x}$ contains a totally isotropic $s$-space, and $C_V(\hat{x})$ (that is, the 1-eigenspace of $\hat{x}$) contains a totally isotropic $(m-s)$-space. Since the two eigenspaces of $\hat{x}$ are nondegenerate, of dimension $i$ and $n-i$, it follows that a maximal totally isotropic subspace of the $(-1)$-eigenspace of $\hat{x}$ has dimension $\lfloor i/2 \rfloor$, and a maximal totally isotropic subspace of $C_V(\hat{x})$ has dimension $\lfloor (n-i)/2 \rfloor$. Therefore, a $t_i$ involution is a derangement if and only if $i$ is odd and $m = n/2$, and such elements lie in $T$ if and only if $(q+1)_2 > (n)_2$ (see Proposition 3.3.6).

For the remainder, let us assume $r > 2$. Set $i = \Phi(r,q)$ and let $j$ be the largest integer in the range $n - 2m + 1 \leqslant j \leqslant n$ such that $r$ divides $q^j - (-1)^j$. Note that $r$ divides $q^{2j} - 1$, so $i$ divides $2j$. By Lemma A.3 (see Appendix A), $n - j = k < 2m$ and $c$ divides $j$.

*Case 2. $r > 2$, $i \not\equiv 2 \pmod 4$.* Set $\hat{x} = [(\Lambda, \Lambda^{-q})^{j/c}, I_{n-j}]$ as in Proposition 3.3.2(i), with $|\Lambda| = b$. Note that $C_V(\hat{x})$ is nondegenerate (or trivial). Since $n - j < 2m$, $C_V(\hat{x})$ does not contain a totally isotropic $m$-space, so if $b > m$ then $x$ is a derangement by Lemma 4.2.2. Similarly, if $b \leqslant m$ and $k < 2\ell$ then $C_V(\hat{x})$ does not contain a totally isotropic $\ell$-space, so once again Lemma 4.2.2 implies that $x$ is a derangement. Now assume that $b \leqslant m$ and $k \geqslant 2\ell$. Here Proposition 3.3.2 implies that $\dim C_V(\hat{y}) \geqslant k$ for every element $y = \hat{y}Z \in T$ of order $r$. Therefore $C_V(\hat{y})$ contains a totally isotropic $\ell$-space, and thus Lemma 4.2.2 implies that $y$ has fixed points.

*Case 3. $r > 2$, $i \equiv 2 \pmod 4$, $i > 2$.* Next assume $i \equiv 2 \pmod 4$ and $i > 2$. Set $\hat{x} = [\Lambda^{2j/i}, I_{n-j}]$ as in Proposition 3.3.2(ii), with $|\Lambda| = b = i/2 = c$. As before, $C_V(\hat{x})$ does not contain a totally isotropic $m$-space, so Lemma 4.2.4 implies that $x$ is a derangement if $b > m$. Similarly, if $b \leqslant m$ and $k < 2\ell$ then $C_V(\hat{x})$ does not contain a totally isotropic $\ell$-space, so once again Lemma 4.2.4 implies that $x$ is a derangement. Therefore, we may assume that $b \leqslant m$ and $k \geqslant 2\ell$. Write $n = n'b + k$ and $m = m'b + \ell$ and note that $m' \geqslant 1$ and $n' \geqslant 2m'$ since $m \geqslant b$, $n \geqslant 2m$ and $b > k \geqslant 2\ell$. Set $s = (r-1)/b$.

If $n' - 2m' < s$ then Lemma 4.2.4 implies that $x$ is a derangement, where

$$\hat{x} = [\Lambda_1^{2m'-1}, \Lambda_2, \ldots, \Lambda_{n'-2m'+1}, I_{k+b}]$$

(Note that $\dim C_V(\hat{x}) < 2b \leqslant 2m$, so $C_V(\hat{x})$ does not contain a totally isotropic $m$-space. Also note that $a_0 = 0$ in the notation of Lemma 4.2.4(ii).). Now assume $n' - 2m' \geqslant s$ and let $y = \hat{y}Z \in T$ be an element of order $r$. By Proposition 3.3.2(ii), we may write

$$\hat{y} = [\Lambda_1^{a_1}, \ldots, \Lambda_s^{a_s}, I_{bd+k}]$$

for some integers $a_j, d \geqslant 0$. Consider the $n'$ blocks

$$\underbrace{\Lambda_1, \dots, \Lambda_1}_{a_1}, \dots, \underbrace{\Lambda_s, \dots, \Lambda_s}_{a_s}, \underbrace{I_b, \dots, I_b}_{d}$$

of size $b$ comprising $\hat{y}$, and note that $n' = d + \sum_{j=1}^{s} a_j$. Since $n' - 2m' \geqslant s$, Lemma 4.2.5 implies that there is a collection of $m'$ pairs of repeated blocks (note that we are colouring $n'$ blocks of size $b$ from the palette $\{\Lambda_1, \dots, \Lambda_s, I_b\}$ of size $s + 1$). Write $m' = m'_0 + m'_1$, where $m'_0 \leqslant \lfloor d/2 \rfloor$ is the number of pairs $\{I_b, I_b\}$ in this collection. Since $\sum_{j=1}^{s} \lfloor a_j/2 \rfloor \geqslant m'_1$, it follows that

$$b \left( \lfloor d/2 \rfloor + \sum_{j=1}^{s} \lfloor a_j/2 \rfloor \right) \geqslant b \left( m'_0 + \sum_{j=1}^{s} \lfloor a_j/2 \rfloor \right) \geqslant bm' = m - \ell$$

We may write $C_V(\hat{y}) = X \perp Y$, where $X$ and $Y$ are nondegenerate spaces of dimension $k$ and $bd$, respectively. Since $k \geqslant 2\ell$, it follows that $X$ contains a totally isotropic $\ell$-space. Similarly, $Y$ contains a totally isotropic $bm'_0$-space and thus $C_V(\hat{y})$ contains a totally isotropic $(bm'_0 + \ell)$-space. Therefore, the conditions in Lemma 4.2.4(ii) are satisfied (taking $a_0 = 2m'_0$) and we conclude that $y$ fixes a totally isotropic $m$-space. In particular, $T$ is $r$-elusive.

*Case 4. $r > 2$, $i = 2$.* Finally, suppose that $i = 2$, so $b = c = 1$ and $k = \ell = 0$. The elements $x \in T$ of order $r$ are described in parts (i) and (ii) of Proposition 3.3.3. Clearly, elements of type (ii) have fixed points if and only if $r$ divides $m$ (see Remark 3.3.5), and $T$ contains such elements if and only if $(n)_r > (q+1)_r$.

Finally, let us assume $x$ is of type (i) (that is, $\hat{x}$ is diagonalisable), say

$$\hat{x} = [\mu I_{a_1}, \dots, \mu^{r-1} I_{a_{r-1}}, I_{a_0}]$$

for some primitive $r$th root of unity $\mu \in \mathbb{F}_{q^2}$. By Lemma 4.2.4, $x$ fixes a totally isotropic $m$-space if and only if it has $m$ pairs of repeated eigenvalues. Therefore Lemma 4.2.5 implies that $\mathrm{PGU}_n(q)$ (and thus $T$ also) contains no derangements of type (i) if $n - 2(m-1) > r$ (that is, if $n - 2m \geqslant s$ where $s = (r-1)/b$).

Now assume $n - 2(m-1) \leqslant r$. Here Lemma 4.2.4 implies that $\mathrm{PGU}_n(q)$ contains derangements of order $r$; more precisely, $x$ is a derangement if and only if $\sum_j \lfloor a_j/2 \rfloor < m$. We need to determine whether or not such elements lie in $T$. Of course, if $(n, r) = 1$ then every element of order $r$ in $\mathrm{PGU}_n(q)$ lies in $T$. Now assume $r$ divides $n$, say $n = dr$. Note that

$$(d-1)r/2 + 1 \leqslant m \leqslant dr/2$$

since $n - 2(m-1) \leqslant r$. Set $\hat{x} = [\mu, \mu^2, \ldots, \mu^{r-1}, I_{n-r+1}]$. Then $\det(\hat{x}) = 1$ and

$$\left\lfloor \frac{n-r+1}{2} \right\rfloor \leqslant \frac{1}{2}((d-1)r+1) < m$$

Therefore $x \in T$ is a derangement of order $r$. □

**Proposition 4.4.2** *In Case III of Table 4.1.1, $T$ is $r$-elusive if and only if one of the following holds:*

(i) *$r = p > 2$ and every partition of $n$ with parts of size at most $p$ contains a partition of $m$.*
(ii) *$r = p = 2$ and either $n$ is odd or $m$ is even.*
(iii) *$r = 2$, $p \neq 2$ and either $n$ is odd, or $m$ is even, or $(q+1)_2 \geqslant (n)_2$ and $q \equiv 3 \pmod 4$.*
(iv) *$r \neq p$, $r > 2$ and one of the following holds:*
    (a) *$i \neq 2$, $c \leqslant m$ and $k \geqslant \ell'$;*
    (b) *$i = 2$ and either $r$ divides $m$, or $(q+1)_r \geqslant (n)_r$.*

*Proof* Here $T = \mathrm{PSU}_n(q)$ and $\Omega$ is the set of nondegenerate $m$-dimensional subspaces of $V$, with $m < n/2$. The degree $|\Omega|$ is given in Table 4.1.2; it follows that $r = p$, or $r$ divides $q^j - (-1)^j$ for some $j$ in the range $n - m + 1 \leqslant j \leqslant n$. Let $x = \hat{x}Z \in T$ be an element of order $r$.

First assume $r = p$. If $\hat{x}$ fixes a nondegenerate $m$-space $U$ then $\hat{x}$ also fixes $U^\perp$ (see Lemma 2.2.16), so $\hat{x} \in \mathrm{GU}_m(q) \times \mathrm{GU}_{n-m}(q) =: L$. Moreover, $\hat{x}$ is $L$-conjugate to $[J_p^{a_p}, \ldots, J_1^{a_1}, J_p^{b_p}, \ldots, J_1^{b_1}]$ by Proposition 3.3.7, where $\sum_v v a_v = m$ and $\sum_v v b_v = n - m$. Therefore $\hat{x}$ is $\mathrm{GU}_n(q)$-conjugate to $[J_p^{a_p+b_p}, \ldots, J_1^{a_1+b_1}]$ and it follows that $T$ is $p$-elusive if and only if every partition of $n$ with parts of size at most $p$ contains a partition of $m$. In particular, if $p = 2$ then $T$ is 2-elusive if and only if $n$ is odd or $m$ is even (see Remark 4.3.5 for further comments).

Next suppose $r = 2$ and $p \neq 2$. Involutions of type $t_i$, with $1 \leqslant i \leqslant n/2$, clearly fix a nondegenerate $m$-space, so let us assume $n$ is even and $x$ is of type $t'_{n/2}$. As described in Section 3.3.2.2, we can take $\hat{x} = [\xi I_{n/2}, \xi^{-q} I_{n/2}]$ (with respect to an hermitian basis $\{e_1, \ldots, e_{n/2}, f_1, \ldots, f_{n/2}\}$), where $\xi \in \mathbb{F}_{q^2}$ has order $2(q+1)_2$. If $U$ is a nondegenerate $m$-space fixed by $x$ then $\xi$ and $\xi^{-q}$ are the eigenvalues of $\hat{x}|_U$, with equal multiplicities. Therefore, $x$ has fixed points if and only if $m$ is even, and we note that $x \in T$ if and only if $q \equiv 1 \pmod 4$ or $(n)_2 > (q+1)_2$.

For the remainder we may assume $r \neq p$ is odd. Set $i = \Phi(r,q)$ and let $j$ be the largest integer in the range $n - m + 1 \leqslant j \leqslant n$ such that $r$ divides $q^j - (-1)^j$. Note that $i$ divides $2j$. By Lemma A.3, $k = n - j < m$ and $c$ divides $j$.

First assume $i \not\equiv 2 \pmod 4$, so $c = 2b$. As in the proof of the previous proposition, set $\hat{x} = [(\Lambda, \Lambda^{-q})^{j/c}, I_{n-j}]$. Since $n - j < m$, $C_V(\hat{x})$ does not contain an $m$-space and thus Lemma 4.2.3 implies that $x$ is a derangement if $c > m$. The same conclusion holds if $c \leqslant m$ and $k < \ell'$ (again via Lemma 4.2.3), so we may assume that $c \leqslant m$ and $k \geqslant \ell'$. Let $y = \hat{y}Z \in T$ be an element of order $r$. By Proposition 3.3.2, $C_V(\hat{y})$ is nondegenerate and $\dim C_V(\hat{y}) \geqslant k \geqslant \ell'$, so it contains a nondegenerate $\ell'$-space. Therefore, Lemma 4.2.3 implies that $y$ fixes an element of $\Omega$ and we conclude that $T$ is $r$-elusive.

Next suppose $i \equiv 2 \pmod 4$ and $i > 2$, so $b = c = i/2$. Set $\hat{x} = [\Lambda^{2j/i}, I_{n-j}]$ and note that $\dim C_V(\hat{x}) < m$. By applying Lemma 4.2.1, we deduce that $x$ is a derangement if $c > m$, or if $c \leqslant m$ and $k < \ell'$. On the other hand, if $c \leqslant m$ and $k \geqslant \ell'$ then Proposition 3.3.2 and Lemma 4.2.1 imply that every element in $T$ of order $r$ has fixed points.

Finally, let us assume that $i = 2$, so $c = 1$ and $k = \ell' = 0$. The possibilities for $x$ are described in parts (i) and (ii) of Proposition 3.3.3. Evidently, $x$ fixes a nondegenerate $m$-space if it is of type (i). However, if $x$ is of type (ii) then $x$ is a derangement if and only if $(m,r) = 1$, and we note that $T$ contains such elements if and only if $(n)_r > (q+1)_r$. $\qquad\square$

## 4.5 Symplectic groups

In this section we deal with the cases labelled IV (totally isotropic $m$-spaces) and V (nondegenerate $m$-spaces) in Table 4.1.1, so $G \leqslant \mathrm{Sym}(\Omega)$ is a primitive group with socle $T = \mathrm{PSp}_n(q)$. Let $Z$ denote the centre of $\mathrm{GSp}_n(q)$.

Let $r \neq p$ be a prime divisor of $|\Omega|$ and set $i = \Phi(r,q)$ and

$$c = c(i) = \begin{cases} i & i \text{ even} \\ 2i & i \text{ odd} \end{cases} \tag{4.5.1}$$

Write

$$n \equiv k \pmod c, \quad m \equiv \ell \pmod i, \quad m \equiv \ell' \pmod c$$

where $0 \leqslant k, \ell' < c$ and $0 \leqslant \ell < i$. Note that if $x = \hat{x}Z \in T$ has order $r \neq p$, then Proposition 3.4.3 implies that the irreducible blocks of $\hat{x}$ are $i$-dimensional, and $c$ is the minimal dimension of an $\hat{x}$-invariant nondegenerate subspace on which $\hat{x}$ acts nontrivially.

Table 4.5.1 *r-elusive subspace actions, $T = \mathrm{PSp}_n(q)$, $r \neq p$, $r > 2$*

| Case | $i$ | Conditions |
|---|---|---|
| IV | odd | $i \leqslant m$ and $k \geqslant 2\ell$ |
|  | even | $i \leqslant m$ and $k \geqslant 2\ell$ and $n' - 2m' \geqslant t$ |
| V | all | $c \leqslant m$ and $k \geqslant \ell'$ |

$n \equiv k \pmod c$, $m \equiv \ell \pmod i$, $m \equiv \ell' \pmod c$
$n' = \lfloor n/i \rfloor$, $m' = \lfloor m/i \rfloor$, $t = (r-1)/i$

**Proposition 4.5.1** *In Case IV of Table 4.1.1, T is r-elusive if and only if one of the following holds:*

(i) *$r = 2$, $p \neq 2$ and either m is even or $q \equiv 1 \pmod 4$.*

(ii) *$r \neq p$, $r > 2$, $i \leqslant m$, $k \geqslant 2\ell$ and either i is odd or $n' - 2m' \geqslant t$, where $n' = \lfloor n/i \rfloor$, $m' = \lfloor m/i \rfloor$ and $t = (r-1)/i$.*

*Proof* Here $T = \mathrm{PSp}_n(q)$ and $\Omega$ is the set of totally isotropic $m$-dimensional subspaces of $V$ (with $m \leqslant n/2$). The degree $|\Omega|$ is given in Table 4.1.2. Since $r$ divides $|\Omega|$, it follows that $r \neq p$ and $r$ divides $q^j - 1$ for some even integer $j$ in the range $n - 2m + 2 \leqslant j \leqslant n$.

Suppose $r = 2$, so $p$ is odd. The conjugacy classes of involutions in $T$ are described in Section 3.4.2 and we use the notation therein for class representatives. Involutions of type $t_i$, with $1 \leqslant i \leqslant n/4$, clearly fix a totally isotropic $m$-space, as do those of type $t_{n/2}$. Involutions of type $t'_{n/2}$ fix even-dimensional totally isotropic subspaces, but they do not fix any odd-dimensional subspaces. Since the latter involutions belong to $T$ if and only if $q \equiv 3 \pmod 4$, we conclude that $T$ is 2-elusive if and only if $m$ is even or $q \equiv 1 \pmod 4$.

For the remainder, let us assume $r > 2$. Let $x = \hat{x}Z \in T$ be an element of order $r$. Let $j$ be the largest even integer such that $n - 2m + 2 \leqslant j \leqslant n$ and $r$ divides $q^j - 1$. Note that $n - j = k$ and $c$ divides $j$ (see Lemma A.2).

First assume $i$ is odd, so $c = 2i$. We set

$$\hat{x} = [(\Lambda, \Lambda^{-1})^{j/2i}, I_{n-j}] \tag{4.5.2}$$

as in Proposition 3.4.3(i). Note that $C_V(\hat{x})$ is nondegenerate (or trivial). In particular, $C_V(\hat{x})$ does not contain a totally isotropic $m$-space since $n - j < 2m$, so Lemma 4.2.2 implies that $x$ is a derangement if $i > m$. Now assume $i \leqslant m$. If $k < 2\ell$ then $C_V(\hat{x})$ does not contain a totally isotropic $\ell$-space, and so once again Lemma 4.2.2 implies that $x$ is a derangement. On the other hand, if $i \leqslant m$ and $k \geqslant 2\ell$ then by applying Proposition 3.4.3 we deduce that $\dim C_V(\hat{y}) \geqslant k$

for every $y = \hat{y}Z \in T$ of order $r$. Therefore Lemma 4.2.2 implies that every element in $T$ of order $r$ fixes a totally isotropic $m$-space.

Now assume $i$ is even, so $c = i$. Set

$$\hat{x} = [\Lambda^{j/i}, I_{n-j}] \tag{4.5.3}$$

as in Proposition 3.4.3(ii), so the irreducible blocks of $\hat{x}$ are nondegenerate $i$-spaces. As before, $C_V(\hat{x})$ does not contain a totally isotropic $m$-space, so $x$ is a derangement if $i > m$ (see Lemma 4.2.4). Now assume $i \leqslant m$. If $k < 2\ell$ then $C_V(\hat{x})$ does not contain a totally isotropic $\ell$-space, so again we deduce that $x$ is a derangement via Lemma 4.2.4.

Finally, suppose that $i \leqslant m$ and $k \geqslant 2\ell$. Write $n = n'i + k$ and $m = m'i + \ell$, and note that $m' \geqslant 1$ and $n' \geqslant 2m'$ since $i \leqslant m$ and $n \geqslant 2m$. Set $t = (r-1)/i$. If $n' - 2m' < t$ then set

$$\hat{x} = [\Lambda_1^{2m'-1}, \Lambda_2, \dots, \Lambda_{n'-2m'+1}, I_{k+i}] \tag{4.5.4}$$

where the $\Lambda_s$ are distinct. Note that $\dim C_V(\hat{x}) = k + i < 2i \leqslant 2m$, so $C_V(\hat{x})$ does not contain a totally isotropic $m$-space. Moreover, if $a_0 \geqslant 0$ is an even integer such that $C_V(\hat{x})$ contains a totally isotropic $(ia_0/2 + \ell)$-space, then $a_0 = 0$ and thus Lemma 4.2.4 implies that $x$ is a derangement.

Now assume that $n' - 2m' \geqslant t$. We claim that $T$ is $r$-elusive. To see this, let $y = \hat{y}Z \in T$ be an element of order $r$, where

$$\hat{y} = [\Lambda_1^{a_1}, \dots, \Lambda_t^{a_t}, I_{id+k}]$$

Consider the $n'$ blocks of size $i$ comprising $\hat{y}$, namely

$$\underbrace{\Lambda_1, \dots, \Lambda_1}_{a_1}, \dots, \underbrace{\Lambda_t, \dots, \Lambda_t}_{a_t}, \underbrace{I_i, \dots, I_i}_{d} \tag{4.5.5}$$

Since $n' - 2m' \geqslant t$, Lemma 4.2.5 implies that $\hat{y}$ has $m'$ pairs of repeated blocks of size $i$ (we are colouring $n'$ blocks of size $i$ by the $(r-1)/i+1$ distinct colours of type $\Lambda_s$ and $I_i$). Write $m' = m_0' + m_1'$, where $m_0'$ is the number of pairs of the form $\{I_i, I_i\}$ in this collection. Since $m_0' \leqslant \lfloor d/2 \rfloor$ and $k \geqslant 2\ell$, it is easy to see that $C_V(\hat{y})$ contains a totally isotropic $(im_0' + \ell)$-space. Moreover,

$$i\left(m_0' + \sum_{s=1}^{t} \lfloor a_s/2 \rfloor\right) \geqslant im' = m - \ell \tag{4.5.6}$$

and thus Lemma 4.2.4 implies that $y$ fixes a totally isotropic $m$-space. This justifies the claim. $\qquad\square$

**Proposition 4.5.2** *In Case V of Table 4.1.1, $T$ is $r$-elusive if and only if one of the following holds:*

(i) $r = p > 2$ and every partition of $n$ with parts of size at most $p$ and all odd parts with even multiplicity contains a partition of $m$ with all odd parts of even multiplicity.

(ii) $r = p = 2$ and either $n \equiv 2 \pmod{4}$ or $m \equiv 0 \pmod{4}$.

(iii) $r = 2$, $p \neq 2$.

(iv) $r \neq p$, $r > 2$, $c \leqslant m$ and $k \geqslant \ell'$.

*Proof* We have $T = \mathrm{PSp}_n(q)$ and $\Omega$ is the set of nondegenerate $m$-dimensional subspaces of $V$ (so $m$ is even and we assume $m < n/2$). Let $r$ be a prime divisor of $|\Omega|$, where $|\Omega|$ is given in Table 4.1.2. If $r \neq p$ then $r$ divides $q^j - 1$ for an even integer $j$ in the range $n - m + 2 \leqslant j \leqslant n$. Let $x = \hat{x}Z \in T$ be an element of order $r$.

First assume $r = p > 2$. By Proposition 3.4.10, there is a bijection between the set of $T$-classes of elements of order $p$ in $T$, and the set of nontrivial *signed partitions* of $n$ of the form

$$(p^{a_p}, \varepsilon_{p-1}(p-1)^{a_{p-1}}, \ldots, \varepsilon_2 2^{a_2}, 1^{a_1})$$

where $\varepsilon_i = \pm$ for all even $i$, and $a_i$ is even for all odd $i$. From the description of these elements in Proposition 3.4.10, it follows that if $x, y \in T$ are elements of order $p$ with the same Jordan form on $V$, then $x$ fixes a nondegenerate $m$-space if and only if $y$ does. In particular, the signs in the partition corresponding to $x$ play no role in determining whether or not $x$ fixes a nondegenerate $m$-space. Now, if $x$ fixes a nondegenerate $m$-space $U$ then $x$ also fixes $U^\perp$, so $\hat{x} \in \mathrm{Sp}_m(q) \times \mathrm{Sp}_{n-m}(q)$ and thus the partition of $n$ corresponding to $x$ contains a partition of $m$ such that all odd parts have even multiplicity. We conclude that $T$ is $p$-elusive if and only if the condition in (i) holds, and we refer the reader to Remark 4.5.3 for further comments on this number-theoretic criterion.

Next assume that $r = p = 2$. Let $x \in T$ be an involution of type $a_s$, $b_s$ or $c_s$ (see Section 3.4.4). From the definition of these elements, it is clear that involutions of type $b_s$ and $c_s$ fix a nondegenerate $m$-space, as do elements of type $a_s$ with $s < n/2$. Finally, suppose $n \equiv 0 \pmod{4}$ and $x \in T$ is an involution of type $a_{n/2}$. If $m \equiv 0 \pmod{4}$ then $x$ has fixed points, so assume $m \equiv 2 \pmod{4}$. If $x$ fixes a nondegenerate $m$-space $U$ then Lemma 3.4.14 implies that the restriction $x|_U \in \mathrm{Sp}(U)$ is of type $a_{m/2}$, but this is not possible since $m/2$ is odd. We conclude that involutions of type $a_{n/2}$ are derangements if and only if $m \equiv 2 \pmod{4}$.

Next suppose that $r = 2$ and $p$ is odd. Recall the labelling of involution class representatives described in Section 3.4.2. It is easy to see that involutions of type $t_i$, with $1 \leqslant i \leqslant n/4$, fix a nondegenerate $m$-space, as do those of type $t_{n/2}$. As explained in Section 3.4.2.5, involutions of type $t'_{n/2}$ fix an orthogonal decomposition of $V$ into $n/2$ distinct 1-dimensional $\mathbb{F}_{q^2}$-spaces, which are nondegenerate with respect to an hermitian form, so they fix an orthogonal decomposition of $V$ into nondegenerate 2-dimensional $\mathbb{F}_q$-spaces (see Construction 2.4.3). It follows that these involutions also have fixed points on $\Omega$, and we conclude that $T$ is 2-elusive if $p$ is odd.

For the remainder, let us assume $r \neq p$ and $r > 2$. Set $i = \Phi(r,q)$ and choose $j$ to be the largest even integer such that $r$ divides $q^j - 1$ and $n - m + 2 \leqslant j \leqslant n$. By Lemma A.2, $n - j = k$ and $c$ divides $j$, where $c$ is defined in (4.5.1).

First assume $i$ is odd, so $c = 2i$. Define $\hat{x}$ as in (4.5.2). Since $n - j < m$, $C_V(\hat{x})$ does not contain an $m$-space and thus Lemma 4.2.3 implies that $x$ is a derangement if $c > m$. Similarly, if $k < \ell'$ then $C_V(\hat{x})$ does not contain an $\ell'$-space, so once again Lemma 4.2.3 implies that $x$ is a derangement. Now assume that $c \leqslant m$ and $k \geqslant \ell'$. By Proposition 3.4.3(i), if $y = \hat{y}Z \in T$ has order $r$ then $\dim C_V(\hat{y}) \geqslant k \geqslant \ell'$, so $C_V(\hat{y})$ contains a nondegenerate $\ell'$-space (note that $m$ and $c$ are both even, so $\ell'$ is also even). Hence Lemma 4.2.3 implies that $y$ fixes an element of $\Omega$.

Finally, let us assume $i$ is even, so $c = i$. Here we set $\hat{x}$ as in (4.5.3). Since $\dim C_V(\hat{x}) = n - j < m$, Lemma 4.2.1 implies that $x$ is a derangement if $c > m$, or if $c \leqslant m$ and $k < \ell'$. Now assume $c \leqslant m$ and $k \geqslant \ell'$. If $y = \hat{y}Z \in T$ has order $r$ then Proposition 3.4.3(ii) implies that $\dim C_V(\hat{y}) \geqslant k \geqslant \ell'$, so $y$ has fixed points on $\Omega$ by Lemma 4.2.1. $\qquad\square$

**Remark 4.5.3** Consider the number-theoretic condition in part (i) of Proposition 4.5.2 (this is denoted by P2 in the statement of Theorem 4.1.4). We can obtain a more concrete criterion for $p$-elusivity when $p = 3$. Indeed, we claim that if $p = 3$ then $T$ is 3-elusive if and only if one of the following holds:

(a) $m \equiv 0 \pmod 6$;
(b) $n \equiv 4 \pmod 6$;
(c) $n \equiv 2 \pmod 6$ and $m \equiv 2 \pmod 6$.

To see this, first observe that $n \geqslant 6$ (since $2 \leqslant m < n/2$). It is easy to see that every partition of $n$ of the form $(3^{a_3}, 2^{a_2}, 1^{a_1})$, where $a_1$ and $a_3$ are even, contains a partition of 6 (with the property that all odd parts have an even multiplicity). Therefore $T$ is 3-elusive if $m$ or $n - m$ is divisible by 6, so let us assume otherwise.

If $n \equiv 0 \pmod 6$ and $m \not\equiv 0 \pmod 6$, then any element in $T$ with Jordan form $[J_3^{n/3}]$ is a derangement. Similarly, if $n \equiv 2 \pmod 6$ and $m \equiv 4 \pmod 6$, then $[J_3^{(n-2)/3}, J_2]$ is a derangement. However, if $n \equiv 4 \pmod 6$ and $m \equiv 2 \pmod 6$ then every valid partition $\lambda$ of $n$ (that is, $\lambda = (3^{a_3}, 2^{a_2}, 1^{a_1})$ with $a_1$ and $a_3$

even) contains a partition of 8 in which all odd parts have an even multiplicity, and since $m - 8$ is divisible by 6 we deduce that $T$ is 3-elusive. This justifies the claim.

## 4.6 Orthogonal groups

In this final section we turn our attention to the remaining cases in Table 4.1.1, and we complete the proofs of Theorems 4.1.4, 4.1.6 and 4.1.7. Here $G \leqslant \mathrm{Sym}(\Omega)$ is a primitive group with socle $T = \mathrm{P}\Omega_n^\varepsilon(q)$, and the relevant cases are labelled VI–XVI in Table 4.1.1. Let $r$ be a prime divisor of $|\Omega|$ and let $Z$ denote the centre of $\mathrm{GO}_n^\varepsilon(q)$.

If $r \neq p$ then set $i = \Phi(r, q)$ and define the integer $c = c(i)$ as in (4.5.1). As before, we write

$$n \equiv k \pmod{c}, \quad m \equiv \ell \pmod{i}, \quad m \equiv \ell' \pmod{c}$$

where $0 \leqslant k, \ell' < c$ and $0 \leqslant \ell < i$. Note that if $x = \hat{x}Z \in T$ has order $r$, then Proposition 3.5.4 implies that the irreducible blocks of $\hat{x}$ are $i$-dimensional, and $c$ is the minimal dimension of a nondegenerate $\hat{x}$-invariant subspace on which $\hat{x}$ acts nontrivially.

**Remark 4.6.1** The relevant $r$-elusive actions of $T$ (with $r \neq p$ and $r > 2$) are recorded in Table 4.6.1. For example, if $i$ is even and $k > 0$ in Case X, then the table indicates that $T$ is $r$-elusive if and only if $i \leqslant m$ and either $k > \ell' > 0$, or $k = \ell' > 0$ and $(n + m - 2k)/i$ is even, or $k > \ell' = 0$ and $m/i$ is even. Note that Case IX does not appear in Table 4.6.1; in this case, $T$ contains derangements of order $r$ for every odd prime divisor $r \neq p$ of $|\Omega|$.

**Proposition 4.6.2** *In Case VI of Table 4.1.1, $T$ is $r$-elusive if and only if one of the following holds:*

(i) *$r = 2$, $p \neq 2$ and one of the following holds:*
  (a) *$q \equiv 1 \pmod 4$;*
  (b) *$q \equiv 3 \pmod 4$, $m \leqslant n/2 - 2$ and either $m$ is even or $n \equiv 2 \pmod 4$.*
(ii) *$r \neq p$, $r > 2$, $i$ is odd, $i \leqslant m$ and $k \geqslant 2\ell$.*
(iii) *$r \neq p$, $r > 2$, $i$ is even, $i \leqslant m$ and $n' - 2m' \geqslant t$ and one of the following holds:*
  (a) *$i \geqslant 2\ell$, $k = 0$, $n/i$ is odd and $n' - 2m' > t$;*
  (b) *$k \geqslant 2\ell$ and $n'$ is even;*
  (c) *$k > 2\ell$ and $n'$ is odd,*
  *where $n' = \lfloor n/i \rfloor$, $m' = \lfloor m/i \rfloor$ and $t = (r-1)/i$.*

Table 4.6.1 *r-elusive subspace actions,* $T = \mathrm{P}\Omega_n^\varepsilon(q)$, $r \neq p$, $r > 2$

| Case | $i$ | Conditions |
|------|-----|------------|
| VI | odd | $i \leqslant m$ and $k \geqslant 2\ell$ |
| | even | $i \leqslant m$ and $n' - 2m' \geqslant t$, $\begin{cases} i \geqslant 2\ell, k = 0, n/i \text{ odd}, n' - 2m' > t \\ k \geqslant 2\ell, n' \text{ even} \\ k > 2\ell, n' \text{ odd} \end{cases}$ |
| VII | odd | $i \leqslant m$ and either $k = 0$ or $k > 2\ell$ |
| | even | $i \leqslant m$ and $n' - 2m' \geqslant t$, $\begin{cases} i \geqslant 2\ell, k = 0, n/i \text{ even}, n' - 2m' > t \\ k \geqslant 2\ell, n' \text{ odd} \\ k > 2\ell, n' \text{ even} \end{cases}$ |
| VIII | odd | $i \leqslant m$ and $k > 2\ell$ |
| | even | $i \leqslant m$ and $k > 2\ell$ and $n' - 2m' \geqslant t$ |
| X | odd | $c \leqslant m$ and $k \geqslant \ell'$ |
| | even | $\begin{cases} k = 0 \text{ and either } n/i \text{ odd, or } \ell' = 0 \text{ and } m/i \text{ even} \\ i \leqslant m, \begin{cases} k > \ell' > 0 \\ k = \ell' > 0, (n+m-2k)/i \text{ even} \\ k > \ell' = 0, m/i \text{ even} \end{cases} \end{cases}$ |
| XI | odd | $c \leqslant m$ and $k > \ell' > 0$ |
| | even | $\begin{cases} k = 0 \text{ and either } n/i \text{ odd, or } \ell' = 0 \text{ and } m/i \text{ odd} \\ i \leqslant m, \begin{cases} k > \ell' > 0 \\ k = \ell' > 0, (n+m-2k)/i \text{ odd} \\ k > \ell' = 0, m/i \text{ odd} \end{cases} \end{cases}$ |
| XII | odd | $c \leqslant m$ and $k > \ell' > 0$ |
| | even | $k = 0$ and $n/i$ odd, or $i \leqslant m$ and $k > \ell'$ |
| XIII | odd | $k = 0$, or $c \leqslant m$ and $k \geqslant \ell' > 0$ |
| | even | $\begin{cases} k = 0 \text{ and either } n/i \text{ even, or } \ell' = 0 \text{ and } m/i \text{ odd} \\ i \leqslant m, \begin{cases} k > \ell' > 0 \\ k = \ell' > 0, (n+m-2k)/i \text{ even} \\ k > \ell' = 0, m/i \text{ odd} \end{cases} \end{cases}$ |
| XIV | odd | $k = 0$, or $c \leqslant m$ and $k > \ell' > 0$ |
| | even | $\begin{cases} k = 0, n/i \text{ even} \\ i \leqslant m \text{ and } k > \ell' > 0 \end{cases}$ |
| XV | odd | $c \leqslant m$ and $k > \ell' \geqslant 0$ |
| | even | $i \leqslant m$ and $k > \ell' \geqslant 0$, with $\ell' = 0$ if and only if $m/i$ even |
| XVI | odd | $c \leqslant m$ and $k > \ell' > 0$ |
| | even | $i \leqslant m$ and $k > \ell' \geqslant 0$, with $\ell' = 0$ if and only if $m/i$ odd |

$n \equiv k \pmod{c}$, $m \equiv \ell \pmod{i}$, $m \equiv \ell' \pmod{c}$
$n' = \lfloor n/i \rfloor$, $m' = \lfloor m/i \rfloor$, $t = (r-1)/i$

*Proof* Here $T = \mathrm{P}\Omega_n^+(q)$ and $\Omega$ is a $T$-orbit of totally singular $m$-spaces with $m \leqslant n/2$. Recall that $T$ acts transitively on the set of all totally singular $m$-spaces if $m < n/2$, and there are precisely two orbits if $m = n/2$, which are fused under the action of $\mathrm{PO}_n^+(q)$ (see Proposition 2.5.4, and recall that two totally singular $\frac{n}{2}$-spaces $U, W$ lie in the same $T$-orbit if and only if the codimension of $U \cap W$ in $U$ is even). The degree $|\Omega|$ is given in Table 4.1.2. Since $r$ divides $|\Omega|$, it follows that $r \neq p$ and $r$ divides $q^j - 1$, where $j$ is an even integer in the range $n - 2m \leqslant j \leqslant n$. Let $x = \hat{x}Z \in T$ be an element of order $r$.

*Case 1.* $r = 2$. First assume $r = 2$, so $p$ is odd. We adopt the notation for semisimple involutions given in Section 3.5.2. Let $x \in T$ be an involution of type $t_i$ or $t_i'$, where $i \leqslant n/4$. Note that if $x$ fixes a totally singular $m$-space $W$, then we obtain a decomposition $W = W_1 \oplus W_{-1}$, where $W_\alpha$ denotes the $\alpha$-eigenspace of $\hat{x}$ on $W$. If $x = t_i$ then a maximal totally singular subspace in the $(-1)$-eigenspace of $\hat{x}$ on $V$ has dimension $i$, and an analogous subspace in the $1$-eigenspace of $\hat{x}$ has dimension $n/2 - i$. Therefore, $x$ fixes a totally singular $m$-space in $\Omega$ (moreover, if $m = n/2$ then we can choose the fixed $m$-space to be in either of the two $T$-orbits). However, if $x = t_i'$ then the respective dimensions are $i - 1$ and $n/2 - i - 1$, so $\dim W \leqslant n/2 - 2$ and thus $x$ is a derangement if and only if $m = n/2 - 1$ or $n/2$. As noted in Proposition 3.5.10, $T$ contains one of these involutions if and only if $q \equiv 3 \pmod 4$.

Now, if $n \equiv 2 \pmod 4$ then involutions of type $t_{n/2}$ fix an element of $\Omega$ (note that if $m = n/2$ then the two eigenspaces of $t_{n/2}$ represent the two distinct $T$-orbits on totally singular $m$-spaces). Similarly, if $n \equiv 0 \pmod 4$, then the elements $t_{n/2}$ and $t_{n/2-1}$ have fixed points. By Proposition 3.5.10, the only other involutions in $T$ are those of type $t_{n/2}'$ or $t_{n/2-1}'$, where $n \equiv 0 \pmod 4$ and $q \equiv 3 \pmod 4$. From the description of these elements in Section 3.5.2.11, it quickly follows that they are derangements if $m$ is odd, and they have fixed points if $m < n/2$ is even. Finally, if $m = n/2$ then either $t_{n/2}'$ or $t_{n/2-1}'$ is a derangement (the particular class of derangements depends on the specific $T$-orbit of totally singular $\frac{n}{2}$-spaces comprising $\Omega$).

For the remainder, let us assume $r$ is odd and $x = \hat{x}Z \in T$ has order $r$. Set $i = \Phi(r, q)$ (note that $i < n$), define $c$ as in (4.5.1) and let $j$ be the largest even integer such that $n - 2m \leqslant j \leqslant n$ and $r$ divides $q^j - 1$. Note that $n - j = k \leqslant 2m$ and $c$ divides $j$ (see Lemma A.2).

We claim that $j = n - 2m$ only if $i$ is even, $j/i$ is odd and $i > m$. To see this, first observe that the condition $j = n - 2m$ implies that $r$ divides $q^{(n-2m)/2} + 1$, but not $q^{(n-2m)/2} - 1$. If $i$ is odd, then $2i$ divides $j$, so $i$ divides $j/2$ and thus $r$ divides $q^{(n-2m)/2} - 1$, which is a contradiction. Now assume $i$ is even. If $j/i$ is even then we reach the same contradiction as before, so let us assume that $j/i$ is odd. If $i \leqslant m$ then $2m = n - j = k < i \leqslant m$, which is absurd, so we must have $i > m$. This justifies the claim.

*Case 2.* $r > 2$, $i$ odd. First assume $i$ is odd, so $c = 2i$ and $n - j = k < 2m$ (as explained above). Set $\hat{x}$ as in (4.5.2), and note that $C_V(\hat{x})$ is hyperbolic (or trivial) and $\dim C_V(\hat{x}) = k = n - j < 2m$. Since $C_V(\hat{x})$ does not contain a totally singular $m$-space, Lemma 4.2.2 implies that $x$ is a derangement if $i > m$. Now assume $i \leqslant m$. If $k < 2\ell$ then $C_V(\hat{x})$ does not contain a totally singular $\ell$-space, and so once again Lemma 4.2.2 implies that $x$ is a derangement.

Now assume $i$ is odd, with $i \leqslant m$ and $k \geqslant 2\ell$. Suppose $y = \hat{y}Z \in T$ has order $r$. By Proposition 3.5.4(i), $\hat{y}$ has the form

$$\hat{y} = [(\Lambda_1, \Lambda_1^{-1})^{a_1}, \ldots, (\Lambda_{t/2}, \Lambda_{t/2}^{-1})^{a_{t/2}}, I_{n-2iu}]$$

where $u = \sum_v a_v$ and $t = (r-1)/i$, and we note that $C_V(\hat{y})$ is hyperbolic (or trivial). Since $\dim C_V(\hat{y}) \geqslant k \geqslant 2\ell$, $C_V(\hat{y})$ contains a totally singular $\ell$-space, so $\hat{y}$ fixes a totally singular $m$-space $W$, by Lemma 4.2.2. The same lemma also implies that $\hat{y}$ fixes a totally singular $m$-space $W'$ such that $W \cap W'$ has codimension $i$ in $W$. Since $i$ is odd, it follows that if $m = n/2$ then the two subspaces $W$ and $W'$ lie in different $T$-orbits. We conclude that $T$ is $r$-elusive if $i$ is odd, $i \leqslant m$ and $k \geqslant 2\ell$.

*Case 3. $r > 2$, $i$ even.* To complete the proof of the proposition, we may assume that $i$ is even, so $c = i$ and the irreducible blocks of $\hat{x}$ are elliptic $i$-spaces (see Proposition 3.5.4(ii)).

First assume that $j = n - 2m$, so $j/i$ is odd and $i > m$ (see the paragraph preceding Case 2 above). Set $x = \hat{x}Z \in T$, where $\hat{x} = [\Lambda^{j/i}, I_{n-j}]$. Here $C_V(\hat{x})$ is an elliptic $2m$-space, which does not contain a totally singular $m$-space, so $x$ is a derangement by Lemma 4.2.4.

For the remainder, we may assume that $j > n - 2m$. For now, let us assume that $n/i$ is even if $j = n$, so as above we can take $x = \hat{x}Z \in T$, where $\hat{x} = [\Lambda^{j/i}, I_{n-j}]$. Here $n - j = k < 2m$, so $C_V(\hat{x})$ does not contain a totally singular $m$-space. If $i > m$ then Lemma 4.2.4 implies that $x$ is a derangement, so let us assume that $i \leqslant m$. If $j/i$ is even then $C_V(\hat{x})$ is hyperbolic (or trivial) and thus $x$ is a derangement if $k < 2\ell$ (see Lemma 4.2.4). Similarly, if $j/i$ is odd then $j \leqslant n - 2$, $C_V(\hat{x})$ is elliptic and the same lemma implies that $x$ is a derangement if $k \leqslant 2\ell$. Therefore, in order to complete the proof we may assume that one of the following holds:

(1) $j = n$ and $n/i$ is odd;
(2) $j/i$ is even, $i \leqslant m$ and $k \geqslant 2\ell$;
(3) $j/i$ is odd, $i \leqslant m$ and $k > 2\ell$.

Note that we allow $k = 0$ in case (2).

Suppose (1) holds. Then $i$ does not divide $n/2$ and we deduce that $i \leqslant 2m$ since $r$ divides $|\Omega|$. If $i = 2m$ then $q^{(n-2m)/2} + 1$ is the only term in the numerator of the expression for $|\Omega|$ that is divisible by $r$, so $i$ divides $n - 2m$ but not $(n-2m)/2$, which is a contradiction since $n/i$ is odd. Therefore $i < 2m$. Set $x = \hat{x}Z \in T$, where $\hat{x} = [\Lambda^{n/i-1}, I_i]$, and note that $C_V(\hat{x})$ is a hyperbolic $i$-space that does not contain a totally singular $m$-space. If $i > m$ then Lemma 4.2.4 implies that $x$ is a derangement. Similarly, if $i \leqslant m$ and $i < 2\ell$ then once again Lemma 4.2.4 implies that $x$ is a derangement. Therefore, we have reduced case (1) to the following:

(1)′ $j = n$, $n/i$ is odd and $2\ell \leqslant i \leqslant m$.

Let $y = \hat{y}Z \in T$ be an element of order $r$. Note that if either (1)′, (2) or (3) holds, then $C_V(\hat{y})$ contains a totally singular $\ell$-space. Write $n = n'i + k$ and $m = m'i + \ell$, and note that $m' \geqslant 1$ and $2m' \leqslant n'$. Also note that $n' = j/i$. Set $t = (r-1)/i$ as before. If $n' - 2m' < t$ then Lemma 4.2.4 implies that the element defined in (4.5.4) is a derangement.

Now assume that $n' - 2m' \geqslant t$. Let $y = \hat{y}Z \in T$ be an element of order $r$, so $\hat{y}$ is of the form

$$\hat{y} = [\Lambda_1^{a_1}, \ldots, \Lambda_t^{a_t}, I_{id+k}]$$

Note that $C_V(\hat{y})$ is a nondegenerate space of type $(-)^{n'+d}$. Consider the $n'$ blocks of size $i$ comprising $\hat{y}$ (see (4.5.5)). Since $n' - 2m' \geqslant t$, Lemma 4.2.5 implies that there exists a collection of $m'$ pairs of repeated blocks (we are colouring $n'$ blocks with $t + 1$ distinct colours of type $\Lambda_s$ and $I_i$). Write $m' = m'_0 + m'_1$, where $m'_0$ is the number of pairs $\{I_i, I_i\}$ in this collection. Then $m'_0 \leqslant \lfloor d/2 \rfloor$ and (4.5.6) holds. We now consider cases (1)′, (2) and (3) in turn (in reverse order).

First assume (3) holds. We may write $C_V(\hat{y}) = X \perp Y$, where $X$ is an elliptic $k$-space and $Y$ is a nondegenerate $id$-space of type $(-)^d$ (of course, $Y$ is trivial if $d = 0$). Now $X$ contains a totally singular $\ell$-space (since $k > 2\ell$) and $Y$ contains a totally singular $i\lfloor d/2 \rfloor$-space, so $C_V(\hat{y})$ contains a totally singular $(im'_0 + \ell)$-space. Therefore, Lemma 4.2.4 implies that $y$ fixes a totally singular $m$-space.

Case (2) is similar. Here $C_V(\hat{y}) = X \perp Y$, where $X$ is a hyperbolic $k$-space and $Y$ is a nondegenerate $id$-space of type $(-)^d$ (note that $X$ is trivial if $k = 0$, and $Y$ is trivial if $d = 0$). Again, we deduce that $C_V(\hat{y})$ contains a totally singular $(im'_0 + \ell)$-space and thus $y$ fixes a totally singular $m$-space.

Finally, suppose that (1)′ holds, so $k = 0$, $d > 0$, $n' = n/i$ is odd and $C_V(\hat{y})$ is of type $(-)^{d+1}$. We may write $C_V(\hat{y}) = X \perp Y$, where $X$ is a hyperbolic $i$-space and $Y$ is a nondegenerate $i(d-1)$-space of type $(-)^{d+1}$. Note that $X$ contains a totally singular $\ell$-space since $i \geqslant 2\ell$.

First assume that $n' - 2m' > t$. We claim that $y$ fixes a totally singular $m$-space. If $d$ is odd, then $Y$ contains a totally singular $i\lfloor d/2 \rfloor$-space and thus

Lemma 4.2.4 implies that $y$ fixes a totally singular $m$-space. Now assume that $d$ is even, so $Y$ contains a totally singular $(i(d-1)/2-1)$-space. Since

$$i(d-1)/2 - 1 \geqslant i(d/2-1)$$

it follows that $C_V(\hat{y})$ contains a totally singular $(\ell + i(d/2-1))$-space and thus Lemma 4.2.4 implies that $y$ fixes a totally singular $m$-space when

$$d/2 + \sum_{s=1}^{t} \lfloor a_s/2 \rfloor > m' \tag{4.6.1}$$

In fact, we claim that this inequality is a consequence of the condition $n' - 2m' > t$, so $y$ always fixes a totally singular $m$-space. Seeking a contradiction, suppose that (4.6.1) is false, so

$$di + 2i\sum_{s=1}^{t} \lfloor a_s/2 \rfloor \leqslant 2m'i < n - ti$$

where the final inequality holds since $n' - 2m' > t$. Without loss of generality, we may assume that $a_1, \ldots, a_v$ are odd and $a_{v+1}, \ldots, a_t$ are even (for some $v \geqslant 0$), so

$$2i\sum_{s=1}^{t} \lfloor a_s/2 \rfloor = i\sum_{s=1}^{t} a_s - vi$$

and thus

$$di + (t-v)i + i\sum_{s=1}^{t} a_s < n$$

This is a contradiction since $di + i\sum_{s=1}^{t} a_s = n$. This justifies the claim, so $T$ is $r$-elusive if $(1)'$ holds and $n' - 2m' > t$.

Finally, let us assume that $(1)'$ holds and $n' - 2m' = t$. We claim that $\hat{x} = [\Lambda_1, \ldots, \Lambda_t, I_{2m'i}]$ is a derangement. To see this, first observe that $t$ is odd (since $n'$ is odd), so $C_V(\hat{x})$ is elliptic and a maximal totally singular subspace of $C_V(\hat{x})$ has dimension $m'i - 1 = m - \ell - 1$. In particular, if $a_0 \geqslant 0$ is an even integer such that $C_V(\hat{x})$ contains a totally singular $(ia_0/2 + \ell)$-space, then

$$i\sum_{s=0}^{t} \lfloor a_s/2 \rfloor = ia_0/2 \leqslant m - 2\ell - 1 < m - \ell$$

and thus Lemma 4.2.4 implies that $x$ is a derangement.

To complete the proof of the proposition, it remains to show that if $i$ is even and $T$ is $r$-elusive on an orbit of maximal totally singular subspaces of $V$, then $T$ is also $r$-elusive on the other $T$-orbit of such subspaces. Let $y \in T$ be an element of order $r$ and note that $m = n/2$ and $\ell > 0$ (if $\ell = 0$ then $k = 0$ and $n' = 2m'$, which is a contradiction since $n' - 2m' \geqslant t$). In particular, we are in one of the cases labelled (2) and (3) above. In the above analysis of these cases, we constructed a $y$-invariant totally singular $m$-space $W$ by choosing a totally singular $\ell$-space $U$ in $C_V(\hat{y})$. If we now repeat this construction, this time choosing a totally singular $\ell$-space $U'$ in $C_V(\hat{y})$ such that $U \cap U'$ has codimension 1 in $U$, then we deduce that $y$ also fixes a totally singular $m$-space $W'$ such that $W \cap W'$ has codimension 1 in $W$. Therefore, $y$ fixes subspaces in both $T$-orbits on maximal totally singular spaces. We conclude that $T$ is $r$-elusive.  □

**Proposition 4.6.3** *In Case VII of Table 4.1.1, $T$ is $r$-elusive if and only if one of the following holds:*

(i) $r = 2$, $p \neq 2$ *and either* $n \equiv 0 \pmod 4$, *or $m$ is even, or $q \not\equiv 7 \pmod 8$.*
(ii) $r \neq p$, $r > 2$, $i$ *is odd*, $i \leqslant m$ *and either* $k = 0$ *or* $k > 2\ell$.
(iii) $r \neq p$, $r > 2$, $i$ *is even*, $i \leqslant m$ *and* $n' - 2m' \geqslant t$ *and one of the following holds:*
   (a) $i \geqslant 2\ell$, $k = 0$, $n/i$ *is even and* $n' - 2m' > t$;
   (b) $k \geqslant 2\ell$ *and $n'$ is odd;*
   (c) $k > 2\ell$ *and $n'$ is even,*
   *where* $n' = \lfloor n/i \rfloor$, $m' = \lfloor m/i \rfloor$ *and* $t = (r-1)/i$.

*Proof* This is very similar to the proof of the previous proposition. Here we have $T = P\Omega_n^-(q)$ and $\Omega$ is the set of totally singular $m$-dimensional subspaces of $V$, where $m < n/2$. The degree $|\Omega|$ is given in Table 4.1.2. It follows that if $r$ is a prime dividing $|\Omega|$, then $r \neq p$ and $r$ divides $q^j - 1$ for some even integer $j$ in the range $n - 2m \leqslant j \leqslant n$.

*Case 1. $r = 2$.* First assume $r = 2$, so $p$ is odd. We use the labelling of involution class representatives given in Section 3.5.2. Involutions of type $t_i$ and $t_i'$, where $i < n/4$, clearly fix a totally singular $m$-space, as do those of type $t_{n/4}$. If $n \equiv 2 \pmod 4$ and $q \equiv 7 \pmod 8$ then there exist additional involutions in $T$ of type $t_{n/2}$, and from their description in Section 3.5.2.13, we deduce that these elements are derangements if and only if $m$ is odd.

For the remainder, let us assume $r$ is odd. As usual, set $i = \Phi(r, q)$, define $c$ as in (4.5.1) and let $j$ be the largest even integer in the range $n - 2m \leqslant j \leqslant n$ such that $r$ divides $q^j - 1$. Note that $n - j = k \leqslant 2m$ and $c$ divides $j$ (see Lemma A.2).

*Case 2. $r > 2$, $i$ odd.* First assume $i$ is odd, so $c = 2i$. To begin with, let us assume that $j = n$, so $i$ divides $n$. Here $i$ divides $n/2$, so $r$ divides $q^{n/2} - 1$, but not $q^{n/2} + 1$. Since $r$ divides $|\Omega|$, it follows that $i \leqslant m$. Indeed, the given expression for $|\Omega|$ in Table 4.1.2 indicates that $r$ divides $q^{n-2e} - 1$ for some $e \in \{1, \dots, m\}$, so $r$ divides $q^{2e} - 1$ and thus $i$ divides $2e$. But $i$ is odd, so $i$

divides $e$ and thus $i \leqslant m$ as required. Now Proposition 3.5.4(i) (and Remark 3.5.5) implies that $\dim C_V(\hat{y}) \geqslant 2i > 2\ell$ for each $y = \hat{y}Z \in T$ of order $r$, and so Lemma 4.2.2 implies that $y$ fixes a totally singular $m$-space. Thus $T$ is $r$-elusive.

Now assume that $i$ is odd and $j < n$. Set $\hat{x} = [(\Lambda, \Lambda^{-1})^{j/2i}, I_{n-j}]$. Here $C_V(\hat{x})$ is a nontrivial elliptic subspace of dimension $k \leqslant 2m$ (see Remark 3.5.5(iii)), so it does not contain a totally singular $m$-space. Therefore, Lemma 4.2.2 implies that $x$ is a derangement if either $i > m$, or $i \leqslant m$ and $k \leqslant 2\ell$. On the other hand, if $i \leqslant m$ and $k > 2\ell$ then Proposition 3.5.4(i) implies that $\dim C_V(\hat{y}) \geqslant k$ for any element $y = \hat{y}Z \in T$ of order $r$, so $C_V(\hat{y})$ contains a totally singular $\ell$-space and thus $y$ fixes a totally singular $m$-space by Lemma 4.2.2.

*Case 3. $r > 2$, $i$ even.* For the remainder, we may assume $i$ is even. First assume $j = n - 2m$, so $2m = k < i$ and thus $q^{(n-2m)/2} - 1$ is the only term in the numerator of the expression for $|\Omega|$ that is divisible by $r$, so $i$ divides $(n - 2m)/2$ and thus $(n - 2m)/i$ is even. In particular, if we set $\hat{x} = [\Lambda^{(n-2m)/i}, I_{2m}]$ then $C_V(\hat{x})$ is an elliptic space that does not contain a totally singular $m$-space. Since $i > 2m$, Lemma 4.2.4 implies that $x$ is a derangement.

From now on we may assume that $j > n - 2m$ and thus $k < 2m$. We define an element $x = \hat{x}Z \in T$ of order $r$ as follows:

$$\hat{x} = \begin{cases} [\Lambda^{j/i-1}, I_i] & j = n \text{ and } n/i \text{ even} \\ [\Lambda^{j/i}, I_{n-j}] & \text{otherwise} \end{cases}$$

Suppose $\hat{x} = [\Lambda^{j/i}, I_{n-j}]$. Here $\dim C_V(\hat{x}) < 2m$ and thus $C_V(\hat{x})$ does not contain a totally singular $m$-space, so in this situation Lemma 4.2.4 implies that $x$ is a derangement if $i > m$. The same lemma also implies that $x$ is a derangement if $j/i$ is odd and $k < 2\ell$, or if $j/i$ is even and $0 < k \leqslant 2\ell$.

Now assume $\hat{x} = [\Lambda^{j/i-1}, I_i]$, so $j = n$, $n/i$ is even and $C_V(\hat{x})$ is hyperbolic. Here $r$ divides $q^{n/2} - 1$, but not $q^{n/2} + 1$, and we quickly deduce that $i \leqslant 2m$ since $r$ divides $|\Omega|$. If $i = 2m$ then $r$ divides $q^{(n-2m)/2} - 1$, but this implies that $i$ divides $m = i/2$ (since $i$ divides $n/2$), which is absurd. Therefore $i < 2m$, so $C_V(\hat{x})$ does not contain a totally singular $m$-space and thus Lemma 4.2.4 implies that $x$ is a derangement if $i > m$ or $2\ell > i$.

In order to complete the proof, we may assume that $i$ is even, $j > n - 2m$ and one of the following holds:

(1) $j = n$, $n/i$ is even and $2\ell \leqslant i \leqslant m$;
(2) $j/i$ is odd, $i \leqslant m$ and $k \geqslant 2\ell$;
(3) $j/i$ is even, $i \leqslant m$ and $k > 2\ell$.

Note that in case (2) we allow $k = 0$. In addition, Proposition 3.5.4(ii) implies that $C_V(\hat{y})$ contains a totally singular $\ell$-space for any element $y = \hat{y}Z \in T$ of order $r$.

Write $n = n'i + k$ and $m = m'i + \ell$, and note that $m' \geqslant 1$ and $2m' \leqslant n'$. Set $t = (r-1)/i$. As in the proof of the previous proposition, if $n' - 2m' < t$ then Lemma 4.2.4 implies that the element defined in (4.5.4) is a derangement, so for the remainder we may assume that $n' - 2m' \geqslant t$.

Let $y = \hat{y}Z \in T$ be an element of order $r$, say $\hat{y} = [\Lambda_1^{a_1}, \ldots, \Lambda_t^{a_t}, I_{id+k}]$. Note that $C_V(\hat{y})$ is a nondegenerate space of type

$$(-)^{1+\sum_{s=1}^{t} a_s} = (-)^{1+(n-di-k)/i} = (-)^{n'+d+1}$$

As in the proof of the previous proposition, $\hat{y}$ contains a collection of $m'$ pairs of repeated blocks of size $i$ and we can write $m' = m'_0 + m'_1$, where $m'_0 \leqslant \lfloor d/2 \rfloor$ is the number of pairs $\{I_i, I_i\}$ in this collection. Note that (4.5.6) holds.

We now consider cases (1), (2) and (3); the arguments are entirely similar to those in the proof of the previous proposition. For example, suppose (3) holds. Here $C_V(\hat{y}) = X \perp Y$, where $X$ is an elliptic $k$-space and $Y$ is a nondegenerate $id$-space of type $(-)^d$ (if $d = 0$ then $Y$ is trivial). Then $X$ contains a totally singular $\ell$-space (since $k > 2\ell$) and $Y$ contains a totally singular $i\lfloor d/2 \rfloor$-space, so $C_V(\hat{y})$ contains a totally singular $(im'_0 + \ell)$-space and thus Lemma 4.2.4 implies that $y$ has fixed points. Similarly, $T$ is $r$-elusive if (2) holds.

Finally, suppose that (1) holds. If $n' - 2m' > t$ then the argument in the proof of the previous proposition (see the analysis of case (1)') goes through and we conclude that $T$ is $r$-elusive. However, if $n' - 2m' = t$ then $t$ is even (since $n'$ is even) and $[\Lambda_1, \ldots, \Lambda_t, I_{2m'i}]$ is a derangement. $\qquad\square$

**Proposition 4.6.4** *In Case VIII of Table 4.1.1, $T$ is $r$-elusive if and only if one of the following holds:*

(i) $r = 2$ and either $m < (n-1)/2$ or $q \equiv 1 \pmod 4$.

(ii) $r \neq p$, $r > 2$, $i \leqslant m$ and $k > 2\ell$ and either $i$ is odd or $n' - 2m' \geqslant t$, where $n' = \lfloor n/i \rfloor$, $m' = \lfloor m/i \rfloor$ and $t = (r-1)/i$.

*Proof* Here $T = \Omega_n(q)$, $nq$ is odd and $\Omega$ is the set of totally singular $m$-dimensional subspaces of $V$, where $m < n/2$. As usual, $|\Omega|$ is given in Table 4.1.2. Let $r$ be a prime divisor of $|\Omega|$. Then $r \neq p$ and $r$ divides $q^j - 1$ for some even integer $j$ in the range $n - 2m + 1 \leqslant j \leqslant n - 1$.

If $r = 2$ then all involutions $x \in T$ are of the form $t_i$ or $t'_i$ with $i \leqslant (n-1)/2$, as described in Sections 3.5.2.1 and 3.5.2.2. If $x$ fixes a totally singular $m$-space $W$, then $W = W_1 \oplus W_{-1}$ where $W_\alpha$ denotes the $\alpha$-eigenspace of $x$ on

$W$. If $x = t_i$ then a maximal totally singular subspace in the $(-1)$-eigenspace of $x$ on $V$ has dimension $i$, and an analogous subspace in the 1-eigenspace has dimension $(n - 2i - 1)/2$. Thus $x$ fixes totally singular subspaces of all possible dimensions. However, if $x = t_i'$ then the respective dimensions are $i - 1$ and $(n - 2i - 1)/2$, so $\dim W \leqslant (n - 3)/2$. Thus involutions of type $t_i'$ are derangements if and only if $m = (n - 1)/2$, and we note that $T$ contains such an element if and only if $q \equiv 3 \pmod 4$.

For the remainder, we may assume $r \neq p$ is odd and $x \in T$ has order $r$. Note that $k$ is odd since $n$ is odd and $c$ is even. Let $j$ be the largest even integer in the range $n - 2m + 1 \leqslant j \leqslant n - 1$ such that $r$ divides $q^j - 1$. Note that $i$ divides $j$, and $k = n - j < 2m$.

First assume $i$ is odd, so $2i$ divides $j$. Set $x = [(\Lambda, \Lambda^{-1})^{j/2i}, I_{n-j}]$ as in Proposition 3.5.4(i), and note that $C_V(x)$ is a parabolic space that does not contain a totally singular $m$-space. Therefore, Lemma 4.2.2 implies that $x$ is a derangement if $i > m$, or if $i \leqslant m$ and $k < 2\ell$. On the other hand, if $i \leqslant m$ and $k > 2\ell$, then $C_V(y)$ contains a totally singular $\ell$-space for every $y \in T$ of order $r$ (see Proposition 3.5.4(i)), so $y$ fixes a totally singular $m$-space by Lemma 4.2.2.

Now suppose $i$ is even. Set $x = [\Lambda^{j/i}, I_{n-j}]$ and note that $C_V(x)$ is a parabolic space of dimension $k < 2m$. In particular, $C_V(x)$ does not contain a totally singular $m$-space and so by Lemma 4.2.4, $x$ is a derangement if $i > m$, or if $k < 2\ell$.

Finally, let us assume that $i$ is even, $i \leqslant m$ and $k > 2\ell$. Write $n = n'i + k$ and $m = m'i + \ell$, and note that $m' \geqslant 1$ and $n' \geqslant 2m'$. Set $t = (r - 1)/i$. If $n' - 2m' < t$ then Lemma 4.2.4 implies that the element $x$ defined in (4.5.4) is a derangement.

Now assume that $n' - 2m' \geqslant t$. We claim that $T$ is $r$-elusive. To see this, let $y = [\Lambda_1^{a_1}, \ldots, \Lambda_t^{a_t}, I_{id+k}] \in T$ be an element of order $r$. By Lemma 4.2.5, $y$ contains a collection of $m'$ pairs of repeated blocks of size $i$ and we may write $m' = m_0' + m_1'$, where $m_0'$ is the number of pairs $\{I_i, I_i\}$ in this collection. Note that $m_0' \leqslant \lfloor d/2 \rfloor$. Now $C_V(y) = X \perp Y$, where $X$ is a parabolic $k$-space and $Y$ is a hyperbolic $id$-space ($Y$ is trivial if $d = 0$). Since $X$ contains a totally singular $\ell$-space, it follows that $C_V(y)$ contains a totally singular $(im_0' + \ell)$-space and thus Lemma 4.2.4 implies that $y$ fixes a totally singular $m$-space (note that (4.5.6) holds). The result follows. $\qquad\square$

**Proposition 4.6.5** *In Case IX of Table 4.1.1, $T$ is $r$-elusive if and only if $r = 2$ and either $\varepsilon = -$ or $n \equiv 2 \pmod 4$.*

*Proof* Here $T = \Omega_n^{\varepsilon}(q)$, $n$ is even, $p = 2$ and $\Omega$ is the set of 1-dimensional nonsingular subspaces of $V$. The degree $|\Omega|$ is given in Table 4.1.2. Note that if $r$ is a prime divisor of $|\Omega|$, then either $r = 2$ or $r$ divides $q^{n/2} - \varepsilon$.

First assume $r = 2$ and set $L = O_2^{\varepsilon'}(q)$. Let $W$ be the natural $L$-module. Now $W$ has exactly $q + 1$ subspaces of dimension one, which is an odd number, so every involution $x \in L$ fixes a 1-dimensional subspace of $W$. Moreover, if $\varepsilon' = -$ then this subspace is nonsingular (there are no singular 1-spaces in an elliptic 2-space). On the other hand, a hyperbolic 2-space only contains two singular 1-spaces, which are interchanged by $x$, and so once again we deduce that $x$ fixes a nonsingular 1-space when $\varepsilon' = +$.

From the description of unipotent involutions in Section 3.5.4, it follows that all involutions of type $b_s$ and $c_s$ have fixed points on $\Omega$. Similarly, if $s < n/2$ is even then an involution of type $a_s$ centralises a nondegenerate subspace, and so these elements also fix nonsingular 1-spaces. Finally, suppose that $\varepsilon = +$, $n \equiv 0 \pmod 4$ and $x \in T$ is an involution of type $a_{n/2}$. Then $C_V(x)$ is a totally singular $\frac{n}{2}$-space and thus $x$ is a derangement.

Now assume that $r$ divides $q^{n/2} - \varepsilon$. As usual, let $x \in T$ be an element of order $r$ and set $i = \Phi(r, q)$. Suppose $\varepsilon = +$ and note that $i$ divides $n/2$. If $i$ is even then $n/i$ is even and $x = [\Lambda^{n/i}]$ is a derangement in $T$. Similarly, if $i$ is odd then $x = [(\Lambda, \Lambda^{-1})^{n/2i}]$ is also a derangement. Finally, suppose $\varepsilon = -$ and $r$ divides $q^{n/2} + 1$. Note that $i$ divides $n$, but not $n/2$, so $i$ is even and $n/i$ is odd. As before, $x = [\Lambda^{n/i}]$ is a derangement. $\square$

In order to complete the analysis of subspace actions, we may assume that $T = \text{P}\Omega_n^{\varepsilon}(q)$ and $\Omega$ is a set of isometric nondegenerate $m$-dimensional subspaces of $V$. For each $m$ there are two isometry types of nondegenerate $m$-spaces, and we require versions of Lemmas 4.2.1 and 4.2.3 that are specifically tailored for orthogonal groups.

Recall that a nondegenerate subspace $U$ of $V$ is either hyperbolic, elliptic or parabolic, and we say that $U$ is of type $+$, $-$ or $\circ$, respectively. By Corollary 2.2.12, a nondegenerate space contains proper nondegenerate subspaces of all isometry types, and we will frequently use this fact, without reference, for the remainder of this chapter.

We can define a natural multiplication on the set of symbols $\{+, -, \circ\}$ as follows:

|   | $+$ | $-$ | $\circ$ |
|---|-----|-----|---------|
| $+$ | $+$ | $-$ | $\circ$ |
| $-$ | $-$ | $+$ | $\circ$ |
| $\circ$ | $\circ$ | $\circ$ | $+$ |

We use this multiplication in the statement of the next lemma.

**Lemma 4.6.6** *Let $x \in O_n^{\varepsilon}(q)$ be a semisimple element of odd prime order $r$ such that $i = \Phi(r, q)$ is even. Let $V$ be the natural $O_n^{\varepsilon}(q)$-module and let $m < n$ be a positive integer such that $V$ contains a nondegenerate $m$-space of type*

$\varepsilon'$. *Write* $m \equiv \ell \pmod{i}$, *with* $0 \leqslant \ell < i$, *and fix* $\varepsilon'' \in \{+, -, \circ\}$ *such that* $\varepsilon' = \varepsilon''(-)^{(m-\ell)/i}$.

*Then* $x$ *fixes a nondegenerate* $m$*-space of type* $\varepsilon'$ *if and only if one of the following holds:*

(i) $C_V(x)$ *contains a nondegenerate* $m$*-space of type* $\varepsilon'$;
(ii) $i \leqslant m$ *and* $\ell > 0$ *and* $C_V(x)$ *contains a nondegenerate* $\ell$*-space of type* $\varepsilon''$;
(iii) $i \leqslant m$ *and* $\ell = 0$ *and either* $(-)^{m/i} = \varepsilon'$, *or* $C_V(x)$ *contains a hyperbolic* $i$*-space.*

*Proof*  By Proposition 3.5.4(ii), $x$ fixes an orthogonal decomposition

$$V = U_1 \perp \ldots \perp U_s \perp C_V(x)$$

where each $U_j$ is an elliptic $i$-space on which $x$ acts irreducibly.

Clearly, if (i) holds then $x$ fixes a nondegenerate $m$-space of type $\varepsilon'$. Next assume that (ii) or (iii) holds. If $\ell > 0$ let $U$ be a nondegenerate $\ell$-space of type $\varepsilon''$ in $C_V(x)$, and set $U = \{0\}$ if $\ell = 0$. Since $i$ is even, if $m$ is odd then $\ell$ is also odd.

Suppose $is \geqslant m$. Then $x$ fixes the $m$-space $W$ spanned by $U$ and $(m - \ell)/i$ of the $U_j$. Since each $U_j$ is elliptic, $W$ is of type $\varepsilon''(-)^{(m-\ell)/i}$ if $\ell > 0$, and of type $(-)^{m/i}$ if $\ell = 0$. If either $\ell > 0$, or $\ell = 0$ and $(-)^{m/i} = \varepsilon'$, then $W$ is of type $\varepsilon'$ as required. It remains to consider the case where $\ell = 0$, $(-)^{m/i} = -\varepsilon'$ and $C_V(x)$ contains a hyperbolic $i$-space $U'$. Note that $i \leqslant m$ since (iii) holds. If $i = m$ then $\varepsilon' = +$ and $U'$ is the required space fixed by $x$. On the other hand, if $i < m$ then $x$ fixes the $m$-space spanned by $U'$ and $m/i - 1 \geqslant 1$ of the $U_j$, which is of type $(-)^{m/i-1} = \varepsilon'$ as required.

Now assume $is < m$. Let

$$U' := U_1 \perp \ldots \perp U_s$$

which is a nondegenerate $is$-space of type $(-)^s$. Since $n - is > m - is > 0$, it follows that $C_V(x)$ contains a nondegenerate subspace $U''$ of dimension $m - is$ and type $\varepsilon'''$ such that $\varepsilon' = (-)^s \varepsilon'''$. Then $W = U' \perp U''$ is a nondegenerate $m$-space of type $\varepsilon'$ fixed by $x$.

Conversely, suppose that $x$ fixes a nondegenerate $m$-space $W$ of type $\varepsilon'$. If $W$ is not contained in $C_V(x)$ then $x$ acts nontrivially on $W$ and thus we obtain a decomposition

$$W = W_1 \perp \ldots \perp W_d \perp C_W(x)$$

where $d \geqslant 1$ and each $W_j$ is an elliptic $i$-space. Note that $i \leqslant m$ and $\dim C_W(x) \geqslant \ell$. If $C_W(x) = \{0\}$ then $\ell = 0$ and $(-)^{m/i} = \varepsilon'$, so (iii) holds. Now assume that $C_W(x)$ is nontrivial, so $m - id > 0$ and $C_W(x)$ is a nondegenerate space of type

$\varepsilon'''$, where $\varepsilon' = (-)^d \varepsilon'''$. If $m - id = \ell$ then $\ell > 0$ and (ii) holds (note that $C_W(x)$ is the required subspace of $C_V(x)$), so we may assume that

$$\dim C_W(x) = m - id > \ell$$

If $\ell > 0$ then $C_W(x)$ contains a nondegenerate $\ell$-space of type $\varepsilon''$. Hence (ii) again holds. Finally, if $\ell = 0$ then $\dim C_W(x) \geqslant i$. If $\dim C_W(x) = i$ and $C_W(x)$ is elliptic then $W$ has type $(-)^{m/i} = \varepsilon'$ and case (iii) holds. If $\dim C_W(x) = i$ and $C_W(x)$ is hyperbolic, or if $\dim C_W(x) > i$, then $C_V(x)$ contains a hyperbolic $i$-space and once again (iii) holds. □

**Lemma 4.6.7** *Let* $x \in O_n^\varepsilon(q)$ *be a semisimple element of odd prime order* $r$ *such that* $i = \Phi(r,q)$ *is odd. Let* $V$ *be the natural* $O_n^\varepsilon(q)$-*module and let* $m < n$ *be a positive integer such that* $V$ *contains a nondegenerate* $m$-*space of type* $\varepsilon'$. *Write* $m \equiv \ell \pmod{2i}$ *with* $0 \leqslant \ell < 2i$. *Then* $x$ *fixes a nondegenerate* $m$-*space of type* $\varepsilon'$ *if and only if one of the following holds:*

(i) $C_V(x)$ *contains a nondegenerate* $m$-*space of type* $\varepsilon'$;
(ii) $2i \leqslant m$, $\ell > 0$ *and* $C_V(x)$ *contains a nondegenerate* $\ell$-*space of type* $\varepsilon'$;
(iii) $2i \leqslant m$, $\ell = 0$ *and either* $\varepsilon' = +$, *or* $\varepsilon' = -$ *and* $C_V(x)$ *contains an elliptic* $2i$-*space.*

*Proof* By Proposition 3.5.4(i), $x$ fixes an orthogonal decomposition

$$V = (U_1 \oplus U_1^*) \perp (U_2 \oplus U_2^*) \perp \ldots \perp (U_t \oplus U_t^*) \perp C_V(x)$$

such that each $\{U_j, U_j^*\}$ is a dual pair of totally singular $i$-spaces and $x$ acts irreducibly on each $U_j$ and $U_j^*$.

Clearly, if $C_V(x)$ contains a nondegenerate $m$-space $W$, then $x$ fixes $W$. Now assume that (ii) or (iii) holds, so $2i \leqslant m$. First assume $2it \geqslant m$. Here $x$ fixes the hyperbolic space $U$ spanned by $(m - \ell)/2i$ of the hyperbolic $2i$-spaces $U_j \oplus U_j^*$. In particular, if $\ell = 0$ then $x$ fixes a hyperbolic $m$-space. Also, if $\ell > 0$ and $C_V(x)$ contains a nondegenerate $\ell$-space $U'$ of type $\varepsilon'$, then $x$ fixes the nondegenerate $m$-space $U \perp U'$ of type $\varepsilon'$. Similarly, if $\ell = 0$, $\varepsilon' = -$ and $C_V(x)$ contains an elliptic $2i$-space $U'$, then $x$ fixes the elliptic $m$-space $U'' \perp U'$, where $U''$ is the hyperbolic $(m - 2i)$-space spanned by $m/2i - 1$ of the spaces $U_j \oplus U_j^*$.

Now assume that $2it < m$. Observe that

$$U' := (U_1 \oplus U_1^*) \perp \ldots \perp (U_t \oplus U_t^*)$$

is a hyperbolic $2it$-space. Since $n - 2it > m - 2it > 0$, it follows that $C_V(x)$ contains a nondegenerate subspace $U''$ of dimension $m - 2it$ and type $\varepsilon'$. Then $x$ fixes the nondegenerate $m$-space $U' \perp U''$, which has type $\varepsilon'$.

Conversely, now assume that $x$ fixes a nondegenerate $m$-space $W$ of type $\varepsilon'$. If $W$ is contained in $C_V(x)$ then (i) holds, so assume otherwise. Then $x$ acts nontrivially on $W$ and we obtain an orthogonal decomposition of $W$ as a sum of $C_W(x)$ and nondegenerate hyperbolic $2i$-spaces of the form $X \oplus X^*$, with $x$ acting irreducibly on each $X$ and $X^*$. In particular, $2i \leqslant m$. If $\ell > 0$ then $C_W(x)$ has dimension at least $\ell$ and is of type $\varepsilon'$, so (ii) holds. On the other hand, if $\ell = 0$ then $m$ is even and so $\varepsilon' = \pm$. If $\varepsilon' = +$ then (iii) holds. Similarly, if $\varepsilon' = -$ then $C_W(x)$ is elliptic and $\dim C_W(x) \geqslant 2i$, so once again (iii) holds.                                                                                                □

**Proposition 4.6.8** *Consider Cases X, XI and XII of Table 4.1.1, and let $\varepsilon' \in \{+, -, \circ\}$ denote the type of $m$-spaces comprising $\Omega$.*

*If $r = p > 2$ then $T$ is $r$-elusive only if $\varepsilon' = +$ and every partition of $n$ with parts of size at most $p$ and all even parts with even multiplicity contains a partition of $m$ with all even parts of even multiplicity.*

*In the remaining cases, $T$ is $r$-elusive if and only if one of the following holds:*

(i) $r = p = 2$, $\varepsilon' = +$ *and either* $n \equiv 2 \pmod 4$ *or* $m \equiv 0 \pmod 4$.
(ii) $r = 2$, $p \neq 2$ *and one of the following holds:*
    (a) $\varepsilon' = +$ *and either* $n \equiv 2 \pmod 4$ *or* $q \equiv 1 \pmod 4$ *or* $m \equiv 0 \pmod 4$;
    (b) $\varepsilon' = -$ *and either* $n \equiv 0 \pmod 4$, $q \equiv 3 \pmod 4$ *and* $m \equiv 2 \pmod 4$, *or* $n \equiv 2 \pmod 4$ *and* $q \not\equiv 1 \pmod 8$;
    (c) $\varepsilon' = \circ$, $n \equiv 2 \pmod 4$ *and* $q \not\equiv 1 \pmod 8$.
(iii) $r \neq p$, $r > 2$, $i$ *is odd,* $c \leqslant m$ *and either* $k > \ell' > 0$, *or* $\varepsilon' = +$ *and* $k = \ell'$ *or* $k > \ell' = 0$.
(iv) $r \neq p$, $r > 2$, $i$ *is even and one of the following holds:*
    (a) $k = 0$ *and either* $n/i$ *is odd, or* $\ell' = 0$ *and* $\varepsilon' = (-)^{m/i}$;
    (b) $i \leqslant m$ *and either* $k > \ell' > 0$, *or* $k = \ell' > 0$ *and* $\varepsilon' = (-)^{(m+n-2k)/i}$, *or* $k > \ell' = 0$ *and* $\varepsilon' = (-)^{m/i}$.

*Proof* In this case we have $T = \mathrm{P}\Omega_n^+(q)$ and $\Omega$ is a $T$-orbit of hyperbolic, elliptic or parabolic $m$-dimensional subspaces of $V$, in Cases X, XI and XII, respectively (consequently, we will say that $\Omega$ is hyperbolic, etc.). Let $\varepsilon' \in \{+, -, \circ\}$ denote the type of the $m$-spaces comprising $\Omega$. Note that $m < n/2$ in all three cases, and $mq$ is odd in Case XII. For each fixed $m$, recall that $T$ has precisely two orbits on the set of nondegenerate $m$-spaces. Indeed, if $m$ is even then the orbits coincide with the hyperbolic and elliptic $m$-spaces, and if $m$ is odd they correspond to the two isometry types of parabolic $m$-spaces. Note that $|\Omega|$ is given in Table 4.1.2. In particular, if $r$ is a prime divisor of

$|\Omega|$ then either $r = p$, or $r$ divides $q^j - 1$ for some even integer $j$ in the range $n - m \leqslant j \leqslant n$.

*Case 1.* $r = p = 2$. We first deal with involutions. Suppose that $r = p = 2$, so $m$ is even. From the description of unipotent involutions in Section 3.5.4, it is clear that all involutions of type $a_s$ (with $s < n/2$) and $c_s$ fix a hyperbolic $m$-space.

Suppose $n \equiv 0 \pmod 4$ and $x \in T$ is an involution of type $a_{n/2}$ that fixes a nondegenerate $m$-space $U$. By Lemma 3.4.14, $x|_U \in O(U)$ must be of type $a_{m/2}$, so $m \equiv 0 \pmod 4$ and $U$ is hyperbolic (recall that there are no elements of type $a_{m/2}$ in $O_m^-(q)$). We conclude that all involutions $x \in T$ fix a hyperbolic $m$-space, unless $n \equiv 0 \pmod 4$, $m \equiv 2 \pmod 4$ and $x$ is of type $a_{n/2}$. Moreover, for any even integer $m$, involutions of type $a_{n/2}$ are derangements on the set of elliptic $m$-spaces. Now assume $n \equiv 2 \pmod 4$ and consider an involution $x$ of type $a_{(n-2)/2}$. If $x$ fixes an elliptic $m$-space $U$ then Lemma 3.4.14 implies that $m \equiv 2 \pmod 4$ and $x|_U$ is of type $a_{(m-2)/2}$. Thus $x|_{U^\perp}$ is of type $a_{(n-m)/2}$, but this is not possible since $x|_{U^\perp} \in O_{n-m}^-(q)$. We conclude that $x = a_{(n-2)/2}$ does not fix an elliptic $m$-space, for any $m$.

*Case 2.* $r = 2$, $p \neq 2$. Next suppose that $r = 2$ and $p \neq 2$. Let $x \in T$ be an involution of type $t_i$ or $t_i'$, where $i \leqslant n/4$. Then $x$ centralises a nondegenerate $\frac{n}{2}$-space, so it fixes an element of $\Omega$ since $m < n/2$. Clearly, if $x$ is an involution of type $t_{n/2}$ (or of type $t_{n/2-1}$ if $n \equiv 0 \pmod 4$) then $x$ fixes a hyperbolic $m$-space. However, $x$ is diagonalisable over $\mathbb{F}_q$ and the two eigenspaces are complementary maximal totally singular subspaces, whence $x$ does not fix any elliptic or parabolic $m$-spaces. If $n \equiv 0 \pmod 4$ then $t_{n/2}, t_{n/2-1} \in T$ if and only if $q \equiv 1 \pmod 4$, and if $n \equiv 2 \pmod 4$ then $t_{n/2} \in T$ if and only if $q \equiv 1 \pmod 8$ (see Proposition 3.5.10).

The only additional involutions in $T$ are those of type $t_{n/2}'$ and $t_{n/2-1}'$, with $q \equiv 3 \pmod 4$ and $n \equiv 0 \pmod 4$. As explained in Section 3.5.2.11, such an involution $x$ is diagonalisable over $\mathbb{F}_{q^2}$ but not over $\mathbb{F}_q$, so $x$ is a derangement if $m$ is odd. Moreover, if $m$ is even then any nondegenerate $m$-space fixed by $x$ can be viewed as a nondegenerate $\frac{m}{2}$-space over $\mathbb{F}_{q^2}$ with respect to an hermitian form. In particular, $x$ fixes a nondegenerate $m$-space of type $\varepsilon'$ if and only if $\varepsilon' = (-)^{m/2}$ (see Construction 2.5.14).

*Case 3.* $r = p > 2$. Let us now turn to unipotent elements of order $p > 2$. We refer the reader to Section 3.5.3 for detailed information on the conjugacy classes of such elements. In particular, recall that if $x \in O_a^{\varepsilon'}(q)$ has order $p$ and Jordan form $[J_p^{a_p}, \ldots, J_1^{a_1}]$, then $a_i$ is even for all even $i$, and if $\varepsilon = -$ then $a_i > 0$ for some odd $i$ (see Proposition 3.5.12).

In Cases XI and XII, take $x \in T$ to be an element with the following Jordan form:

| | $n \equiv 0 \pmod 4$ | $n \equiv 2 \pmod 4$ |
|---|---|---|
| $m \equiv 0 \pmod 2$ | $[J_2^{n/2}]$ | $[J_2^{n/2-1}, J_1^2]$ |
| $m \equiv 1 \pmod 4$ | $[J_2^{n/2}]$ | $[J_3^2, J_2^{n/2-3}]$ |
| $m \equiv 3 \pmod 4$ | $[J_2^{n/2}]$ | $[J_2^{n/2-1}, J_1^2]$ |

In all six cases, it is easy to see that $x$ is a derangement. (Note that in each case, *any* element $x \in T$ with the given Jordan form is a derangement; there is no need to consider the labelled partitions referred to in Proposition 3.5.12.) Finally, let us consider Case X, so $\Omega$ is hyperbolic. Here the necessary condition for $r$-elusivity given in the statement of the proposition is clear, but we have not determined a necessary and sufficient condition for $r$-elusivity in all cases (note that if $n \equiv 0 \pmod 4$ and $m \equiv 2 \pmod 4$, then any element $x \in T$ with Jordan from $[J_2^{n/2}]$ is a derangement).

To complete the proof of the proposition, we may assume $r \ne p$ and $r > 2$. Let $x = \hat{x}Z \in T$ be an element of order $r$. Set $i = \Phi(r, q)$ and let $j$ be the largest even integer in the range $n - m \leqslant j \leqslant n$ such that $r$ divides $q^j - 1$. Then by Lemma A.2, $k = n - j \leqslant m$ and $c$ divides $j$.

To get started, let us briefly consider the special case $j = n - m$. Note that $m$ is even and $i > m$ by the maximality of $j$. If $\varepsilon' = +$ then $q^{(n-m)/2} - 1$ is a term in the denominator of the expression for $|\Omega|$ in Table 4.1.2, so $i$ does not divide $(n - m)/2$ and thus $(n - m)/i$ is odd. Similarly, if $\varepsilon' = -$ then the denominator contains a $q^{(n-m)/2} + 1$ term, which implies that $(n - m)/i$ is even.

*Case 4. $r \ne p$, $r > 2$, $i$ odd.* First assume $i$ is odd, so $c = 2i$. Set $\hat{x} = [(\Lambda, \Lambda^{-1})^{j/2i}, I_{n-j}]$ and note that $C_V(\hat{x})$ is hyperbolic (or trivial). Moreover, $\dim C_V(\hat{x}) = k \leqslant m$, with equality only if $m$ is even, $j = n - m$ and $(n - m)/i$ is even, so our above comments imply that $\varepsilon' = -$ if equality holds. Therefore, in all cases $C_V(\hat{x})$ does not contain an $m$-space in $\Omega$, so Lemma 4.6.7 implies that $x$ is a derangement if $c > m$. Now assume $c \leqslant m$. If $k < \ell'$, or if $k = \ell' > 0$ and $\varepsilon' = -$, then $C_V(\hat{x})$ does not contain a nondegenerate $\ell'$-space of type $\varepsilon'$, so once again Lemma 4.6.7 implies that $x$ is a derangement. Similarly, $x$ is a derangement if $\ell' = 0$ and $\varepsilon' = -$. Therefore, to complete the analysis when $i$ is odd, we may assume that $c \leqslant m$ and $k \geqslant \ell'$, with $k = \ell'$ or $\ell' = 0$ only if $\varepsilon' = +$.

Let $y = \hat{y}Z \in T$ be an element of order $r$, and note that $C_V(\hat{y})$ is hyperbolic (or trivial) and $\dim C_V(\hat{y}) \geqslant k$. In particular, if $\ell' > 0$ then $C_V(\hat{y})$ contains a nondegenerate $\ell'$-space of type $\varepsilon'$, so Lemma 4.6.7 implies that $y$ fixes an $m$-space in $\Omega$. Similarly, if $\ell' = 0$ then the same lemma implies that $y$ fixes a hyperbolic $m$-space.

*Case 5.* $r \neq p, r > 2, i \ even.$ Now assume $i$ is even. First we consider the special case $j = n$, so $i$ divides $n$. Suppose $n/i$ is even and set $\hat{x} = [\Lambda^{n/i}]$ as in Proposition 3.5.4(ii). Then Lemma 4.6.6 implies that $x$ is a derangement if $\ell' > 0$, or if $\ell' = 0$ and $\varepsilon' = (-)^{m/i+1}$ (note that if $\ell' = 0$ then $m$ is even, so Case XII does not arise). Moreover, if $\ell' = 0$ then Lemma 4.6.6(iii) implies that any element in $T$ of order $r$ fixes a nondegenerate $m$-space of type $(-)^{m/i}$. Now assume $n/i$ is odd. If $y = \hat{y}Z \in T$ has order $r$, then Proposition 3.5.4(ii) implies that $\dim C_V(\hat{y}) \geqslant i$, with equality only if $C_V(\hat{y})$ is hyperbolic. Therefore, Lemma 4.6.6 implies that $y$ fixes nondegenerate $m$-spaces of all types.

Now let us assume $i$ is even and $j < n$. Recall that $c = i$. By Lemma A.2 we have $0 < n - j = k \leqslant m$. Set

$$\hat{x} = [\Lambda^{j/i}, I_{n-j}] \qquad (4.6.2)$$

as in Proposition 3.5.4(ii). By Remark 3.5.5, if $j/i$ is even then $C_V(\hat{x})$ is hyperbolic, and $C_V(\hat{x})$ is elliptic if $j/i$ is odd. Moreover, we have already noted that if $j = n - m$ (in which case $m$ is even) then $(n - m)/i$ is odd if $\varepsilon' = +$, and thus $C_V(\hat{x})$ is elliptic in this situation, and similarly $(n - m)/i$ is even if $\varepsilon' = -$, and thus $C_V(\hat{x})$ is hyperbolic. Therefore, $C_V(\hat{x})$ does not contain an $m$-space in $\Omega$, so Lemma 4.6.6 implies that $x$ is a derangement if $i > m$.

For the remainder, we may assume that $i \leqslant m$ and $j < n$. Since $n - j = k < i \leqslant m$, it follows that $j > n - m$. There are three cases to consider, according to the value of $k$.

If $k < \ell'$ then Lemma 4.6.6 implies that the element $\hat{x}$ in (4.6.2) is a derangement since $C_V(\hat{x})$ does not contain a nondegenerate $\ell'$-space.

Next assume $k = \ell'$. Since $k$ is even, it follows that $m$ is even and thus $\varepsilon' = \pm$. Also recall that $k > 0$, so $\ell' > 0$. Define $x = \hat{x}Z \in T$ with $\hat{x}$ as in (4.6.2) (note that $C_V(\hat{x})$ is a nondegenerate $\ell'$-space of type $(-)^{j/i}$). By Lemma 4.6.6, $x$ fixes an $m$-space in $\Omega$ if and only if

$$\varepsilon' = (-)^{(m-\ell'+j)/i}$$

Furthermore, the same lemma implies that $T$ is $r$-elusive if and only if this condition holds. For example, suppose $\varepsilon' = +$ and $(m - \ell' + j)/i$ is even. Let $y = \hat{y}Z \in T$ be an element of order $r$, so we may write $\hat{y} = [\Lambda_1^{a_1}, \ldots, \Lambda_t^{a_t}, I_e]$, where $e = di + \ell'$ with $d \geqslant 0$. If $d > 0$ then $C_V(\hat{y})$ contains nondegenerate $\ell'$-spaces of each isometry type, so $y$ has fixed points by Lemma 4.6.6(ii). Now assume $d = 0$, so $\sum_v a_v = j/i$ and $C_V(\hat{y})$ is of type $(-)^{j/i}$. Since $(m - \ell' + j)/i$ is even we have $(-)^{j/i} = (-)^{(m-\ell')/i}$, so Lemma 4.6.6(ii) implies that $y$ has fixed points. The case $\varepsilon' = -$ is similar.

Finally, let us assume that $k > \ell'$. Suppose $\ell' > 0$ and let $y = \hat{y}Z \in T$ be an element of order $r$. Here $\dim C_V(\hat{y}) \geqslant k$ and $C_V(\hat{y})$ contains a nondegenerate

$\ell'$-space of every possible isometry type, so Lemma 4.6.6(ii) implies that $y$ has fixed points on $\Omega$. Finally suppose that $k > \ell' = 0$. Define $x = \hat{x}Z \in T$ as in (4.6.2), and note that $m$ is even. Since $k < i$, $C_V(\hat{x})$ does not contain a hyperbolic $i$-space and so Lemma 4.6.6 implies that $x$ fixes an $m$-space in $\Omega$ if and only if

$$\varepsilon' = (-)^{m/i}$$

Moreover, it is easy to see that the same lemma implies that $T$ is $r$-elusive if and only if this condition holds. □

**Proposition 4.6.9** *Consider Cases XIII and XIV of Table 4.1.1.*

*If $r = p > 2$ then $T$ is $r$-elusive only if Case XIII holds, $n \equiv m \pmod 4$ and every partition of $n$ with parts of size at most $p$ and all even parts with even multiplicity contains a partition of $m$ with all even parts of even multiplicity.*

*In the remaining cases, $T$ is $r$-elusive if and only if one of the following holds:*

(i) *$r = p = 2$ and either $n \equiv 0 \pmod 4$ or $m \equiv 2 \pmod 4$.*
(ii) *$r = 2$, $p \neq 2$ and either $n \equiv 0 \pmod 4$ or $q \not\equiv 7 \pmod 8$ or $m \equiv 2 \pmod 4$.*
(iii) *$r \neq p$, $r > 2$, $i$ is odd and either $k = 0$, or $c \leqslant m$ and $k \geqslant \ell' > 0$.*
(iv) *$r \neq p$, $r > 2$, $i$ is even and one of the following holds:*
    (a) *$k = 0$ and either $n/i$ is even, or $\ell' = 0$ and $m/i$ is odd;*
    (b) *$k > 0$, $i \leqslant m$ and either $k > \ell' > 0$, or $\ell' = 0$ and $m/i$ is odd, or $k = \ell'$ and $(m+n-2k)/i$ is even.*

*Proof* Here $T = P\Omega_n^-(q)$ and $\Omega$ is a $T$-orbit of $m$-dimensional elliptic or parabolic subspaces of $V$. In the parabolic case, we assume $p$ is odd and $m < n/2$ (note that if $\Omega$ is elliptic then we impose no such condition on $m$; see Remark 4.1.3). Recall that $T$ has two orbits on parabolic $m$-spaces, where the subspaces in distinct orbits are similar but not isometric. Subspaces in the two orbits can be distinguished by considering the discriminant of the restriction of the underlying quadratic form $Q$ on $V$, so we refer to *square-parabolic* and *nonsquare-parabolic* spaces accordingly. The actions of $T$ on the two orbits are equivalent (see Section 2.2.6).

We recall that $|\Omega|$ is recorded in Table 4.1.2. In particular, if $r$ is a prime divisor of $|\Omega|$ then either $r = p$, or $r$ divides $q^j - 1$ for some even integer $j$ in the range $n - m \leqslant j \leqslant n$.

*Case 1. $r = p = 2$.* First assume $r = p = 2$, so $\Omega$ is elliptic. As recorded in Proposition 3.5.16, the involutions $x \in T$ are of type $a_s$ and $c_s$, where $s \leqslant n/2$ is even (and $s < n/2$ if $x = a_s$). From the definition of these elements in Section

3.5.4, it is clear that all $c_s$-involutions fix an elliptic $m$-space, and so do involutions of type $a_s$ if $s \leqslant (n-4)/2$. If $n \equiv 0 \pmod 4$, then there are no further involutions in $T$. Now, if $n \equiv 2 \pmod 4$ and $x$ is an involution of type $a_{(n-2)/2}$ then $x$ centralises an elliptic 2-space and therefore fixes an elliptic $m$-space if $m \equiv 2 \pmod 4$. However, if $m \equiv 0 \pmod 4$ then $x$ does not fix an elliptic $m$-space $U$; if it did, then the restriction $x|_U \in O(U)$ would either be of type $a_s$ with $s \leqslant (m-4)/2$ even, or type $b_s$ or $c_s$. But each of these possibilities is incompatible with the fact that $x$ is of type $a_{(n-2)/2}$ (see Lemma 3.4.14, which is also valid for orthogonal groups, as noted in Remark 3.5.17).

*Case 2.* $r = 2$, $p \neq 2$. Next assume $r = 2$ and $p \neq 2$. Clearly, involutions of type $t_i$ and $t_i'$ (with $i < n/4$) have fixed points, and so do involutions of type $t_{n/4}$ (with $n \equiv 0 \pmod 4$). The only other involutions in $T$ are those of type $t_{n/2}$, with $n \equiv 2 \pmod 4$ and $q \equiv 7 \pmod 8$. Let $x$ be such an involution. From the description of these elements in Section 3.5.2.13, it is clear that $x$ does not fix an odd-dimensional subspace of $V$, so $x$ is a derangement if $m$ is odd. Similarly, if $m$ is even then the only nondegenerate $m$-spaces fixed by $x$ are of type $(-)^{m/2}$ (see Construction 2.5.14).

*Case 3.* $r = p > 2$. Next suppose that $r = p > 2$. Let $D(Q)$ denote the discriminant of the underlying quadratic form $Q$. First assume $m$ is odd. According to the congruences of $n$ and $m$ modulo 4, take $x \in T$ to be an element with the following Jordan form:

|  | $n \equiv 0 \pmod 4$ | $n \equiv 2 \pmod 4$ |
| --- | --- | --- |
| $m \equiv 1 \pmod 4$ | $[J_3, J_2^{n/2-2}, J_1]$ | $[J_3^2, J_2^{n/2-3}]$ |
| $m \equiv 3 \pmod 4$ | $[J_3, J_2^{n/2-2}, J_1]$ | $[J_2^{n/2-1}, J_1^2]$ |

If $n \equiv 2 \pmod 4$, then *any* $x \in T$ with the given Jordan form is a derangement. Now assume $n \equiv 0 \pmod 4$, so $D(Q) = \boxtimes$ (see Lemma 2.2.9(ii)). Here we take $x \in T$ to be a unipotent element corresponding to the following *labelled partition* of $n$ (see Proposition 3.5.12), which depends on both $m$ and the isometry type of the $m$-spaces comprising $\Omega$:

$$(\square 3, 2^{n/2-2}, \boxtimes 1) \quad \begin{cases} m \equiv 1 \pmod 4 \text{ and } \Omega \text{ square-parabolic, or} \\ m \equiv 3 \pmod 4 \text{ and } \Omega \text{ nonsquare-parabolic} \end{cases}$$

$$(\boxtimes 3, 2^{n/2-2}, \square 1) \quad \begin{cases} m \equiv 3 \pmod 4 \text{ and } \Omega \text{ square-parabolic, or} \\ m \equiv 1 \pmod 4 \text{ and } \Omega \text{ nonsquare-parabolic} \end{cases}$$

It is easy to check that in each case $x$ is a derangement. For example, suppose $m \equiv 1 \pmod 4$ and $x$ corresponds to $(\square 3, 2^{n/2-2}, \boxtimes 1)$. If $x$ fixes a nondegenerate $m$-space $U$, then $x|_U$ and $x|_{U^\perp}$ must correspond to the labelled partitions $(2^{(m-1)/2}, \boxtimes 1)$ and $(\square 3, 2^{(n-m-3)/2})$. This implies that $U$ is a nonsquare-parabolic $m$-space, so $x$ has no fixed points on the set of square-parabolic $m$-spaces. The other cases are entirely similar.

Now assume $m$ is even, so $\Omega$ is elliptic. If $n \equiv 0 \pmod 4$ and $m \equiv 2 \pmod 4$ then any element $x \in T$ with Jordan form $[J_3, J_2^{n/2-2}, J_1]$ is a derangement. Similarly, $[J_2^{n/2-1}, J_1^2]$ is a derangement if $n \equiv 2 \pmod 4$ and $m \equiv 0 \pmod 4$. Finally, if $n \equiv m \pmod 4$ then the necessary condition in the statement of the proposition is clear, but we have not determined a necessary and sufficient condition for $r$-elusivity in this situation.

For the remainder, we may assume that $r \neq p$ is odd. Set $i = \Phi(r, q)$ and note that $i$ divides $j$, where $j$ is the largest even integer in the range $n - m \leqslant j \leqslant n$ such that $r$ divides $q^j - 1$. By Lemma A.2, $k = n - j \leqslant m$ and $c$ divides $j$. Let $x = \hat{x}Z \in T$ be an element of order $r$.

*Case 4.* $r \neq p$, $r > 2$, $i$ odd. First assume $i$ is odd, so $c = 2i$. Consider the special case $j = n$. According to Proposition 3.5.4(i) and Remark 3.5.5(iii), $\hat{x}$ is of the form

$$\hat{x} = [(\Lambda_1, \Lambda_1^{-1})^{a_1}, \ldots, (\Lambda_{t/2}, \Lambda_{t/2}^{-1})^{a_{t/2}}, I_{n-2iu}]$$

where $t = (r - 1)/i$, $u = \sum_v a_v$ and $C_V(\hat{x})$ is elliptic with $\dim C_V(\hat{x}) \geqslant 2i$. Therefore, Lemma 4.6.7 implies that $x$ fixes an $m$-space in $\Omega$.

Next assume $i$ is odd and $j < n$. Set

$$\hat{x} = [(\Lambda, \Lambda^{-1})^{j/2i}, I_{n-j}] \tag{4.6.3}$$

as in Proposition 3.5.4(i), and note that $C_V(\hat{x})$ is nontrivial and elliptic. If $n - j = m$ then $m$ is even, so $\Omega$ is elliptic and $2i$ divides $j = n - m$. In particular, $r$ divides $q^{(n-m)/2} - 1$, but not $q^{(n-m)/2} + 1$, so $(q^{n-m} - 1)_r$ divides $q^{(n-m)/2} - 1$ but this contradicts the fact that $r$ divides $|\Omega|$ (see Table 4.1.2). Therefore $k = n - j < m$, and by applying Lemma 4.6.7 we deduce that $x$ is a derangement if $c > m$. Now assume $c \leqslant m$. If $k < \ell'$ then $C_V(\hat{x})$ does not contain a nondegenerate $\ell'$-space and so Lemma 4.6.7 once again implies that $x$ is a derangement. If $\ell' = 0$ then $m$ is even (so $\Omega$ is elliptic) and $C_V(\hat{x})$ does not contain a nondegenerate subspace of dimension $2i > k$, so Lemma 4.6.7 implies that $x$ is also a derangement in this case.

Finally, let us assume that $c \leqslant m$ and $k \geqslant \ell' > 0$. Let $y = \hat{y}Z \in T$ be an element of order $r$. Then by Proposition 3.5.4(i), $\hat{y}$ has the form

$$\hat{y} = [(\Lambda_1, \Lambda_1^{-1})^{a_1}, \ldots, (\Lambda_{t/2}, \Lambda_{t/2}^{-1})^{a_{t/2}}, I_{n-2iu}]$$

where $t = (r-1)/i$, $u = \sum_v a_v$ and $C_V(\hat{y})$ is a nontrivial elliptic space. Since $\dim C_V(\hat{y}) \geqslant k \geqslant \ell'$ and $k$ is even, it follows that $C_V(\hat{y})$ contains an elliptic $\ell'$-space if $\ell'$ is even, and parabolic $\ell'$-spaces of both isometry types if $\ell'$ is odd, and so Lemma 4.6.7 implies that $y$ fixes an $m$-space in $\Omega$.

*Case 5.* $r \neq p$, $r > 2$, $i$ even. To complete the proof, we may assume $i$ is even, so $c = i$. To begin with, let us assume that $j = n$ and $n/i$ is odd. Set $\hat{x} = [\Lambda^{n/i}]$. Then Lemma 4.6.6 implies that $x$ is a derangement unless $\ell' = 0$ (so $m$ is even) and $m/i$ is odd, in which case the same lemma implies that every element in $T$ of order $r$ fixes an elliptic $m$-space. If $j = n$ and $n/i$ is even, then Proposition 3.5.4(ii) implies that $\dim C_V(\hat{y}) \geqslant i > \ell'$ for each $y = \hat{y}Z \in T$ of order $r$. Further, if $\dim C_V(\hat{y}) = i$ then $C_V(\hat{y})$ is hyperbolic. Therefore, $C_V(\hat{y})$ always contains a hyperbolic $i$-space and a nondegenerate $\ell'$-space (if $\ell' > 0$) of every isometry type, so Lemma 4.6.6 implies that $y$ fixes an $m$-space in $\Omega$.

Finally, let us assume $i$ is even and $j < n$. Note that $k = n - j > 0$ is even. Set

$$\hat{x} = [\Lambda^{j/i}, I_{n-j}] \tag{4.6.4}$$

as in Proposition 3.5.4(ii), and note that $C_V(\hat{x})$ is of type $(-)^{j/i+1}$. Recall that $n - j \leqslant m$.

First assume $n - j = m$, so $\Omega$ is elliptic and the maximality of $j$ implies that $i > m$. If $(n-m)/i$ is even then $i$ divides $(n-m)/2$, so $(q^{n-m} - 1)_r$ divides $q^{(n-m)/2} - 1$, but this contradicts the fact that $r$ divides $|\Omega|$. Therefore $(n-m)/i$ is odd, $C_V(\hat{x})$ is hyperbolic and thus Lemma 4.6.6 implies that $x$ is a derangement.

Now assume $n - j < m$. If $i > m$ then Lemma 4.6.6 once again implies that $x$ is a derangement, so let us assume $i \leqslant m$. We consider three cases, according to the value of $k$.

If $k < \ell'$ then Lemma 4.6.6 implies that $x$ is a derangement since $C_V(\hat{x})$ does not contain a nondegenerate $\ell'$-space.

Next suppose $k = \ell'$, so $m$ is even and $\ell' > 0$. Since $C_V(\hat{x})$ is of type $(-)^{j/i+1}$, Lemma 4.6.6 implies that $x$ has fixed points if and only if

$$- = (-)^{(m-\ell'+j+i)/i}$$

or equivalently, if and only if $(m + n - 2k)/i$ is even. Moreover, it is easy to check that $T$ is $r$-elusive if and only if this condition holds.

Finally, let us assume $k > \ell'$. Suppose $\ell' = 0$, so $\Omega$ is elliptic. If $x$ is the element defined in (4.6.4), then $\dim C_V(\hat{x}) = k < i \leqslant m$ and thus $C_V(\hat{x})$ does not contain a hyperbolic $i$-space. In particular, if $m/i$ is even then Lemma 4.6.6 implies that $x$ is a derangement. However, if $m/i$ is odd then condition (iii) in Lemma 4.6.6 implies that every element in $T$ of order $r$ has fixed points. Finally, suppose $k > \ell' > 0$. Let $y = \hat{y}Z \in T$ be an element of order $r$, so $\hat{y}$ has the form

$$\hat{y} = [\Lambda_1^{a_1}, \ldots, \Lambda_t^{a_t}, I_{n-iu}]$$

with $t = (r-1)/i$ and $u = \sum_v a_v \leqslant (n-k)/i$. Since $n - iu \geqslant k > \ell'$, the subspace $C_V(\hat{y})$ contains a nondegenerate $\ell'$-space of every possible isometry type, so Lemma 4.6.6(ii) implies that $y$ has fixed points.                                              □

Our final result handles the two remaining cases in Table 4.1.1.

**Proposition 4.6.10** *Consider Cases XV and XVI of Table 4.1.1, and let $\varepsilon' \in \{+, -\}$ denote the type of m-spaces comprising $\Omega$.*

*If $r = p$ then $T$ is $r$-elusive only if $\varepsilon' = +$, $m \equiv 0$ (mod 4) and every partition of $n$ with parts of size at most $p$ and all even parts with even multiplicity contains a partition of $m$ with all even parts of even multiplicity.*

*In the remaining cases, $T$ is $r$-elusive if and only if $r = 2$, or $r \neq p$ and one of the following holds:*

(i) *$i$ is odd, $c \leqslant m$ and either $k > \ell' > 0$, or $\ell' = 0$ and $\varepsilon' = +$;*
(ii) *$i$ is even, $i \leqslant m$ and $k > \ell' \geqslant 0$, with $\ell' = 0$ if and only if $\varepsilon' = (-)^{m/i}$.*

*Proof* In this case $T = \Omega_n(q)$, $nq$ is odd and $\Omega$ is the set of hyperbolic or elliptic $m$-dimensional subspaces of $V$ (we will write $\varepsilon' \in \{+, -\}$ to denote the type of subspaces in $\Omega$). In both cases, $|\Omega|$ is recorded in Table 4.1.2.

First assume $r = p$. If $\varepsilon' = -$ then let $x \in T$ be an element with Jordan form $[J_2^{(n-1)/2}, J_1]$ if $n \equiv 1$ (mod 4), and $[J_3, J_2^{(n-3)/2}]$ if $n \equiv 3$ (mod 4); in both cases, it is clear that $x$ is a derangement (here there is no need to consider labelled partitions). Now assume $\varepsilon' = +$. Suppose $m \equiv 2$ (mod 4). If $n \equiv 1$ (mod 4) then set $x = [J_2^{(n-1)/2}, J_1]$, and if $n \equiv 3$ (mod 4) take $x = [J_3, J_2^{(n-3)/2}]$. Again, it is easy to see that $x$ is a derangement. Finally, if $m \equiv 0$ (mod 4) then the necessary condition in the statement of the proposition is clear, but we have not determined a necessary and sufficient condition for $r$-elusivity in this case.

Next suppose $r = 2$. It is clear that the involutions $t_i, t_i' \in T$ (with $1 \leqslant i \leqslant (n-1)/2$) fix elliptic and hyperbolic $m$-spaces, so $T$ is 2-elusive. For the remainder we may assume $r \neq p$ is odd. Since $r$ divides $|\Omega|$, we may define $j$ to be the largest even integer in the range $n - m + 1 \leqslant j < n$ such that $r$ divides $q^j - 1$. Since $n$ is odd and $c$ is even it follows that $k$ is odd and $\ell'$ is even. In particular,

Lemma A.2 implies that $n - j = k > 0$ and $c$ divides $j$. Let $x = \hat{x}Z \in T$ be an element of order $r$.

First assume $i$ is odd, so $c = 2i$. Define $\hat{x}$ as in (4.6.3) and note that $0 < n - j = k < m$. If $c > m$ then Lemma 4.6.7 implies that $x$ is a derangement, so assume $c \leqslant m$. If $k < \ell'$ then once again Lemma 4.6.7 implies that $x$ is a derangement since $C_V(\hat{x})$ does not contain a nondegenerate $\ell'$-space, so assume $k > \ell'$ (recall that $k$ is odd and $\ell'$ is even, so $k \neq \ell'$). Now $C_V(\hat{x})$ has dimension $n - j < 2i$, so it does not contain an elliptic $2i$-space and thus $x$ does not fix an elliptic $m$-space if $\ell' = 0$. However, if $\ell' = 0$ then Lemma 4.6.7 implies that every element in $T$ of order $r$ fixes a hyperbolic $m$-space. Finally, if $k > \ell' > 0$ then Proposition 3.5.4(i) implies that $\dim C_V(\hat{y}) \geqslant k$ for every element $y = \hat{y}Z \in T$ of order $r$, so $C_V(\hat{y})$ contains a nondegenerate $\ell'$-space of each isometry type, and thus Lemma 4.6.7 implies that $y$ fixes an $m$-space in $\Omega$.

To complete the proof, let us assume $i$ is even. Define $\hat{x}$ as in (4.6.4) and note that $C_V(\hat{x})$ is parabolic and $0 < n - j < m$. In particular, Lemma 4.6.6 implies that $x$ is a derangement if $i > m$, or if $i = m$ and $\varepsilon' = +$. On the other hand, if $i = m$ and $\Omega$ is elliptic then Lemma 4.6.6(iii) implies that every element in $T$ of order $r$ has fixed points. Finally, suppose that $i < m$. Recall that $\ell'$ is even and $k = n - j$ is odd. If $k < \ell'$ then $C_V(\hat{x})$ does not contain a nondegenerate $\ell'$-space, so Lemma 4.6.6 implies that $x$ is a derangement. Now assume $k > \ell'$. If $\ell' = 0$ then Lemma 4.6.6 implies that $x$ has fixed points if and only if

$$\varepsilon' = (-)^{m/i}$$

More generally, if $\ell' = 0$ then it is easy to check that $T$ is $r$-elusive if and only if this condition holds. Finally, if $k > \ell' > 0$ then by applying Proposition 3.5.4(ii) and Lemma 4.6.6 we deduce that $T$ is $r$-elusive. $\qquad\square$

This completes our analysis of derangements of prime order in subspace actions of classical groups. In particular, we have completed the proofs of Theorems 4.1.4, 4.1.6 and 4.1.7, and this establishes Theorem 1.5.1 for subspace actions.

# 5

# Non-subspace actions

## 5.1 Introduction

Let $G$ be a primitive almost simple classical group over $\mathbb{F}_q$ with point stabiliser $H$, socle $T$ and natural module $V$ with $\dim V = n$. Write $q = p^f$, where
$p$ is a prime. We may assume that $n$ and $q$ satisfy the conditions recorded in
Table 3.0.1. As described in Section 2.6, Aschbacher's subgroup structure theorem [3] implies that $H$ belongs to one of ten subgroup collections, denoted by
$\mathscr{C}_1, \ldots, \mathscr{C}_8, \mathscr{S}$ and $\mathscr{N}$.

In the previous chapter we studied derangements of prime order in the case
where $H \in \mathscr{C}_1$ (the so-called *subspace actions*) and now we turn our attention
to the remaining *non-subspace actions* of finite classical groups. More precisely, in this chapter we will assume that $H$ belongs to one of the geometric
subgroup collections $\mathscr{C}_i$ with $2 \leqslant i \leqslant 8$ (see Table 5.1.1 for a brief description of these subgroups). We adopt the precise definition of the $\mathscr{C}_i$ collections
used by Kleidman and Liebeck [86], and the specific permutation groups with
$H \in \mathscr{C}_i$ will be handled in Section 5.*i*. In addition, in Section 5.9 we consider
the small collection $\mathscr{N}$ of *novelty* subgroups that arises when $T = \mathrm{P}\Omega_8^+(q)$ or
$\mathrm{Sp}_4(q)'$ (with $q$ even).

As noted in Chapter 1, rather different techniques are needed to handle the
almost simple irreducible subgroups comprising the collection $\mathscr{S}$, and we will
deal with them in a separate paper.

We will continue to adopt the notation introduced in Chapters 2 and 3. In
particular, recall that if $x \in G \cap \mathrm{PGL}(V)$ and $\hat{x} \in \mathrm{GL}(\overline{V})$ is a preimage of $x$,
where $\overline{V} = V \otimes \mathbb{F}$ and $\mathbb{F}$ is the algebraic closure of $\mathbb{F}_q$, then we define

$$\nu(x) = \min\{\dim[\overline{V}, \lambda \hat{x}] \mid \lambda \in \mathbb{F}^\times\}$$

where $[\overline{V}, \lambda \hat{x}]$ is the subspace of $\overline{V}$ spanned by the vectors $v - v^{\lambda \hat{x}}$ (with $v \in
\overline{V}$); see Definition 3.1.4. Observe that $\nu(x)$ is the codimension of the largest
eigenspace of $\hat{x}$ on $\overline{V}$. Lower bounds on this parameter for $x \in H$ will play an

Table 5.1.1 *Non-subspace subgroups $H \in \mathscr{C}_i$, $2 \leqslant i \leqslant 8$*

| Collection | Description |
|---|---|
| $\mathscr{C}_2$ | Stabilisers of decompositions $V = \bigoplus_{i=1}^{t} V_i$, where $\dim V_i = a$ |
| $\mathscr{C}_3$ | Stabilisers of prime degree extension fields of $\mathbb{F}_q$ |
| $\mathscr{C}_4$ | Stabilisers of decompositions $V = V_1 \otimes V_2$ |
| $\mathscr{C}_5$ | Stabilisers of prime index subfields of $\mathbb{F}_q$ |
| $\mathscr{C}_6$ | Normalisers of symplectic-type $r$-groups, $r \neq p$ |
| $\mathscr{C}_7$ | Stabilisers of decompositions $V = \bigotimes_{i=1}^{t} V_i$, where $\dim V_i = a$ |
| $\mathscr{C}_8$ | Stabilisers of nondegenerate forms on $V$ |

important role in our study of derangements in this chapter (see Lemmas 5.3.3, 5.4.2, 5.6.3 and 5.7.3, for instance).

Throughout this chapter we will frequently refer to the *type* of $H$, which provides an approximate description of the group-theoretic structure of $H \cap \mathrm{PGL}(V)$. For example, if $G = \mathrm{PSL}_6(q)$ and $H \in \mathscr{C}_2$ is the stabiliser of a decomposition $V = V_1 \oplus V_2 \oplus V_3$, where each $V_i$ is 2-dimensional, then $H$ is a subgroup of type $\mathrm{GL}_2(q) \wr S_3$ (the precise structure of $H$ is given in [86, Proposition 4.2.9]). This convenient terminology is adopted by Kleidman and Liebeck (see [86, Section 3.1]).

Let $r \neq p$ be an odd prime. Recall that we define $\Phi(r,q)$ to be the smallest positive integer $i$ such that $r$ divides $q^i - 1$ (see (3.1.2)). As explained in Section 3.1.1, this integer plays an important role in the study of semisimple elements in $G$ of order $r$. As before, in this chapter it will be convenient to define an integer $c = c(i)$ as follows:

**Definition 5.1.1** Let $r \neq p$ be an odd prime and set $i = \Phi(r,q)$.

(a) If $T = \mathrm{PSL}_n(q)$ then $c(i) = i$.
(b) If $T = \mathrm{PSU}_n(q)$ then $c(i)$ is defined as in (4.4.2), so

$$c(i) = \begin{cases} i & i \equiv 0 \ (\mathrm{mod}\ 4) \\ i/2 & i \equiv 2 \ (\mathrm{mod}\ 4) \\ 2i & i \ \mathrm{odd} \end{cases}$$

(c) If $T = \mathrm{PSp}_n(q)$ or $\mathrm{P\Omega}_n^{\varepsilon}(q)$ then $c(i)$ is defined as in (4.5.1), that is

$$c(i) = \begin{cases} i & i \ \mathrm{even} \\ 2i & i \ \mathrm{odd} \end{cases}$$

Our main results on the $r$-elusivity of $\mathscr{C}_i$-actions (for $i \geqslant 2$) will be presented in Section 5.$i$, with reference to specific tables. The following theorem provides a summary of our results in the special case $r = 2$. Note that in Table 5.1.2, Case 2.I refers to the case $H \in \mathscr{C}_2$ that is labelled I in Table 5.2.1, and Case

Table 5.1.2 *2-elusive actions, $H \in \bigcup_{i \geqslant 2} \mathscr{C}_i \cup \mathscr{N}$*

| $T$ | Case | Type of $H$ | Conditions |
|---|---|---|---|
| $\mathrm{PSL}_n^\varepsilon(q)$ | 2.I | $\mathrm{GL}_a^\varepsilon(q) \wr S_t$ | always |
| | 2.VI | $\mathrm{GL}_{n/2}(q^2).2$ | $1 < (q+1)_2 \leqslant (n)_2$ |
| | 3.I | $\mathrm{GL}_{n/2}(q^2)$ | $n = 2;\ 1 < (q-1)_2 \leqslant (n)_2$ |
| | 4.I | $\mathrm{GL}_2^\varepsilon(q) \otimes \mathrm{GL}_{n/2}^\varepsilon(q)$ | $1 < (q-\varepsilon)_2 \leqslant (n)_2$ |
| | 5.I | $\mathrm{GL}_n^\varepsilon(q_0)$ | $p = 2;\ q = q_0^2,\ (n)_2 \leqslant (q-1)_2$ |
| | 5.V, 8.I | $\mathrm{Sp}_n(q)$ | $(q-\varepsilon)_2 \leqslant (n)_2$ |
| | 5.VI, 8.II | $\mathrm{O}_n^{\varepsilon'}(q)$ | always |
| | 6.I | $3^2.\mathrm{Sp}_2(3)$ | $n = 3$ |
| | 6.II | $2^4.\mathrm{Sp}_4(2)$ | $n = 4$ |
| | 6.III | $2^2.\mathrm{O}_2^-(2)$ | $n = 2$ |
| | 8.III | $\mathrm{GU}_n(q^{1/2})$ | $p = 2;\ (n)_2 \leqslant (q-1)_2$ |
| $\mathrm{PSp}_n(q)'$ | 2.II | $\mathrm{Sp}_a(q) \wr S_t$ | always |
| | 2.VII | $\mathrm{GL}_{n/2}(q).2$ | always |
| | 3.II | $\mathrm{Sp}_2(q^2)$ | $n = 4,\ q$ odd |
| | 3.V | $\mathrm{GU}_{n/2}(q)$ | always |
| | 4.II | $\mathrm{Sp}_2(q) \otimes \mathrm{O}_{n/2}^\varepsilon(q)$ | always |
| | 5.II | $\mathrm{Sp}_n(q_0)$ | $p = 2;\ q = q_0^2$ |
| | 6.IV | $2^4.\mathrm{O}_4^-(2)$ | $n = 4$ |
| | 7.II | $\mathrm{Sp}_2(q) \wr S_3$ | $n = 8$ |
| | 8.IV | $\mathrm{O}_n^\varepsilon(q)$ | $\varepsilon = +;\ n \equiv 2 \pmod 4$ |
| | 9.I | $\mathrm{O}_2^\varepsilon(q) \wr S_2$ | $n = 4,\ p = 2$ |
| $\mathrm{P}\Omega_n^\varepsilon(q)$ | 2.III | $\mathrm{O}_1(q) \wr S_n$ | always |
| | 2.IV | $\mathrm{O}_a^{\varepsilon'}(q) \wr S_t$ | see Remark 5.1.3 |
| | 2.V | $\mathrm{O}_{n/2}(q)^2$ | $q \not\equiv \varepsilon \pmod 8$ |
| | 2.VIII | $\mathrm{GL}_{n/2}(q).2$ | $n \equiv 2 \pmod 4,\ q \equiv 1 \pmod 4$ |
| | 3.III | $\mathrm{O}_{n/2}^\varepsilon(q^2)$ | $q$ odd |
| | 3.IV | $\mathrm{O}_{n/2}(q^2)$ | $q$ odd |
| | 3.VI | $\mathrm{GU}_{n/2}(q)$ | $q \equiv 3 \pmod 4$, with $n \equiv 4 \pmod 8$ if $\varepsilon = +$ |
| | 5.III | $\mathrm{O}_n^\varepsilon(q_0)$ | $p = 2;\ q = q_0^2$ |
| | 5.IV | $\mathrm{O}_n^-(q^{1/2})$ | $n \equiv 2 \pmod 4$ |
| | 9.IV | $\mathrm{GL}_3^\varepsilon(q) \times \mathrm{GL}_1^\varepsilon(q)$ | $(n,\varepsilon) = (8,+),\ q \equiv \varepsilon \pmod 4$ |
| | 9.V | $\mathrm{O}_2^-(q^2) \times \mathrm{O}_2^-(q^2)$ | $(n,\varepsilon) = (8,+),\ q$ odd |
| | 9.VII | $[2^9].\mathrm{SL}_3(2)$ | $(n,\varepsilon) = (8,+),\ q \equiv \pm 1 \pmod 8$ |
| | 9.VIII | $P_{1,3,4}$ | $(n,\varepsilon) = (8,+),\ q \equiv 1 \pmod 4$ |

9.IV is the case $H \in \mathscr{N}$ labelled IV in Table 5.9.1, etc. Also note that the conditions listed in the final column are additional to the relevant conditions listed in the final column of Tables $5.i.1,\ 2 \leqslant i \leqslant 9$. (In the final column, we adopt the notation described in Remark 4.1.8, so a comma means 'and' and a semicolon means 'or'.)

**Theorem 5.1.2** *Let* $G \leqslant \mathrm{Sym}(\Omega)$ *be a primitive almost simple classical group with socle* $T$ *and point stabiliser* $H \in \bigcup_{i \geqslant 2} \mathscr{C}_i \cup \mathscr{N}$. *Then* $T$ *is* 2-*elusive if and only if* $|\Omega|$ *is even and* $(G, H)$ *is one of the cases listed in Table 5.1.2.*

**Remark 5.1.3** Consider Case IV in Table 5.2.1, where $T = \mathrm{P}\Omega_n^{\varepsilon}(q)$ and $H$ is a $\mathscr{C}_2$-subgroup of type $\mathrm{O}_a^{\varepsilon'}(q) \wr S_t$ with $a \geqslant 2$. The conditions for 2-elusivity in this case are rather complicated, so we record them here: $T$ is 2-elusive unless $q$ is even, $\varepsilon' = -$, $a \equiv 0 \pmod 4$ and either $a = 4$ or $t$ is even.

Finally, some words on the organisation of this chapter. For $2 \leqslant i \leqslant 8$, the groups $G$ with $H \in \mathscr{C}_i$ will be handled in Section 5.$i$, and the possibilities for $H$ will be presented in Table 5.$i$.1. In the final column of these tables we record conditions that are necessary for the existence and/or maximality of $H$ in $G$. (We are not claiming that these conditions are always sufficient.) For $i \geqslant 3$, the $r$-elusive examples with $H \in \mathscr{C}_i$ will be recorded in Table 5.$i$.2. Finally, the primitive groups with $H \in \mathscr{N}$ will be studied in Section 5.9.

## 5.2 $\mathscr{C}_2$: Imprimitive subgroups

We begin by considering the imprimitive subgroups comprising Aschbacher's $\mathscr{C}_2$ collection. As described in Section 2.6.2.2, these subgroups are the stabilisers in $G$ of appropriate direct sum decompositions

$$V = V_1 \oplus \cdots \oplus V_t \qquad (5.2.1)$$

of the natural $T$-module $V$, where $t \geqslant 2$ and $\dim V_i = a$ for all $i$, so $n = at$. If $T = \mathrm{PSL}_n(q)$, then there are no additional conditions on the subspaces in (5.2.1). In the remaining cases, one of the following holds:

  (i) The $V_i$ are isometric nondegenerate subspaces, and the decomposition in (5.2.1) is orthogonal.

  (ii) $t = 2$ and $V_1$ and $V_2$ are maximal totally singular subspaces.

  (iii) $t = 2$, $T = \mathrm{P\Omega}_n^{\pm}(q)$, $nq/2$ is odd and $V_1$ and $V_2$ are orthogonal nonisometric nondegenerate subspaces of dimension $n/2$.

The specific cases that we will consider in this section are listed in Table 5.2.1; the conditions recorded in the final column are needed for the existence and maximality of $H$ in $G$ (see [86, Tables 3.5.A–H and 4.2.A] and [13, Section 8]). We refer the reader to [86, Section 4.2] for further details.

Let us now state our main results for $\mathscr{C}_2$-actions, starting with a theorem on derangements of order $p > 2$. In order to state this result, we need to introduce the following number-theoretic conditions on partitions of $n$ (recall that a partition $\lambda$ of $n$ *contains* a partition $\lambda'$ of $m < n$ if and only if each part of $\lambda'$ is also a part of $\lambda$).

P1$'$  Every partition $(p^{a_p}, \ldots, 1^{a_1})$ of $n$, with $a_p < a$, contains a partition of $a$.

P2$'$  Every partition $(p^{a_p}, \ldots, 1^{a_1})$ of $n$, such that $a_p < a$ and $a_i$ is even for all odd $i$, contains a partition of $a$ with the property that all odd parts have even multiplicity.

P3$'$  Every partition $(p^{a_p}, \ldots, 1^{a_1})$ of $n$, such that $a_p < a$ and $a_i$ is even for all even $i$, contains a partition of $a$ with the property that all even parts have even multiplicity.

**Theorem 5.2.1** *Let $G \leqslant \mathrm{Sym}(\Omega)$ be a primitive almost simple classical group with socle $T$ and point stabiliser $H \in \mathscr{C}_2$. Suppose $p > 2$ and $p$ divides $|\Omega|$. Then $T$ is $p$-elusive only if one of the following holds:*

  (i) *P1$'$ holds in Case I of Table 5.2.1.*

  (ii) *P2$'$ holds in Case II.*

  (iii) *P3$'$ holds in Case IV, with $\varepsilon = \varepsilon' = +$ and $a \equiv 0 \pmod 4$.*

Table 5.2.1 *The collection $\mathscr{C}_2$*

| Case | $T$ | Type of $H$ | Conditions |
|------|-----|-------------|------------|
| I | $\mathrm{PSL}_n^\varepsilon(q)$ | $\mathrm{GL}_a^\varepsilon(q) \wr S_t$ | |
| II | $\mathrm{PSp}_n(q)$ | $\mathrm{Sp}_a(q) \wr S_t$ | |
| III | $\mathrm{P}\Omega_n^\varepsilon(q)$ | $\mathrm{O}_1(q) \wr S_n$ | $q = p \geqslant 3$, with $\varepsilon = (-)^{n(q-1)/4}$ if $n$ even |
| IV | $\mathrm{P}\Omega_n^\varepsilon(q)$ | $\mathrm{O}_a^{\varepsilon'}(q) \wr S_t$ | $a \geqslant 2$, and $q$ odd if $a$ odd, and $(\varepsilon')^t = \varepsilon$ if $a$ even |
| V | $\mathrm{P}\Omega_n^\varepsilon(q)$ | $\mathrm{O}_{n/2}(q)^2$ | $n \equiv 2 \pmod 4$, $q \equiv -\varepsilon \pmod 4$ |
| VI | $\mathrm{PSU}_n(q)$ | $\mathrm{GL}_{n/2}(q^2).2$ | $n \geqslant 4$ even |
| VII | $\mathrm{PSp}_n(q)$ | $\mathrm{GL}_{n/2}(q).2$ | $q$ odd |
| VIII | $\mathrm{P}\Omega_n^+(q)$ | $\mathrm{GL}_{n/2}(q).2$ | |

**Remark 5.2.2** We present more detailed results in Propositions 5.2.12, 5.2.15 and 5.2.17. For instance, in Case I we prove that $T$ is $p$-elusive (for $p > 2$) only if one of the following holds:

(a) $p = 3$ and either $a \equiv 0 \pmod 6$, or $a \equiv \pm 2 \pmod 6$, $a \geqslant 8$ and $t \equiv 1 \pmod 3$;

(b) $p \geqslant 5$, $m$ divides $a$ for every positive integer $m < p$, and either $p$ divides $a$, or $t \equiv 1 \pmod p$.

In fact, if $p = 3$ then the necessary conditions in (a) are also sufficient.

Next we present our main result on semisimple derangements of odd prime order. Note that in Table 5.2.2 we adopt the usual notation $i = \Phi(r,q)$ (see (3.1.2)) and we define the integer $c$ as in Definition 5.1.1.

**Theorem 5.2.3** *Let $G \leqslant \mathrm{Sym}(\Omega)$ be a primitive almost simple classical group with socle $T$ and point stabiliser $H \in \mathscr{C}_2$. Let $r \neq p$ be an odd prime. Then $T$ is $r$-elusive if and only if $(G,H,r)$ is one of the cases recorded in Table 5.2.2, and $r$ divides $|\Omega|$.*

**Remark 5.2.4** The relevant conditions for $r$-elusivity are recorded in the second column of Table 5.2.2, which are additional to the conditions appearing in the final column of Table 5.2.1. For instance, Table 5.2.2 indicates that $T$ is $r$-elusive in Case I if either $1 = a < c = r - 1$ and $n \leqslant kr$, or if $c \leqslant a$ and $k = t\ell$. (As recorded in the table, $k$ and $\ell$ are integers such that $n \equiv k \pmod c$, $a \equiv \ell \pmod c$ and $0 \leqslant k, \ell < c$.)

Finally, our main result on derangements of order 2 is the following theorem.

Table 5.2.2  *r-elusive $\mathscr{C}_2$-actions, $r \neq p$, $r > 2$*

| Case | Conditions |
|------|------------|
| I | $\begin{cases} 1 = a < c = r-1, n \leqslant kr \\ c \leqslant a, k = t\ell \end{cases}$ |
| II | $c \leqslant a, k = t\ell$ |
| III | $c = r-1, \begin{cases} n \leqslant kr \\ k = 0, n \leqslant cr, \varepsilon = (-)^i, \text{ with } n/i \text{ odd if } \varepsilon = + \end{cases}$ |
| IV | $\begin{cases} \ell = 0, \varepsilon' = (-)^{a/i} \\ \ell > 0, c < a, \begin{cases} k = t\ell, a \not\equiv 0 \ (\mathrm{mod}\ i) \\ k = 0, c = t\ell, \varepsilon = (-)^{n/i+1} \end{cases} \end{cases}$ |
| V | $\begin{cases} c < n/2, k = 2\ell \\ i < n/2, k = 0, c = 2\ell, \varepsilon = (-)^i \end{cases}$ |

$n \equiv k \ (\mathrm{mod}\ c), a \equiv \ell \ (\mathrm{mod}\ c)$

Table 5.2.3  *2-elusive $\mathscr{C}_2$-actions*

| Case | Conditions |
|------|------------|
| I, II, III, VII | always |
| IV | see Remark 5.1.3 |
| V | $q \not\equiv \varepsilon \ (\mathrm{mod}\ 8)$ |
| VI | $1 < (q+1)_2 \leqslant (n)_2$ |
| VIII | $n \equiv 2 \ (\mathrm{mod}\ 4), q \equiv 1 \ (\mathrm{mod}\ 4)$ |

**Theorem 5.2.5** *Let $G \leqslant \mathrm{Sym}(\Omega)$ be a primitive almost simple classical group with socle $T$ and point stabiliser $H \in \mathscr{C}_2$. Then $T$ is 2-elusive if and only if $|\Omega|$ is even and $(G, H)$ is one of the cases in Table 5.2.3.*

## 5.2.1 Preliminaries

Here we record some preliminary results that will be needed in the proofs of our main theorems for $\mathscr{C}_2$-actions.

**Lemma 5.2.6** *Let $x \in \mathrm{GL}_n(\mathbb{F})$ be an element of prime order $r$, where $\mathbb{F}$ is an algebraically closed field of characteristic $p$. Suppose $x$ cyclically permutes the subspaces $\{V_1, \ldots, V_r\}$ in a decomposition*

$$V = V_1 \oplus \cdots \oplus V_r$$

*of the natural $\mathrm{GL}_n(\mathbb{F})$-module, where each $V_i$ is a-dimensional.*

(i) *If $r \neq p$ then each $r$th root of unity occurs as an eigenvalue of $x$ with multiplicity $a$.*

(ii) *If $r = p$ then $x$ has Jordan form $[J_p^a]$.*

*Proof* This follows immediately from the fact that the characteristic polynomial of $x$ is $(t^r - 1)^a$. Alternatively, we can argue as follows. Let $\{e_{1,1}, \dots, e_{a,1}\}$ be a basis for $V_1$ and set $e_{i,j+1} = (e_{i,1})^{x^j}$, where $1 \leqslant i \leqslant a$ and $1 \leqslant j < r$ (note that $e_{i,j+1} \in V_{j+1}$). Then $x$ induces a permutation of cycle-shape $(r^a)$ on the vectors in the basis

$$\{e_{1,1}, e_{1,2}, \dots, e_{1,r}, e_{2,1}, \dots, e_{2,r}, \dots, e_{a,1}, \dots, e_{a,r}\}$$

of $V$. The result follows.                                                  $\square$

**Corollary 5.2.7** *Suppose $x \in \mathrm{GL}_n(\mathbb{F})$ has prime order $r$ and stabilises a decomposition $V = V_1 \oplus \cdots \oplus V_t$, where $t \geqslant 2$ and each $V_i$ is $a$-dimensional.*

(i) *Suppose $r = p$ and $x$ has Jordan form $[J_p^{a_p}, \dots, J_1^{a_1}]$. If $a_p < a$ then $x$ fixes each $V_i$.*

(ii) *Suppose $r \neq p$, say $x = [I_{a_0}, \mu I_{a_1}, \dots, \mu^{r-1} I_{a_{r-1}}]$, where $\mu \in \mathbb{F}$ is a primitive $r$th root of unity. If $a_j < a$ for any $j$, then $x$ fixes each $V_i$.*

In the next two lemmas we have

$$G \in \{\mathrm{GU}_n(q), \mathrm{Sp}_n(q), \mathrm{O}_n^\varepsilon(q)\}$$

and $V$ is the natural module for $G$. Let $x \in G$ be a semisimple element of odd prime order $r$ and let $i = \Phi(r, q)$ (see (3.1.2)). We set

$$b = b(i) = \begin{cases} i/2 & G = \mathrm{GU}_n(q) \text{ and } i \text{ even} \\ i & \text{otherwise} \end{cases} \tag{5.2.2}$$

As noted in Remark 2.2.6, if $V$ is equipped with a nondegenerate sesquilinear form then it is convenient to use the term *totally singular* to describe a totally isotropic subspace of $V$. Also recall that a pair $\{U, U^*\}$ of complementary totally singular subspaces of $V$ is a *dual pair* if $U \oplus U^*$ is nondegenerate.

**Lemma 5.2.8** *Let $x \in G$ be a semisimple element of odd prime order $r$ with $i = \Phi(r, q)$, where $i$ is odd if $G$ is symplectic or orthogonal, and $i \not\equiv 2 \pmod 4$ if $G$ is unitary. Let $a$ be a positive integer such that $a \geqslant 2b$ and $V$ contains a nondegenerate $a$-space, where $b$ is the integer defined in (5.2.2). Write*

$$n \equiv k \pmod{2b}, \quad a \equiv \ell \pmod{2b}, \quad e = \dim C_V(x)$$

*where $0 \leqslant k, \ell < 2b$.*

*Then x fixes componentwise an orthogonal decomposition*

$$V = V_1 \perp \ldots \perp V_t$$

*into nondegenerate a-spaces if and only if $e \geqslant t\ell$.*

*In particular, if a is even, $G = \mathrm{O}_n^\varepsilon(q)$ and each $V_j$ is of type $\varepsilon'$, then x fixes such a decomposition componentwise if and only if one of the following holds:*

(i) *$\ell > 0$ and $e \geqslant t\ell$;*
(ii) *$\ell = 0$ and either $\varepsilon' = +$, or $\varepsilon' = -$ and $e \geqslant 2tb$.*

*Proof*  In view of Propositions 3.3.2(i), 3.4.3(i) and 3.5.4(i), we observe that $x$ fixes an orthogonal decomposition of the form

$$V = (U_1 \oplus U_1^*) \perp (U_2 \oplus U_2^*) \perp \ldots \perp (U_s \oplus U_s^*) \perp C_V(x)$$

where each $\{U_j, U_j^*\}$ is a dual pair of totally singular $b$-spaces, and $x$ acts irreducibly on each $U_j$ and $U_j^*$.

Suppose $x$ fixes an appropriate decomposition

$$V = V_1 \perp V_2 \perp \ldots \perp V_t \tag{5.2.3}$$

componentwise and consider $x|_{V_j}$, which is the restriction of $x$ to $V_j$. Under the action of $x$, each $V_j$ admits a decomposition into irreducible dual pairs of totally singular $b$-spaces, together with the 1-eigenspace $C_{V_j}(x)$. Hence $\dim C_{V_j}(x) \geqslant \ell$ and thus $e \geqslant t\ell$. Moreover, if $G = \mathrm{O}_n^\varepsilon(q)$, $\ell = 0$ and the $V_j$ are of type $\varepsilon' = -$, then each $C_{V_j}(x)$ must be nontrivial and elliptic (since each $U_d \oplus U_d^*$ is hyperbolic), so $\dim C_{V_j}(x) \geqslant 2b$ and thus $e \geqslant 2tb$ in this case.

Conversely, suppose that $e \geqslant t\ell$. Then there exists an orthogonal decomposition

$$C_V(x) = X \perp W_1 \perp \ldots \perp W_t \tag{5.2.4}$$

where $X$ is a nondegenerate $(e - t\ell)$-space, and each $W_j$ is a nondegenerate $\ell$-space (if $\ell = 0$, then $X = C_V(x)$). For each $1 \leqslant j \leqslant t$, let $Y_j$ be the span of $W_j$ and a nondegenerate $(a - \ell)$-space spanned by a combination of dual pairs $\{U_d, U_d^*\}$ and appropriate nondegenerate $2b$-dimensional subspaces of $X$ such that the nondegenerate spaces $Y_j$ constructed in this way are pairwise orthogonal. Then $x$ fixes the decomposition $V = Y_1 \perp \ldots \perp Y_t$ componentwise.

To complete the proof, let us assume that $G = \mathrm{O}_n^\varepsilon(q)$, $e \geqslant t\ell$, $a$ is even and $V$ admits an orthogonal decomposition as in (5.2.3), where each $V_j$ is a nondegenerate $a$-space of type $\varepsilon'$. Note that $\varepsilon = (\varepsilon')^t$. There are several cases to consider.

First observe that $C_V(x)$ is either trivial (in which case $\varepsilon = +$), or it is nondegenerate of type $\varepsilon$. Suppose $\ell > 0$. Since the $V_j$ in (5.2.3) are of type

$\varepsilon'$, in (5.2.4) we may assume that $X$ is hyperbolic and each $W_j$ has type $\varepsilon'$. Therefore, by taking $2b$-dimensional hyperbolic subspaces of $X$ we may assume that the $(a - \ell)$-subspaces constructed above are hyperbolic, and thus the nondegenerate spaces $Y_j$ in the above decomposition $V = Y_1 \perp \ldots \perp Y_t$ are of type $\varepsilon'$.

Finally, let us assume that $\ell = 0$. Note that $e$ is divisible by $2b$. If $\varepsilon' = +$ then $\varepsilon = +$, $C_V(x)$ is hyperbolic (or trivial) and it is easy to construct an appropriate orthogonal decomposition that is fixed (componentwise) by $x$. Now assume $\varepsilon' = -$ and $e \geqslant 2tb$. Since $C_V(x)$ is of type $\varepsilon = (-)^t$, we can write

$$C_V(x) = X' \perp W_1' \perp \ldots \perp W_t'$$

where $X'$ is hyperbolic and each $W_j'$ is an elliptic $2b$-space. Take $Y_j$ to be the span of $W_j'$ and an appropriate combination of dual pairs $\{U_d, U_d^*\}$ and hyperbolic $2b$-dimensional subspaces of $X'$ so that $V = Y_1 \perp \ldots \perp Y_t$. Then $x$ fixes this decomposition (componentwise), and each $Y_j$ is elliptic as required. $\qquad\square$

**Lemma 5.2.9** *Let $x \in G$ be a semisimple element of odd prime order $r$ with $i = \Phi(r,q)$, where $i$ is even if $G$ is symplectic or orthogonal, and $i \equiv 2 \pmod{4}$ if $G$ is unitary. Let $a$ be a positive integer such that $a \geqslant b$ and $V$ contains a nondegenerate $a$-space, where $b$ is the integer defined in (5.2.2). Write*

$$n \equiv k \pmod{b}, \quad a \equiv \ell \pmod{b}, \quad e = \dim C_V(x)$$

*where $0 \leqslant k, \ell < b$.*

*Then $x$ fixes componentwise an orthogonal decomposition*

$$V = V_1 \perp \ldots \perp V_t$$

*into nondegenerate $a$-spaces if and only if $e \geqslant t\ell$.*

*In particular, if $a$ is even, $G = O_n^\varepsilon(q)$ and each $V_j$ is of type $\varepsilon'$, then $x$ fixes such a decomposition componentwise if and only if one of the following holds:*

  (i) *$\ell > 0$ and $C_V(x)$ contains a nondegenerate $t\ell$-space that admits an orthogonal decomposition into $t$ nondegenerate $\ell$-spaces of type $\varepsilon''$, where $\varepsilon' = \varepsilon''(-)^{(a-\ell)/b}$.*

  (ii) *$\ell = 0$ and either $\varepsilon' = (-)^{a/b}$, or $C_V(x)$ contains a hyperbolic $tb$-space.*

*Proof* By Propositions 3.3.2(ii), 3.4.3(ii) and 3.5.4(ii), it follows that $x$ fixes an orthogonal decomposition

$$V = U_1 \perp U_2 \perp \ldots \perp U_s \perp C_V(x)$$

where each $U_j$ is a nondegenerate $b$-space on which $x$ acts irreducibly. Furthermore, if $G = O_n^\varepsilon(q)$ then each $U_j$ is elliptic.

Suppose $x$ fixes componentwise an appropriate decomposition as in (5.2.3). Under the action of $x$, each $V_j$ admits an orthogonal decomposition into irreducible nondegenerate $b$-spaces, together with the 1-eigenspace $C_{V_j}(x)$. Hence $\dim C_{V_j}(x) \geqslant \ell$ and thus $e \geqslant t\ell$.

In addition, let us assume that $a$ is even, $G = \mathrm{O}_n^\varepsilon(q)$ and each $V_j$ is of type $\varepsilon'$. Recall that each $U_j$ is an elliptic $b$-space. If $\ell = 0$ and $(-)^{a/b} = -\varepsilon'$, then $\dim C_{V_j}(x) \geqslant b$. More precisely, $C_{V_j}(x)$ contains a hyperbolic $b$-space $U$ such that $V_j = U \perp U'$ with $U'$ of type $\varepsilon'$ (or trivial), whence $C_V(x)$ contains a hyperbolic $tb$-space. Now assume that $\ell > 0$. Then $\dim C_{V_j}(x) \geqslant \ell$ and $C_{V_j}(x)$ contains a nondegenerate $\ell$-space of type $\varepsilon''$ such that $\varepsilon' = \varepsilon''(-)^{(a-\ell)/b}$. Therefore $C_V(x)$ contains a nondegenerate $t\ell$-space that admits an orthogonal decomposition into $t$ nondegenerate $\ell$-spaces of type $\varepsilon''$.

For the converse, let us assume that $e \geqslant t\ell$. Now $C_V(x)$ admits an orthogonal decomposition as in (5.2.4), with $X$ a nondegenerate $(e - t\ell)$-space, and each $W_j$ a nondegenerate $\ell$-space (note that $X = C_V(x)$ if $\ell = 0$). As in the proof of the previous lemma, for each $1 \leqslant j \leqslant t$, let $Y_j$ be the span of $W_j$ and a nondegenerate $(a - \ell)$-space spanned by a combination of irreducible blocks $U_d$ and appropriate nondegenerate $b$-dimensional subspaces of $X$ such that the $Y_j$ constructed in this way are pairwise orthogonal. Then each $Y_j$ is a nondegenerate $a$-space, and $x$ fixes the decomposition $V = Y_1 \perp \ldots \perp Y_t$ componentwise.

To complete the proof, we may assume that $a$ is even, $G = \mathrm{O}_n^\varepsilon(q)$ and $V$ admits an orthogonal decomposition as in (5.2.3), where each $V_j$ is a nondegenerate $a$-space of type $\varepsilon'$. We need to show that $x$ fixes such a decomposition componentwise if the conditions in (i) or (ii) hold.

First assume that (i) holds, so $\ell > 0$ and $C_V(x)$ contains a nondegenerate $t\ell$-space that admits an orthogonal decomposition into $t$ nondegenerate $\ell$-spaces of type $\varepsilon''$, where $\varepsilon' = \varepsilon''(-)^{(a-\ell)/b}$. In particular, in (5.2.4) we may assume that each $W_j$ has type $\varepsilon''$. If $X = \{0\}$ (in (5.2.4)) then each $Y_j$ constructed above has type $\varepsilon''(-)^{(a-\ell)/b} = \varepsilon'$ and the result follows. Now assume that $X \neq \{0\}$, in which case $X$ has type $\varepsilon'''$, where $\varepsilon'''(\varepsilon'')^t(-)^s = \varepsilon = (\varepsilon')^t$. Since $\varepsilon' = \varepsilon''(-)^{(a-\ell)/b}$ we have

$$\varepsilon'''(\varepsilon'')^t(-)^s = (\varepsilon''(-)^{(a-\ell)/b})^t$$

and thus

$$\varepsilon''' = (-)^{(a-\ell)t/b-s} = (-)^{(n-t\ell-sb)/b} = (-)^{(\dim X)/b}$$

Therefore, $X$ admits an orthogonal decomposition into elliptic $b$-spaces. If we now use these $b$-spaces in the above construction of the $Y_j$, then each $Y_j$ has type $\varepsilon'$ and thus $x$ fixes an orthogonal decomposition of the desired form.

Finally, let us assume that (ii) holds, so $\ell = 0$ and $X = C_V(x)$ in (5.2.4). Suppose that $\varepsilon' = (-)^{a/b}$. If $X = \{0\}$ then it is clear that the $Y_j$ constructed above have type $\varepsilon'$, so let us assume that $X \neq \{0\}$. If $\varepsilon'''$ denotes the type of $X$, then

$$\varepsilon'''(-)^s = \varepsilon = (\varepsilon')^t = (-)^{n/b}$$

and we deduce that $\varepsilon''' = (-)^{(\dim X)/b}$. Therefore, $X$ admits an orthogonal decomposition into elliptic $b$-spaces and once again we see that the $Y_j$ have type $(-)^{a/b} = \varepsilon'$. Finally, let us assume that $(-)^{a/b} = -\varepsilon'$ and $X$ contains a hyperbolic $tb$-space. In this situation we have an orthogonal decomposition

$$C_V(x) = X = X' \perp W_1' \perp \ldots \perp W_t'$$

where each $W_j'$ is a hyperbolic $b$-space. As above, we deduce that $X'$ has type $(-)^{(\dim X')/b}$, so $X'$ can be decomposed as an orthogonal sum of elliptic $b$-spaces. If we now use these elliptic $b$-spaces in the construction of the $Y_j$ above, then each $Y_j$ will be of type $(-)^{(a-b)/b} = \varepsilon'$ and the result follows. $\quad\square$

**Remark 5.2.10** Let $x \in \mathrm{GL}_n(q)$ be a semisimple element of odd prime order $r$ and set $i = \Phi(r,q)$. Then the argument in the proof of Lemma 5.2.9 shows that $x$ fixes componentwise a decomposition of the natural $\mathrm{GL}_n(q)$-module $V$ into $t$ subspaces of dimension $a$ if and only if $e \geqslant t\ell$, where $e = \dim C_V(x)$ and $a \equiv \ell$ (mod $i$) with $0 \leqslant \ell < i$.

The final result of this section is a technical lemma that we will need in the analysis of involutions in the proof of Proposition 5.2.23.

**Lemma 5.2.11** Let $V = V_1 \oplus V_2$ be the natural $\mathrm{O}_{2a}^+(q)$-module, where $V_1$ and $V_2$ are complementary totally singular $a$-spaces. Assume $q \equiv 1$ (mod 4) and set $x = [\lambda I_a, -\lambda I_a] \in \mathrm{O}_{2a}^+(q)$ with respect to this decomposition, where $\lambda \in \mathbb{F}_q$ has order 4. Then $B(v, vx) = 0$ for all $v \in V$, where $B$ is the corresponding nondegenerate symmetric bilinear form on $V$.

*Proof* Let $v \in V$. The conclusion is clear if $v \in V_1$ or $v \in V_2$, so we may assume that $v = v_1 + v_2$ for some nonzero vectors $v_i \in V_i$. Then $vx = \lambda v_1 - \lambda v_2$ and thus

$$B(v, vx) = B(v_1 + v_2, \lambda v_1 - \lambda v_2) = -\lambda B(v_1, v_2) + \lambda B(v_2, v_1) = 0$$

since $B$ is symmetric. $\quad\square$

## 5.2.2 Unipotent elements

In this section we focus on derangements of order $p$ and we establish Theorem 5.2.1. For $p > 2$, we adopt the parameterisations of unipotent classes in symplectic and orthogonal groups described in Sections 3.4.3 and 3.5.3. Similarly, for $p = 2$ we use the labelling of unipotent involutions presented in Sections 3.4.4 and 3.5.4 (this agrees with the notation introduced originally by Aschbacher and Seitz [5]). In particular, $J_i$ denotes a standard (lower-triangular) unipotent Jordan block of size $i$, and it will be convenient to interpret $J_0$ as the 'empty' Jordan block of dimension zero. We will consider each of the cases labelled I–VIII in Table 5.2.1 in turn, starting with Case I.

**Proposition 5.2.12** *Consider Case I in Table 5.2.1.*

(i) *If $p = 2$ then $T$ is $p$-elusive.*

(ii) *If $p = 3$ then $T$ is $p$-elusive if and only if $a \equiv 0 \pmod 6$, or $a \equiv \pm 2$ $\pmod 6$, $a \geqslant 8$ and $t \equiv 1 \pmod 3$.*

(iii) *If $p \geqslant 5$ then $T$ is $p$-elusive only if $m$ divides $a$ for every positive integer $m < p$, and either $p$ divides $a$, or $t \equiv 1 \pmod p$.*

*Proof* Here $T = \mathrm{PSL}_n^\varepsilon(q)$ and $H$ is of type $\mathrm{GL}_a^\varepsilon(q) \wr S_t$, which is the stabiliser in $G$ of a decomposition $V = V_1 \oplus \cdots \oplus V_t$, where $\dim V_i = a$ for all $i$ (and the $V_i$ are nondegenerate and pairwise orthogonal if $\varepsilon = -$). By Propositions 3.2.6 and 3.3.7, there is a bijection between the set of $\mathrm{PGL}_n^\varepsilon(q)$-classes of elements of order $p$ in $T$ and the set of nontrivial partitions of $n$ with parts of size at most $p$. Let $x = \hat{x} Z \in T$ be an element of order $p$.

First assume $p = 2$. We claim that $T$ is 2-elusive (so this case is recorded in Tables 5.1.2 and 5.2.3). It is clear that every involution fixes an appropriate decomposition of $V$ componentwise if $a$ is even, so let us assume $a$ is odd. In view of Lemma 5.2.6 and Corollary 5.2.7, it suffices to show that any element $\hat{x} = [J_2^\ell, J_1^{n-2\ell}]$ with $\ell < a$ fixes a decomposition componentwise. This is obvious if $\ell \leqslant a/2$, so let us assume $a/2 < \ell < a$. Here

$$\hat{x} = [J_2^{(a-1)/2}, J_1] \oplus [J_2^{\ell-(a-1)/2}, J_1^{2a-2\ell-1}] \oplus [J_1^a]^{t-2}$$

and we conclude that $T$ is 2-elusive as claimed.

Now assume $p > 2$. If $n = 2$ then $[J_2]$ is a derangement, so assume $n > 2$. If $p > n$ then part (i) of Corollary 5.2.7 implies that $[J_{n-1}, J_1]$ is a derangement if $a = 1$, and $[J_n]$ is a derangement if $a > 1$. Similarly, if $n \geqslant p > a$ then $[J_{p-1}, J_1^{n-p+1}]$ is a derangement if $a = 1$, and $[J_p, J_1^{n-p}]$ is a derangement if $a > 1$. Therefore, for the remainder we may assume that $p \leqslant a$. Notice that if

$m < p$ is a positive integer that does not divide $a$, then $[J_m^c, J_d]$ is a derangement, where $n = cm + d$ with $0 \leqslant d < m$.

Next suppose $p$ does not divide $a$, but $p$ does divide $t$, say $t = \ell p$ with $\ell \geqslant 1$. We claim that $\hat{x} = [J_p^{n/p-1}, J_{p-1}, J_1]$ is a derangement. Now, if $x$ fixes an appropriate decomposition $V = V_1 \oplus \cdots \oplus V_t$ then it must induce a nontrivial permutation of the $V_i$, with cycle-shape $(p^h, 1^{t-hp})$ for some $1 \leqslant h < \ell$ (note that $h < \ell$ since $x$ has Jordan blocks $J_i$ with $i < p$). Let $U$ be the subspace spanned by the $t - hp$ summands fixed by $x$, so $\hat{x}|_U = [J_p^{a(\ell-h)-1}, J_{p-1}, J_1]$. Now $t - hp = p(\ell - h) \geqslant p \geqslant 3$ since $\ell > h$, so $\hat{x}|_U$ cannot fix each of the $t - hp$ subspaces in the corresponding decomposition of $U$, which is a contradiction. This justifies the claim.

Now assume $p$ fails to divide both $a$ and $t$. Write $n = cp + d$, where $1 \leqslant d < p$, and set $\hat{x} = [J_p^c, J_d]$. We claim that $x$ is a derangement if $t \not\equiv 1 \pmod{p}$. To see this, suppose $t \not\equiv 1 \pmod{p}$ and $x$ fixes a decomposition. Then as before, $x$ must act nontrivially on the $t$ summands, with cycle-shape $(p^h, 1^{t-hp})$ for some $h \geqslant 1$. If $U$ is the subspace spanned by the summands fixed by $x$, then $\hat{x}|_U = [J_p^{c-ah}, J_d]$. But visibly, $\hat{x}|_U$ can only fix each of the $t - hp$ subspaces in the decomposition of $U$ if $t - hp = 1$, a contradiction since $t \not\equiv 1 \pmod{p}$. This justifies the claim.

To summarise, for $p > 2$ we have now shown that $T$ is $p$-elusive only if one of the following holds:

(a) $m$ divides $a$ for every positive integer $m \leqslant p$, or
(b) $m$ divides $a$ for every positive integer $m < p$, and we have $p < a$, $(a, p) = 1$ and $t \equiv 1 \pmod{p}$.

To complete the proof of the proposition, it remains to show that if $p = 3$ then $T$ is 3-elusive in cases (a) and (b) (with $a \neq 4$). First consider (a). Here $a$ is divisible by 6, so $n = 6k$ for some integer $k \geqslant 2$ and it is easy to see that every partition of $n$ of the form $(3^{a_3}, 2^{a_2}, 1^{a_1})$ contains a partition of 6. Therefore every element in $T$ of order 3 fixes a decomposition componentwise, whence $T$ is 3-elusive if (a) holds.

Now consider (b). Here $a \geqslant 4$ and $t \equiv 1 \pmod{3}$. If $a = 4$ then $[J_3^3, J_2^{n/2-5}, J_1]$ is a derangement, so assume $a > 4$. Suppose $a \equiv b \pmod{6}$, say $a = 6k + b$ where $b = 2$ or 4 (since $a$ is even and indivisible by 3). In order to deduce that $T$ is 3-elusive, it suffices to show that the specific elements

$$\hat{x} = [J_3^{2c}, J_2^{n/2-3c}], \quad \hat{y} = [J_3^{2d-1}, J_2^{n/2-3d+1}, J_1]$$

have fixed points, where $c \geqslant 0$ and $d \geqslant 1$ are integers such that $\max\{2c, 2d - 1\} < a$. First consider $\hat{x}$. If $6c \leqslant a$ then it is clear that $\hat{x}$ has fixed points, so assume otherwise. Then $2c > a/3 > 2k$ and we may write $2c = 2ke + f$, where $0 \leqslant f < 2k$. Then

$$\hat{x} = [J_3^{2k}, J_2^{b/2}]^e \oplus [J_3^f, J_2^{a/2-3f/2}] \oplus [J_2^{a/2}]^{t-e-1}$$

Similarly, $\hat{y}$ has fixed points if $3(2d-1) < a$, so let us assume that $2d-1 \geqslant a/3 > 2k$. Write $2d-1 = 2kg+h$, where $1 \leqslant h < 2k$. Then

$$\hat{y} = [J_3^{2k}, J_2^{b/2}]^g \oplus [J_3^h, J_2^{a/2-3h/2-1/2}, J_1] \oplus [J_2^{a/2}]^{t-g-1}$$

The result follows.                                                              □

**Remark 5.2.13** If $p \geqslant 5$ then the given conditions in part (iii) of Proposition 5.2.12 are necessary for $p$-elusivity, but they are not always sufficient. For example, if $(a,p) = (12,5)$ and $t \equiv 1 \pmod 5$ then the second condition in (iii) is satisfied, but $[J_5^{11}, J_3^{4t-19}, J_2]$ is a derangement.

**Remark 5.2.14** We can replace the necessary conditions in part (iii) of Proposition 5.2.12 by a more succinct number-theoretic condition that is also necessary for $p$-elusivity. Indeed, for $p > 2$ it is clear that $T$ is $p$-elusive only if every partition $(p^{a_p}, \ldots, 1^{a_1})$ of $n$ with $a_p < a$ contains a partition of $a$. This is the condition P1$'$ referred to in the statement of Theorem 5.2.1.

**Proposition 5.2.15** *Consider Case II in Table 5.2.1.*

(i) *If $p = 2$ then $T$ is p-elusive.*

(ii) *If $p = 3$ then $T$ is p-elusive if and only if $a \not\equiv 4 \pmod 6$, or $a \geqslant 10$ and $t \equiv 1 \pmod 3$.*

(iii) *If $p \geqslant 5$ then $T$ is p-elusive only if $m$ divides $a$ for every positive integer $m < p$, and either $p$ divides $a$, or $t \equiv 1,2 \pmod p$.*

*Proof* In this case $T = \mathrm{PSp}_n(q)$ and $H$ is of type $\mathrm{Sp}_a(q) \wr S_t$. Here $H$ stabilises an orthogonal decomposition $V = V_1 \perp \ldots \perp V_t$, where each $V_i$ is a nondegenerate $a$-space.

First assume $p = 2$. We claim that $T$ is 2-elusive. We will use the notation for involution class representatives given in Section 3.4.4. In view of Lemma 3.4.14, and the description of involutions of type $b_s$ and $c_s$, it is easy to see that these elements fix an orthogonal decomposition of $V$ (componentwise) into nondegenerate 2-spaces, so they fix an appropriate decomposition in $\Omega$. Similarly, we observe that $a_s$-type involutions also have fixed points if $a \equiv 0 \pmod 4$.

To complete the analysis of unipotent involutions, we may assume that $a \equiv 2 \pmod 4$ and $x$ is an $a_s$-type involution (note that $s$ is even). First observe that if $y \in S_t < H$ has cycle-shape $(2^h, 1^{t-2h})$, then Lemma 5.2.6(ii) implies that $y$ has Jordan form $[J_2^{ah}, J_1^{n-2ah}]$ on $V$. Moreover, we deduce that $y$ is an $a$-type involution since $B(v,vy) = 0$ for all $v \in V$ (where $B$ denotes the underlying alternating form on $V$). Write $s = a\ell + b$, where $0 \leqslant b < a$ and $b$ is even. Also write $b = a'j + c$, where $0 \leqslant c < a' = a/2 - 1$ and $c$ is even. Note that $j \leqslant 2$, with equality only if $c = 0$. Also note that $t \geqslant 2\ell$.

If $t = 2\ell$ then $s = a\ell$, $b = 0$ and $x$ fixes a decomposition in $\Omega$ by inducing a permutation of cycle-shape $(2^{t/2})$ on the component spaces. Similarly, if $t = 2\ell + 1$ then $b \leqslant a'$ and thus $(j,c) = (0,b)$ or $(1,0)$. Here $x$ fixes a decomposition by inducing a permutation of cycle-shape $(2^\ell)$ on the first $2\ell$ component spaces, and an involution of type $a_b$ on the final (fixed) component. Finally, if $t \geqslant 2\ell + 2$ then $x$ fixes an appropriate decomposition $V = V_1 \perp \ldots \perp V_t$ by permuting $V_1, \ldots, V_{2\ell}$ as before, and acting as an involution of type $a_{a'}$ on the spaces $V_{2\ell+1}, \ldots, V_{2\ell+j}$, and of type $a_c$ on $V_{2\ell+j+1}$ if $c > 0$ (recall that $c = 0$ if $j = 2$). This justifies the claim.

For the remainder we may assume that $p > 2$. By Proposition 3.4.10, there is a bijection between the set of $T$-classes of elements of order $p$ in $T$, and the set of nontrivial *signed partitions* of $n$ of the form

$$(p^{a_p}, \varepsilon_{p-1}(p-1)^{a_{p-1}}, \ldots, \varepsilon_2 2^{a_2}, 1^{a_1})$$

where $\varepsilon_i = \pm$ for each even $i$, and $a_i$ is even for all odd $i$. It is important to note that if $x, y \in T$ are elements of order $p$ with the same Jordan block structure on $V$, then $x$ fixes an appropriate decomposition of $V$ if and only if $y$ does (this is clear from the description of unipotent elements in Section 3.4.3). Therefore, in this situation, signed partitions play no role in our analysis of unipotent elements. Let $x = \hat{x}Z \in T$ be an element of order $p$.

Suppose $2p > n$. If $n = 4$ then $a = t = 2$ and $T$ is $p$-elusive if and only if $p = 3$ (note that $[J_4]$ is a derangement of order $p$ if $p \geqslant 5$). Now assume that $n > 4$. If $a < n/2$ then $[J_{n/2}^2]$ is a derangement, and if $a = n/2$ we can take $[J_{n/2-1}^2, J_2]$. For the remainder, we may assume that $2p \leqslant n$.

Suppose $2p > a$. If $a \geqslant 4$ then $[J_p^2, J_1^{n-2p}]$ is a derangement, so let us assume $a = 2$. If $p \geqslant 5$ then $[J_{p-2}^2, J_1^{n-2p+4}]$ is a derangement. If $(a,p) = (2,3)$ then

$$\hat{x} = [J_3^{2\ell}, J_2^m, J_1^{n-6\ell-2m}]$$

for some $\ell, m \geqslant 0$, and it is clear that $x$ fixes a decomposition by inducing a permutation with cycle-shape $(3^\ell, 1^{t-3\ell})$ on the $t$ component spaces in the decomposition. Therefore, $T$ is 3-elusive if $(a,p) = (2,3)$.

Now assume $2p \leqslant a$. If there exists a positive even integer $m < p$ that does not divide $a$, then $[J_m^c, J_d]$ is a derangement, where $n = cm + d$ with $0 \leqslant d < m$. Next suppose $m < p$ is odd and $m$ does not divide $a$. Write $n = 2cm + d$ with $0 \leqslant d < 2m$, and note that $d \leqslant 2p - 6$ is even. If $t \geqslant 3$ then $[J_m^{2c}, J_{d/2}^2]$

is a derangement of order $p$, so let us assume $t = 2$ (in which case $a \equiv d/2$ (mod $m$)). If $d < p$ then $[J_m^{2c}, J_d]$ is a derangement, and $[J_m^{2c}, J_{p-1}, J_{d+1-p}]$ has the same property if $d \geqslant p$ (note that $p - 1 \neq d + 1 - p$ since $d \leqslant 2p - 6$).

Next suppose $p$ divides $t$, but not $a$. First assume $p = 3$ and $a \equiv 2$ (mod 6). We have already observed that $T$ is 3-elusive when $a = 2$, so let us assume $a \geqslant 8$. Again, we claim that $T$ is 3-elusive. To see this, write $a = 6k + 2$, where $k \geqslant 1$. In view of Lemma 5.2.6(ii), it suffices to show that $\hat{x} = [J_3^{2\ell}, J_2^{n/2-3\ell}]$ has fixed points, where $2\ell < a$. Write $2\ell = 2kb + c$, where $0 \leqslant c < 2k$ and $c$ is even. Since $2\ell < a$ we have $b \leqslant 3$, with equality only if $c = 0$. If $b = 3$ then $n/2 - 3\ell \geqslant 3$ (since $t \geqslant 3$) and thus

$$\hat{x} = [J_3^{2k}, J_2]^3 \oplus [J_2^{a/2}]^{t-3}$$

Similarly, if $b < 3$ then $n/2 - 3\ell \geqslant b + a/2 - 3c/2$ and

$$\hat{x} = [J_3^{2k}, J_2]^b \oplus [J_3^c, J_2^{a/2-3c/2}] \oplus [J_2^{a/2}]^{t-b-1}$$

This justifies the claim.

Now assume $p \geqslant 5$, or $p = 3$ and $a \equiv 4$ (mod 6) (we continue to assume that $p$ divides $t$, but not $a$). Write $t = \ell p$, where $\ell \geqslant 1$. We claim that $\hat{x} = [J_p^{n/p-2}, J_{p-2}^2, J_2^2]$ is a derangement. Seeking a contradiction, suppose $\hat{x}$ fixes a decomposition. Then $\hat{x}$ must induce a nontrivial permutation on the $t$ component spaces in this decomposition, of cycle-shape $(p^h, 1^{t-hp})$ for some $1 \leqslant h < \ell$. Let $U$ be the subspace spanned by the $t - hp$ subspaces fixed by $\hat{x}$, so $\hat{x}|_U = [J_p^{a(\ell-h)-2}, J_{p-2}^2, J_2^2]$. Now $t - hp = p(\ell - h) \geqslant p \geqslant 3$ since $\ell > h$, so $\hat{x}|_U$ cannot fix each of the $t - hp$ subspaces in the decomposition of $U$, a contradiction. This justifies the claim.

Now assume $n = at$ is indivisible by $p$. Write $n = 2cp + d$, where $d$ is even and $2 \leqslant d < 2p$. Suppose $t \not\equiv 1, 2$ (mod $p$), so $p \geqslant 5$. We claim that $\hat{x} = [J_p^{2c}, J_{d/2}^2]$ is a derangement. Seeking a contradiction, suppose $\hat{x}$ fixes a decomposition. Then as before, $\hat{x}$ must act nontrivially on the set of $t$ summands of the decomposition, say with cycle-shape $(p^h, 1^{t-hp})$ for some $h \geqslant 1$. Then $\hat{x}|_U = [J_p^{2c-ah}, J_{d/2}^2]$, where $U$ is the subspace spanned by the component spaces fixed by $\hat{x}$. But visibly, $\hat{x}|_U$ can only fix each of the $t - hp$ subspaces in the decomposition of $U$ if $t - hp = 1$ or 2, which is not possible since $t \not\equiv 1, 2$ (mod $p$).

To summarise, for $p > 2$ we have now shown that $T$ is $p$-elusive only if one of the following holds:

- $p = 3$ and $a \neq 4$, with $a \not\equiv 4$ (mod 6) if $t \equiv 0$ (mod 3);
- $p \geqslant 5$, $m$ divides $a$ for every positive integer $m < p$, and either $p$ divides $a$, or $t \equiv 1, 2$ (mod $p$).

In order to complete the proof of the proposition, we may assume $p = 3$. Recall that we have already shown that $T$ is 3-elusive if $a = 2$, or if $a \equiv 2$ (mod 6) and $t \equiv 0$ (mod 3). Now, if $a \equiv 4$ (mod 6) and $t \equiv 2$ (mod 3) then $n \equiv 2$ (mod 6), say $n = 6k + 2$, and by repeating the argument above we deduce that $[J_3^{2k}, J_2]$ is a derangement. Therefore, to complete the proof we need to show that $T$ is 3-elusive if $a \geqslant 6$ and one of the following holds:

(a) $a \equiv 0$ (mod 6);
(b) $a \equiv 2$ (mod 6) and $t \not\equiv 0$ (mod 3);
(c) $a \equiv 4$ (mod 6) and $t \equiv 1$ (mod 3).

This is straightforward. In case (a), every suitable partition of $n$ contains a partition of 6 (with all odd parts occurring with even multiplicity), so every element of order 3 fixes a decomposition componentwise. Next consider (b). Write $a = 6k + 2$ with $k \geqslant 1$. Now, if $t \geqslant 4$ then it suffices to show that $\hat{x} = [J_3^{6k}, J_2^{n/2-9k}]$ has fixed points, which is clear since

$$\hat{x} = [J_3^{2k}, J_2]^3 \oplus [J_2^{a/2}]^{t-3}$$

Similarly, if $t = 2$ then we only need to show that $[J_3^{4k}, J_2^2]$ has fixed points, and once again this is clear. Finally, if (c) holds then $a = 6k + 4 \geqslant 10$ and $t \geqslant 4$, and we need to show that $\hat{x} = [J_3^{6k+2}, J_2^{n/2-9k-3}]$ has fixed points. Here

$$\hat{x} = [J_3^{2k}, J_2^2]^3 \oplus [J_3^2, J_2^{a/2-3}] \oplus [J_2^{a/2}]^{t-4}$$

and the result follows. $\qquad\qquad\qquad\qquad\qquad\qquad\qquad\qquad\qquad\quad\square$

**Remark 5.2.16** We can replace the necessary conditions in part (iii) of Proposition 5.2.15 by an alternative number-theoretic condition that is also necessary for $p$-elusivity. Indeed, in the proof of Proposition 5.2.15 we noted that signed partitions play no role in the analysis, so for $p > 2$ it follows that $T$ is $p$-elusive only if every partition $(p^{a_p}, \ldots, 1^{a_1})$ of $n$, where $a_p < a$ and $a_i$ is even for all odd $i$, contains a partition of $a$ such that all odd parts have even multiplicity. This is the condition P2$'$ referred to in Theorem 5.2.1.

**Proposition 5.2.17** *Consider Cases III, IV and V in Table 5.2.1, so $T = \mathrm{P\Omega}_n^\varepsilon(q)$ and $H$ is the stabiliser of a decomposition $V = V_1 \perp \ldots \perp V_t$, where each $V_j$ is a nondegenerate $a$-space of type $\varepsilon'$.*

(i) *If $p = 2$ then $T$ is $p$-elusive if and only if one of the following holds:*
   (a) $\varepsilon' = +$;
   (b) $\varepsilon' = -$ *and* $a \equiv 2$ (mod 4);
   (c) $\varepsilon' = -$, $a \equiv 0$ (mod 4), $a \geqslant 8$ *and* $t$ *is odd.*

(ii) *If $p = 3$ then $T$ is $p$-elusive only if $\varepsilon = \varepsilon' = +$, $a \equiv 0 \pmod 4$ and $a \neq 8$.*

(iii) *If $p \geqslant 5$ then $T$ is $p$-elusive only if $\varepsilon = \varepsilon' = +$, $a \equiv 0 \pmod 4$, $m$ divides $a$ for every positive integer $m < p$, and either $p$ divides $a$, or $t \equiv 1,2 \pmod p$.*

*Proof* Here $T = \mathrm{P}\Omega_n^\varepsilon(q)$, where $n \geqslant 7$. In Cases III and IV, $H$ is of type $\mathrm{O}_a^{\varepsilon'}(q) \wr S_t$ (with $a = 1$ in Case III), and in Case V, $H$ is of type $\mathrm{O}_{n/2}(q)^2$. We will deal with all three cases simultaneously. As in the statement of the proposition, let $\varepsilon' \in \{+,-,\circ\}$ denote the type of nondegenerate spaces in the orthogonal decomposition stabilised by $H$. Note that $\varepsilon = (\varepsilon')^t$ if $a$ is even.

First assume $p = 2$, in which case $n$ and $a$ are even, so we are in Case IV. We will use the description of unipotent involutions from Section 3.5.4. As in the proof of Proposition 5.2.15, it is easy to see that every $c_s$-type involution fixes (componentwise) an orthogonal decomposition of $V$ into nondegenerate 2-spaces of all possible isometry types. Therefore, $c_s$-type involutions have fixed points on $\Omega$. Now assume that $x$ is an involution of type $a_s$ (recall that there are no $b_s$-type involutions in $T$). If $\varepsilon' = +$ then the argument in the proof of Proposition 5.2.15 goes through unchanged, and similarly if $\varepsilon' = -$ and $a \equiv 2 \pmod 4$; in both cases, we deduce that $T$ is 2-elusive. (In particular, if $\varepsilon = +$ and $n \equiv 0 \pmod 4$ then involutions of type $a_{n/2}$ and $a'_{n/2}$ both have fixed points.)

Finally, let us assume that $\varepsilon' = -$ and $a \equiv 0 \pmod 4$. If $t$ is even, then it is easy to check that $a_{n/2-2}$-type involutions are derangements (here we use the fact that $\mathrm{O}_a^-(q)$ does not contain involutions of type $a_{a/2}$; see Section 3.5.4). Now assume $t$ is odd, so $\varepsilon = -$ and $s \leqslant n/2 - 2$. If $a = 4$ then $a_2$-type involutions are derangements, so let us assume that $a \geqslant 8$. Let $x \in T$ be an $a_s$-type involution and write $s = a\ell + b$ and $b = a'j + c$, where $0 \leqslant b < a$, $0 \leqslant c < a' = a/2 - 2$ and $b,c$ are even. Note that $j \leqslant 3$, with equality only if $c = 0$. Also note that $t \geqslant 2\ell + 1$. We claim that $x$ has fixed points.

If $t = 2\ell + 1$ then $b \leqslant a'$ and $x$ fixes a decomposition in $\Omega$ by inducing a permutation of cycle-shape $(2^\ell)$ on the first $2\ell$ component spaces, and an involution of type $a_b$ on the final (fixed) component. Finally, if $t \geqslant 2\ell + 3$ then $x$ fixes an appropriate decomposition $V = V_1 \perp \ldots \perp V_t$ by permuting $V_1, \ldots, V_{2\ell}$ as before, and acting as an involution of type $a_{a'}$ on $V_{2\ell+1}, \ldots, V_{2\ell+j}$, and of type $a_c$ on $V_{2\ell+j+1}$ (recall that $c = 0$ if $j = 3$). This justifies the claim.

For the remainder, let us assume $p > 2$. Let $\delta \in \{\square, \boxtimes\}$ be the discriminant of the underlying quadratic form $Q$ on $V$. As in Section 3.5.3, let $\mathscr{Q}^\delta(n,p)$ be the set of nontrivial *labelled partitions* of $n$ of the form

$$(\delta_p p^{a_p}, (p-1)^{a_{p-1}}, \ldots, 2^{a_2}, \delta_1 1^{a_1})$$

where $a_i$ is even for all even $i$, $\delta_{2j+1} \in \{\square, \boxtimes\}$ and $\prod_j \delta_{2j+1} = \delta$ (with the additional condition that if $\varepsilon = -$ then $a_i > 0$ for some odd $i$). Then Proposition 3.5.12 gives a bijection from $\mathscr{Q}^\delta(n,p)$ to the set of $\mathrm{PO}_n^\varepsilon(q)$-classes of elements in $T$ of order $p$. Let $x = \hat{x}Z \in T$ be an element of order $p$.

Suppose $n$ is odd, so $a$ and $t$ are also odd. Let $\hat{x} = [J_2^{(n-1)/2}, J_1]$ if $n \equiv 1$ (mod 4), otherwise set $\hat{x} = [J_3, J_2^{(n-3)/2}]$. Then it is easy to see that $x$ is a derangement. (Note that *any* element in $T$ with the given Jordan form is a derangement; there is no need to consider labelled partitions in this situation.)

Now assume $n$ is even. There are several cases to consider. To begin with, let us assume $a$ is odd, so $t$ is even. If $\varepsilon = +$ then define $\hat{x}$ as follows:

|  | $t \equiv 0 \pmod 4$ | $t \equiv 2 \pmod 4$ |
|---|---|---|
| $a \equiv 1 \pmod 4$ | $[J_2^{n/2}]$ | $[J_3^2, J_2^{n/2-3}]$ |
| $a \equiv 3 \pmod 4$ | $[J_2^{n/2}]$ | $[J_2^{n/2-1}, J_1^2]$ |

Similarly, if $\varepsilon = -$ we choose $\hat{x}$ as follows:

|  | $t \equiv 0 \pmod 4$ | $t \equiv 2 \pmod 4$ |
|---|---|---|
| $a \equiv 1 \pmod 4$ | $[J_3^4, J_2^{n/2-6}]$ | $[J_3^2, J_2^{n/2-3}]$ |
| $a \equiv 3 \pmod 4$ | $[J_2^{n/2-2}, J_1^4]$ | $[J_2^{n/2-1}, J_1^2]$ |

In each of these cases, it is easy to check that $x$ is a derangement. (Note that if $\varepsilon = -$, $a \equiv 1 \pmod 4$ and $t \equiv 0 \pmod 4$ then $a \geqslant 5$ since we may assume that $n \equiv 2 \pmod 4$ if $\varepsilon = -$ and $a = 1$; see Table 5.2.1.)

For the remainder let us assume $n$ and $a$ are both even. If $\varepsilon' = -$ then set $\hat{x} = [J_3, J_2^{n/2-2}, J_1]$ if $n \equiv 0 \pmod 4$, and $\hat{x} = [J_2^{n/2-1}, J_1^2]$ if $n \equiv 2 \pmod 4$. Since there are no elements in $\mathrm{O}_m^-(q)$ with Jordan form $[J_2^{m/2}]$, we quickly deduce that $x$ is a derangement.

Now assume that $\varepsilon = \varepsilon' = +$. Suppose $a \equiv 2 \pmod 4$. Set $\hat{x} = [J_2^{n/2-1}, J_1^2]$ if $t$ is odd, and $\hat{x} = [J_3, J_2^{n/2-2}, J_1]$ if $t$ is even. Once again, it is easy to check that $x$ is a derangement.

Finally, let us assume that $\varepsilon = \varepsilon' = +$ and $a \equiv 0 \pmod 4$. If $(p, a) = (3, 8)$ then $[J_3^4, J_2^{n/2-6}]$ is a derangement. In order to complete the proof of the proposition, we may assume that $p \geqslant 5$. We need to show that $T$ is $p$-elusive only if $m$ divides $a$ for every positive integer $m < p$, and either $p$ divides $a$, or $t \equiv 1, 2 \pmod p$.

If $p > n$ then $[J_{n/2}^2]$ is a derangement, so we may assume that $p \leqslant n$. If $p > a$ then $[J_p, J_1^{n-p}]$ is a derangement, so let us assume $p \leqslant a$.

Suppose $m < p$ is an odd integer that does not divide $a$. Write $n = mc + d$, where $0 \leqslant d < m$. If $d$ is odd then $[J_m^c, J_d]$ is a derangement, and similarly we see that $[J_m^c, J_{d-1}, J_1]$ is a derangement if $d$ is even and $(t, d) \neq (2, 2)$. Finally, if $(t, d) = (2, 2)$ then $[J_{m+2}, J_m^{c-1}]$ is a derangement (note that $3 \leqslant m < p$ is odd, so $m + 2 \leqslant p$ and this element has order $p$).

Similarly, suppose $m < p$ is an even integer that does not divide $a$. Write $n = 2mc + d$ with $0 \leqslant d < 2m$, and note that $d \equiv 0 \pmod 4$. Then $[J_m^{2c}, J_{d/2}^2]$ is a derangement. Therefore, for the remainder we may assume that $a$ is divisible by every positive integer $m < p$.

Suppose $p$ does not divide $a$. If $p$ divides $t$, say $t = \ell p$, then it is straightforward to see that $\hat{x} = [J_p^{n/p-1}, J_{p-2}, J_1^2]$ is a derangement. Indeed, suppose $\hat{x}$ has fixed points. Since $t \geqslant 5$, $\hat{x}$ must induce a nontrivial permutation on the $t$ subspaces in a fixed decomposition, say with cycle-shape $(p^h, 1^{t-hp})$ for some $1 \leqslant h < \ell$. If $U$ denotes the subspace spanned by the $t - hp$ summands fixed by $\hat{x}$, then $\hat{x}|_U = [J_p^{a(\ell-h)-1}, J_{p-2}, J_1^2]$, but $t - hp \geqslant p \geqslant 5$, so $\hat{x}|_U$ cannot fix each of the $t - hp$ subspaces in the decomposition of $U$. This is a contradiction.

Finally, suppose that $p$ fails to divide both $a$ and $t$. It remains to show that $T$ contains derangements of order $p$ if $t \not\equiv 1, 2 \pmod p$. Suppose $t \not\equiv 1, 2 \pmod p$ and write $n = 2cp + d$ with $2 \leqslant d < 2p$, so $d$ is even. By arguing as in the proof of Proposition 5.2.15, it is easy to check that $[J_p^{2c}, J_{d/2}^2]$ is a derangement.  $\square$

**Remark 5.2.18** We can replace the necessary condition in parts (ii) and (iii) of Proposition 5.2.17 with a simplified number-theoretic condition. Indeed, if $p > 2$ then $T$ is $p$-elusive only if every partition $(p^{a_p}, \ldots, 1^{a_1})$ of $n$, such that $a_p < a$ and $a_i$ is even for all even $i$, contains a partition of $a$ with the property that all even parts have even multiplicity. This is the condition P3′ referred to in part (iii) of Theorem 5.2.1. In addition, the proof of Proposition 5.2.17 shows that $\varepsilon = \varepsilon' = +$ and $a \equiv 0 \pmod 4$.

**Proposition 5.2.19** *In Cases VI–VIII of Table 5.2.1, $T$ is not $p$-elusive.*

*Proof*  In each of these cases, $H$ is the stabiliser in $G$ of a decomposition $V = V_1 \oplus V_2$, where $V_1$ and $V_2$ are maximal totally singular subspaces. Let $x = \hat{x}Z \in T$ be an element of order $p$, and let $\lambda = (p^{a_p}, \ldots, 1^{a_1})$ be the partition of $n$ corresponding to the Jordan form of $\hat{x}$ on $V$.

From the definition of $H$, if $x$ interchanges $V_1$ and $V_2$ then $p = 2$ and $\lambda = (2^{n/2})$, and if $x$ fixes both $V_1$ and $V_2$ then each $a_i$ in $\lambda$ is even (cf. Lemma 2.2.17, noting that if $A \in \mathrm{GL}(U)$ is unipotent then $A$ and $A^{-\mathsf{T}}$ have the same Jordan form). In particular, any element with Jordan form $[J_2, J_1^{n-2}]$ is a derangement in Cases VI and VII. Now consider Case VIII. If $p > 2$ then $[J_3, J_1^{n-3}]$ is a derangement. Finally, if $p = 2$ then let $x \in T$ be a $c_2$-type involution. Here $x$ has two nontrivial Jordan blocks, both acting on hyperbolic 2-spaces, so $x$ does not fix $V_1$ and $V_2$, and it does not interchange them since $n \geqslant 8$. Therefore, $x$ is a derangement.                                                                      $\square$

### 5.2.3 Semisimple elements

Now let us turn to Theorem 5.2.3. Let $r \neq p$ be a prime divisor of $|\Omega|$. If $r > 2$ then set $i = \Phi(r,q)$ as in (3.1.2) (so $i \geqslant 1$ is minimal such that $r$ divides $q^i - 1$) and write

$$n \equiv k \ (\mathrm{mod}\ c), \quad a \equiv \ell \ (\mathrm{mod}\ c) \tag{5.2.5}$$

where $0 \leqslant k, \ell < c$, and the integer $c = c(i)$ is given in Definition 5.1.1. (Here $a = 1$ in Case III of Table 5.2.1, and $a = n/2$ in Cases V–VIII.) Note that $k \leqslant t\ell$.

The $r$-elusive cases are recorded in Tables 5.2.2 (for $r > 2$) and 5.2.3 (for $r = 2$).

**Proposition 5.2.20** *In Case I of Table 5.2.1, if $r \neq p$ then $T$ is $r$-elusive if and only if $r = 2$, or $r > 2$ and one of the following holds:*

(i) $1 = a < c = r - 1$ *and* $n \leqslant kr$;
(ii) $c \leqslant a$ *and* $k = t\ell$.

*Proof* Here $T = \mathrm{PSL}_n^\varepsilon(q)$ and $H$ is of type $\mathrm{GL}_a^\varepsilon(q) \wr S_t$. Set $\mathbb{F} = \mathbb{F}_{q^u}$, where $u = 1$ if $\varepsilon = +$, otherwise $u = 2$, and let $\overline{\mathbb{F}}$ denote the algebraic closure of $\mathbb{F}$. Let $Z$ be the centre of $\mathrm{GL}_n^\varepsilon(q)$. Let $r \neq p$ be a prime dividing $|\Omega|$, and let $x = \hat{x}Z \in T$ be an element of order $r$.

*Case 1. $\varepsilon = +$, $r = 2$.* First assume $\varepsilon = +$ and $r = 2$, so $q$ is odd. We use the notation for involutions presented in Section 3.2.2 (recall that this agrees with the notation introduced by Gorenstein, Lyons and Solomon [67]). If $x$ is of type $t_i$, with $1 \leqslant i \leqslant n/2$, then $\hat{x}$ is diagonalisable over $\mathbb{F}_q$ and thus $x$ visibly fixes an appropriate decomposition of $V$. Now assume $n$ is even and $x = t'_{n/2}$. Here $\hat{x}$ is a block-diagonal matrix of the form $[A^{n/2}]$, where

$$A = \begin{pmatrix} 0 & 1 \\ \xi^2 & 0 \end{pmatrix} \tag{5.2.6}$$

and $\xi \in \mathbb{F}_{q^2}$ has order $2(q-1)_2$ (where $(q-1)_2$ is the largest 2-power dividing $q - 1$). Clearly, $x$ fixes a decomposition (componentwise) if $a$ is even. Now, if $a$ is odd then $t$ is even (since $n$ is even) and $a$ divides $n/2$. Moreover, if $\{v_1, \ldots, v_{n/2}\}$ is an $\mathbb{F}_{q^2}$-basis for $V$ then

$$\{v_1, \xi v_1, v_2, \xi v_2, \ldots, v_{n/2}, \xi v_{n/2}\}$$

is an $\mathbb{F}_q$-basis for $V$ on which $\hat{x}$ acts via scalar multiplication by $\xi$. Therefore, $\hat{x}$ interchanges the $a$-spaces $\langle v_1, \ldots, v_a \rangle$ and $\langle \xi v_1, \ldots, \xi v_a \rangle$, and since $a$ divides $n/2$ it follows that $x$ stabilises a suitable direct sum decomposition of $V$. To be precise, $x$ induces a permutation of cycle-shape $(2^{t/2})$ on the $t$ spaces

$$\{\langle v_{a\ell+1}, \ldots, v_{a\ell+a} \rangle, \langle \xi v_{a\ell+1}, \ldots, \xi v_{a\ell+a} \rangle \mid 0 \leqslant \ell < t/2\}$$

We conclude that $T$ is 2-elusive.

*Case 2.* $\varepsilon = -$, $r = 2$. Next assume $\varepsilon = -$ and $r = 2$. Clearly, if $x = t_i$ with $i \leqslant n/2$ then $x$ fixes an orthogonal decomposition of $V$ into nondegenerate 1-spaces (note that the two eigenspaces of $\hat{x}$ on $V$ are nondegenerate), so $x$ has fixed points. Now assume $n$ is even and $x \in T$ is of type $t'_{n/2}$, in which case $q \equiv 1 \pmod 4$ or $(n)_2 > (q+1)_2$ (see Section 3.3.2.2). Let $\{e_1, \ldots, e_{n/2}, f_1, \ldots, f_{n/2}\}$ be an hermitian basis for $V$. Here $\hat{x}$ is contained in the centre of a subgroup of type $\mathrm{GL}_{n/2}(q^2)$, and with respect to the given basis we have $\hat{x} = [\xi I_{n/2}, -\xi I_{n/2}]$, where $\xi \in \mathbb{F}_{q^2}$ has order $2(q+1)_2$. Clearly, $x$ fixes a decomposition of $V$ into nondegenerate 2-spaces, so $x$ has fixed points if $a$ is even. Now assume $a$ is odd, so $t$ is even. With respect to the orthogonal basis

$$\{e_1 + f_1, e_1 - f_1, \ldots, e_{n/2} + f_{n/2}, e_{n/2} - f_{n/2}\}$$

we have $\hat{x} = [A^{n/2}]$, where

$$A = \begin{pmatrix} 0 & \xi \\ \xi & 0 \end{pmatrix}$$

Therefore, $x$ normalises an orthogonal decomposition of $V$ into nondegenerate 1-spaces (by inducing a permutation of cycle-shape $(2^{n/2})$ on summands), so by taking a suitable combination of these 1-spaces we deduce that $x$ has fixed points on $\Omega$, whence $T$ is 2-elusive.

For the remainder we may assume that $r > 2$. Set $i = \Phi(r, q)$ and define the integer $c = c(i)$ as in Definition 5.1.1. Also define $k$ and $\ell$ as in (5.2.5). We will use the notation for semisimple elements introduced in Sections 3.2.1 and 3.3.1.

*Case 3.* $r > 2$, $c = 1$. First assume $c = 1$, so $k = \ell = 0$ and either $(i, \varepsilon) = (1, +)$ or $(2, -)$. The possibilities for $x$ (up to $T$-conjugacy) are described in Propositions 3.2.2 and 3.3.3. Clearly, if $\hat{x}$ is diagonalisable over $\mathbb{F}$ then $x$ fixes a decomposition in $\Omega$, so let us assume otherwise (note that if $\varepsilon = -$ and $\hat{x}$ is diagonalisable then $\hat{x}$ fixes (componentwise) an orthogonal decomposition of

$V$ into nondegenerate 1-spaces). In this situation, $r$ divides $n$ and either $\varepsilon = +$ and $\hat{x}$ is diagonalisable over $\mathbb{F}_{q^r}$ with no eigenvalues in $\mathbb{F}_q$, or $\varepsilon = -$ and $\hat{x}$ is diagonalisable over $\mathbb{F}_{q^{2r}}$ with no eigenvalues in $\mathbb{F}_{q^2}$. In both cases, $x$ fixes a direct sum decomposition of $V$ into $r$-spaces (which are nondegenerate if $\varepsilon = -$), whence $x$ has fixed points if $r$ divides $a$.

Now suppose $r$ does not divide $a$, so $r$ divides $t$, and $a$ divides $n/r$. First assume $\varepsilon = +$ and let $\{v_1, \ldots, v_{n/r}\}$ be a basis for $V$ over $\mathbb{F}_{q^r}$. Let $\xi \in \mathbb{F}_{q^r}$ be an element of order $r(q-1)_r$ and note that the minimal polynomial of $\xi$ over $\mathbb{F}_q$ has degree $r$, so $\{1, \xi, \xi^2, \ldots, \xi^{r-1}\}$ is an $\mathbb{F}_q$-basis for $\mathbb{F}_{q^r}$ and thus

$$\{v_1, \xi v_1, \ldots, \xi^{r-1} v_1, v_2, \xi v_2, \ldots, \xi^{r-1} v_2, \ldots, v_{n/r}, \xi v_{n/r}, \ldots, \xi^{r-1} v_{n/r}\}$$

is a basis for $V$ over $\mathbb{F}_q$. Then $\hat{x}$ acts on this basis via scalar multiplication by $\xi$, hence $x$ induces a permutation of cycle-shape $(r^{t/r})$ on the $t$ spaces

$$\{\langle \xi^i v_{a\ell+1}, \ldots, \xi^i v_{a\ell+a} \rangle \mid 0 \leqslant i < r,\ 0 \leqslant \ell < t/r\}$$

Therefore $x$ has fixed points. A similar argument applies when $\varepsilon = -$. (Note that if we take an orthonormal basis $\{v_1, \ldots, v_{n/r}\}$ for $V$ over $\mathbb{F}_{q^{2r}}$ then each $\mathbb{F}_{q^2}$-space $\langle v_{a\ell+1}, \ldots, v_{a\ell+a} \rangle$ is nondegenerate, and so is the image of this subspace under the isometry $\hat{x}$.)

*Case 4.* $r > 2$, $c > a$. For the remainder we may assume that $c > 1$. First assume $c > a$ and set

$$\hat{x} = \begin{cases} [\Lambda, \Lambda^{-q}, I_{n-c}] & \text{if } \varepsilon = -, i \not\equiv 2 \pmod 4 \\ [\Lambda, I_{n-c}] & \text{otherwise} \end{cases}$$

Suppose $x$ fixes a decomposition in $\Omega$. Since $r$ does not divide $|\mathrm{GL}_a^\varepsilon(q)|$, $x$ must induce a nontrivial permutation on the component spaces, so Lemma 5.2.6 implies that each $r$th root of unity occurs as an eigenvalue of $\hat{x}$ (on $V \otimes \overline{\mathbb{F}}$) with multiplicity at least $a$. Therefore, $x$ is a derangement if $a > 1$, or if $a = 1$ and $c < r - 1$.

Now assume $a = 1$ and $c = r - 1$, so $n = t$. Set $x = \hat{x} Z \in T$, where

$$\hat{x} = \begin{cases} [(\Lambda, \Lambda^{-q})^{(n-k)/c}, I_k] & \varepsilon = -, i \not\equiv 2 \pmod 4 \\ [\Lambda^{(n-k)/c}, I_k] & \text{otherwise} \end{cases} \tag{5.2.7}$$

If $x$ fixes a decomposition in $\Omega$ then it must induce a permutation with $(n-k)/c$ orbits of length $r$ on the set of component spaces, so Lemma 5.2.6 implies that the multiplicity of each $r$th root of unity as an eigenvalue of $\hat{x}$ (on $V \otimes \overline{\mathbb{F}}$) is at least $(n-k)/c$. Since $\dim C_V(\hat{x}) = k$, it follows that $x$ is a derangement if $(n-k)/c > k$. On the other hand, if $(n-k)/c \leqslant k$ then $x$ fixes an appropriate decomposition $V = V_1 \oplus \cdots \oplus V_n$ by inducing a permutation of shape

$(r^{(n-k)/c}, 1^{n-r(n-k)/c})$ on the $n$ summands. In fact, if $y = \hat{y}Z \in T$ is any element of order $r$ then Propositions 3.2.2 and 3.3.3 imply that $\dim C_V(\hat{y}) \geqslant k$ and each primitive $r$th root of unity has multiplicity $s \leqslant k$ as an eigenvalue of $\hat{y}$. Therefore $y$ fixes an appropriate decomposition by inducing a permutation of shape $(r^s, 1^{t-rs})$ on the component spaces. We conclude that $T$ is $r$-elusive in this situation. (Note that the condition $(n-k)/c \leqslant k$, with $c = r-1$, is equivalent to the condition $n \leqslant kr$.)

*Case 5.* $r > 2$, $1 < c \leqslant a$. Finally, let us assume that $1 < c \leqslant a$. Recall that $k \leqslant t\ell$. If $k < t\ell$ then a combination of Lemmas 5.2.8 and 5.2.9 (also see Remark 5.2.10) implies that the element $x \in T$ defined in (5.2.7) is a derangement (note that $\dim C_V(\hat{x}) = k < c \leqslant a$), so let us assume $k = t\ell$. By Propositions 3.2.1 and 3.3.2, we have $\dim C_V(\hat{y}) \geqslant k = t\ell$ for any element $y = \hat{y}Z \in T$ of order $r$. Hence by applying Lemmas 5.2.8 and 5.2.9, we deduce that $y$ fixes a decomposition in $\Omega$. Thus $T$ is $r$-elusive in this case. $\qquad\square$

**Proposition 5.2.21** *In Case II of Table 5.2.1, if $r \neq p$ then $T$ is $r$-elusive if and only if $r = 2$, or $r > 2$, $c \leqslant a$ and $k = t\ell$.*

*Proof* Here $T = \mathrm{PSp}_n(q)$ and $H$ is of type $\mathrm{Sp}_a(q) \wr S_t$. Let $r \neq p$ be a prime dividing $|\Omega|$ and let $x = \hat{x}Z \in T$ be an element of order $r$, where $Z$ denotes the centre of $\mathrm{GSp}_n(q)$.

First assume that $r = 2$. Involutions of type $t_i$ with $i \leqslant n/4$ clearly fix a decomposition of $V$ into nondegenerate 2-spaces, so they have fixed points on $\Omega$. If $x \in T$ is an involution of type $t_{n/2}$ then $q \equiv 1 \pmod 4$ and we may take $\hat{x} = [\lambda I_{n/2}, \lambda^{-1}I_{n/2}]$ with respect to a symplectic basis $\{e_1, \ldots, e_{n/2}, f_1, \ldots, f_{n/2}\}$, where $\lambda \in \mathbb{F}_q$ has order 4. Once again, these elements have fixed points since they fix a decomposition of $V$ into nondegenerate 2-spaces. Finally, suppose $x$ is an involution of type $t'_{n/2}$. As described in Section 3.4.2.5, $x$ arises from an element in the centre of $\Delta U_{n/2}(q)$ and so it fixes an orthogonal decomposition of $V$ into $n/2$ nondegenerate 1-dimensional subspaces over $\mathbb{F}_{q^2}$, and therefore a decomposition into $n/2$ nondegenerate 2-spaces over $\mathbb{F}_q$. Therefore, $x$ fixes a decomposition in $\Omega$ componentwise, whence $T$ is 2-elusive.

For the remainder, let us assume $r > 2$. Set $i = \Phi(r, q)$ and define $c = c(i)$ as in Definition 5.1.1. Set

$$\hat{x} = \begin{cases} [\Lambda, \Lambda^{-1}, I_{n-c}] & i \text{ odd} \\ [\Lambda, I_{n-c}] & i \text{ even} \end{cases} \qquad (5.2.8)$$

Now, if $c > a$ then $r$ does not divide $|\mathrm{GSp}_a(q)|$, and so if $x$ fixes a decomposition in $\Omega$ then it must induce a nontrivial permutation on the component spaces. Therefore, Lemma 5.2.6 implies that every $r$th root of unity occurs as an eigenvalue of $\hat{x}$ (on $V \otimes \overline{\mathbb{F}}_q$) with multiplicity at least $a$. But $a \geqslant 2$ and every nontrivial eigenvalue of $\hat{x}$ has multiplicity 1, whence $x$ is a derangement.

Now assume $c \leqslant a$. Define $k$ and $\ell$ as in (5.2.5), and note that $k \leqslant t\ell$. Suppose $k < t\ell$ and set

$$\hat{x} = \begin{cases} [(\Lambda, \Lambda^{-1})^{(n-k)/c}, I_k] & i \text{ odd} \\ [\Lambda^{(n-k)/c}, I_k] & i \text{ even} \end{cases}$$

Since $\dim C_V(\hat{x}) = k < t\ell$, Lemmas 5.2.8 and 5.2.9 imply that $x$ does not fix a decomposition in $\Omega$ componentwise, so Lemma 5.2.6 implies that $x$ is a derangement since $k < c \leqslant a$. Finally, let us assume $k = t\ell$. By Proposition 3.4.3, we have $\dim C_V(\hat{y}) \geqslant k = t\ell$ for any element $y = \hat{y}Z \in T$ of order $r$, so in view of Lemmas 5.2.8 and 5.2.9 we deduce that $y$ fixes a decomposition in $\Omega$. Therefore, $T$ is $r$-elusive.                                                    □

Let $T = \mathrm{P}\Omega_n^\varepsilon(q)$ be an orthogonal group, where $q$ is odd, and suppose that

$$V = V_1 \perp V_2 \perp \ldots \perp V_t \tag{5.2.9}$$

is an orthogonal decomposition of $V$ into isometric parabolic $a$-spaces. Let $Q$ be the underlying quadratic form on $V$, and let $D(Q|_{V_i}) \in \{\square, \boxtimes\}$ be the discriminant of the restriction of $Q$ to $V_i$ (see Section 2.2.6). We say that (5.2.9) is a *square-parabolic $a$-decomposition* if $D(Q|_{V_i}) = \square$ for each $i$, otherwise it is a *nonsquare-parabolic $a$-decomposition*.

**Lemma 5.2.22** *Let $T = \mathrm{P}\Omega_n^\varepsilon(q)$, where $n$ is even and $q$ is odd, and let $x \in T$ be an element of odd prime order $r \neq p$. Let $a \geqslant 1$ be an odd integer. Then $x$ fixes a square-parabolic $a$-decomposition of $V$ if and only if $x$ fixes a nonsquare-parabolic $a$-decomposition.*

*Proof* Suppose that $x$ fixes a square-parabolic $a$-decomposition of $V$ and let $g \in \mathrm{PGO}_n^\varepsilon(q)\backslash\mathrm{PO}_n^\varepsilon(q)$. In view of Lemma 2.5.2 (and Remark 2.5.3), we see that $g$ maps the set of square-parabolic $a$-decompositions of $V$ to the set of nonsquare-parabolic $a$-decompositions of $V$ (and vice versa), so the conjugate element $x^g$ fixes a nonsquare-parabolic $a$-decomposition. By Proposition 3.5.8(i), $x$ and $x^g$ are $\mathrm{PO}_n^\varepsilon(q)$-conjugate and thus $x$ also fixes a nonsquare-parabolic $a$-decomposition.

Conversely, this argument also shows that if $x$ fixes a nonsquare-parabolic $a$-decomposition, then it fixes a square-parabolic $a$-decomposition.                □

**Proposition 5.2.23** *In Cases III and IV of Table 5.2.1, if $r \neq p$ and $a$ is odd then $T$ is $r$-elusive if and only if $r = 2$, or $r > 2$ and one of the following holds:*

(i) $a = 1$, $c = r - 1$ and one of the following holds:
  (a) $n \leqslant kr$;
  (b) $k = 0$, $n \leqslant cr$, $\varepsilon = (-)^i$ and $n/i$ is odd if $\varepsilon = +$.
(ii) $c < a$, $k = t\ell$ and $a \not\equiv 0 \pmod{i}$.
(iii) $c < a$, $k = 0$, $c = t\ell$ and $\varepsilon = (-)^{n/i+1}$.

*Proof*  Here $T = \mathrm{P\Omega}_n^\varepsilon(q)$ and $H$ is of type $\mathrm{O}_a(q) \wr S_t$, where $aq$ is odd. Let $Q$ be the underlying nondegenerate quadratic form on $V$, and let $B$ be the corresponding symmetric bilinear form. If $n$ is odd then we may assume that $D(Q) = \square$ (see Remark 2.2.10). Similarly, if $n$ is even then $t$ must also be even, whence $D(Q) = \square$ since the component subspaces in any decomposition in $\Omega$ are isometric (see Lemma 2.2.11). Therefore, $D(Q) = \square$ for all $n$. In particular, Lemma 2.2.9 implies that $\varepsilon = +$ if $t \equiv 0 \pmod 4$, and $q \equiv \varepsilon \pmod 4$ if $t \equiv 2 \pmod 4$.

Let $V = V_1 \perp \ldots \perp V_t$ be a decomposition in $\Omega$. If $t$ is odd then Lemma 2.2.11 implies that $D(Q|_{V_i}) = \square$ for all $i$. Now assume $t$ is even, in which case $T$ has two orbits on the set of orthogonal decompositions of $V$ into isometric parabolic $a$-spaces, which are distinguished by the discriminant $D(Q|_{V_i})$. These two orbits are fused by a similarity of $Q$, and so the two actions of $T$ are equivalent. Therefore, for any $t$ we may assume that $D(Q|_{V_i}) = \square$ for all $i$. In terms of our earlier terminology, this means that $\Omega$ is the set of square-parabolic $a$-decompositions of $V$.

Let $r$ be a prime divisor of $|\Omega|$ and let $x = \hat{x}Z \in T$ be an element of order $r$. We begin with the case $r = 2$.

*Case 1. $r = 2$.*  First assume $x$ is an involution of type $t_i$ or $t_i'$, where $1 \leqslant i \leqslant n/4$ if $n$ is even, and $1 \leqslant i \leqslant (n-1)/2$ if $n$ is odd. Let $U$ be the $(-1)$-eigenspace of $\hat{x}$ on $V$ and set $W = C_V(\hat{x})$. The discriminants $D(Q|_U)$ and $D(Q|_W)$ are determined by $q$ and $i$, and we note that $D(Q|_U) = D(Q|_W)$ since $D(Q) = \square$ (see Lemma 2.2.11). Moreover, the $(-1)$-eigenspace of an involution of type $t_i$ has the opposite discriminant to the $(-1)$-eigenspace of a $t_i'$ involution, and the respective 1-eigenspaces have the same property.

Suppose that $D(Q|_U) = D(Q|_W) = \square$. Then $U$ and $W$ admit orthogonal decompositions into parabolic 1-spaces $X_j$ such that $D(Q|_{X_j}) = \square$. Therefore, by taking a suitable combination of these 1-spaces, we see that $x$ fixes a decomposition in $\Omega$ componentwise.

Now assume that $D(Q|_U) = D(Q|_W) = \boxtimes$. Then $U$ admits an orthogonal decomposition into parabolic 1-spaces such that all but one have discriminant a square and the remaining one, $\langle v_1 \rangle$ say, has discriminant a nonsquare. Similarly for $W$, where $\langle v_2 \rangle$ is a subspace with nonsquare discriminant. If $a \geqslant 3$

then by combining these 1-spaces into $a$-spaces, keeping the two nonsquare discriminant 1-spaces together, we deduce that $x$ fixes a decomposition in $\Omega$ componentwise. Finally, let us assume $a = 1$. Fix $u \in \langle v_1 \rangle$ and $w \in \langle v_2 \rangle$ such that $B(u,u) + B(w,w)$ is a square. (This is possible since we can always find two nonsquares in $\mathbb{F}_q$ whose sum is a square; if not, then the set of all nonsquares, together with 0, would form an additive subgroup of $\mathbb{F}_q$ of order $(q+1)/2$, which is absurd.) Then the 1-spaces $\langle u + w \rangle$ and $\langle -u + w \rangle$ both have square discriminants, and they are interchanged by $x$. Therefore, $x$ fixes a decomposition in $\Omega$ by interchanging two of the component spaces and fixing the rest. We conclude that $T$ is 2-elusive when $n$ is odd.

Next suppose $\varepsilon = +$ and $x$ is an involution of type $t_{n/2}$ or $t_{n/2-1}$. By Lemma 5.2.11, $\hat{x}$ maps each nonsingular vector $v \in V$ into $\langle v \rangle^{\perp}$. In particular, $x$ fixes a decomposition of $V$ into square-parabolic 1-spaces by inducing a permutation of shape $(2^{n/2})$ on the components. By taking a suitable combination of these 1-spaces we see that $x$ fixes a decomposition in $\Omega$ by inducing a permutation of shape $(2^{t/2})$ on the $t$ summands.

To complete the analysis of involutions, we may assume that $\varepsilon = +$ and $x = t'_{n/2}$ or $t'_{n/2-1}$, or $\varepsilon = -$ and $x = t_{n/2}$. As noted in Sections 3.5.2.11 and 3.5.2.13, $x$ maps each vector $v \in V$ to a vector in $\langle v \rangle^{\perp}$. Therefore, as in the previous paragraph, we deduce that $x$ fixes a decomposition in $\Omega$ by inducing a permutation of shape $(2^{t/2})$ on components. We conclude that $T$ is 2-elusive.

For the remainder, let us assume $r$ is odd. As usual, set $i = \Phi(r,q)$ and define $c = c(i)$ as in Definition 5.1.1. Also define $k, \ell$ as in (5.2.5), and note that $k \leqslant t\ell$, $c$ is even and $\ell$ is odd. By Lemma 5.2.22, an element $x \in T$ of order $r$ fixes a decomposition in $\Omega$ if and only if it fixes an orthogonal decomposition of $V$ into isometric parabolic $a$-spaces.

*Case 2.* $r > 2$, $c > a$. First assume $c > a$. If we define $\hat{x}$ as in (5.2.8), then it is clear that $x$ is a derangement if $a \geqslant 3$, so we may assume $a = 1$. Clearly, the same element $x$ can only fix a decomposition in $\Omega$ if it induces a nontrivial permutation of the component spaces, so Lemma 5.2.6 implies that $x$ is a derangement if $c < r - 1$.

Now assume $a = 1$ and $c = r - 1$. Write $n = n'c + k$, so $n' = \lfloor n/c \rfloor$. Note that $n' \geqslant 1$ since we are assuming that $T$ contains elements of order $r$. Also note that $n' \leqslant n/r$ if and only if $n \leqslant kr$. We claim that $T$ is $r$-elusive if and only if one of the following holds:

(a) $n$ is odd and $n \leqslant kr$;
(b) $\varepsilon = +$ and either $n \leqslant kr$, or $k = 0$, $i$ is even, $n/i$ is odd and $n \leqslant ir$;
(c) $\varepsilon = -$ and either $n \leqslant kr$, or $k = 0$, $i$ is odd and $n \leqslant 2ir$.

To see this, first assume $n \leqslant kr$, so $n' \leqslant n/r$. Let $d$ be an integer such that $1 \leqslant d \leqslant n'$. By [86, Proposition 4.2.15], the alternating group $A_t$ is contained in $H \cap T$, so there exists an element $x = \hat{x}Z \in T$ of order $r$ that fixes a decomposition in $\Omega$ by inducing a permutation of shape $(r^d, 1^{n-dr})$ on the $n$ component spaces. Since $a = 1$, $x$ acts trivially on each fixed component and thus Lemma 5.2.6 implies that

$$\hat{x} = \begin{cases} [(\Lambda, \Lambda^{-1})^d, I_{n-dc}] & i \text{ odd} \\ [\Lambda^d, I_{n-dc}] & i \text{ even} \end{cases}$$

up to conjugacy. By Proposition 3.5.4, any element in $T$ of order $r$ is $\mathrm{PO}_n^\varepsilon(q)$-conjugate to $x$ (for some $d$), whence $T$ is $r$-elusive.

On the other hand, if $n > kr$ then $n' > n/r > k$ and we set

$$\hat{x} = \begin{cases} [(\Lambda, \Lambda^{-1})^{n'}, I_k] & i \text{ odd} \\ [\Lambda^{n'}, I_k] & i \text{ even} \end{cases} \tag{5.2.10}$$

If $k > 0$ then $x \in T$ and Lemma 5.2.6 implies that $x$ is a derangement. This justifies the claim when $n$ is odd.

Now assume $k = 0$, so $n$ is even. If $\varepsilon = +$ then the element defined in (5.2.10) is in $T$ (and is a derangement), unless $i$ is even and $n/i$ is odd. In the latter situation, if $r < n/i$ then $[\Lambda^{n/i-1}, I_{r-1}]$ is a derangement. However, if $r \geqslant n/i$ then every element $y = \hat{y}Z \in T$ of order $r$ is of the form $\hat{y} = [\Lambda^d, I_{n-di}]$ with $1 \leqslant d < n/i$ and $n - di \geqslant d$. By arguing as above we deduce that $y$ fixes a decomposition in $\Omega$ by inducing a permutation of shape $(r^d, 1^{n-dr})$ on components. This justifies claim (b) above.

Finally, let us assume that $k = 0$ and $\varepsilon = -$. Here the element $x$ in (5.2.10) is a derangement in $T$ unless $i$ is odd, or both $i$ and $n/i$ are even. We can eliminate the latter possibility since $n \equiv 2 \pmod 4$ (see Table 5.2.1). If $i$ is odd then $T$ is $r$-elusive if $r \geqslant n/2i$, otherwise $[(\Lambda, \Lambda^{-1})^{n'-1}, I_{r-1}]$ is a derangement. This justifies (c).

*Case 3.* $r > 2$, $c < a$, $a \equiv 0 \pmod i$. To complete the proof, we may assume that $c < a$, so $a \geqslant 3$. First assume $i$ divides $a$, so $\ell = i$ and $a/i$ are both odd. Clearly, if $\varepsilon = +$ then $[(\Lambda, \Lambda^{-1})^{n/2i}]$ is a derangement. Similarly, if $n$ is odd (in which case $t \geqslant 3$ is also odd) then $\hat{x} = [(\Lambda, \Lambda^{-1})^{(n-i)/2i}, I_i]$ is a derangement. Indeed, in the latter case we note that $\dim C_V(\hat{x}) < c < a$, so if $x$ fixes a decomposition in $\Omega$ then it must fix each component. Moreover, $C_V(\hat{x})$ has to be contained in one of the component spaces, but this is not possible.

Next suppose $i$ divides $a$ and $\varepsilon = -$. Note that $2i$ divides $n$, so $k = 0$. First consider the case $t = 2$, so $n \equiv 2 \pmod 4$ (since $n = 2a$) and $q \equiv 3 \pmod 4$

(since $D(Q) = \square$). In particular, $n/2i$ is odd. We claim that $T$ is $r$-elusive. To see this, let $y = \hat{y}Z \in T$ be an element of order $r$, so

$$\hat{y} = [(\Lambda_1, \Lambda_1^{-1})^{a_1}, \dots, (\Lambda_{s/2}, \Lambda_{s/2}^{-1})^{a_{s/2}}, I_{2i(d+1)}]$$

for some $a_j, d \geqslant 0$, where $s = (r-1)/i$. The elliptic space $C_V(\hat{y})$ admits a decomposition $C_V(\hat{y}) = X \perp Y$, where $X$ is an elliptic $2i$-space and $Y$ is a hyperbolic $2di$-space (so $Y$ is trivial if $d = 0$). Since $D(Q|_X) = \square$ (see Lemma 2.2.9) we may write $X = X_1 \perp X_2$, where $X_1, X_2$ are square-parabolic $i$-spaces. Since $C_V(\hat{y})^\perp$ and $Y$ both admit orthogonal decompositions into hyperbolic $2i$-spaces that are fixed by $y$, it follows that $y$ fixes a decomposition of $V$ into isometric parabolic $a$-spaces and thus $y$ has fixed points on $\Omega$. This justifies the claim.

Now assume $t \geqslant 4$ (we continue to assume that $\varepsilon = -$ and $i$ divides $a$). Set $\hat{x} = [(\Lambda, \Lambda^{-1})^{n/2i-1}, I_{2i}]$ and note that $C_V(\hat{x})$ is an elliptic space. Suppose $x$ fixes a decomposition $V = V_1 \perp \dots \perp V_t$ in $\Omega$. If $x$ fixes $V_j$ then the only possibility is $\hat{x}|_{V_j} = [(\Lambda, \Lambda^{-1})^{(a-i)/2i}, I_i]$, so $x$ can only fix at most two components. In particular, $x$ must induce a nontrivial permutation on the component spaces, so Lemma 5.2.6 implies that $\dim C_V(\hat{x}) = 2i \geqslant a$. But $2i = c < a$, a contradiction, so $x$ is a derangement. We conclude that if $c < a$ and $i$ divides $a$ then $T$ is $r$-elusive if and only if $\varepsilon = -$ and $t = 2$ (note that $k = 0$, $\ell = i$ and $c = t\ell$ in this situation).

*Case 4.* $r > 2$, $c < a$, $a \not\equiv 0 \pmod{i}$, $k = t\ell$. For the remainder, we may assume that $c < a$ and $i$ does not divide $a$. Recall that $k \leqslant t\ell$. Let us consider the special case $k = t\ell$. We claim that every element $x \in T$ of order $r$ fixes a decomposition in $\Omega$ componentwise, whence $T$ is $r$-elusive. To see this, it suffices to show that $x = \hat{x}Z \in T$ has the desired property, where

$$\hat{x} = \begin{cases} [(\Lambda, \Lambda^{-1})^d, I_{n-dc}] & i \text{ odd} \\ [\Lambda^d, I_{n-dc}] & i \text{ even} \end{cases}$$

and $1 \leqslant d \leqslant n'$ with $n' = \lfloor n/c \rfloor$. Set $a' = \lfloor a/c \rfloor$ and write $d = \alpha a' + \beta$, where $0 \leqslant \beta < a'$ and $0 \leqslant \alpha \leqslant t$. Set $U = C_V(\hat{x})$.

First let us consider the case $\alpha = t$, so $d = ta'$ and $\dim U = t\ell$. We claim that $U$ admits an orthogonal decomposition into isometric parabolic $\ell$-spaces, which implies that $x$ fixes (componentwise) an orthogonal decomposition of $V$ into isometric parabolic $a$-spaces. In particular, $x$ fixes a decomposition in $\Omega$. For instance, if $i$ is even then

$$\hat{x} = [\Lambda^{ta'}, I_{t\ell}] = [\Lambda^{a'}, I_\ell]^t$$

To justify the claim, first observe that if $D(Q|_U) = \square$ then $U$ admits an orthogonal decomposition into square-parabolic $\ell$-spaces, so we may assume that $D(Q|_U) = \boxtimes$. If $n$, and hence $t$, is odd then there exists an orthogonal decomposition of $U$ into nonsquare-parabolic $\ell$-spaces. Finally, suppose that $n$, and hence $t$, is even. Here $d$ is even, so $U^\perp$ is hyperbolic and $\dim U^\perp \equiv 0 \pmod 4$, so Lemma 2.2.9 implies that $D(Q|_{U^\perp}) = \square$. Since $D(Q) = \square$, it follows that $D(Q|_U) = \square$, which is a contradiction.

Now assume that $\alpha < t$. We claim that $x$ fixes a square-parabolic decomposition in $\Omega$. To see this, first observe that $U = C_V(\hat{x})$ admits an orthogonal decomposition

$$U = X_1 \perp \dots \perp X_\alpha \perp Y \perp Z_1 \perp \dots \perp Z_{t-\alpha-1} \qquad (5.2.11)$$

into parabolic spaces, where each $X_j$ is an $\ell$-space, $Y$ is an $(a - \beta c)$-space, and each $Z_j$ is an $a$-space. To justify the claim, we need to show that the isometry types of the spaces in this decomposition can be chosen in such a way that we can construct an appropriate decomposition in $\Omega$ that is fixed by $x$.

To do this, let us first assume $i$ is even and $D(Q|_U) = \boxtimes$, so $d$ is odd since $D(Q) = \square$. In (5.2.11), we may choose the component spaces so that

$$D(Q|_{X_j}) = \begin{cases} \square & a' \text{ even} \\ \boxtimes & a' \text{ odd} \end{cases} \quad D(Q|_Y) = \begin{cases} \square & \beta \text{ even} \\ \boxtimes & \beta \text{ odd} \end{cases} \quad D(Q|_{Z_j}) = \square$$

for all $j$. Indeed, the fact that $d$ is odd implies that either $a'$ or $\beta$ is odd, and if $a'$ is odd then $\alpha$ is even if and only if $\beta$ is odd, so in all cases we have

$$D(Q|_{X_j})^\alpha D(Q|_Y) D(Q|_{Z_j})^{t-\alpha-1} = \boxtimes$$

as required. Now write

$$\hat{x} = [\Lambda^{a'}, I_\ell]^\alpha \oplus [\Lambda^\beta, I_{a-\beta c}] \oplus [I_a]^{t-\alpha-1}$$

where the 1-eigenspaces of the blocks correspond to the component spaces in (5.2.11). Therefore, $x$ fixes a parabolic $a$-decomposition $V = V_1 \perp \dots \perp V_t$. In fact, our specific choice of the component spaces in (5.2.11) implies that each $V_j$ is square-parabolic. For instance, $\hat{x}|_{V_1} = [\Lambda^{a'}, I_\ell]$ so

$$V_1 = U_1 \perp \dots \perp U_{a'} \perp X_1$$

where each $U_j$ is an elliptic $i$-space and the $\ell$-space $X_1$ has been chosen as above. Then $D(Q|_{U_1}) = \boxtimes$ (since $D(Q|_U) = \boxtimes$ and $D(Q) = \square$), so

$$D(Q|_{V_1}) = D(Q|_{U_1})^{a'} D(Q|_{X_1}) = \square$$

as required. Therefore $x$ fixes a decomposition in $\Omega$ componentwise.

The other cases are very similar. For example, suppose $i$ is odd and $D(Q|_U) = \square$, so either $d$ is even or $q \equiv 1 \pmod 4$. If $q \equiv 1 \pmod 4$ then we may choose each component space in (5.2.11) to be square-parabolic, in which case

$$\hat{x} = [(\Lambda, \Lambda^{-1})^{a'}, I_\ell]^\alpha \oplus [(\Lambda, \Lambda^{-1})^\beta, I_{a-\beta c}] \oplus [I_a]^{t-\alpha-1}$$

and the result follows (note that if $W$ is a hyperbolic $2i$-space then the condition $q \equiv 1 \pmod{4}$ implies that $D(Q|_W) = \square$). Now assume that $q \equiv 3 \pmod{4}$. Since $d$ is even, we may choose the component spaces so that

$$D(Q|_{x_j}) = \begin{cases} \square & a' \text{ even} \\ \boxtimes & a' \text{ odd} \end{cases} \quad D(Q|_Y) = \begin{cases} \square & \beta \text{ even} \\ \boxtimes & \beta \text{ odd} \end{cases} \quad D(Q|_{z_j}) = \square$$

for all $j$, and the result follows as above.

*Case 5.* $r > 2$, $c < a$, $a \not\equiv 0 \pmod{i}$, $k < t\ell$. To complete the proof, we may assume that $k < t\ell$ (we continue to assume that $c < a$ and $i$ does not divide $a$). First assume $i$ is odd, so $c = 2i$. If $(\varepsilon, k) \neq (-, 0)$ then $\hat{x} = [(\Lambda, \Lambda^{-1})^{(n-k)/c}, I_k]$ is a derangement of order $r$ (indeed, $x$ cannot fix a decomposition componentwise since $k < t\ell$, and it cannot permute components since $\dim C_V(\hat{x}) = k < c < a$). Now assume $\varepsilon = -$ and $k = 0$. Here $c$ divides $t\ell$, so we have $c \leqslant t\ell$. Set $\hat{x} = [(\Lambda, \Lambda^{-1})^{n/c-1}, I_c]$. Since $\dim C_V(\hat{x}) = c < a$, if $x$ fixes a decomposition in $\Omega$ then it must fix each component, so $x$ is a derangement if $c < t\ell$. However, if $c = t\ell$ then we can repeat the arguments in Case 4 to show that every element of order $r$ in $T$ fixes a decomposition in $\Omega$ componentwise. We omit the details.

Finally, let us assume that $k < t\ell$ and $i$ is even, so $c = i$. If $k > 0$ then $[\Lambda^{(n-k)/c}, I_k]$ is a derangement. Similarly, if $k = 0$ and $\varepsilon = (-)^{n/c}$ then $[\Lambda^{n/c}]$ is a derangement, so to complete the proof we may assume that $k = 0$ and $\varepsilon = (-)^{n/c+1}$. Set $\hat{x} = [\Lambda^{n/c-1}, I_c]$. If $x$ fixes a decomposition in $\Omega$ then it must fix each component space, so $x$ is a derangement if $c < t\ell$. Finally, if $c = t\ell$ then we can proceed as in Case 4 to deduce that $T$ is $r$-elusive. $\qquad\square$

Next we deal with the remaining cases arising in Case IV of Table 5.2.1. Here $T = \mathrm{P}\Omega_n^\varepsilon(q)$ and $H$ is of type $\mathrm{O}_a^{\varepsilon'}(q) \wr S_t$, where $a$ is even and $\varepsilon = (\varepsilon')^t$. We will consider the cases $(\varepsilon, \varepsilon') = (+, +)$, $(+, -)$ and $(-, -)$ in the next three propositions. We refer the reader to Remark 5.2.27 for a general result on $r$-elusivity (with $r \neq p$) in Case IV.

**Proposition 5.2.24** *In Case IV of Table 5.2.1, if $r \neq p$ and $\varepsilon = \varepsilon' = +$ then $T$ is $r$-elusive if and only if $r = 2$, or $r > 2$ and one of the following holds:*

(i) $\ell = 0$ *and* $a/i$ *is even.*
(ii) $\ell > 0$, $c < a$ *and* $k = t\ell$.
(iii) $\ell > 0$, $c < a$, $k = 0$, $n/i$ *is odd and* $i = t\ell$.

*Proof*   Here $T = \mathrm{P}\Omega_n^+(q)$ and $H$ is of type $\mathrm{O}_a^+(q) \wr S_t$. Let $r$ be a prime divisor of $|\Omega|$ and let $x = \hat{x}Z \in T$ be an element of order $r$.

*Case 1. $r = 2$.*   First assume $r = 2$. If $x = t_i$, with $1 \leqslant i \leqslant n/4$, then $x$ fixes a decomposition of $V$ into hyperbolic 2-spaces, hence $x$ fixes a decomposition in $\Omega$ componentwise. Next suppose $x = t_i'$ and $1 \leqslant i \leqslant n/4$. Let $U$ be the $(-1)$-eigenspace of $\hat{x}$ and set $W = C_V(\hat{x})$, so $U$ and $W$ are elliptic spaces and $x$ fixes the decomposition $V = U \perp W$. Now $U = U' \perp U_1$ and $W = W' \perp W_1$, where $U_1, W_1$ are elliptic 2-spaces and $U', W'$ are hyperbolic (or trivial) spaces. By Lemma 2.2.13(ii), there exist hyperbolic 2-spaces $X_1, X_2$ such that $U_1 \perp W_1 = X_1 \perp X_2$ and $\dim(U_1 \cap X_i) = \dim(W_1 \cap X_i) = 1$ for $i = 1, 2$. Thus $\hat{x}|_{X_i} = [-I_1, I_1]$, so $x$ fixes each subspace in the decomposition $V = U' \perp X_1 \perp X_2 \perp W'$, and therefore $x$ fixes (componentwise) a decomposition of $V$ into hyperbolic 2-spaces. We conclude that $x$ fixes a decomposition in $\Omega$.

Next suppose that $x$ is of type $t_{n/2}$ or $t_{n/2-1}$. Here $x$ fixes a pair of complementary maximal totally singular subspaces of $V$, so $x$ fixes a decomposition of $V$ into hyperbolic 2-spaces and thus once again $x$ fixes a decomposition in $\Omega$ componentwise.

To complete the analysis of involutions, we may assume that $x$ is of type $t_{n/2}'$ or $t_{n/2-1}'$, in which case $n \equiv 0 \pmod 4$ and $q \equiv 3 \pmod 4$. As explained in Section 3.5.2.11, we can take $\hat{x} = [A^{n/2}]$ with respect to a hyperbolic basis

$$\{e_1, \ldots, e_{n/2}, f_1, \ldots, f_{n/2}\}$$

where $A$ is given in (5.2.6) (note that $A = -A^\mathsf{T}$ since $q \equiv 3 \pmod 4$). Clearly, $x$ fixes a decomposition of $V$ into hyperbolic 4-spaces, so $x$ has fixed points on $\Omega$ if $a \equiv 0 \pmod 4$. On the other hand, if $a \equiv 2 \pmod 4$ then $t$ is even (since $n \equiv 0 \pmod 4$) and we observe that $x$ induces a permutation of shape $(2^{n/4})$ on the set of hyperbolic 2-spaces

$$\{\langle e_\ell, f_\ell \rangle \mid 1 \leqslant \ell \leqslant n/2\}$$

Therefore, $x$ fixes a decomposition in $\Omega$ by inducing a permutation of shape $(2^{t/2})$ on the component spaces. We conclude that $T$ is 2-elusive.

For the remainder, let us assume $r > 2$. Set $i = \Phi(r, q)$ and define $c = c(i)$ as in Definition 5.1.1. Note that $i < n$ since we are assuming that $T$ contains an element of order $r$.

*Case 2. $r > 2$, $c > a$.*   If $c > a$ then the usual argument implies that $x \in T$ is a derangement, where $\hat{x} = [\Lambda, \Lambda^{-1}, I_{n-c}]$ if $i$ is odd, and $\hat{x} = [\Lambda, I_{n-c}]$ if $i$ is even (see the proof of Proposition 5.2.21, for example).

*Case 3.* $r > 2$, $c \leqslant a$. For the remainder, we may assume that $c \leqslant a$. Define $k, \ell$ as in (5.2.5), and note that $\ell$ is even and $k \leqslant t\ell$.

First assume that $\ell = 0$. If $i$ is odd then Lemma 5.2.8 implies that every element in $T$ of order $r$ fixes a decomposition in $\Omega$ and so $T$ is $r$-elusive. Similarly, if $i$ is even then Lemma 5.2.9 implies that $T$ is $r$-elusive if $a/i$ is even, so let us assume that $i$ is even and $a/i$ is odd. If $n/i$ is even then the element $x = \hat{x}Z \in T$ with $\hat{x} = [\Lambda^{n/i}]$ is a derangement (indeed, $x$ cannot fix a decomposition in $\Omega$ componentwise because $a/i$ is odd, and it cannot permute components because $C_V(\hat{x})$ is trivial). Now assume $n/i$ is odd (so $t$ is odd) and set $\hat{x} = [\Lambda^{n/i-1}, I_i]$. Suppose that $x$ fixes a decomposition $V = V_1 \perp \ldots \perp V_t$ in $\Omega$. Since $a/i$ is odd, if $x$ fixes each $V_j$ then $\dim C_{V_j}(\hat{x}) \geqslant i$ for each $j$, whence $\dim C_V(\hat{x}) \geqslant ti$. This is a contradiction. Similarly, if $x$ induces a nontrivial permutation on the $V_j$, then Lemma 5.2.6 implies that each $r$th root of unity occurs as an eigenvalue of $\hat{x}$ with multiplicity at least $a \geqslant i$. Therefore, $x$ is a derangement unless $a = i = r - 1$. However, in this situation it is easy to see that $[\Lambda, I_{n-a}]$ is a derangement.

To complete the proof we may assume that $\ell > 0$, so $c < a$. For now, let us assume that we are not in the case where $k = 0$ and $n/i$ is odd.

First assume that $k < t\ell$ and set

$$\hat{x} = \begin{cases} [(\Lambda, \Lambda^{-1})^{(n-k)/c}, I_k] & i \text{ odd} \\ [\Lambda^{(n-k)/c}, I_k] & i \text{ even} \end{cases}$$

Since $\dim C_V(\hat{x}) < t\ell$, Lemmas 5.2.8 and 5.2.9 imply that $x$ does not fix a decomposition in $\Omega$ componentwise. Furthermore, in view of Lemma 5.2.6, we deduce that $x$ is a derangement since $k < c < a$.

Now let us assume $k = t\ell$. We claim that $T$ is $r$-elusive. By Proposition 3.5.4, we have $\dim C_V(\hat{y}) \geqslant k = t\ell$ for all $y = \hat{y}Z \in T$ of order $r$. Therefore, if $i$ is odd, Lemma 5.2.8 implies that $y$ fixes a decomposition in $\Omega$ and so $T$ is $r$-elusive. Now suppose $i$ is even. If $(a - \ell)/i$ is even then $(n - k)/i$ is even and so either $\dim C_V(\hat{y}) > t\ell$ or $C_V(\hat{y})$ is a hyperbolic $t\ell$-space. In both cases, it follows that $C_V(\hat{y})$ contains a nondegenerate $t\ell$-space that admits a decomposition into $t$ hyperbolic $\ell$-spaces, so Lemma 5.2.9 implies that $y$ fixes a decomposition in $\Omega$ and thus $T$ is $r$-elusive. If $(a - \ell)/i$ is odd and $t$ is odd then either $\dim C_V(\hat{y}) > t\ell$ or $C_V(\hat{y})$ is an elliptic $t\ell$-space. In both cases, $C_V(\hat{y})$ contains a nondegenerate $t\ell$-space that admits a decomposition into $t$ elliptic $\ell$-spaces, so once again we conclude that $T$ is $r$-elusive. Finally, if $(a - \ell)/i$ is odd and $t$ is even then either $\dim C_V(\hat{y}) > t\ell$ or $C_V(\hat{y})$ is a hyperbolic $t\ell$-space. Again, it follows that $C_V(\hat{y})$ contains a nondegenerate $t\ell$-space that admits a decomposition into $t$ elliptic $\ell$-spaces, whence $T$ is $r$-elusive.

It remains to consider the case where $k = 0$ and $n/i$ is odd. Note that $i$ is even and $i$ divides $t\ell$, so $i \leqslant t\ell$. If $i < t\ell$ then Lemmas 5.2.6 and 5.2.9 imply that $[\Lambda^{n/i-1}, I_i]$ is a derangement since $i < a$. Now assume that $i = t\ell$ and let $y = \hat{y}Z \in T$ be an element of order $r$. Then either $\dim C_V(\hat{y}) > t\ell$ or $C_V(\hat{y})$ is a hyperbolic $t\ell$-space. Therefore, if $(a - \ell)/i$ is even then $C_V(\hat{y})$ contains a non-degenerate $t\ell$-space that admits an orthogonal decomposition into hyperbolic $\ell$-spaces. Similarly, if $(a - \ell)/i$ is odd then $t$ is even and $C_V(\hat{y})$ contains a non-degenerate $t\ell$-space that admits a decomposition into elliptic $\ell$-spaces. In both cases, Lemma 5.2.9 implies that $y$ fixes a decomposition in $\Omega$ and thus $T$ is $r$-elusive. $\qquad\qquad\square$

**Proposition 5.2.25** *In Case IV of Table 5.2.1, if $r \neq p$ and $(\varepsilon, \varepsilon') = (+, -)$ then $T$ is $r$-elusive if and only if $r = 2$, or $r > 2$ and one of the following holds:*

(i) *$\ell = 0$ and $a/i$ is odd.*
(ii) *$\ell > 0$, $c < a$ and $k = t\ell$.*
(iii) *$\ell > 0$, $c < a$, $k = 0$, $n/i$ is odd and $i = t\ell$.*

*Proof* Here $T = \mathrm{P}\Omega_n^+(q)$ and $H$ is of type $\mathrm{O}_a^-(q) \wr S_t$, where $t$ is even (so $n \equiv 0$ (mod 4)). Let $r$ be a prime divisor of $|\Omega|$ and let $x = \hat{x}Z \in T$ be an element of order $r$.

*Case 1. $r = 2$.* First assume $r = 2$. Suppose $x = t_i$ with $1 \leqslant i \leqslant n/4$. Then $x$ fixes each component in a decomposition $V = X_1 \perp X_2 \perp \ldots \perp X_{n/2}$ into hyperbolic 2-spaces, where $\hat{x}|_{X_j} = I_2$ or $-I_2$ for each $j$. By Lemma 2.2.13(ii) we can write $X_{2j-1} \perp X_{2j} = Y_{2j-1} \perp Y_{2j}$, where $Y_{2j-1}$ and $Y_{2j}$ are elliptic 2-spaces such that $\dim(X_\alpha \cap Y_\beta) = 1$ for each $\alpha, \beta \in \{2j-1, 2j\}$. Since each nonzero vector in $X_j$ is an eigenvector of $\hat{x}$, it follows that $x$ fixes each $Y_j$. Moreover, since $t$ is even, $x$ fixes each subspace in the orthogonal decomposition

$$V = Y_1 \perp \ldots \perp Y_t \perp X_{t+1} \perp \ldots \perp X_{n/2}$$

By combining each $Y_j$ with an appropriate number of the $X_k$, we deduce that $x$ fixes a decomposition in $\Omega$ componentwise.

Next suppose $x = t_i'$ with $1 \leqslant i \leqslant n/4$. Since $n \equiv 0$ (mod 4), it follows that $i$ is odd and $q \equiv 3$ (mod 4) (see Section 3.5.2.4). Since $i$ and $n/2 - i$ are both odd, the two eigenspaces of $\hat{x}$ on $V$ admit orthogonal decompositions into elliptic 2-spaces, so $x$ fixes componentwise a decomposition $V = Y_1 \perp \ldots \perp Y_{n/2}$ into elliptic 2-spaces, with $\hat{x}|_{Y_j} = -I_2$ or $I_2$. In particular, $x$ has fixed points if $a \equiv 2$ (mod 4). As in the previous paragraph, we can write $Y_{2j-1} \perp Y_{2j} = X_{2j-1} \perp X_{2j}$, where $X_{2j-1}$ and $X_{2j}$ are hyperbolic 2-spaces such that $\dim(X_\alpha \cap Y_\beta) = 1$ for each $\alpha, \beta \in \{2j-1, 2j\}$. Since each nonzero vector in $Y_j$ is an eigenvector of $\hat{x}$, it follows that $x$ fixes each subspace in the decomposition

$$V = Y_1 \perp \ldots \perp Y_t \perp X_{t+1} \perp \ldots \perp X_{n/2}$$

and thus $x$ also fixes a decomposition in $\Omega$ when $a \equiv 0 \pmod 4$.

Now let us turn to the involutions $x \in T$ of type $t_{n/2}$ or $t_{n/2-1}$, so $q \equiv 1$ (mod 4) (see Section 3.5.2.10). With respect to a hyperbolic basis

$$\{e_1, \ldots, e_{n/2}, f_1, \ldots, f_{n/2}\} \tag{5.2.12}$$

we can take $\hat{x} = [\lambda I_{n/2}, \lambda^{-1} I_{n/2}]$, where $\lambda \in \mathbb{F}_q$ has order 4. Fix $\mu \in \mathbb{F}_q$ such that $-\mu$ is a nonsquare. By Lemma 2.2.13(ii),

$$\langle e_1, e_2, f_1, f_2 \rangle = \langle e_1 + f_1, e_2 + \mu f_2 \rangle \perp \langle e_1 - f_1, e_2 - \mu f_2 \rangle$$

is a decomposition into elliptic 2-spaces that are interchanged by $x$. Therefore, since $n \equiv 0 \pmod 4$, we can construct a decomposition $V = Y_1 \perp \ldots \perp Y_{n/2}$ into elliptic 2-spaces so that $x$ induces a permutation of shape $(2^{n/4})$ on the component spaces. In particular, if $a \equiv 2 \pmod 4$ then by combining together $a/2$ of the $Y_i$ ($t$ times), in a suitable manner, we obtain an orthogonal decomposition of $V$ into elliptic $a$-spaces on which $x$ induces a permutation of shape $(2^{t/2})$. Therefore, $x$ has fixed points if $a \equiv 2 \pmod 4$. Now assume that $a \equiv 0 \pmod 4$. One can check that

$$\langle e_1, e_2, f_1, f_2 \rangle = \langle e_1 + f_1, e_2 + f_2 \rangle \perp \langle e_1 - f_1, e_2 - f_2 \rangle$$

is a decomposition into hyperbolic 2-spaces that are interchanged by $x$ (note that $-1 \in \mathbb{F}_q$ is a square since $q \equiv 1 \pmod 4$), so by arguing as before we can construct a decomposition in $\Omega$ such that $x$ induces a permutation of shape $(2^{t/2})$ on the components. We conclude that $x$ has fixed points for any $a$.

To complete the analysis of involutions, we may assume that $x = t'_{n/2}$ or $t'_{n/2-1}$, so $q \equiv 3 \pmod 4$ (see Section 3.5.2.11). In terms of the hyperbolic basis in (5.2.12), we can take $\hat{x} = [A^{n/2}]$, where $A$ is given in (5.2.6). Since $q \equiv 3 \pmod 4$, Lemma 2.2.13 implies that

$$\langle e_1, e_2, f_1, f_2 \rangle = \langle e_1 + f_1, e_2 + f_2 \rangle \perp \langle e_1 - f_1, e_2 - f_2 \rangle$$

is a decomposition into elliptic 2-spaces that is fixed componentwise by $x$. Therefore, we can construct a decomposition $V = Y_1 \perp \ldots \perp Y_{n/2}$ into elliptic 2-spaces with the property that $x$ fixes each $Y_j$. We deduce that $x$ has fixed points on $\Omega$ if $a \equiv 2 \pmod 4$.

Now assume $a \equiv 0 \pmod 4$. Fix $\zeta \in \mathbb{F}_q$ so that $\mathbf{x}^2 + \mathbf{x} + \zeta \in \mathbb{F}_q[\mathbf{x}]$ is irreducible. Then $x$ interchanges the hyperbolic 2-spaces in the decomposition

$$\langle e_1, e_2, f_1, f_2 \rangle = \langle e_1, f_1 \rangle \perp \langle e_2, f_2 \rangle$$

and it also interchanges the elliptic 2-spaces in the decomposition

$$\langle e_1, e_2, f_1, f_2 \rangle = \langle e_1 + \zeta f_1, e_2 + f_1 + f_2 \rangle \perp \langle e_2 + \zeta f_2, -e_1 - f_1 + f_2 \rangle$$

Therefore, by taking suitable combinations of these spaces, we deduce that $x$ also has fixed points if $a \equiv 0 \pmod 4$ (more precisely, $x$ fixes an appropriate decomposition by inducing a permutation of shape $(2^{t/2})$ on the summands). We conclude that $T$ is 2-elusive.

*Case 2. $r > 2$.* For the remainder, let us assume that $r > 2$. Set $i = \Phi(r, q)$ and define $c = c(i)$ as in Definition 5.1.1. In the usual way, it is easy to see that $T$ contains a derangement of order $r$ if $c > a$, so let us assume that $c \leqslant a$. Define $k, \ell$ as in (5.2.5) and note that $\ell$ is even and $k \leqslant t\ell$.

First assume that $\ell = 0$. If $i$ is odd, then by combining Lemmas 5.2.6 and 5.2.8 we deduce that $[(\Lambda, \Lambda^{-1})^{n/2i}]$ is a derangement. Now assume $i$ is even. Suppose that $a/i$ is even and set $\hat{x} = [\Lambda^{n/i}]$, noting that $x \in T$ since $n/i$ is even. Since $(-)^{a/i} = -\varepsilon'$, Lemma 5.2.9 implies that $x$ does not fix a decomposition in $\Omega$ componentwise, and thus $x$ is a derangement by Lemma 5.2.6. On the other hand, if $a/i$ is odd then Lemma 5.2.9 implies that every element in $T$ of order $r$ fixes componentwise a decomposition in $\Omega$, so $T$ is $r$-elusive in this situation.

Now let us assume that $\ell > 0$, so $c < a$. To begin with, we exclude the case where $k = 0$ and $n/i$ is odd.

First assume that $k < t\ell$ and set

$$\hat{x} = \begin{cases} [(\Lambda, \Lambda^{-1})^{(n-k)/c}, I_k] & i \text{ odd} \\ [\Lambda^{(n-k)/c}, I_k] & i \text{ even} \end{cases}$$

Since $\dim C_V(\hat{x}) < t\ell$, Lemmas 5.2.8 and 5.2.9 imply that $x$ does not fix a decomposition in $\Omega$ componentwise, so Lemma 5.2.6 implies that $x$ is a derangement since $k < c < a$.

Now assume $k = t\ell$. According to Proposition 3.5.4, if $y = \hat{y}Z \in T$ has order $r$ then $\dim C_V(\hat{y}) \geqslant k = t\ell$. Consequently, if $i$ is odd then Lemma 5.2.8 implies that $y$ fixes a decomposition in $\Omega$ componentwise and so $T$ is $r$-elusive. Now assume $i$ is even. Since $t$ is even, $(n-k)/i = (a-\ell)t/i$ and $i$ divides $a - \ell$, it follows that $(n-k)/i$ is even and so either $\dim C_V(\hat{y}) > t\ell$ or $C_V(\hat{y})$ is a hyperbolic $t\ell$-space. Therefore, $C_V(\hat{y})$ contains a nondegenerate $t\ell$-space that admits a decomposition into $t$ elliptic $\ell$-spaces, and also a decomposition into $t$ hyperbolic $\ell$-spaces. By Lemma 5.2.9, we conclude that $y$ fixes a decomposition in $\Omega$ and thus $T$ is $r$-elusive.

To complete the proof, we may assume that $k = 0$ and $n/i$ is odd. Note that $i$ is even and $i$ divides $t\ell$, so $i \leqslant t\ell$. If $i < t\ell$ then by applying Lemmas

5.2.6 and 5.2.9 we deduce that $[\Lambda^{n/i-1}, I_i]$ is a derangement (note that $i < a$). Now assume $i = t\ell$. If $y = \hat{y}Z \in T$ has order $r$ then either $\dim C_V(\hat{y}) > t\ell$ or $C_V(\hat{y})$ is a hyperbolic $t\ell$-space. Since $t$ is even, it follows that $C_V(\hat{y})$ contains a nondegenerate $t\ell$-space that admits a decomposition into hyperbolic $\ell$-spaces, and also a decomposition into elliptic $\ell$-spaces. As before, by applying Lemma 5.2.9, we conclude that $T$ is $r$-elusive. $\qquad\qquad\square$

**Proposition 5.2.26** *In Case IV of Table 5.2.1, if $r \neq p$ and $\varepsilon = \varepsilon' = -$ then $T$ is $r$-elusive if and only if $r = 2$, or $r > 2$ and one of the following holds:*

  (i) $\ell = 0$ and $a/i$ is odd.
 (ii) $\ell > 0$, $c < a$ and $k = t\ell$.
(iii) $\ell > 0$, $c < a$, $k = 0$, $n/i$ is even and $c = t\ell$.

*Proof*   Here $T = P\Omega_n^-(q)$ and $H$ is of type $O_a^-(q) \wr S_t$, so $t \geqslant 3$ is odd. Let $r$ be a prime divisor of $|\Omega|$ and let $x = \hat{x}Z \in T$ be an element of order $r$.

*Case 1. $r = 2$.* First assume $r = 2$. Suppose $x \in T$ is an involution of type $t_i$ with $1 \leqslant i \leqslant n/4$, or type $t'_i$ with $1 \leqslant i < n/4$. From the description of these elements in Section 3.5.2, it follows that $x$ fixes a decomposition $V = X_1 \perp \ldots \perp X_{n/2-1} \perp Y_0$ where each $X_j$ is a hyperbolic 2-space such that $\hat{x}|_{X_j} = -I_2$ or $I_2$, and $Y_0$ is an elliptic 2-space with $\hat{x}|_{Y_0} = I_2$ if $x = t_i$, and $\hat{x}|_{Y_0} = -I_2$ if $x = t'_i$. Since $t$ is odd, we can use Lemma 2.2.13(ii) to construct an orthogonal decomposition

$$V = X_1 \perp \ldots \perp X_{n/2-t} \perp Y_0 \perp Y_1 \perp \ldots \perp Y_{t-1}$$

where each $Y_j$ is an elliptic 2-space, and each subspace in this decomposition is fixed by $x$. Therefore, by combining each $Y_j$ with an appropriate number of the $X_k$, we obtain a decomposition in $\Omega$ that is fixed componentwise by $x$.

To complete the analysis of involutions we may assume $x = t_{n/2}$, in which case $n \equiv 2 \pmod 4$ and $q \equiv 7 \pmod 8$. In particular, $a \equiv 2 \pmod 4$. These involutions arise from elements in the centre of $\Delta U_{n/2}(q)$ (see Section 3.5.2.13). If $\{v_1, \ldots, v_{n/2}\}$ is an orthonormal basis for $V$ over $\mathbb{F}_{q^2}$ with respect to an hermitian form $\bar{B}$, then each $\langle v_i \rangle_{\mathbb{F}_{q^2}}$ is an elliptic 2-space over $\mathbb{F}_q$ with respect to the quadratic form $Q$ on the $\mathbb{F}_q$-space $V$ defined by $Q(v) = \bar{B}(v, v)$ (see Construction 2.5.14). Since $a \equiv 2 \pmod 4$ and $x$ fixes each of these 2-spaces, by taking suitable combinations of these subspaces we deduce that $x$ fixes a decomposition in $\Omega$ componentwise. We conclude that $T$ is 2-elusive.

*Case 2. $r > 2$.* For the remainder, let us assume $r > 2$. Set $i = \Phi(r, q)$ and define $c = c(i)$ as in Definition 5.1.1. As before, it is easy to see that $T$ contains derangements of order $r$ if $c > a$, so let us assume $c \leqslant a$. Define $k, \ell$ as in (5.2.5) and note that $\ell$ is even and $k \leqslant t\ell$.

First assume that $\ell = 0$. If $a/i$ is odd then $i$ is even, $(-)^{a/i} = \varepsilon'$ and Lemma 5.2.9 implies that every element in $T$ of order $r$ fixes a decomposition in $\Omega$ componentwise. Hence $T$ is $r$-elusive in this case. Now assume $a/i$ is even. If $i$ is also even then $\hat{x} = [\Lambda^{n/i-1}, I_i]$ is a derangement (indeed, Lemma 5.2.6 implies that $x$ cannot permute components because $\dim C_V(\hat{x}) = i < a$, and by Lemma 5.2.9 it cannot fix componentwise because $(-)^{a/i} = -\varepsilon'$ and $\dim C_V(\hat{x}) = i < ti$). Finally, suppose that $a/i$ is even and $i$ is odd. Let $\hat{x} = [(\Lambda, \Lambda^{-1})^{n/2i-1}, I_{2i}]$ and assume that $x$ fixes a decomposition in $\Omega$. By Lemma 5.2.8, $x$ must induce a nontrivial permutation on the component spaces since $\dim C_V(\hat{x}) = 2i < 2ti$. Therefore, Lemma 5.2.6 implies that $2i \geqslant a$ and thus $2i = a = r - 1$ is the only possibility. In this situation, it is easy to see that $y = \hat{y}Z \in T$ is a derangement, where $\hat{y} = [\Lambda, \Lambda^{-1}, I_{n-a}]$. Indeed, $y$ does not fix a decomposition in $\Omega$ componentwise, and it does not permute component spaces since the eigenvalues of $\hat{y}$ are incompatible with Lemma 5.2.6.

Now assume $\ell > 0$, so $c < a$. Recall that $k \leqslant t\ell$. If $k = t\ell$ and $i$ is odd then Proposition 3.5.4 implies that $\dim C_V(\hat{y}) \geqslant k = t\ell$ for all $y = \hat{y}Z \in T$ of order $r$, so Lemma 5.2.8 implies that $y$ fixes a decomposition in $\Omega$ componentwise. Therefore, $T$ is $r$-elusive in this situation. Now assume $k = t\ell$ and $i$ is even. Once again we have $\dim C_V(\hat{y}) \geqslant k = t\ell$ for every element $y = \hat{y}Z \in T$ of order $r$. Moreover, if $(a - \ell)/i$ is odd then so is $(n - k)/i = (a - \ell)t/i$, hence $C_V(\hat{y})$ contains a hyperbolic $t\ell$-space that admits a decomposition into $t$ hyperbolic $\ell$-spaces. Therefore, Lemma 5.2.9 implies that $T$ is $r$-elusive. Similarly, if $(a - \ell)/i$ is even then so is $(n - k)/i$ and thus $C_V(\hat{y})$ contains an elliptic $t\ell$-space. Since $t$ is odd, this $t\ell$-space admits a decomposition into $t$ elliptic $\ell$-spaces, and once again we conclude that $T$ is $r$-elusive.

It remains to consider the case $k < t\ell$. First assume $k > 0$. If $i$ is odd then $[(\Lambda, \Lambda^{-1})^{(n-k)/2i}, I_k]$ is a derangement by the usual argument, using Lemmas 5.2.6 and 5.2.8. Similarly, if $i$ is even then the same argument shows that $[\Lambda^{(n-k)/i}, I_k]$ is a derangement.

Finally, suppose that $k = 0$. Note that $c$ divides $t\ell$, so $c \leqslant t\ell$. First we assume $i$ is odd. Set $\hat{x} = [(\Lambda, \Lambda^{-1})^{n/2i-1}, I_{2i}]$. Since $c = 2i < a$, the element $x$ cannot permute the component spaces in a decomposition in $\Omega$, and thus Lemma 5.2.8 implies that $x$ is a derangement if $c < t\ell$. On the other hand, if $c = t\ell$ then the same lemma implies that $x$, and in fact any element in $T$ of order $r$, fixes a decomposition in $\Omega$ and so $T$ is $r$-elusive. Now assume $i$ is even. If $n/i$ is odd then $[\Lambda^{n/i}]$ is a derangement by Lemmas 5.2.6 and 5.2.9, so we may assume that $n/i$ is even. Set $\hat{x} = [\Lambda^{n/i-1}, I_i]$ and note that $C_V(\hat{x})$ is hyperbolic. If $i < t\ell$ then $x$ is a derangement by Lemmas 5.2.6 and 5.2.9. However, if $i = t\ell$ then

$(a - \ell)/i$ is odd and the same lemmas imply that $x$ (and indeed any element in $T$ of order $r$) fixes a decomposition in $\Omega$, whence $T$ is $r$-elusive in this situation. This completes the proof of the proposition.                                    $\square$

**Remark 5.2.27** For $r \neq p$, let us summarise the conditions for $r$-elusivity in Case IV of Table 5.2.1 that we have derived in Propositions 5.2.23–5.2.26. As before, write $n \equiv k \pmod{c}$ and $a \equiv \ell \pmod{c}$, where $0 \leqslant k, \ell < c$ and $c$ is given in Definition 5.1.1. In addition, let us assume that $a \geqslant 2$. Then $T$ is $r$-elusive if and only if $r = 2$, or $r > 2$ and one of the following holds:

(i) $\ell = 0$ and $\varepsilon' = (-)^{a/i}$.
(ii) $\ell > 0$, $c < a$ and either
   (a) $k = t\ell$ and $a \not\equiv 0 \pmod{i}$, or
   (b) $k = 0$, $c = t\ell$ and $\varepsilon = (-)^{n/i+1}$.

Note that $a \not\equiv 0 \pmod{i}$ if $\ell > 0$ and $a$ is even, so in this situation the second condition in (ii)(a) always holds.

**Proposition 5.2.28** *In Case V of Table 5.2.1, if $r \neq p$ then $T$ is $r$-elusive if and only if $r = 2$ and $q \not\equiv \varepsilon \pmod{8}$, or $r > 2$ and one of the following holds:*

(i) $c < n/2$ and $k = 2\ell$.
(ii) $i < n/2$, $k = 0$, $c = 2\ell$ and $\varepsilon = (-)^i$.

*Proof*  Here $T = \mathrm{P\Omega}_n^\varepsilon(q)$ and $H$ is of type $\mathrm{O}_{n/2}(q)^2$, so $n \equiv 2 \pmod{4}$ and $q$ is odd. We can view $\Omega$ as the set of orthogonal decompositions $V = V_1 \perp V_2$, where $V_1$ is a square-parabolic $\frac{n}{2}$-space and $V_2$ is a nonsquare-parabolic $\frac{n}{2}$-space (that is, if $Q$ is the underlying quadratic form on $V$, then $D(Q|_{V_1}) = \square$ and $D(Q|_{V_2}) = \boxtimes$). In particular, $V_1$ and $V_2$ are similar, but not isometric, and we have $D(Q) = \boxtimes$ (see Lemma 2.2.9). Note that an element $x \in T$ fixes a decomposition in $\Omega$ if and only if it fixes a parabolic $\frac{n}{2}$-space (indeed, if $x$ fixes a parabolic $\frac{n}{2}$-space $U$ then it also fixes its orthogonal complement $U^\perp$, and $D(Q|_U) \neq D(Q|_{U^\perp})$ since $D(Q) = \boxtimes$). Let $r$ be a prime divisor of $|\Omega|$ and let $x = \hat{x}Z \in T$ be an element of order $r$.

First assume $r = 2$. Suppose $x \in T$ is an involution of type $t_i$ or $t_i'$, where $1 \leqslant i < n/4$. Then $\hat{x}$ centralises a nondegenerate $(n - 2i)$-space, and since $n - 2i > n/2$ it follows that $x$ fixes a square-parabolic $\frac{n}{2}$-space (see Corollary 2.2.12) and thus $x$ fixes a decomposition in $\Omega$.

To complete the analysis of involutions, we may assume that $q \equiv \varepsilon \pmod{8}$ and $x \in T$ is of type $t_{n/2}$. As noted in Section 3.5.2.12, if $\varepsilon = +$ then we can take $\hat{x} = [\lambda I_{n/2}, \lambda^{-1} I_{n/2}]$ (with respect to a suitable hyperbolic basis), where

$\lambda \in \mathbb{F}_q$ has order 4. Clearly, $x$ does not fix a parabolic subspace of $V$, so $x$ is a derangement. Similarly, if $\varepsilon = -$ then there exists an elliptic basis of $V$ such that $\hat{x}$ is a block-diagonal matrix, with $n/2$ irreducible blocks of size two (see Section 3.5.2.13). Once again we deduce that $x$ is a derangement.

For the remainder we may assume that $r > 2$. Set $i = \Phi(r,q)$ and define $c = c(i)$ as in Definition 5.1.1. It is easy to see that $T$ contains derangements of order $r$ if $c > n/2$, so let us assume $c < n/2$. Define $k, \ell$ as in (5.2.5) (with $a = n/2$) and note that $\ell$ is odd and $k \leqslant 2\ell$.

First assume $i$ divides $n/2$ (so $i$ and $n/2i$ are odd, $k = 0$, $\ell = i$ and $c = 2\ell$). If $\varepsilon = +$ then $[(\Lambda, \Lambda^{-1})^{n/2i}]$ is a derangement, so let us assume $\varepsilon = -$, in which case $i < n/2$ (since we are assuming that $T$ contains elements of order $r$). We claim that $T$ is $r$-elusive. To see this, let $y = \hat{y}Z \in T$ be an element of order $r$. Recall that it suffices to show that $y$ fixes a parabolic $\frac{n}{2}$-space. Since $C_V(\hat{y})$ is elliptic, we may write $C_V(\hat{y}) = X \perp Y$, where $X$ is an elliptic $2i$-space and $Y$ is a hyperbolic $2bi$-space for some $b \geqslant 0$. In particular,

$$C_V(\hat{y}) = X_1 \perp X_2 \perp Y_1 \perp \ldots \perp Y_b$$

where $X_1$ and $X_2$ are parabolic $i$-spaces and the $Y_j$ are hyperbolic $2i$-spaces. In view of Proposition 3.5.4, it follows that $y$ fixes a decomposition

$$V = X_1 \perp X_2 \perp Z_1 \perp \ldots \perp Z_{n/2i-1}$$

where the $Z_j$ are hyperbolic $2i$-spaces. In particular, $y$ fixes a parabolic $\frac{n}{2}$-space and thus $y$ has fixed points on $\Omega$. This justifies the claim.

Now assume $i$ does not divide $n/2$, noting that $k > 0$ if $i$ is odd. If $k = 2\ell$ and $x \in T$ has order $r$ then $C_V(\hat{x})$ contains a parabolic $\ell$-space, and by applying Lemmas 4.6.6 and 4.6.7 we deduce that $x$ fixes a parabolic $\frac{n}{2}$-space. Therefore $x$ fixes a decomposition in $\Omega$ and we conclude that $T$ is $r$-elusive.

Finally, suppose $k < 2\ell$. If $i$ is odd then $k > 0$ and $[(\Lambda, \Lambda^{-1})^{(n-k)/2i}, I_k]$ is a derangement, so let us assume $i$ is even. If $k > 0$ then $[\Lambda^{(n-k)/i}, I_k]$ is a derangement (see Lemmas 4.6.6 and 4.6.7), so assume $k = 0$, in which case $i$ divides $n$, and $n/i$ is odd. If $\varepsilon = -$ then $[\Lambda^{n/i}]$ is a derangement, so we reduce to the case $\varepsilon = +$. Now $i$ divides $2\ell$, and it is clear that $[\Lambda^{n/i-1}, I_i]$ is a derangement if $i < 2\ell$. However, if $i = 2\ell$ and $y \in T$ has order $r$ then $C_V(\hat{y})$ contains a parabolic $\ell$-space, so $y$ has fixed points and thus $T$ is $r$-elusive. $\qquad\square$

**Proposition 5.2.29** *In Case VI of Table 5.2.1, if $r \neq p$ then $T$ is $r$-elusive if and only if $r = 2$ and $(q+1)_2 \leqslant (n)_2$.*

*Proof* Here $T = \mathrm{PSU}_n(q)$ and $H$ is of type $\mathrm{GL}_{n/2}(q^2).2$, with $n \geqslant 4$. We may identify $\Omega$ with the set of decompositions $V = V_1 \oplus V_2$, where $V_1$ and $V_2$ are maximal totally isotropic subspaces of $V$. We calculate that

$$|\Omega| = \frac{1}{2}q^{\frac{1}{4}n^2}\prod_{k=1}^{n/2}(q^{2k-1}+1)$$

(see [86, Proposition 4.2.4]). Let $r$ be a prime divisor of $|\Omega|$ and let $x = \hat{x}Z \in T$ be an element of order $r$.

First assume $r = 2$. The involution classes in $T$ are described in Section 3.3.2 and we adopt the notation therein. It is clear that involutions of type $t'_{n/2}$ and $t_i$, with $i \leqslant n/2$ even, fix a decomposition in $\Omega$ componentwise. For example, if $x = t_i$ and $i$ is even then the $(-1)$-eigenspace of $\hat{x}$ decomposes into two totally isotropic $\frac{i}{2}$-spaces, and the 1-eigenspace decomposes into two totally isotropic $\frac{1}{2}(n-i)$-spaces, so by taking an appropriate combination of these spaces we deduce that $x$ fixes a decomposition in $\Omega$. Finally, suppose $x = t_i$ and $i < n/2$ is odd. By Lemma 5.2.6, if $x$ fixes a decomposition $V = V_1 \oplus V_2$ in $\Omega$ then it must fix $V_1$ and $V_2$. In particular, $\dim C_{V_1}(\hat{x}) = \dim C_{V_2}(\hat{x})$ (see Lemma 2.2.17), but this implies that $\dim C_V(\hat{x})$ is even, which is a contradiction. Therefore, $x$ is a derangement, and we note that these elements belong to $T$ if and only if $(q+1)_2 > (n)_2$. We conclude that $T$ is 2-elusive if and only if $(q+1)_2 \leqslant (n)_2$.

To complete the proof, let us assume $r > 2$. Set $i = \Phi(r,q)$ and note that $i \equiv 2 \pmod 4$ since $r$ divides $|\Omega|$. Now, if $x \in T$ is an element of order $r$ that fixes a decomposition in $\Omega$ then Lemma 2.2.17 implies that every eigenvalue of $\hat{x}$ (on $V \otimes \overline{\mathbb{F}}_{q^2}$) has even multiplicity. In particular, if $i \neq 2$ then $[\Lambda, I_{n-i/2}]$ is a derangement, and $[\Lambda, \Lambda^{-1}, I_{n-2}]$ is a derangement if $i = 2$. $\square$

**Proposition 5.2.30** *In Case VII of Table 5.2.1, if $r \neq p$ then $T$ is $r$-elusive if and only if $r = 2$.*

*Proof* Here $T = \mathrm{PSp}_n(q)$, $H$ is of type $\mathrm{GL}_{n/2}(q).2$ and $q$ is odd. We may identify $\Omega$ with the set of decompositions $V = V_1 \oplus V_2$, where $V_1$ and $V_2$ are maximal totally isotropic subspaces of $V$. From [86, Proposition 4.2.5] we calculate that

$$|\Omega| = \frac{1}{2}q^{\frac{1}{8}n(n+2)}\prod_{k=1}^{n/2}(q^k+1)$$

Let $r$ be a prime divisor of $|\Omega|$ and let $x = \hat{x}Z \in T$ be an element of order $r$.

First assume $r = 2$. The involutions in $T$ are described in Section 3.4.2, and it is easy to see that $x$ fixes a decomposition in $\Omega$ componentwise, unless $n \equiv 2 \pmod 4$ and $x$ is of type $t'_{n/2}$. Suppose $n \equiv 2 \pmod 4$ and $x$ is of type $t'_{n/2}$, so $q \equiv 3 \pmod 4$. Let

$$\{e_1,\ldots,e_{(n-2)/4},f_1,\ldots,f_{(n-2)/4},x'\}$$

be an hermitian $\mathbb{F}_{q^2}$-basis for $V$ as in Construction 2.4.3. As explained in Section 3.4.2.5, $x$ interchanges the two totally isotropic $\mathbb{F}_q$-spaces in the decomposition $V = U \oplus W$, where

$$U = \langle e_1, \ldots, e_{(n-2)/4}, f_1, \ldots, f_{(n-2)/4}, x' \rangle$$
$$W = \langle \xi e_1, \ldots, \xi e_{(n-2)/4}, \xi^{-1} f_1, \ldots, \xi^{-1} f_{(n-2)/4}, \xi x' \rangle$$

and $\xi \in \mathbb{F}_{q^2}$ has order 4. We conclude that $T$ is 2-elusive.

Finally, suppose $r > 2$. Set $i = \Phi(r,q)$ and note that $i$ is even since $r$ divides $|\Omega|$. By Lemma 2.2.17, if $x$ has fixed points on $\Omega$ then every eigenvalue of $\hat{x}$ (on $V \otimes \overline{\mathbb{F}}_q$) has even multiplicity. Therefore, $[\Lambda, I_{n-i}]$ is a derangement. $\qquad \square$

**Proposition 5.2.31** *In Case VIII of Table 5.2.1, if $r \neq p$ then $T$ is $r$-elusive if and only if $r = 2$, $n \equiv 2$ (mod 4) and $q \equiv 1$ (mod 4).*

*Proof* Here $T = \mathrm{P}\Omega_n^+(q)$ and $H$ is of type $\mathrm{GL}_{n/2}(q).2$. We may identify $\Omega$ with a $T$-orbit on the set of decompositions of the form $V = V_1 \oplus V_2$, where $V_1$ and $V_2$ are maximal totally singular subspaces. Recall that if $n/2$ is even then there are two such orbits (see Proposition 2.5.4), whereas if $n/2$ is odd then there is a unique orbit and there is no element in $T$ that interchanges the spaces $V_1$ and $V_2$ in such a decomposition. Let $r$ be a prime divisor of $|\Omega|$ and let $x = \hat{x}Z \in T$ be an element of order $r$.

First assume $r > 2$ and set $i = \Phi(r,q)$. Note that $i$ is even since $r$ divides $\prod_{k=1}^{n/2-1}(q^k + 1)$ (see [86, Proposition 4.2.7]). In particular, by arguing as in the proof of Proposition 5.2.30 we deduce that $[\Lambda, I_{n-i}]$ is a derangement.

Now assume $r = 2$. Suppose $q \equiv 3$ (mod 4) and let $x \in T$ be an involution of type $t_1'$. Assume $x$ fixes a decomposition $V = V_1 \oplus V_2$ in $\Omega$. Since $n \geqslant 8$ and $x$ centralises a subspace of dimension $n - 2$, it follows that $x$ fixes $V_1$ and $V_2$. However, this contradicts the fact that the $(-1)$-eigenvectors of $\hat{x}$ are nonsingular, whence $x$ is a derangement.

Finally, suppose $r = 2$ and $q \equiv 1$ (mod 4). First assume that $n \equiv 0$ (mod 4). Let $x \in T$ be an involution of type $t_{n/2-1}$ or $t_{n/2}$, as described in Section 3.5.2.10. Since 1 is not an eigenvalue of $\hat{x}$, it can only fix a decomposition in $\Omega$ componentwise. By considering the construction of these involutions, it is clear that one of them fixes a decomposition in $\Omega$, while the other fixes a decomposition in the other $T$-orbit of decompositions. Therefore, one of these involutions is a derangement.

To complete the proof, we may assume that $r = 2$, $q \equiv 1$ (mod 4) and $n \equiv 2$ (mod 4). We claim that $T$ is 2-elusive. To see this, first observe that if $x = t_i \in T$

and $i < n/4$ then the two eigenspaces of $\hat{x}$ are hyperbolic, whence $x$ fixes an appropriate decomposition in $\Omega$. Finally, if $x$ is an involution of type $t_{n/2}$ then the description of these involutions in Section 3.5.2.12 clearly shows that $x$ fixes a decomposition in $\Omega$ componentwise. □

This completes the proof of Theorems 5.2.1, 5.2.3 and 5.2.5.

## 5.3 $\mathscr{C}_3$: Field extension subgroups

The subgroups in the collection $\mathscr{C}_3$ arise from field extensions of prime degree $k$, where $k$ divides the dimension $n$ of the natural $T$-module $V$. The cases we will consider in this section are listed in Table 5.3.1 (see [86, Table 4.3.A]). We refer the reader to [86, Section 4.3] for additional information on these subgroups.

**Theorem 5.3.1** *Let $G \leqslant \mathrm{Sym}(\Omega)$ be a primitive almost simple classical group with socle $T$ and point stabiliser $H \in \mathscr{C}_3$. Let $r$ be a prime divisor of $|\Omega|$. Then $T$ is $r$-elusive if and only if $(G, H, r)$ is one of the cases in Table 5.3.2.*

In the first row of Table 5.3.2, the integer $c = c(i)$ (with $i = \Phi(r, q)$) is defined in Definition 5.1.1.

### 5.3.1 Preliminaries

Let $G = \mathrm{GL}_n(q)$, where $q = p^f$ for a prime $p$ and positive integer $f$. Let $V$ be the natural $G$-module and set $L = \mathrm{GL}_{n/k}(q^k)$, where $k$ is a prime divisor of $n$.

Table 5.3.1 *The collection $\mathscr{C}_3$*

| Case | $T$ | Type of $H$ | Conditions |
|------|-----|-------------|------------|
| I | $\mathrm{PSL}_n^\varepsilon(q)$ | $\mathrm{GL}_{n/k}^\varepsilon(q^k)$ | $k$ odd if $\varepsilon = -$ |
| II | $\mathrm{PSp}_n(q)$ | $\mathrm{Sp}_{n/k}(q^k)$ | $n/k$ even |
| III | $\mathrm{P}\Omega_n^\varepsilon(q)$ | $\mathrm{O}_{n/k}^\varepsilon(q^k)$ | $n/k \geqslant 4$ even |
| IV | $\mathrm{P}\Omega_n^\varepsilon(q)$ | $\mathrm{O}_{n/k}(q^k)$ | $n/k \geqslant 3$ odd, $q$ odd |
| V | $\mathrm{PSp}_n(q)$ | $\mathrm{GU}_{n/2}(q)$ | $q$ odd |
| VI | $\mathrm{P}\Omega_n^\varepsilon(q)$ | $\mathrm{GU}_{n/2}(q)$ | $\varepsilon = (-)^{n/2}$ |

Table 5.3.2 *$r$-elusive $\mathscr{C}_3$-actions*

| Case | $r$ | Conditions |
|------|-----|------------|
| I | $k$ | $n = k > 3, k \neq p, c = n - 1$ |
|  | $3$ | $n = k = 3, q \equiv 4\varepsilon, 7\varepsilon, 8\varepsilon \pmod 9$ |
|  | $2$ | $k = 2, \begin{cases} n = 2 \\ 1 < (q-1)_2 \leqslant (n)_2 \end{cases}$ |
| II | $2$ | $(n, k) = (4, 2), q$ odd |
| III | $2$ | $k = 2, q$ odd |
| IV | $2$ | $k = 2$ |
| V | $2$ | always |
| VI | $2$ | $q \equiv 3 \pmod 4$, with $n \equiv 4 \pmod 8$ if $\varepsilon = +$ |

Let $V_\#$ be the natural $L$-module over $\mathbb{F}_{q^k}$, and consider the natural embedding of $L$ in $G$, with respect to a fixed $\mathbb{F}_{q^k}$-basis $\{v_1, \ldots, v_{n/k}\}$ for $V_\#$ (see Construction 2.1.4). Let $\phi$ denote the standard field automorphism of $L$ of order $k$ with respect to this basis (that is, $(a_{ij})^\phi = (a_{ij}^q)$ for $(a_{ij}) \in L$) and set

$$H = L.C \leqslant \Gamma L_{n/k}(q^k)$$

where $C = \langle \phi \rangle = C_k$. Note that $\phi$ acts as an $\mathbb{F}_q$-linear map on $V$, so we may view $H$ as a subgroup of $G$.

Let $r \neq p$ be a prime. Set $i = \Phi(r, q)$ and $i_0 = \Phi(r, q^k)$ (see (3.1.2)), and observe that

$$i_0 = \begin{cases} i/k & \text{if } k \text{ divides } i \\ i & \text{otherwise} \end{cases} \tag{5.3.1}$$

As in Section 3.1.1, let $\mathscr{S}_r$ be the set of $r$th roots of unity in $\mathbb{F}_{q^i}$ and let $\{\Lambda_1, \ldots, \Lambda_t\}$ be the distinct orbits of $\sigma_0$ on $\mathscr{S}_r$, where $\sigma_0$ is the permutation $\sigma_0 : \lambda \mapsto \lambda^{q^k}$. For each positive integer $\ell$, set

$$\Lambda_j^{q^\ell} = \{\lambda^{q^\ell} \mid \lambda \in \Lambda_j\}$$

For $1 \leqslant \ell < k$, note that $\Lambda_j^{q^\ell} = \Lambda_j$ if and only if $i = i_0$. In particular, if $\sigma$ is the permutation of $\mathscr{S}_r$ defined by $\sigma : \lambda \mapsto \lambda^q$, then either each $\sigma$-orbit is a union of $k$ distinct $\sigma_0$-orbits, or $i = i_0$ and $\sigma_0$-orbits and $\sigma$-orbits coincide.

In the next result we use the notation for semisimple and unipotent elements introduced in Sections 3.1.1 and 3.1.2.

**Lemma 5.3.2** *Let $x \in H$ be an element of prime order $r$.*

(i) *If $r = p$ and $x = [J_p^{a_p}, \ldots, J_1^{a_1}] \in L$, then $x$ has Jordan form $[J_p^{ka_p}, \ldots, J_1^{ka_1}]$ on $V$.*

(ii) *If $r \neq p$ and $x = [\Lambda_1^{a_1}, \ldots, \Lambda_t^{a_t}, I_e] \in L$, then, up to conjugacy,*

$$x = [\Lambda_1^{a_1}, (\Lambda_1^q)^{a_1}, \ldots, (\Lambda_1^{q^{k-1}})^{a_1}, \ldots, \Lambda_t^{a_t}, (\Lambda_t^q)^{a_t}, \ldots, (\Lambda_t^{q^{k-1}})^{a_t}, I_{ke}]$$

*on $V$. In particular, if $i = i_0$ then $x$ is conjugate to $[\Lambda_1^{ka_1}, \ldots, \Lambda_t^{ka_t}, I_{ke}]$.*

(iii) *If $x \in H \setminus L$, then $r = k$ and $x$ has characteristic polynomial $(\mathbf{x}^k - 1)^{n/k}$, so either $k = p$ and $x$ has Jordan form $[J_k^{n/k}]$ on $V$, or $k \neq p$ and every $k$th root of unity is an eigenvalue of $x$ (on $V \otimes \overline{\mathbb{F}}_q$) with multiplicity $n/k$.*

*Proof*  Parts (i) and (ii) are an easy exercise. Now consider (iii). Since $k$ is a prime number, $r = k$ is the only possibility. We may assume that $x \in L\phi$. The *Shintani correspondence* (see [56, Theorem 2.10], for example) gives a bijection from the set of $H$-classes in the coset $L\phi$ to the set of conjugacy classes in $\mathrm{GL}_{n/k}(q)$. Moreover, this bijection has the property that any $H$-class of elements of order $k$ in $L\phi$ maps to the trivial conjugacy class in $\mathrm{GL}_{n/k}(q)$. It follows that there is a unique $H$-class of elements of order $k$ in $L\phi$, so without loss of generality we may assume that $x = \phi$.

Fix an $\mathbb{F}_{q^k}$-basis $\{v_1, \ldots, v_{n/k}\}$ for $V_\#$ and let $\lambda \in \mathbb{F}_{q^k} \setminus \mathbb{F}_q$ be a scalar such that

$$\{\lambda v_1, \lambda^q v_1, \ldots, \lambda^{q^{k-1}} v_1, \ldots, \lambda v_{n/k}, \lambda^q v_{n/k}, \ldots, \lambda^{q^{k-1}} v_{n/k}\}$$

is an $\mathbb{F}_q$-basis for $V$ (such a scalar exists by the Normal Basis Theorem). In terms of this basis we calculate that $x = [A^{n/k}]$, where $A \in \mathrm{GL}_k(q)$ is a permutation matrix with characteristic polynomial $\mathbf{x}^k - 1$. The result follows. $\qquad\square$

For $x \in L$, we define $\nu(x)$ and $\nu_0(x)$ with respect to the natural modules for $G$ and $L$, respectively (see Definition 3.1.4). The following result will play a key role in our analysis of $\mathscr{C}_3$-actions.

**Lemma 5.3.3** *Let $x \in H$ be an element of prime order $r$.*

(i) *If $x \in L$ then $\nu(x) \geqslant k\nu_0(x)$, with equality if $r = p$.*
(ii) *Either $\nu(x) = 0$, or $\nu(x) \geqslant k - \delta_{n,k}$.*

*Proof* Part (i) follows from the proof of [97, Lemma 4.2] (alternatively, it also follows from Lemma 5.3.2).

Now consider (ii). First assume $x \in L$. The result is clear if $\nu_0(x) > 0$, so let us assume that $\nu_0(x) = 0$, in which case $x = [\lambda I_{n/k}]$ for some $\lambda \in \mathbb{F}_{q^k}^\times$. If $\nu(x) > 0$ then $\lambda \notin \mathbb{F}_q$ and thus

$$\{\lambda, \lambda^q, \ldots, \lambda^{q^{k-1}}\}$$

are the distinct eigenvalues of $x$ on $V \otimes \overline{\mathbb{F}}_q$, each occurring with multiplicity $n/k$. Therefore $\nu(x) = n(1 - 1/k)$ and the desired bound holds (note that if $n > k$ then $\nu(x) \geqslant 2k(1 - 1/k) \geqslant k$). Similarly, if $x \in H \setminus L$ then Lemma 5.3.2(iii) implies that $\nu(x) = n(1 - 1/k)$ and the result follows as before. $\qquad\square$

### 5.3.2 Subgroup structure

Let $G$ be an almost simple classical group with socle $T$ and let $H$ be a maximal $\mathscr{C}_3$-subgroup of $G$. Set $H_0 = H \cap T$. Detailed information on the structure of $H_0$ is provided in [86, Section 4.3], but in some cases we need to determine the precise structure. For easy reference, we do this here. Note that we freely use the notation for involutions defined in Section 3.5.2.

In the next two lemmas, $\phi$ denotes a standard involutory field automorphism of $\mathrm{P}\Omega_{n/2}^{\varepsilon}(q^2)$.

**Lemma 5.3.4** *Let* $T = \mathrm{P}\Omega_n^{\varepsilon}(q)$ *and let* $H$ *be a* $\mathscr{C}_3$-*subgroup of type* $\mathrm{O}_{n/2}^{\varepsilon}(q^2)$ *as in Case III of Table 5.3.1 (with* $k = 2$), *so* $n \equiv 0$ (mod 4).

(i) *If* $p = 2$ *then*

$$H_0 = \begin{cases} \mathrm{O}_{n/2}^{+}(q^2).\langle\phi\rangle & \varepsilon = + \\ \mathrm{O}_{n/2}^{-}(q^2) & \varepsilon = - \end{cases}$$

(ii) *If* $p > 2$ *then*

$$H_0 = \begin{cases} \langle \mathrm{P}\Omega_{n/2}^{+}(q^2), \gamma_1', \phi \rangle & \varepsilon = +,\ q^{n/4} \equiv 1 \ (\mathrm{mod}\ 4) \\ \langle \mathrm{P}\Omega_{n/2}^{+}(q^2), \gamma_1', t_1'\phi \rangle & \varepsilon = +,\ q^{n/4} \equiv 3 \ (\mathrm{mod}\ 4) \\ \mathrm{PO}_{n/2}^{-}(q^2) & \varepsilon = - \end{cases}$$

*In particular, if* $\varepsilon = +$ *then* $H_0$ *is a nonsplit extension of* $\mathrm{P}\Omega_{n/2}^{+}(q^2)$ *by* $C_2 \times C_2$.

*Proof*   Let $V$ and $V_{\#}$ be the natural modules for $T$ and $\mathrm{P}\Omega_{n/2}^{\varepsilon}(q^2)$, respectively.

First assume $p = 2$. Recall that $\mathrm{O}_{n/2}^{\varepsilon}(q^2) = \Omega_{n/2}^{\varepsilon}(q^2).\langle x \rangle$, where $x$ is a $b_1$-type involution (see Proposition 3.5.16). By Lemma 5.3.2(i), $x$ has Jordan form $[J_2^2, J_1^{n-4}]$ on $V$, so $x \in T$ and thus $\mathrm{O}_{n/2}^{\varepsilon}(q^2) \leqslant H_0$. The result now follows from [86, Propositions 4.3.14 and 4.3.16]. For the remainder, we may assume that $p > 2$.

First assume $\varepsilon = -$. Here $\mathrm{PSO}_{n/2}^{-}(q^2) = \mathrm{P}\Omega_{n/2}^{-}(q^2)$ (see Section 2.5) and $\mathrm{PO}_{n/2}^{-}(q^2) = \mathrm{PSO}_{n/2}^{-}(q^2).\langle x \rangle$, where $x$ is an involution of type $[-I_1, I_{n/2-1}]$ on $V_{\#}$. In the terminology of Section 3.5.2, $x$ is an involutory graph automorphism of type $\gamma_1$ or $\gamma_1'$. By [86, Proposition 4.3.16] we have

$$H_0 = \mathrm{P}\Omega_{n/2}^{-}(q^2).2 = \mathrm{PSO}_{n/2}^{-}(q^2).2$$

Now Lemma 5.3.2(ii) implies that the $(-1)$-eigenspace of $x$ on $V$ is a nondegenerate 2-space and thus $x$ embeds in $T$ as an involution of type $t_1$ or $t_1'$, both of which are in $T$ (see Sections 3.5.2.3 and 3.5.2.4). It follows that $H_0 = \mathrm{PO}_{n/2}^{-}(q^2)$ as claimed.

Now assume $\varepsilon = +$. Here [86, Proposition 4.3.14] gives

$$H_0 = N_T(\mathrm{P}\Omega_{n/2}^{+}(q^2)) = \mathrm{P}\Omega_{n/2}^{+}(q^2).[4] < \mathrm{Aut}(\mathrm{P}\Omega_{n/2}^{+}(q^2))$$

where [4] denotes an unspecified group of order 4. Let $\xi \in \mathbb{F}_{q^2}$ be an element of order $2(q-1)_2$ and fix a hyperbolic basis $\{e_1, \ldots, e_{n/4}, f_1, \ldots, f_{n/4}\}$ for $V_{\#}$ (with respect to a quadratic form $\overline{Q}$). As explained in Construction 2.5.13,

$$\left\{ e_1, \xi e_1, \ldots, e_{n/4}, \xi e_{n/4}, \frac{1}{2}f_1, \frac{1}{2}\xi^{-1}f_1, \ldots, \frac{1}{2}f_{n/4}, \frac{1}{2}\xi^{-1}f_{n/4} \right\}$$

is a hyperbolic basis for $V$ over $\mathbb{F}_q$ (with respect to the quadratic form $Q = \text{Tr} \circ \overline{Q}$ on $V$, where $\text{Tr} : \mathbb{F}_{q^2} \to \mathbb{F}_q$ is the trace map $\lambda \mapsto \lambda + \lambda^q$).

First observe that $H_0$ contains some (but not all) involutory graph automorphisms of $\text{P}\Omega_{n/2}^+(q^2)$. For instance, let $\gamma_1$ and $\gamma_1'$ be representatives of the two $\text{P}\Omega_{n/2}^+(q^2)$-classes of graph automorphisms of type $[-I_1, I_{n/2-1}]$ (here we may assume that the $(-1)$-eigenspace of $\gamma_1$ is $\langle v \rangle$, where $\overline{Q}(v) = \square$). By appealing to Table 2.5.1 we see that $\gamma_1$ is a $t_1'$ involution if $q \equiv 1 \pmod 4$, and a $t_1$ involution if $q \equiv 3 \pmod 4$. However, $t_1 \in T$ if and only if $q \equiv 1 \pmod 4$, and similarly $t_1' \in T$ if and only if $q \equiv 3 \pmod 4$, so $\gamma_1 \notin H_0$. On the other hand, $\gamma_1'$ is a $t_1$ involution if $q \equiv 1 \pmod 4$, and a $t_1'$ involution if $q \equiv 3 \pmod 4$, so $\gamma_1' \in H_0$. Similarly, if $\phi \in \text{Aut}(\text{P}\Omega_{n/2}^+(q^2))$ is the standard involutory field automorphism then an easy calculation using the set-up in Construction 2.5.13 reveals that $\phi$ is a $t_{n/4}$ involution, which belongs to $T$ if and only if $q^{n/4} \equiv 1 \pmod 4$. In particular, $\phi \in H_0$ if and only if $q^{n/4} \equiv 1 \pmod 4$. One can check that $[\phi, \gamma_1'] \neq 1$.

The above observations show that if $q^{n/4} \equiv 1 \pmod 4$ then

$$H_0 = \langle \text{P}\Omega_{n/2}^+(q^2), \gamma_1', \phi \rangle$$

is a nonsplit extension of $\text{P}\Omega_{n/2}^+(q^2)$ by $C_2 \times C_2$.

Now assume $q^{n/4} \equiv 3 \pmod 4$, so $n \equiv 4 \pmod 8$ and $q \equiv 3 \pmod 4$. Let $x \in \text{PSO}_{n/2}^+(q^2)$ be an involution of type $t_1'$. Then $x$ embeds in $\text{PSO}_n^+(q) \setminus T$ as a $t_2'$-type involution (see Construction 2.5.13), and we also note that $\phi \in \text{PSO}_n^+(q) \setminus T$. Therefore $x\phi \in T$ and we may assume that $[x, \gamma_1'] = 1$, so in this situation

$$H_0 = \langle \text{P}\Omega_{n/2}^+(q^2), \gamma_1', x\phi \rangle$$

is a nonsplit extension of $\text{P}\Omega_{n/2}^+(q^2)$ by $C_2 \times C_2$.                    □

**Lemma 5.3.5** *Let $T = \text{P}\Omega_n^\varepsilon(q)$ and let $H$ be a $\mathcal{C}_3$-subgroup of type $\text{O}_{n/2}(q^2)$ as in Case IV of Table 5.3.1 (with $k = 2$), so $n \geqslant 10$ and $n \equiv 2 \pmod 4$. Then*

$$H_0 = \begin{cases} \text{SO}_{n/2}(q^2) & q \equiv -\varepsilon \pmod 4 \\ \Omega_{n/2}(q^2).\langle \phi \rangle & q \equiv \varepsilon \pmod 8 \\ \langle \Omega_{n/2}(q^2), t_1'\phi \rangle & \text{otherwise} \end{cases}$$

*Proof* As in the previous lemma, let $V$ and $V_\#$ be the natural modules for $T$ and $\Omega_{n/2}(q^2)$, respectively. Similarly, let $Q$ and $\overline{Q}$ be the respective quadratic forms on $V$ and $V_\#$, with corresponding symmetric bilinear forms $B$ and $\overline{B}$. As in Construction 2.5.13, we have $Q = \mathrm{Tr} \circ \overline{Q}$ and $B = \mathrm{Tr} \circ \overline{B}$. By [86, Proposition 4.3.20], $H_0$ is an almost simple group with structure

$$H_0 = \Omega_{n/2}(q^2).2$$

Let $x \in \mathrm{SO}_{n/2}(q^2) \setminus \Omega_{n/2}(q^2)$ be an involution of type $t_1'$ (see Section 3.5.2.2). The $(-1)$-eigenspace of $x$ is a 2-dimensional elliptic space, so $x$ embeds in $\mathrm{PSO}_n^\varepsilon(q)$ as an involution of type $t_2'$ (see Construction 2.5.13). These involutions belong to $T$ if and only if $q \equiv -\varepsilon \pmod 4$, whence $H_0 = \mathrm{SO}_{n/2}(q^2)$ if $q \equiv -\varepsilon \pmod 4$.

For the remainder we may assume that $q \equiv \varepsilon \pmod 4$, so $x \in \mathrm{PSO}_n^\varepsilon(q) \setminus T$. Here $D(Q) = \square$ and $D(\overline{Q}) = \boxtimes$ (see Construction 2.5.13), so we can fix an orthogonal basis $\{v_1, \ldots, v_{n/2}\}$ for $\overline{V}$ so that $\overline{B}(v_i, v_j) = \lambda \delta_{i,j}$, where $\mathbb{F}_{q^2}^\times = \langle \lambda \rangle$. Let $\phi$ be the standard involutory field automorphism of $\Omega_{n/2}(q^2)$ with respect to this basis, so

$$(\mu v_i)^\phi = \mu^q \lambda^{(q-1)/2} v_i$$

for all $\mu \in \mathbb{F}_{q^2}$ (see [86, (4.3.27)]). In terms of the $\mathbb{F}_q$-basis

$$\{v_1, \lambda^{(q-1)/2} v_1, \ldots, v_{n/2}, \lambda^{(q-1)/2} v_{n/2}\}$$

for $V$, $\phi$ is represented by the block-diagonal matrix $[A^{n/2}]$, where

$$A = \begin{pmatrix} 0 & 1 \\ -1 & 0 \end{pmatrix}$$

Since this matrix has determinant 1, it follows that the element $\phi$ embeds in $\mathrm{PSO}_n^\varepsilon(q)$ as an involution of type $t_{n/2}$. In particular, if $q \equiv \varepsilon \pmod 8$ then $\phi \in T$ (see Sections 3.5.2.12 and 3.5.2.13), so $H_0 = \Omega_{n/2}(q^2).\langle \phi \rangle$ in this situation. On the other hand, if $q \equiv (4+\varepsilon) \pmod 8$ then $\phi \in \mathrm{PSO}_n^\varepsilon(q) \setminus T$, so $x\phi \in T$ and thus $H_0 = \langle \Omega_{n/2}(q^2), x\phi \rangle$ as required. $\square$

In the statement of the next lemma, if $n \equiv 0 \pmod 4$ we define

$$J = \begin{cases} \mathrm{PGU}_{n/2}(q) & p = 2 \\ \frac{1}{2}\mathrm{PGU}_{n/2}(q) & p > 2 \end{cases} \tag{5.3.2}$$

where $\frac{1}{2}\mathrm{PGU}_{n/2}(q)$ is the unique index-two subgroup of $\mathrm{PGU}_{n/2}(q)$. Note that if $p > 2$ and $x = \hat{x}Z \in \mathrm{PGU}_{n/2}(q)$ (where $Z = Z(\mathrm{GU}_{n/2}(q))$) then

$$x = \hat{x}Z \in \tfrac{1}{2}\mathrm{PGU}_{n/2}(q) \text{ if and only if there exists a scalar} \atop \lambda \in \mathbb{F}_{q^2} \text{ such that } \lambda^{q+1} = 1 \text{ and } \det(\hat{x}) = \lambda^2. \tag{5.3.3}$$

In addition, let $\varphi$ be an involutory graph automorphism of $\mathrm{PSU}_{n/2}(q)$ (arising from the involutory field automorphism of $\mathbb{F}_{q^2}$).

**Lemma 5.3.6** *Let $T = P\Omega_n^\varepsilon(q)$ and let $H$ be a $\mathcal{C}_3$-subgroup of type $\mathrm{GU}_{n/2}(q)$ as in Case VI in Table 5.3.1, so $\varepsilon = (-)^{n/2}$. Then*

$$H_0 = \begin{cases} C_a.J.\langle\varphi\rangle & \varepsilon = + \\ C_a.\mathrm{PGU}_{n/2}(q) & \varepsilon = - \end{cases}$$

*where $a = (q+1)/(3-\varepsilon, q+1)$.*

*Proof* Consider the natural embeddings

$$\mathrm{GU}_{n/2}(q) < \mathrm{GL}_{n/2}(q^2) < \mathrm{GL}_n(q)$$

and observe that $\Gamma\mathrm{U}_{n/2}(q) \cap \mathrm{GL}_n(q) = \Delta\mathrm{U}_{n/2}(q).\langle\varphi\rangle$. Now [86, Proposition 4.3.18] gives

$$H_0 = C_a.\mathrm{PSU}_{n/2}(q).[\ell(n/2, q+1)] \tag{5.3.4}$$

where $\ell = 2$ if $\varepsilon = +$ and $p = 2$, otherwise $\ell = 1$. In particular, we have $H_0 \leqslant C_a.\mathrm{PGU}_{n/2}(q).\langle\varphi\rangle$ (recall that $P\Delta\mathrm{U}_{n/2}(q) = \mathrm{PGU}_{n/2}(q)$). Also note that $\varphi \in H_0$ if and only if $\varepsilon = +$ (see Remark 2.5.15).

First assume that $\varepsilon = +$. If $q$ is even, then by comparing orders we deduce that $H_0 = C_a.\mathrm{PGU}_{n/2}(q).\langle\varphi\rangle$ as claimed. If $q$ is odd then $C_a.J.\langle\varphi\rangle$ is the only index-two subgroup of $C_a.\mathrm{PGU}_{n/2}(q).\langle\varphi\rangle$ that contains $C_a$ and $\varphi$, and again the result follows.

Finally, suppose that $\varepsilon = -$, so $n/2$ is odd and $\varphi \notin H_0$. If $q$ is even, then the result follows from (5.3.4) by comparing orders. Similarly, if $q$ is odd we use the fact that $C_a.\mathrm{PGU}_{n/2}(q)$ is the only index-two subgroup of $C_a.\mathrm{PGU}_{n/2}(q).\langle\varphi\rangle$ that does not contain $\varphi$. $\square$

### 5.3.3 Proof of Theorem 5.3.1

**Proposition 5.3.7** *Theorem 5.3.1 holds in Case I of Table 5.3.1.*

*Proof* Here $T = \mathrm{PSL}_n^\varepsilon(q)$ and $H$ is of type $\mathrm{GL}_{n/k}^\varepsilon(q^k)$, where $k$ is a prime (and $k$ is odd if $\varepsilon = -$). According to [86, Proposition 4.3.6] we have

$$H_0 = H \cap T = L.C$$

where

$$L = C_a.\mathrm{PSL}_{n/k}^\varepsilon(q^k).C_b, \quad C = \langle\psi\rangle = C_k$$

and

$$a = \frac{(n/k, q-\varepsilon)(q^k-\varepsilon)}{(n, q-\varepsilon)(q-\varepsilon)}, \quad b = \frac{(n/k, q^k-\varepsilon)}{(n/k, q-\varepsilon)} \tag{5.3.5}$$

Here $\psi$ acts on $\mathrm{PSL}_{n/k}^{\varepsilon}(q^k)$ as a field automorphism of order $k$, and $L$ is the image (modulo scalars) of $\mathrm{SL}_{n/k}^{\varepsilon}(q^k)Z \cap \mathrm{SL}_n^{\varepsilon}(q)$, where $Z$ is the centre of $\mathrm{GL}_n^{\varepsilon}(q)$. More precisely, we may assume that $\psi$ is a standard field automorphism of order $k$ (with respect to a suitable basis for $V$ over $\mathbb{F}_{q^k}$), unless $k = 2$, $n \equiv 2 \pmod 4$ and $q \equiv 3 \pmod 4$, in which case $\psi$ is an appropriate scalar multiple of a standard involutory field automorphism.

Let $x = \hat{x}Z \in T$ be an element of order $r$, where $r$ is a prime divisor of $|\Omega|$.

*Case 1.* $r = p > 2$. First assume $r = p > 2$. We claim that $\hat{x} = [J_2, J_1^{n-2}]$ is a derangement. Since $\nu(x) = 1$, this is clear if $n > k$ or $n = k \geqslant 3$ (see Lemma 5.3.3). If $n = k = 2$ then $|H_0|$ is indivisible by $p$ and thus $x$ is a derangement.

*Case 2.* $r \neq p$, $r > 2$. Next suppose $r \neq p$ and $r > 2$. Set $i = \Phi(r,q)$ and $i_0 = \Phi(r,q^k)$ (see (3.1.2)). As noted in (5.3.1), we have

$$i_0 = \begin{cases} i/k & \text{if } k \text{ divides } i \\ i & \text{otherwise} \end{cases}$$

We claim that $k$ does not divide $i$. To see this, suppose $\varepsilon = +$ and $k$ divides $i$. Since $r$ divides $|\Omega|$, $r$ must also divide

$$N := \frac{(q^i-1)(q^{2i}-1)\cdots(q^{mi}-1)}{(q^{i_0k}-1)(q^{2i_0k}-1)\cdots(q^{m'i_0k}-1)}$$

where $n = mi+t$ and $n/k = m'i_0+t'$ with $0 \leqslant t < i$ and $0 \leqslant t' < i_0$. But $i = i_0k$, so $m = m'$ and $N = 1$, which is a contradiction. This justifies the claim for $\varepsilon = +$, and a very similar argument applies when $\varepsilon = -$. It follows that $i = i_0$. In addition, we may assume that $n \geqslant 3$ since $r$ does not divide $|H_0|$ if $n = 2$. Define the integer $c = c(i,\varepsilon)$ as follows (this agrees with Definition 5.1.1):

$$c(i,+) = i, \quad c(i,-) = \begin{cases} i & i \equiv 0 \pmod 4 \\ i/2 & i \equiv 2 \pmod 4 \\ 2i & i \text{ odd} \end{cases} \tag{5.3.6}$$

Let $y = \hat{y}Z \in H_0$ be an element of order $r$.

First assume $r \neq k$, so $y \in L$. If $\hat{y}$ has order $r$, then Lemma 5.3.2(ii) implies that $k$ divides the multiplicity of each eigenvalue of $\hat{y}$ on $\overline{V} = V \otimes \overline{\mathbb{F}}_q$ (this follows from the fact that $i = i_0$). On the other hand, if $y$ does not lift to an element of order $r$ in $\mathrm{GL}_n^{\varepsilon}(q)$, then $c = 1$, $r$ divides $n$ and each eigenvalue of $\hat{y}$ on $\overline{V}$ has multiplicity $n/r \geqslant 2$ (see Sections 3.2.1 and 3.3.1). In particular, it is easy to see that $T$ contains derangements of order $r$ if $c > 1$. For example, if $\varepsilon = -$ and $i \not\equiv 2 \pmod 4$ then $[\Lambda, \Lambda^{-q}, I_{n-c}]$ is a derangement. Similarly, if $c = 1$ then $[\Lambda, \Lambda^{-1}, I_{n-2}]$ is a derangement.

Now assume that $r = k > 2$ and $r \neq p$. As in the previous case, if $y \in L$ then either $\hat{y}$ has order $k$ and the multiplicity of each eigenvalue is divisible by $k$, or $c = 1$ and each eigenvalue has multiplicity $n/k$ (and in addition, $C_V(\hat{y})$ is trivial). Similarly, if $y \in H_0 \setminus L$ then by applying Lemma 5.3.2(iii) we deduce that each eigenvalue of $\hat{y}$ on $\overline{V}$ has multiplicity $n/k$. Therefore, if $n/k \geqslant 2$ then it is easy to find derangements of order $r$ as before. For instance, if $\varepsilon = +$ and $i > 1$, then $[\Lambda, I_{n-i}]$ is a derangement.

Finally, let us assume that $n = r = k > 2$. If $1 < c < n - 1$, or if $c = 1$ and $n > 3$, then we can find an element $x \in T$ of order $r$ such that $\dim C_V(\hat{x}) \geqslant 2$ and $\hat{x}$ has an eigenvalue (on $\overline{V}$) with multiplicity 1. From the above description of the elements of order $r$ in $H_0$, it follows that $x$ is a derangement. Now assume $c = 1$ and $n = 3$. If $q \equiv \varepsilon \pmod 9$ then the previous argument implies that $[\Lambda, I_2]$ is a derangement in $T$. However, if $q \equiv 4\varepsilon, 7\varepsilon \pmod 9$ then $T$ has a unique class of elements of order 3 and we quickly deduce that $T$ is 3-elusive. Similarly, if $c = n - 1$ then $T$ has a unique class of elements of order $r$ and once again $T$ is $r$-elusive. These cases are recorded in Table 5.3.2. (Note that if $n = r = k = 3$ and $c = 2$, then $r$ divides $|\Omega|$ if and only if $q \equiv 8\varepsilon \pmod 9$.)

*Case 3.* $r = 2$. For the remainder we may assume that $r = 2$. Suppose $k > 2$. As before, if $p = 2$ then $[J_2, J_1^{n-2}]$ is a derangement. Similarly, if $p \neq 2$ then $[-I_2, I_{n-2}]$ is a derangement. Now assume $(r, k) = (2, 2)$, so $\varepsilon = +$. If $n = 2$ then $T$ has a unique class of involutions and thus $T$ is 2-elusive. Therefore we may assume $n \geqslant 4$. If $p = 2$ then $[J_2, J_1^{n-2}]$ is a derangement (see Lemma 5.3.3), so let us assume $p \neq 2$. We adopt the notation for semisimple involutions introduced in Sections 3.2.2 and 3.3.2.

If $(q - 1)_2 > (n)_2$ then $x = t_1 \in T$ is a derangement since $v(x) = 1$, so let us assume $(q - 1)_2 \leqslant (n)_2$. We claim that $T$ is 2-elusive. To see this, let $x \in T$ be an involution. First assume that $v(x) < n/2$, in which case $v(x)$ is even (see Section 3.2.2). Observe that either the integer $b$ in (5.3.5) is even, or $(q^2 - 1)_2 > (n/2)_2$. Therefore, from our analysis of semisimple involutions in Section 3.2.2, we deduce that $H_0$ contains an involution of type $[-I_\ell, I_{n/2-\ell}]$ for all $1 \leqslant \ell \leqslant n/4$ (with respect to the natural $\frac{n}{2}$-dimensional module), which embeds in $T$ as an involution of type $t_{2\ell}$. We conclude that all involutions $x \in T$ with $v(x) < n/2$ have fixed points.

Finally, let us assume that $1 < (q - 1)_2 \leqslant (n)_2$ and $x \in T$ is an involution with $v(x) = n/2$. Note that the involutory field automorphism $\psi \in H_0$ has the property $v(\psi) = n/2$, so we may as well assume $T$ contains involutions of type $t_{n/2}$ and $t'_{n/2}$, in which case $n \equiv 0 \pmod 4$. Now $q^2 \equiv 1 \pmod 4$, so $H_0$ contains an involution of type $[-I_{n/4}, I_{n/4}]$, which embeds in $T$ as an involution of type $t_{n/2}$. Finally, if $x = t'_{n/2} \in T$ then either $q \equiv 3 \pmod 4$ or $(q-1)_2 < (n)_2$ (see Section 3.2.2.2), so the integer $a$ in (5.3.5) is even and thus $x$ has fixed points since the central involution in $L$ is $T$-conjugate to $x$. $\qquad\square$

**Proposition 5.3.8** *Theorem 5.3.1 holds in Case II of Table 5.3.1.*

*Proof* Here $T = \mathrm{PSp}_n(q)$ and [86, Proposition 4.3.10] implies that

$$H_0 = L.C \leqslant \mathrm{P\Gamma Sp}_{n/k}(q^k)$$

where

$$L = \mathrm{PSp}_{n/k}(q^k), \quad C = \langle \phi \rangle = C_k$$

and $\phi$ is a standard field automorphism of $L$ of order $k$. Let $r$ be a prime divisor of $|\Omega|$. If $r = p$ then Lemma 5.3.3 implies that $[J_2, J_1^{n-2}]$ is a derangement of order $r$, so for the remainder we may assume that $r \neq p$.

Suppose $r > 2$. Define the integers $i$ and $i_0$ as in the proof of the previous proposition, and note that (5.3.1) holds. If $i = k = 2$ then Lemma 5.3.2(ii) implies that $[\Lambda, I_{n-2}]$ is a derangement. In the remaining cases, it is easy to check that $i = i_0$ since $r$ divides $|\Omega|$. Therefore, if we set $\hat{x} = [\Lambda, I_{n-i}]$ if $i$ is even, and $\hat{x} = [\Lambda, \Lambda^{-1}, I_{n-2i}]$ if $i$ is odd, then once again $x$ is a derangement by Lemma 5.3.2.

For the remainder we may assume that $r = 2$ and $p > 2$. If $n > 4$ then $t_1$-type involutions in $T$ are derangements, so we may assume that $(n,k) = (4,2)$. We claim that $T$ is 2-elusive.

Let $\{e_1, f_1\}$ be a symplectic basis for the natural $\mathrm{Sp}_2(q^2)$-module, with respect to a nondegenerate alternating form $\overline{B}$. Let $\xi \in \mathbb{F}_{q^2}$ be an element of order $2(q-1)_2$. Then $\{e_1, \xi e_1, \frac{1}{2}f_1, \frac{1}{2}\xi^{-1}f_1\}$ is a symplectic basis for $V$ with respect to the form $B = \mathrm{Tr} \circ \overline{B}$ on $V$ (see Construction 2.4.2). It is easy to see that the field automorphism $\phi \in H_0$ embeds in $T$ as an involution of type $t_1$.

Let $y$ be a representative of the unique class of involutions in $\mathrm{PSp}_2(q^2)$, where $\hat{y} = [\lambda, \lambda^{-1}]$ and $\lambda \in \mathbb{F}_{q^2}$ has order 4. If $q \equiv 1 \pmod 4$ then $\hat{y}$ acts on $V$ as $[\lambda I_2, \lambda^{-1}I_2]$, whence $y$ is an involution of type $t_2$. On the other hand, if $q \equiv 3 \pmod 4$ then we may assume $\xi = \lambda$ and thus $y$ acts on $V$ as $[A, -A^{\mathrm{T}}]$, where

$$A = \begin{pmatrix} 0 & 1 \\ \xi^2 & 0 \end{pmatrix} \qquad (5.3.7)$$

Therefore, $y$ is a $t'_2$-involution (see Section 3.4.2.5) and we conclude that $T$ is 2-elusive. $\qquad\square$

**Proposition 5.3.9** *Theorem 5.3.1 holds in Case III of Table 5.3.1.*

*Proof* Here $T = \mathrm{P}\Omega_n^\varepsilon(q)$ and $H$ is of type $\mathrm{O}_{n/k}^\varepsilon(q^k)$, where $n/k \geqslant 4$ is even. Set $H_0 = H \cap T$. We refer the reader to Lemma 5.3.4 for the precise structure of $H_0$ when $k = 2$. Let $r$ be a prime divisor of $|\Omega|$.

First assume $r$ is odd. Lemma 5.3.2 implies that $[J_3, J_1^{n-3}]$ is a derangement of order $r = p$, so we may assume that $r \neq p$. Here we define the integers $i$ and $i_0$ as in (5.3.1), and we proceed as in the proof of Proposition 5.3.8. The reader can check that $T$ always contains derangements of order $r$.

Now assume $r = 2$. We adopt the notation for involutions introduced in Sections 3.5.2 (for $p \neq 2$) and 3.5.4 (for $p = 2$). If $p = 2$ then there are two $T$-classes of involutions $x \in T$ with $v(x) = 2$ (represented by the elements $a_2$ and $c_2$), which are clearly derangements if $k > 2$. Similarly, if $k = 2$ then there is a unique class of involutions $y \in H_0$ with $v_0(y) = 1$, so once again we can find derangements in $T$ of order $2$. More precisely, we claim that $a_2$-type involutions are derangements.

To see this, it suffices to show that the $b_1$-type involutions in $H_0$ embed in $T$ as involutions of type $c_2$. Let $V_\#$ be the natural module for $\Omega_{n/2}^\varepsilon(q^2)$ and let $\overline{Q}$ be the corresponding quadratic form on $V_\#$, with associated bilinear form $\overline{B}$. As indicated in [86, Table 4.3.A], we may assume that $Q = \mathrm{Tr} \circ \overline{Q}$ is the defining quadratic form on $V$, in which case $B = \mathrm{Tr} \circ \overline{B}$ is the associated bilinear form. Let $\{e_1, f_1\}$ be a standard basis of a hyperbolic 2-space in $V_\#$ and let $x \in H_0$ be the involution that interchanges $e_1$ and $f_1$, and centralises $\langle e_1, f_1 \rangle^\perp$ (note that $x$ is a $b_1$-type involution). Fix an element $\xi \in \mathbb{F}_{q^2} \setminus \mathbb{F}_q$, so $\mathrm{Tr}(\xi) \neq 0$. As an element of $T$, we see that $x$ interchanges $e_1, f_1$ and $\xi e_1, \xi f_1$, and it centralises the subspace $\langle e_1, f_1, \xi e_1, \xi f_1 \rangle^\perp$ of $V$. Now

$$B(\xi e_1, \xi e_1 x) = B(\xi e_1, \xi f_1) = \mathrm{Tr}(\overline{B}(\xi e_1, \xi f_1)) = \mathrm{Tr}(\xi^2) = \mathrm{Tr}(\xi)^2 \neq 0$$

and thus $x$ is a $c_2$-type involution (see Section 3.5.4). This justifies the claim.

For the remainder we may assume that $r = 2$ and $p > 2$. If $k \geqslant 3$ then any involution $x \in T$ with $v(x) = 2$ is a derangement, so we reduce to the case $(r, k) = (2, 2)$. Here $n \equiv 0 \pmod 4$ and the precise structure of $H_0$ is given in Lemma 5.3.4. We claim that $T$ is 2-elusive. Fix an element $\xi \in \mathbb{F}_{q^2}$ of order $2(q-1)_2$.

First suppose $\varepsilon = -$, so $H_0 = \mathrm{PO}_{n/2}^-(q^2)$ by Lemma 5.3.4. In terms of the notation in Section 3.5.2, the involution classes in $T$ are represented by the elements

$$\{t_i, t_i', t_{n/4} \mid 1 \leqslant i < n/4\}$$

In particular, there is a unique $T$-class with $v(x) = n/2$, so these involutions have fixed points (note that there are always involutions $y \in H_0$ with $v_0(y) = n/4$, and for any such involution we have $v(y) = n/2$).

Similarly, by appealing to Table 2.5.1 in Construction 2.5.13, we see that $t_i$ and $t'_i$ also have fixed points for all $1 \leqslant i < n/4$. For example, observe that $H_0$ has two classes of graph automorphisms of type $[-I_1, I_{n/2-1}]$, labelled $\gamma_1$ and $\gamma'_1$, say. More precisely, if $\langle v \rangle$ and $\langle v' \rangle$ are the respective $(-1)$-eigenspaces (over $\mathbb{F}_{q^2}$), and $\overline{Q}$ is the relevant quadratic form on the natural $\mathrm{P\Omega}^-_{n/2}(q^2)$-module, then $\overline{Q}(v) = \square$ and $\overline{Q}(v') = \boxtimes$, respectively. (Here it is worth noting that $\gamma_1$ and $\gamma'_1$ are conjugate in $\mathrm{Inndiag}(\mathrm{P\Omega}^-_{n/2}(q^2))$, but not in $\mathrm{PO}^-_{n/2}(q^2)$, since the 1-spaces $\langle v \rangle$ and $\langle v' \rangle$ are similar, but not isometric.) According to Table 2.5.1, the $\mathbb{F}_q$-space $\langle v, \xi v \rangle$ is elliptic, in which case $\gamma_1$ is $T$-conjugate to $t_1$, whereas $\langle v', \xi v' \rangle$ is hyperbolic and thus $\gamma'_1$ is a $t'_1$ involution. It follows that $t_1$ and $t'_1$ have fixed points. Similarly, to see that $t_2$ and $t'_2$ have fixed points, first note that $H_0$ contains both classes of involutions of type $[-I_2, I_{n/2-2}]$. As observed in Construction 2.5.13, if $x \in H_0$ is such an involution, with a hyperbolic $(-1)$-eigenspace, then $x$ is $T$-conjugate to an involution of type $t_2$ in $T$, otherwise $x$ is of type $t'_2$. The other cases with $3 \leqslant i < n/4$ are entirely similar, and we conclude that $T$ is 2-elusive.

For the remainder, we may assume that $k = 2$ and $\varepsilon = +$. Now $n \equiv 0 \pmod{4}$ and Lemma 5.3.4 gives

$$H_0 = \begin{cases} \langle \mathrm{P\Omega}^+_{n/2}(q^2), \gamma'_1, \phi \rangle & q^{n/4} \equiv 1 \pmod{4} \\ \langle \mathrm{P\Omega}^+_{n/2}(q^2), \gamma'_1, t'_1\phi \rangle & q^{n/4} \equiv 3 \pmod{4} \end{cases}$$

which is a nonsplit extension of $\mathrm{P\Omega}^+_{n/2}(q^2)$ by $C_2 \times C_2$. Here we recall that $\gamma'_1$ denotes an involutory graph automorphism of $\mathrm{P\Omega}^+_{n/2}(q^2)$ of type $[-I_1, I_{n/2-1}]$, such that the $(-1)$-eigenspace $\langle v \rangle$ has the property $\overline{Q}(v) = \boxtimes$.

Let $x \in T$ be an involution, so $v(x) = 2i$ for some $i \leqslant n/4$. If $i < n/4$ then $T$ has a unique class of such involutions, and they clearly have fixed points, so we may assume that $v(x) = n/2$. Here there are several cases to consider, according to various congruences of $n$ and $q$.

It will be helpful to fix some notation. As before, let $V_\#$ be the natural module for $\mathrm{P\Omega}^+_{n/2}(q^2)$ (over $\mathbb{F}_{q^2}$) and let $\overline{Q}$ be the underlying quadratic form on $V_\#$, with associated bilinear form $\overline{B}$. As in Construction 2.5.13, we may assume that $Q = \mathrm{Tr} \circ \overline{Q}$ is the quadratic form on $V$ defining $T$, which has associated bilinear form $B = \mathrm{Tr} \circ \overline{B}$. Let

$$\{e_1, \ldots, e_{n/4}, f_1, \ldots, f_{n/4}\}$$

be a hyperbolic basis for $V_\#$ and note that

$$\left\{ e_1, \xi e_1, \ldots, e_{n/4}, \xi e_{n/4}, \frac{1}{2}f_1, \frac{1}{2}\xi^{-1}f_1, \ldots, \frac{1}{2}f_{n/4}, \frac{1}{2}\xi^{-1}f_{n/4} \right\}$$

is a hyperbolic basis for $V$, where once again $\xi \in \mathbb{F}_{q^2}$ has order $2(q-1)_2$ (see Construction 2.5.13). Using this basis, it is easy to see that the standard involutory field automorphism $\phi \in \mathrm{Aut}(\mathrm{P}\Omega_{n/2}^+(q^2))$ embeds in $\mathrm{PSO}_n^+(q)$ as an involution of type $t_{n/4}$.

*Case 1.* $q \equiv 1 \pmod 4$. Here $t_{n/4}, t_{n/2}$ and $t_{n/2-1}$ are the relevant involutions in $T$. As remarked above, the element $\phi \in H_0$ is an involution of type $t_{n/4}$. Fix $\lambda \in \mathbb{F}_q$ of order 4. In terms of the hyperbolic bases given above, the involution $x = [\lambda I_{n/4}, \lambda^{-1}I_{n/4}] \in H_0$ (which is of type $t_{n/4}$ or $t_{n/4-1}$ as an element of $H_0$, in the notation of Section 3.5.2) embeds in $T$ as $[\lambda I_{n/2}, \lambda^{-1}I_{n/2}]$, whence $x$ is a $t_{n/2}$ (or $t_{n/2-1}$) involution. Note that

$$U = \langle e_1, \xi e_1, \ldots, e_{n/4}, \xi e_{n/4} \rangle$$

is the $\lambda$-eigenspace of $x$. Since $H_0$ contains involutory graph automorphisms, the two $\mathrm{Inndiag}(\mathrm{P}\Omega_{n/2}^+(q^2))$-classes represented by $t_{n/4}$ and $t_{n/4-1}$ are fused in $H_0$, and hence in $T$. Therefore, in order to deduce that $T$ is 2-elusive we need to find an involution $y \in H_0$ of type $t_{n/2}$ or $t_{n/2-1}$, that is not $T$-conjugate to $x$. We will use the description of these involutions in Section 3.5.2.10 in order to distinguish the two classes; if $W$ denotes the $\lambda$-eigenspace of $y$ (as a subspace of $V$), then $x$ and $y$ are $T$-conjugate if and only if $\dim(U \cap W)$ is even. We consider two subcases, according to the value of $n$ modulo 8.

*Subcase 1(a).* $q \equiv 1 \pmod 4$, $n \equiv 0 \pmod 8$. As above, let $\xi \in \mathbb{F}_{q^2}$ be an element of order $2(q-1)_2$. Note that $\xi$ is an $\mathbb{F}_{q^2}$-nonsquare, so we can take $\gamma_1' \in H_0$ to be the involution whose $(-1)$-eigenspace is $\langle e_1 + \xi f_1 \rangle$. Now

$$W = \langle e_1 + \xi f_1, \xi e_1 - \xi^2 f_1, e_2, \ldots, e_{n/4}, \xi f_2, \ldots, \xi f_{n/4} \rangle$$

is the $\lambda$-eigenspace of $x\gamma_1'\phi \in H_0$ on $V$, which is a maximal totally singular subspace (with respect to the quadratic form $Q$), so this element is a $t_{n/2}$ or $t_{n/2-1}$ involution in $T$. Since $\dim(U \cap W) = n/4 - 1$ is odd, we deduce that $x\gamma_1'\phi$ is not $T$-conjugate to $x$.

*Subcase 1(b).* $q \equiv 1 \pmod 4$, $n \equiv 4 \pmod 8$. Order the above hyperbolic basis as follows:

$$\left\{ e_1, \ldots, e_{n/4}, \frac{1}{2}f_1, \ldots, \frac{1}{2}f_{n/4}, \xi e_1, \ldots, \xi e_{n/4}, \frac{1}{2}\xi^{-1}f_1, \ldots, \frac{1}{2}\xi^{-1}f_{n/4} \right\}$$

and observe that $\phi$ has matrix $[I_{n/2}, -I_{n/2}]$ with respect to this basis. Let $\lambda \in \mathbb{F}_q$ be an element of order 4 and let $y$ be the diagonal matrix

$$y = [C, -C] \in \mathrm{SO}_{n/2}^{+}(q^2)$$

with respect to the basis $\{e_1, \ldots, e_{n/4}, f_1, \ldots, f_{n/4}\}$, where $C = [-\lambda, \lambda I_{n/4-1}]$. Then Lemma 2.5.7 implies that $y \in \Omega_{n/2}^{+}(q^2)$ (since $\det(y) = -(\lambda^{n/4})$ is an $\mathbb{F}_{q^2}$-square), so we may view $y$ as an element of $H_0$. In terms of the above $\mathbb{F}_q$-basis for $V$, $y$ is the diagonal matrix $[C, -C, C, -C]$, so the product $y\phi \in H_0$ has matrix $[C, -C, -C, C]$ and thus $y\phi$ embeds in $T$ as an involution of type $t_{n/2}$ or $t_{n/2-1}$. The $\lambda$-eigenspace of $y\phi$ is

$$W = \langle f_1, e_2, \ldots, e_{n/4}, \xi e_1, \xi f_2, \ldots, \xi f_{n/4} \rangle$$

so $\dim(U \cap W) = n/4$ is odd and thus $x$ and $y\phi$ are not $T$-conjugate.

*Case 2.* $q \equiv 3 \pmod 4$. Here the relevant involutions in $T$ are $t'_{n/2}$, $t'_{n/2-1}$ and either $t_{n/4}$ (if $n \equiv 0 \pmod 8$) or $t'_{n/4}$ (if $n \equiv 4 \pmod 8$). Also note that $\xi$ has order 4, so $\xi^2 = -1$. Let $x = \hat{x}Z \in \mathrm{PSO}_{n/2}^{+}(q^2)$, where $\hat{x} = [\xi I_{n/4}, -\xi I_{n/4}]$. Since $\xi$ is an $\mathbb{F}_{q^2}$-square, Lemma 2.5.7 implies that $\hat{x} \in \Omega_{n/2}^{+}(q^2)$. As an element of $T$, we see that $x$ fixes the following decomposition

$$V = \langle e_1, \xi e_1, \ldots, e_{n/4}, \xi e_{n/4} \rangle \oplus \langle f_1, \xi^{-1} f_1, \ldots, f_{n/4}, \xi^{-1} f_{n/4} \rangle$$

of $V$ into maximal totally singular subspaces, inducing the block-diagonal matrix $[A^{n/4}]$ on both summands, where $A = -A^{\mathsf{T}}$ is the matrix given in (5.3.7). Therefore, $x \in T$ is an involution of type $t'_{n/2}$ or $t'_{n/2-1}$.

As before, our main aim is to find another involution in $H_0$ of this type, which is not $T$-conjugate to $x$. We will use the description of these involutions given in Section 3.5.2.11 in order to distinguish the two relevant $T$-classes. More precisely, it suffices to find an involution in $H_0$ that has no eigenvalues in $\mathbb{F}_q$, and which fixes a maximal totally singular subspace $W$ of $V$ such that $\dim(U \cap W)$ is odd, where

$$U = \langle e_1, \xi e_1, \ldots, e_{n/4}, \xi e_{n/4} \rangle$$

as before. Once again, we consider two subcases.

*Subcase 2(a).* $q \equiv 3 \pmod 4$, $n \equiv 0 \pmod 8$. As previously noted, the field automorphism $\phi \in H_0$ is an involution of type $t_{n/4}$. Therefore, to show that $T$ is 2-elusive we need to find an involution in $H_0$ of type $t'_{n/2}$ or $t'_{n/2-1}$, which is not $T$-conjugate to $x$.

To get started, fix $\mu \in \mathbb{F}_{q^2}$ of order $2(q+1)_2$ and note that $\mu$ is an $\mathbb{F}_{q^2}$-nonsquare. Thus we can take $\gamma_1' \in H_0$ to be the element

$$
\begin{pmatrix}
0 & & -\mu & \\
 & I_{n/4-1} & & \\
\hline
-\mu^{-1} & & 0 & \\
 & & & I_{n/4-1}
\end{pmatrix}
$$

with respect to the $\mathbb{F}_{q^2}$-basis $\{e_1, \ldots, e_{n/4}, f_1, \ldots, f_{n/4}\}$. Similarly, set

$$
y = 
\begin{pmatrix}
 & & 1 & \\
 & & & \xi I_{n/4-1} \\
\hline
1 & & & \\
 & -\xi I_{n/4-1} & &
\end{pmatrix}
$$

By applying Lemmas 2.5.7 and 2.5.8, using the fact that $n \equiv 0 \pmod 8$ and $\pm \xi$ are $\mathbb{F}_{q^2}$-squares, we deduce that $y \in \Omega_{n/2}^{+}(q^2)$. Then $y\gamma_1' \phi \in H_0$ is an involution with no eigenvalues in $\mathbb{F}_q$, which fixes the following decomposition

$$
V = \langle e_1, \xi e_1, e_2, \xi f_2, \ldots, e_{n/4}, \xi f_{n/4} \rangle \oplus \langle f_1, \xi f_1, f_2, \xi e_2, \ldots, f_{n/4}, \xi e_{n/4} \rangle
$$

of $V$ into two maximal totally singular subspaces. Since

$$
\dim(U \cap \langle e_1, \xi e_1, e_2, \xi f_2, \ldots, e_{n/4}, \xi f_{n/4} \rangle) = n/4 + 1
$$

is odd, we conclude that $y\gamma_1' \phi$ is not $T$-conjugate to $x$. Therefore, $T$ is 2-elusive.

*Subcase 2(b).* $q \equiv 3 \pmod 4$, $n \equiv 4 \pmod 8$. Here $\phi \notin H_0$, in contrast to the previous three cases. Let $\gamma \in \mathrm{Aut}(\mathrm{P\Omega}_{n/2}^{+}(q^2))$ be a graph automorphism of type $[-I_{n/4}, I_{n/4}]$ such that both eigenspaces are nonsquare-parabolic. From the discussion in Case 2 of Construction 2.5.13, we deduce that $\gamma$ has elliptic eigenspaces on $V$, so it embeds in $T$ as an involution of type $t_{n/4}'$. Therefore $\gamma \in H_0$ and thus $t_{n/4}'$-type involutions have fixed points. As before, to conclude that $T$ is 2-elusive we need to find an involution in $H_0$ of type $t_{n/2}'$ or $t_{n/2-1}'$, that is not $T$-conjugate to $x$.

Let $\gamma_1 \in \mathrm{Aut}(\mathrm{P\Omega}_{n/2}^{+}(q^2))$ be a graph automorphism of the form $[-I_1, I_{n/2-1}]$, where the $(-1)$-eigenspace $\langle v \rangle$ has the property $\overline{Q}(v) = \square$. More precisely, in terms of the hyperbolic basis $\{e_1, \ldots, e_{n/4}, f_1, \ldots, f_{n/4}\}$, we may assume that $\gamma_1$ interchanges $e_1$ and $-f_1$, and fixes $e_i, f_i$ for all $i \geqslant 2$ (so $v = e_1 + f_1$). As explained in Construction 2.5.13 (see Table 2.5.1), $\gamma_1$ embeds in $\mathrm{PSO}_n^{+}(q)$ as an involution of type $t_1$. Therefore, $\gamma_1, \phi \in \mathrm{PSO}_n^+(q) \setminus \mathrm{P\Omega}_n^+(q)$, so their product $\gamma_1 \phi$ is in $H_0$.

Let $A$ be the matrix defined in (5.3.7) and set

$$
y = \left(
\begin{array}{c|c}
\begin{array}{c} \xi \\ (A^{\mathsf{T}})^{(n-4)/8} \end{array} & \\
\hline
& \begin{array}{c} -\xi \\ (A^{\mathsf{T}})^{(n-4)/8} \end{array}
\end{array}
\right)
$$

with respect to the $\mathbb{F}_{q^2}$-basis $\{e_1, \ldots, e_{n/4}, f_1, \ldots, f_{n/4}\}$. Note that $y \in \Omega_{n/2}^+(q^2)$ since $\xi$ is an $\mathbb{F}_{q^2}$-square (see Lemma 2.5.7). Therefore $y\gamma_1\phi \in H_0$ is an involution with no eigenvalues in $\mathbb{F}_q$. Moreover, this element fixes the following decomposition

$$
V = \langle e_1, \xi f_1, e_2, \xi e_2, \ldots, e_{n/4}, \xi e_{n/4} \rangle \oplus \langle \xi e_1, f_1, f_2, \xi f_2, \ldots, f_{n/4}, \xi f_{n/4} \rangle
$$

of $V$ into maximal totally singular subspaces. Here

$$
\dim(U \cap \langle e_1, \xi f_1, e_2, \xi e_2, \ldots, e_{n/4}, \xi e_{n/4} \rangle) = n/2 - 1
$$

is odd and thus $y\gamma_1\phi$ is not $T$-conjugate to $x$. The result follows.

This completes the proof of Proposition 5.3.9.                    □

**Proposition 5.3.10** *Theorem 5.3.1 holds in Case IV of Table 5.3.1.*

*Proof*  Here $T = \mathrm{P\Omega}_n^\varepsilon(q)$ and $H$ is of type $\mathrm{O}_{n/k}(q^k)$, where $n/k \geqslant 3$ is odd and $q$ is odd. First assume $k \geqslant 3$, so $n$ is odd. Here $\nu(x) \geqslant k \geqslant 3$ for all $x \in H_0$ of prime order (see Lemma 5.3.3), so any involution in $T$ of type $[-I_{n-1}, I_1]$ is a derangement. Similarly, any element with Jordan form $[J_2^2, J_1^{n-4}]$ is a derangement of order $p$. Finally, let $r$ be a prime divisor of $|\Omega|$ such that $r \neq p$ and $r > 2$. Define $i$ and $i_0$ as before, and note that $i = i_0$ since $r$ divides $|\Omega|$. Set $\hat{x} = [\Lambda, I_{n-i}]$ if $i$ is even, and $\hat{x} = [\Lambda, \Lambda^{-1}, I_{n-2i}]$ if $i$ is odd. Then $x = \hat{x}Z \in T$ is a derangement of order $r$.

For the remainder, let us assume $k = 2$, in which case the precise structure of $H_0$ is given in Lemma 5.3.5. By arguing as above, we can quickly reduce to the case $r = 2$. We claim that $T$ is 2-elusive. To see this, let $V_\#$ be the natural module for $\Omega_{n/2}(q^2)$ and let $x \in T$ be an involution.

If $\nu(x) = n/2$ then $q \equiv \varepsilon \pmod 8$ and $x$ is of type $t_{n/2}$ (see Tables B.10 and B.11, for example). Since $H_0 = \Omega_{n/2}(q^2).\langle \phi \rangle$ and $\nu(\phi) = n/2$, we conclude that $x$ has fixed points.

Now assume $\nu(x) < n/2$. As explained in Construction 2.5.13, any hyperbolic (respectively, elliptic) $m$-space in $V_\#$ corresponds to a hyperbolic (respectively, elliptic) $2m$-space in $V$. This easily allows us to find representatives in

$H_0$ of each relevant $T$-class of involutions, and we ask the reader to check the details. For instance, if $\varepsilon = +$ and $q \equiv 3 \pmod 4$ then $H_0 = \mathrm{SO}_{n/2}(q^2)$ and the involutions in $H_0$ of type $t_{(n-2)/4}$ and $t'_{(n-2)/4}$ embed in $T$ as involutions of type $t_1$ and $t'_1$, respectively. The result follows. $\qquad\square$

In order to complete the proof of Theorem 5.3.1, we need to deal with the cases labelled V and VI in Table 5.3.1.

Let $G = \mathrm{GL}_n(q)$ and $H = \mathrm{GU}_{n/2}(q)$. Consider the natural embeddings

$$H < \mathrm{GL}_{n/2}(q^2) < G$$

Let $V$ and $V_{\#}$ be the natural modules for $G$ and $H$, respectively. Let $r \neq p$ be an odd prime and set $i = \Phi(r,q)$ and $i_0 = \Phi(r,q^2)$. Note that (5.3.1) holds (with $k = 2$). As before, let $\mathscr{S}_r$ be the set of $r$th roots of unity in $\mathbb{F}_{q^i}$, and let $\sigma$ and $\sigma_0$ be the permutations of $\mathscr{S}_r$ defined by $\sigma : \lambda \mapsto \lambda^q$ and $\sigma_0 : \lambda \mapsto \lambda^{q^2}$. Set $t = (r-1)/i$, $s = (r-1)/i_0$ and let $\{\Lambda_1,\ldots,\Lambda_s\}$ and $\{\Gamma_1,\ldots,\Gamma_t\}$ be the respective $\sigma_0$-orbits and $\sigma$-orbits on $\mathscr{S}_r$.

As noted in Section 3.3.1, if $i \not\equiv 2 \pmod 4$ then $s$ is even and we may label the $\Lambda_j$ so that $\Lambda_j^{-q} = \Lambda_{s/2+j}$ for all $1 \leqslant j \leqslant s/2$ (where $\Lambda_j^{-q} = \{\lambda^{-q} \mid \lambda \in \Lambda_j\}$). In particular, if $i \equiv 0 \pmod 4$ then we may assume that $\Gamma_j = \Lambda_j \cup \Lambda_j^{-q}$ for all $1 \leqslant j \leqslant s/2$. Similarly, if $i$ is odd then we will assume that $\Gamma_j = \Lambda_j$ and $\Gamma_{s/2+j} = \Lambda_j^{-q}$ for all $1 \leqslant j \leqslant s/2$.

The following lemma will be useful (here we use the notation in Propositions 3.2.1 and 3.3.2 for semisimple elements).

**Lemma 5.3.11** *Let $x \in H$ be an element of prime order $r$, and assume that $i \not\equiv 2 \pmod 4$ if $r \neq p$.*

(i) *If $r = p$ and $x = [J_p^{a_p}, \ldots, J_1^{a_1}]$ on $V_{\#}$, then $x$ has Jordan form $[J_p^{2a_p}, \ldots, J_1^{2a_1}]$ on $V$.*

(ii) *If $r \neq p$ and*

$$x = [(\Lambda_1, \Lambda_1^{-q})^{a_1}, \ldots, (\Lambda_{s/2}, \Lambda_{s/2}^{-q})^{a_{s/2}}, I_e]$$

*on $V_{\#}$, then*

$$x = \begin{cases} [\Gamma_1^{2a_1}, \ldots, \Gamma_{s/2}^{2a_{s/2}}, I_{2e}] & i \equiv 0 \pmod 4 \\ [\Gamma_1^{2a_1}, \ldots, \Gamma_{s/2}^{2a_{s/2}}, \Gamma_{s/2+1}^{2a_1}, \ldots, \Gamma_s^{2a_{s/2}}, I_{2e}] & i \text{ odd} \end{cases}$$

*on $V$ (up to conjugacy).*

*Proof* Part (i) follows immediately from Lemma 5.3.2(i). Similarly, part (ii) follows from Lemma 5.3.2(ii), given the relationship between the $\Lambda_j$ and $\Gamma_j$ described above. □

**Proposition 5.3.12** *Theorem 5.3.1 holds in Case V of Table 5.3.1.*

*Proof* Here $T = \mathrm{PSp}_n(q)$ and $H$ is of type $\mathrm{GU}_{n/2}(q)$, where $q$ is odd. By [86, Proposition 4.3.7] we have

$$H_0 = C_{(q+1)/2}.\mathrm{PGU}_{n/2}(q).\langle \psi \rangle$$

where $\psi$ is an appropriate $\mathbb{F}_{q^2}$-scalar multiple of the involutory automorphism of $\mathrm{PSU}_{n/2}(q)$ induced by the $q$-power map on $\mathbb{F}_{q^2}$ (see (2.2.6)). We refer the reader to Construction 2.4.3 for specific details on the embedding of $\mathrm{GU}_{n/2}(q)$ in $\mathrm{Sp}_n(q)$. Let $r$ be a prime divisor of $|\Omega|$.

First assume $r$ is odd. If $r = p$ then Lemma 5.3.11(i) implies that $[J_2, J_1^{n-2}]$ is a derangement of order $r$. Now assume $r \neq p$. Set $i = \Phi(r, q)$ and observe that $i \not\equiv 2 \pmod 4$ since $r$ divides $|\Omega|$. Let $x = \hat{x}Z \in T$ be an element of order $r$, where $\hat{x} = [\Lambda, I_{n-i}]$ if $i \equiv 0 \pmod 4$, and $\hat{x} = [\Lambda, \Lambda^{-1}, I_{n-2i}]$ if $i$ is odd. Then $x$ is a derangement by Lemma 5.3.11(ii).

Now assume $r = 2$. We claim that $T$ is 2-elusive. Let $\xi \in \mathbb{F}_{q^2}$ be an element of order $2(q-1)_2$. As in Construction 2.4.3, let $\{v_1, \ldots, v_{n/2}\}$ be an orthonormal basis for the underlying $\frac{n}{2}$-dimensional hermitian space over $\mathbb{F}_{q^2}$, corresponding to an hermitian form $\overline{B}$. Then

$$\left\{ v_1, \ldots, v_{n/2}, -\frac{1}{2}\xi^{-1}v_1, \ldots, -\frac{1}{2}\xi^{-1}v_{n/2} \right\}$$

is a symplectic basis for $V$, with respect to the alternating form $B = \mathrm{Tr} \circ (\xi \overline{B})$. In particular, if $x \in \mathrm{PGU}_{n/2}(q)$ is an involution of type $[-I_i, I_{n/2-i}]$, then $x$ embeds in $T$ as a $t_i$ involution, so all involutions $y \in T$ with $\nu(y) < n/2$ have fixed points. It remains to show that the involutions $x \in T$ with $\nu(x) = n/2$ also have fixed points.

If $n \equiv 2 \pmod 4$ then $T$ has a unique class of involutions $x$ with $\nu(x) = n/2$, which have fixed points since $\psi \in H_0$ is an involution with $\nu(\psi) = n/2$. Now assume $n \equiv 0 \pmod 4$. Here there are two such $T$-classes, represented by $t_{n/4}$, and either $t_{n/2}$ (if $q \equiv 1 \pmod 4$) or $t'_{n/2}$ (if $q \equiv 3 \pmod 4$). Clearly, any involution in $H_0$ of type $[-I_{n/4}, I_{n/4}]$ embeds in $T$ as an involution of type $t_{n/4}$. Similarly, if $q \equiv 3 \pmod 4$ and $x \in H_0$ is the central involution in $C_{(q+1)/2}.\mathrm{PGU}_{n/2}(q)$, then $x$ is an involution of type $t'_{n/2}$ in $T$. Finally, if $q \equiv 1 \pmod 4$ then it is easy to check that the eigenvalues of $\psi$ on $V$ are in $\mathbb{F}_q$ and the corresponding eigenspaces are totally isotropic, so $\psi$ embeds in $T$ as an involution of type $t_{n/2}$. This justifies the claim, and we conclude that $T$ is 2-elusive. $\qquad\Box$

**Proposition 5.3.13** *Theorem 5.3.1 holds in Case VI of Table 5.3.1.*

*Proof* Here $T = \mathrm{P}\Omega_n^\varepsilon(q)$ and $H$ is of type $\mathrm{GU}_{n/2}(q)$, where $\varepsilon = (-)^{n/2}$ (see Table 5.3.1). We refer the reader to Construction 2.5.14 for details of the embedding of $\mathrm{GU}_{n/2}(q)$ in $\mathrm{O}_n^\varepsilon(q)$.

Let $V$ be the natural $T$-module and let $\overline{B} : V \times V \to \mathbb{F}_{q^2}$ be a nondegenerate hermitian form on $V$ (where we view $V$ as a $\frac{n}{2}$-dimensional space over $\mathbb{F}_{q^2}$). Let $\{v_1, \ldots, v_{n/2}\}$ be an orthonormal basis for $V$ over $\mathbb{F}_{q^2}$ with respect to $\overline{B}$. Then

$$Q(v) = \overline{B}(v, v) \tag{5.3.8}$$

defines a nondegenerate quadratic form of type $\varepsilon$ on the $\mathbb{F}_q$-space $V$, with associated symmetric bilinear form $B = \mathrm{Tr} \circ \overline{B}$.

Set $H_0 = H \cap T$ and recall that the precise structure of $H_0$ is given in Lemma 5.3.6 (with $J$ defined in (5.3.2)). Let $r$ be a prime divisor of $|\Omega|$.

In the usual manner, we reduce to the case $r = 2$. For instance, if $r = p > 2$ then Lemma 5.3.11(i) implies that $[J_3, J_1^{n-3}]$ is a derangement of order $r$.

Let us consider the case $r = p = 2$. As explained in Section 3.5.4, there are precisely two $T$-classes of involutions in $T$ with Jordan form $[J_2^2, J_1^{n-4}]$, represented by the elements $a_2$ and $c_2$. However, there is a unique $H_0$-class of involutions with Jordan form $[J_2, J_1^{n/2-2}]$ on the natural $\mathrm{PSU}_{n/2}(q)$-module, which implies that $T$ contains derangements of order 2. More precisely, we claim that $c_2$-type involutions are derangements.

To see this, let $\{e_1, f_1\}$ be an hermitian basis of a nondegenerate 2-space in $V$ (over $\mathbb{F}_{q^2}$) and let $x \in H_0$ be the involution that interchanges $e_1$ and $f_1$, and centralises $\langle e_1, f_1 \rangle^\perp$. Let $\mu \in \mathbb{F}_{q^2}$ be an element of order $q+1$ and observe that $x$ (as an element of $T$) interchanges $e_1, f_1$ and $\mu e_1, \mu f_1$, and it centralises the $\mathbb{F}_q$-space $\langle e_1, f_1, \mu e_1, \mu f_1 \rangle^\perp$. Therefore, $x$ has Jordan form $[J_2^2, J_1^{n-4}]$ on $V$, and it is straightforward to verify that $x$ is an $a_2$-type involution, which justifies the claim. For instance,

$$B(e_1, e_1 x) = B(e_1, f_1) = \mathrm{Tr}(\overline{B}(e_1, f_1)) = \mathrm{Tr}(1) = 0$$

and

$$B(\mu e_1, \mu e_1 x) = B(\mu e_1, \mu f_1) = \mathrm{Tr}(\overline{B}(\mu e_1, \mu f_1)) = \mathrm{Tr}(\mu^{q+1} \overline{B}(e_1, f_1))$$
$$= \mathrm{Tr}(1)$$
$$= 0$$

For the remainder, we may assume that $r = 2$ and $p > 2$. Let $\xi \in \mathbb{F}_{q^2}$ be an element of order $2(q-1)_2$, so

$$\{v_1, \xi v_1, v_2, \xi v_2, \dots, v_{n/2}, \xi v_{n/2}\} \tag{5.3.9}$$

is an orthogonal $\mathbb{F}_q$-basis for $V$.

*Case 1.* $\varepsilon = +$. First assume that $\varepsilon = +$, so $n \equiv 0 \pmod 4$ and $H_0 = C_{(q+1)/2}.J.\langle \varphi \rangle$, where $J = \frac{1}{2}\mathrm{PGU}_{n/2}(q)$ and $\varphi$ is the involutory graph automorphism of $\mathrm{PSU}_{n/2}(q)$ induced from the $q$-power map on $\mathbb{F}_{q^2}$.

*Subcase 1(a).* $\varepsilon = +$, $q \equiv 1 \pmod 4$. Here there is no element $\lambda \in \mathbb{F}_{q^2}$ such that $\lambda^{q+1} = 1$ and $\lambda^2 = -1$. Therefore, (5.3.3) indicates that $J$ does not contain an involution of type $[-I_1, I_{n/2-1}]$, and thus every $t_1$ involution in $T$ is a derangement.

*Subcase 1(b).* $\varepsilon = +$, $q \equiv 3 \pmod 4$. We claim that $T$ is 2-elusive if and only if $n \equiv 4 \pmod 8$. To see this, first observe that if $i < n/4$ then there is a unique $T$-class of involutions $x \in T$ with $\nu(x) = 2i$, and all of these elements have fixed points (note that $q+1$ is divisible by 4, so $H_0$ contains involutions of type $[-I_i, I_{n/2-i}]$ with $i$ odd).

Now let $x \in T$ be an involution with $\nu(x) = n/2$, so $x = t_{n/4}$ (or $t'_{n/4}$ if $n \equiv 4 \pmod 8$)), $t'_{n/2}$ or $t'_{n/2-1}$. Note that elements of type $t_{n/4}$ or $t'_{n/4}$ are genuine involutions, in the sense that they lift to involutions in $\Omega_n^+(q)$, while those of type $t'_{n/2}$ and $t'_{n/2-1}$ lift to elements of order 4 (see Sections 3.5.2.5, 3.5.2.6 and 3.5.2.11).

Set $m = n/2$. One can check that there are precisely seven $H_0$-classes of involutions $y \in H_0$ with $\nu(y) = n/2$, which are represented by the following elements:

$$\{z, t_{m/2}, \tau, \tau', zt_{m/2}, z\tau, z\tau'\} \tag{5.3.10}$$

Here $z$ is the central involution, and we take $\tau$ and $\tau'$ to be symplectic and non-symplectic graph automorphisms of $\mathrm{PSU}_m(q)$, respectively (see Remark 3.2.12). Note that elements in $\mathrm{PGU}_m(q)$ of type $t'_{m/2}$ correspond to elements of order 4 in $H_0$. Our aim is to determine the $T$-class of each of the involutions in (5.3.10).

Clearly, if $x \in H_0$ is an involution of type $t_{m/2}$ (that is, $x = [-I_{m/2}, I_{m/2}]$ with nondegenerate eigenspaces) then $x$ embeds in $T$ as an involution with an $m$-dimensional nondegenerate $(-1)$-eigenspace. The isometry type of this space is determined in Construction 2.5.14, and we deduce that $x$ is $T$-conjugate to $t_{n/4}$ if $n/4$ is even, and to $t'_{n/4}$ if $n/4$ is odd.

As explained in Section 3.3.5, we can take

$$
\tau = \begin{pmatrix} \begin{matrix} 0 & 1 \\ -1 & 0 \end{matrix} & & & \\ & \ddots & & \\ & & \begin{matrix} 0 & 1 \\ -1 & 0 \end{matrix} & \end{pmatrix} \varphi, \quad \tau' = \begin{pmatrix} \begin{matrix} 0 & 1 \\ 1 & 0 \end{matrix} & & & \\ & \ddots & & \\ & & \begin{matrix} 0 & 1 \\ 1 & 0 \end{matrix} & \end{pmatrix} \varphi
$$

with respect to an orthonormal basis $\{v_1, \ldots, v_{n/2}\}$ for the hermitian space, where $\varphi$ is the $q$-power map as before. Since $q \equiv 3 \pmod 4$ it follows that $\xi^{q+1} = 1$ and thus $\tau'$ and $z\tau'$ are both genuine involutions in $T$. Therefore, the $H_0$-classes represented by $t_{m/2}, \tau'$ and $z\tau'$ are fused in $T$.

By definition, the central involution $z$ embeds in $T$ as an involution of type $t'_{n/2}$ or $t'_{n/2-1}$ (see Section 3.5.2.11). With respect to the orthogonal $\mathbb{F}_q$-basis in (5.3.9) we have $z = [C^{n/4}]$ where

$$
C = \begin{pmatrix} \begin{matrix} 0 & 1 \\ -1 & 0 \end{matrix} & \\ & \begin{matrix} 0 & 1 \\ -1 & 0 \end{matrix} \end{pmatrix}
$$

Similarly, $\tau = [D^{n/4}]$ and $z\tau = [E^{n/4}]$, where

$$
D = \begin{pmatrix} & \begin{matrix} 1 & 0 \\ 0 & -1 \end{matrix} \\ \begin{matrix} -1 & 0 \\ 0 & 1 \end{matrix} & \end{pmatrix}, \quad E = \begin{pmatrix} & \begin{matrix} 0 & -1 \\ -1 & 0 \end{matrix} \\ \begin{matrix} 0 & 1 \\ 1 & 0 \end{matrix} & \end{pmatrix}
$$

Finally, if we take $\langle v_2, v_4, \ldots, v_{n/2} \rangle$ to be the $(-1)$-eigenspace of an involution in $H_0$ of type $t_{m/2}$, then $zt_{m/2} = [F^{n/4}]$ with

$$
F = \begin{pmatrix} \begin{matrix} 0 & 1 \\ -1 & 0 \end{matrix} & \\ & \begin{matrix} 0 & -1 \\ 1 & 0 \end{matrix} \end{pmatrix}
$$

It follows that the four involutions $z, \tau, z\tau$ and $zt_{m/2}$ are of type $t'_{n/2}$ or $t'_{n/2-1}$ in $T$. We need to determine whether or not all these involutions are $T$-conjugate.

Let $Q$ be the quadratic form on $V$ given in (5.3.8), with associated bilinear form $B$. Since $Q(v_i) = Q(\xi v_i) = 1$, the stabiliser in $O_n^+(q)$ of the orthogonal decomposition

$$
V = \langle v_1 \rangle \perp \langle \xi v_1 \rangle \perp \ldots \perp \langle v_{n/2} \rangle \perp \langle \xi v_{n/2} \rangle
$$

is a $\mathscr{C}_2$-subgroup of type $O_1(q) \wr S_n$ (see Case III in Table 5.2.1).

In the subgroup $O_1(q) \wr S_4 < O_4^+(q)$, we calculate that $C$ is conjugate to $D$, $E$ and $F$ under the permutations $(2,3,4)$, $(1,3,4)$ and $(3,4)$, respectively. Now any transposition in $S_4$ is a reflection of the form $r_u$ for some nonsingular vector $u$ with $B(u,u) = 4$. Hence, every even permutation in $S_4$ has square spinor norm and thus lies in $\Omega_4^+(q)$ (see Lemma 2.5.5). Therefore, $(3,4) \in O_4^+(q) \setminus SO_4^+(q)$, while $(2,3,4) = (3,4)(2,3)$ and $(1,3,4)$ are in $\Omega_4^+(q)$. Thus $z$, $\tau$ and $z\tau$ are $T$-conjugate. By Lemma 2.5.9(iii), if $g \in O_{n/2}^+(q)$ then the block-diagonal matrix $[g,g]$ lies in $\Omega_n^+(q)$, hence $z, \tau, z\tau$ and $zt_{m/2}$ are all $T$-conjugate if $n \equiv 0 \pmod 8$. In particular, in this case $T$ contains a derangement of type $t'_{n/2}$ or $t'_{n/2-1}$. However, if $n \equiv 4 \pmod 8$ then $z$ is conjugate to $zt_{m/2}$ by an element of $O_n^+(q) \setminus SO_n^+(q)$. Such an element must interchange the two $T$-classes $t'_{n/2}$ and $t'_{n/2-1}$, and so $z$ and $zt_{m/2}$ belong to distinct $T$-classes. We conclude that $T$ is 2-elusive in this case.

*Case 2.* $\varepsilon = -$. To complete the proof of the proposition, let us assume $\varepsilon = -$, so $n \equiv 2 \pmod 4$. If $q \equiv 1 \pmod 4$ and $i < n/4$ then $T$ contains two distinct classes of involutions $x$ with $v(x) = 2i$ (labelled $t_i$ and $t'_i$). However, there is a unique class of involutions $y \in H_0$ with $v_0(y) = i$ (with respect to the natural $PSU_{n/2}(q)$-module), so in this situation $T$ contains derangements of order two. Now assume $q \equiv 3 \pmod 4$. It is clear that all involutions $x \in T$ with $v(x) < n/2$ have fixed points (there is a unique class with $v(x) = 2i < n/2$), so let us assume $v(x) = n/2$. If $q \equiv 3 \pmod 8$ then there are no such involutions in $T$, so in this case $T$ is 2-elusive. On the other hand, if $q \equiv 7 \pmod 8$ then there is a unique such $T$-class. These involutions have fixed points since $(q+1)/4$ is even and $v(z) = n/2$ for the central involution $z \in H_0$. We conclude that if $\varepsilon = -$ then $T$ is 2-elusive if and only if $q \equiv 3 \pmod 4$. $\qquad\square$

This completes the proof of Theorem 5.3.1.

## 5.4 $\mathscr{C}_4$: Tensor product subgroups, I

The subgroups in this collection are the stabilisers of appropriate tensor product decompositions $V = V_1 \otimes V_2$ of the natural module $V$. The specific cases that arise are listed in Table 5.4.1, where $\dim V = n = ab$ (see [86, Tables 3.5.H, 4.4.A]).

**Theorem 5.4.1** *Let $G \leqslant \mathrm{Sym}(\Omega)$ be a primitive almost simple classical group with socle $T$ and point stabiliser $H \in \mathscr{C}_4$. Let $r$ be a prime divisor of $|\Omega|$. Then $T$ is $r$-elusive if and only if $r = 2$ and $(G, H)$ is one of the cases in Table 5.4.2.*

In the proof of Theorem 5.4.1, we will frequently appeal to the following lemma, which is [97, Lemma 3.7]. Here $x = x_1 \otimes x_2$ is an element of the central product $\mathrm{GL}_a(q) \otimes \mathrm{GL}_b(q) < \mathrm{GL}_{ab}(q)$, and the parameter $\nu(x)$ is defined in Definition 3.1.4 (with respect to the natural $\mathrm{GL}_{ab}(q)$-module). Note that $\nu(x_1)$ and $\nu(x_2)$ are defined in terms of the respective natural modules for $\mathrm{GL}_a(q)$ and $\mathrm{GL}_b(q)$.

**Lemma 5.4.2** *Let $x = x_1 \otimes x_2 \in \mathrm{GL}_a(q) \otimes \mathrm{GL}_b(q)$ be an element of prime order, where $a \leqslant b$. Then*

$$\nu(x) \geqslant \max\{a\nu(x_2), b\nu(x_1)\}$$

*In particular, if $x$ is a non-scalar then $\nu(x) \geqslant a$.*

Table 5.4.1 *The collection $\mathscr{C}_4$*

| Case | $T$ | Type of $H$ | Conditions |
|---|---|---|---|
| I | $\mathrm{PSL}_n^\varepsilon(q)$ | $\mathrm{GL}_a^\varepsilon(q) \otimes \mathrm{GL}_b^\varepsilon(q)$ | $2 \leqslant a < b$ |
| II | $\mathrm{PSp}_n(q)$ | $\mathrm{Sp}_a(q) \otimes \mathrm{O}_b^\varepsilon(q)$ | $b \geqslant 3$, $q$ odd |
| III | $\mathrm{P}\Omega_n^+(q)$ | $\mathrm{Sp}_a(q) \otimes \mathrm{Sp}_b(q)$ | $2 \leqslant a < b$ |
| IV | $\Omega_n(q)$ | $\mathrm{O}_a(q) \otimes \mathrm{O}_b(q)$ | $3 \leqslant a < b$, $abq$ odd |
| V | $\mathrm{P}\Omega_n^+(q)$ | $\mathrm{O}_a^{\varepsilon_1}(q) \otimes \mathrm{O}_b^{\varepsilon_2}(q)$ | $4 \leqslant a \leqslant b$, $a$ and $b$ even, $q$ odd, $(a, \varepsilon_1) \neq (b, \varepsilon_2)$ |
| VI | $\mathrm{P}\Omega_n^\varepsilon(q)$ | $\mathrm{O}_a^\varepsilon(q) \otimes \mathrm{O}_b(q)$ | $a \geqslant 4$ even, $bq$ odd, $b \geqslant 3$ |

Table 5.4.2 *$r$-elusive $\mathscr{C}_4$-actions*

| Case | $r$ | Conditions |
|---|---|---|
| I | 2 | $a = 2$, $1 < (q - \varepsilon)_2 \leqslant (n)_2$ |
| II | 2 | $a = 2$ |

**Proposition 5.4.3** *Theorem 5.4.1 holds in Case I of Table 5.4.1.*

*Proof*  Here $T = \mathrm{PSL}_n^\varepsilon(q)$ and $H$ is of type $\mathrm{GL}_a^\varepsilon(q) \otimes \mathrm{GL}_b^\varepsilon(q)$, where $n = ab$ and $2 \leqslant a < b$. Let $H_0 = H \cap T$. The structure of $H_0$ is given in [86, Proposition 4.4.10], and we note that

$$\mathrm{PSL}_a^\varepsilon(q) \times \mathrm{PSL}_b^\varepsilon(q) \leqslant H_0 \leqslant \mathrm{PGL}_a^\varepsilon(q) \times \mathrm{PGL}_b^\varepsilon(q)$$

Let $r$ be a prime divisor of $|\Omega|$ and let $x = \hat{x}Z \in T$ be an element of order $r$. First assume $r$ is odd. By Lemma 5.4.2, if $r = p$ then $\hat{x} = [J_2, J_1^{n-2}]$ is a derangement of order $r$ since $\nu(x) = 1$. Now suppose $r \neq p$. As usual, set $i = \Phi(r,q)$ as in (3.1.2) (so $i \geqslant 1$ is minimal such that $r$ divides $q^i - 1$) and define the integer $c = c(i,\varepsilon)$ as in (5.3.6). We may as well assume that $r$ divides $|H_0|$, so $c \leqslant b$.

Suppose $c > a$. If $y \in H_0$ has order $r$, then we may write $y = I_a \otimes y_2$, so the multiplicity of each eigenvalue of $\hat{y}$ on $\overline{V} = V \otimes \overline{\mathbb{F}}_q$ is divisible by $a$. We quickly deduce that $T$ contains derangements of order $r$. For example, take

$$\hat{x} = \begin{cases} [\Lambda, \Lambda^{-q}, I_{n-c}] & \varepsilon = - \text{ and } i \not\equiv 2 \ (\mathrm{mod}\ 4) \\ [\Lambda, I_{n-c}] & \text{otherwise} \end{cases}$$

Similarly, if $c = a$ then each $y \in H_0$ of order $r$ is of the form $I_a \otimes y_2$ or $y_1 \otimes y_2$ with $\nu(y_1) = a - 1$. In particular, the same element $x \in T$ is a derangement (note that if $y = y_1 \otimes y_2$ then $\nu(y) \geqslant (a-1)b > a$ by Lemma 5.4.2).

Finally, suppose $c < a$. If we define $x$ as above then $\nu(x) = c$ and thus Lemma 5.4.2 implies that $x$ is a derangement. Moreover, $x \in T$ if $c > 1$, so let us assume $c = 1$. Here we take $\hat{x} = [\Lambda, \Lambda^{-1}, I_{n-2}]$. Then $x \in T$ and we claim that $x$ is a derangement. This is clear if $a \geqslant 3$ (since $\nu(x) = 2$), so let us assume $(c,a) = (1,2)$. Suppose $y = y_1 \otimes y_2 \in H_0$ has order $r$. If $\nu(y_1) \geqslant 1$ then Lemma 5.4.2 implies that $\nu(y) \geqslant b \geqslant 3$, and if $\nu(y_1) = 0$ then each eigenvalue of $\hat{y}$ on $\overline{V}$ has even multiplicity. We deduce that $x$ is a derangement.

For the remainder we may assume $r = 2$. As above, if $p = 2$ then $[J_2, J_1^{n-2}]$ is a derangement, so let us assume $p > 2$. We will use the notation for involutions introduced in Sections 3.2.2 and 3.3.2. If $a \geqslant 3$ then $x = t_2 \in T$ is a derangement since $\nu(x) = 2$, so we have reduced to the case $a = 2$. If $(q - \varepsilon)_2 > (n)_2$ then $T$ contains $t_1$ involutions, which are clearly derangements, so we may assume that $(q - \varepsilon)_2 \leqslant (n)_2$. Here we claim that $T$ is 2-elusive.

To see this, let $x \in T$ be an involution. If $\nu(x) < n/2$ then $x = t_{2\ell}$ for some $\ell < n/4$, and $x$ has fixed points since $I_2 \otimes [-I_\ell, I_{n/2-\ell}] \in H_0$ is $T$-conjugate to $x$. Now assume $\nu(x) = n/2$. If $n/2$ is odd then the condition $(q - \varepsilon)_2 \leqslant (n)_2$ implies that $q \equiv -\varepsilon \ (\mathrm{mod}\ 4)$, so there is a unique class of such involutions in $T$, which necessarily have fixed points. Finally, suppose $n/2$ is even. Since

$$I_2 \otimes [-I_{n/4}, I_{n/4}] \in H_0$$

is an involution of type $t_{n/2}$, we may as well assume that $(q-\varepsilon)_2 < (n)_2$, so that $T$ also contains involutions of type $t'_{n/2}$. Let $y_1 \in \mathrm{PGL}_2(q)$ be a $t'_1$-type involution. Then $y_1 \otimes I_{n/2} \in H_0$ is conjugate to $t'_{n/2}$ since its eigenvalues on $\overline{V}$ are in $\mathbb{F}_{q^2} \setminus \mathbb{F}_q$. We conclude that $T$ is 2-elusive if $a = 2$ and $1 < (q-\varepsilon)_2 \leqslant (n)_2$, as claimed.                                                                                  $\square$

**Proposition 5.4.4** *Theorem 5.4.1 holds in Case II of Table 5.4.1.*

*Proof*   Here $T = \mathrm{PSp}_n(q)$ and $H$ is of type $\mathrm{Sp}_a(q) \otimes \mathrm{O}_b^{\varepsilon}(q)$, where $n = ab$, $q$ is odd and $b \geqslant 3$. Set $H_0 = H \cap T$. By [86, Proposition 4.4.11],

$$\mathrm{PSp}_a(q) \times \mathrm{PO}_b^{\varepsilon}(q) \leqslant H_0 \leqslant \mathrm{PGSp}_a(q) \times \mathrm{PGO}_b^{\varepsilon}(q)$$

Let $r$ be a prime divisor of $|\Omega|$. In order to prove the proposition, we may as well assume that $r$ also divides $|H_0|$.

First assume $r$ is odd. If $r = p$ then Lemma 5.4.2 implies that $[J_2, J_1^{n-2}]$ is a derangement of order $r$. Next suppose $r \neq p$. Set $i = \Phi(r, q)$ and define $x = \hat{x}Z \in T$, where

$$\hat{x} = \begin{cases} [\Lambda, \Lambda^{-1}, I_{n-2i}] & i \text{ odd} \\ [\Lambda, I_{n-i}] & i \text{ even} \end{cases} \tag{5.4.1}$$

We claim that $x$ is a derangement.

To see this, first assume $i$ is even. If $i > a$ then every element of order $r$ in $H_0$ is of the form $I_a \otimes y_2$, so $x$ is a derangement. An entirely similar argument applies if $i > b$, while Lemma 5.4.2 implies that $x$ is a derangement if $i < a$ and $i < b$. Next suppose $i < a$ and $i = b$. Each $y \in H_0$ of order $r$ is of the form $y_1 \otimes I_b$ or $y_1 \otimes y_2$, where $\nu(y_2) = b - 1$. In the former case, the multiplicity of each eigenvalue of $\hat{y}$ on $\overline{V} = V \otimes \overline{\mathbb{F}}_q$ is divisible by $b$, while in the latter case Lemma 5.4.2 implies that $\nu(y) \geqslant a(b-1) > b = i$. We conclude that $x$ is a derangement. The remaining cases $i = a < b$ and $i = a = b$ are very similar. An entirely similar argument applies if $i$ is odd.

To complete the proof, let us assume $r = 2$. We use the notation for involutions from Section 3.4.2. If $a \geqslant 4$ then $t_1$ involutions are derangements (since $\nu(t_1) = 2$), so we may assume that $a = 2$. We claim that $T$ is 2-elusive.

First observe that if $i < n/4$ then $I_2 \otimes [-I_i, I_{n/2-i}] \in H_0$ embeds in $T$ as an involution of type $t_i$, so it remains to show that each involution $x \in T$ with $\nu(x) = n/2$ has fixed points. Let $y_1 \in \mathrm{PSp}_2(q)$ be an involution. If $q \equiv 1$ (mod 4) then $\hat{y}_1 = [\lambda, \lambda^{-1}]$, where $\lambda \in \mathbb{F}_q$ has order 4, and thus $y_1 \otimes I_{n/2} \in H_0$ is of type $t_{n/2}$. On the other hand, if $q \equiv 3$ (mod 4) then $y_1 \otimes I_{n/2} \in H_0$ is of type $t'_{n/2}$ (since its eigenvalues on $\overline{V}$ are in $\mathbb{F}_{q^2} \setminus \mathbb{F}_q$). This justifies the claim if

$n \equiv 2 \pmod 4$. Finally, if $n \equiv 0 \pmod 4$ then $I_2 \otimes [-I_{n/4}, I_{n/4}] \in H_0$ is of type $t_{n/4}$, and once again the claim follows.                                    □

**Proposition 5.4.5** *Theorem 5.4.1 holds in Case III of Table 5.4.1.*

*Proof* Here $T = P\Omega_n^+(q)$ and $H$ is of type $\mathrm{Sp}_a(q) \otimes \mathrm{Sp}_b(q)$, where $n = ab$ and $2 \leqslant a < b$. Let $r$ be a prime divisor of $|\Omega|$.

If $r = p > 2$ then Lemma 5.4.2 implies that $[J_3, J_1^{n-3}]$ is a derangement of order $r$ (note that $I_2 \otimes [J_p^{a_p}, \ldots, J_1^{a_1}]$ has Jordan form $[J_p^{2a_p}, \ldots, J_1^{2a_1}]$). If $p = 2$ then $c_2$-type involutions are derangements. This follows from Lemma 5.4.2 if $a \geqslant 4$, and for $a = 2$ it is easy to check that $I_2 \otimes b_1$ is an $a_2$-type involution. If $r \neq p$ and $r > 2$ then it is straightforward to verify that the element $x$ defined in (5.4.1) is a derangement of order $r$ (where $i = \Phi(r,q)$). Finally, if $r = 2$ and $p > 2$ then $\nu(y) \geqslant 4$ for every involution $y \in H_0$, so $t_1$ or $t_1'$ involutions in $T$ are derangements.                                    □

**Proposition 5.4.6** *Theorem 5.4.1 holds in Cases IV–VI of Table 5.4.1.*

*Proof* This is very similar to the previous cases. For example, consider Case IV. Here $T = \Omega_n(q)$ and $H$ is of type $O_a(q) \otimes O_b(q)$, where $3 \leqslant a < b$ and $abq$ is odd. In view of Lemma 5.4.2, $[J_3, J_1^{n-3}]$ is a derangement of order $p$, and involutions of type $t_1$ or $t_1'$ are also derangements. Finally, if $r \neq p$ and $r > 2$ then it is easy to check that the element defined in (5.4.1) is a derangement. Cases V and VI are very similar and we ask the reader to check the details.    □

This completes the proof of Theorem 5.4.1.

## 5.5 $\mathscr{C}_5$: Subfield subgroups

Let $G$ be an almost simple classical group over $\mathbb{F}_q$ with socle $T$, where $q = p^f$ and $p$ is a prime. Set $\mathbb{F} = \mathbb{F}_{q^u}$, where $u = 2$ if $T = \mathrm{PSU}_n(q)$, otherwise $u = 1$. The $\mathscr{C}_5$-subgroups of $G$ arise from prime index subfields $\mathbb{F}_{q_0}$ of $\mathbb{F}$, so $q^u = q_0^k$ and $k$ is a prime divisor of $uf$. The specific cases that arise are listed in Table 5.5.1 (see [86, Table 4.5.A]), and we refer the reader to [86, Section 4.5] for further details. Note that $q = q_0$ in Cases V and VI.

**Theorem 5.5.1** *Let $G \leqslant \mathrm{Sym}(\Omega)$ be a primitive almost simple classical group with socle $T$ and point stabiliser $H \in \mathscr{C}_5$. Let $r$ be a prime divisor of $|\Omega|$. Then $T$ is $r$-elusive if and only if $(G, H, r)$ is one of the cases in Table 5.5.2.*

Table 5.5.1 *The collection $\mathscr{C}_5$*

| Case | $T$ | Type of $H$ | Conditions |
|------|-----|-------------|------------|
| I | $\mathrm{PSL}_n^\varepsilon(q)$ | $\mathrm{GL}_n^\varepsilon(q_0)$ | $k > 2$ if $\varepsilon = -$ |
| II | $\mathrm{PSp}_n(q)$ | $\mathrm{Sp}_n(q_0)$ | |
| III | $\mathrm{P\Omega}_n^\varepsilon(q)$ | $\mathrm{O}_n^\varepsilon(q_0)$ | $k > 2$ if $\varepsilon = -$ |
| IV | $\mathrm{P\Omega}_n^+(q)$ | $\mathrm{O}_n^-(q_0)$ | $k = 2$ |
| V | $\mathrm{PSU}_n(q)$ | $\mathrm{Sp}_n(q)$ | |
| VI | $\mathrm{PSU}_n(q)$ | $\mathrm{O}_n^\varepsilon(q)$ | $q$ odd |

Table 5.5.2 *$r$-elusive $\mathscr{C}_5$-actions*

| Case | $r, p$ | Conditions |
|------|--------|------------|
| I | $r = 2$ | $k = 2$ and $(n)_2 \leqslant (q-1)_2$ if $q$ odd |
| | $r = p > 2$ | see Remark 5.5.2(i) |
| | $r \neq p, r > 2$ | $\begin{cases} k = 2, r = n \in \{3,5\}, (q_0+1)_r = r \\ r = k \end{cases}$ |
| II, III | $r = 2$ | $k = 2$ if $q$ odd |
| | $r = p > 2$ | $k$ odd |
| | $r \neq p, r > 2$ | $r = k$ |
| IV | $r = 2$ | $n \equiv 2 \pmod 4$ |
| V | $r = 2$ | $(q+1)_2 \leqslant (n)_2$ |
| VI | $r = 2$ | always |
| | $r \neq p, r > 2$ | $r = n \in \{3,5\}, (q+1)_r = r$ |

**Remark 5.5.2** Some comments on Case I in Table 5.5.1.

(i) In Case I, if $r = p > 2$ then $T$ is $r$-elusive if and only if every nontrivial partition $\lambda = (p^{a_p}, \ldots, 1^{a_1})$ of $n$ satisfies the following conditions:
  (a) $(v, q - \varepsilon) = (v, q_0 - \varepsilon)$, where $v = \gcd\{j \mid a_j \neq 0\}$, and
  (b) if $(v, q - \varepsilon) = (v, q_0 - \varepsilon) > 1$ then either $k \geqslant p$, or $(n, q_0 - \varepsilon)$ is indivisible by $k$.

(ii) Suppose $r = p = 3$ in Case I, and consider the above conditions (a) and (b) for $r$-elusivity. First note that either $(v, q - \varepsilon) = 1$, or $\lambda = (2^{n/2})$ and $(v, q - \varepsilon) = 2$, so (a) always holds. In particular, if $n$ is odd then $(v, q - \varepsilon) = 1$ and thus $T$ is 3-elusive. If $n$ is even and $(v, q - \varepsilon) = (v, q_0 - \varepsilon) = 2$ then $(n, q_0 - \varepsilon)$ is even, so we require $k > 2$ for 3-elusivity. Therefore, $T$ is 3-elusive if and only if $n$ is odd, or $n$ is even and $k > 2$.
As we will explain in the proof of Proposition 5.5.7, if $n$ is even and $k = 2$ then the two $\mathrm{PSL}_n(q_0)$-classes of elements with Jordan form $[J_2^{n/2}]$ are fused in $T$, so there are derangements in $T$ of type $[J_2^{n/2}]$.

(iii) More generally, if $k = 2$ and $r = p > 3$ in Case I then $T$ is $p$-elusive only if $n$ is odd. However, this is not always a sufficient condition. For example, if $(n, k, q_0) = (3, 2, 5)$ and $\lambda = (3)$, then $(v, q_0 - 1) = 1$, $(v, q - 1) = 3$ and we deduce that $T$ contains derangements of order 5.

### 5.5.1 Subgroup structure

Let $H$ be a maximal $\mathscr{C}_5$-subgroup of $G$ and set $H_0 = H \cap T$. In our analysis of Cases III, IV and VI in Table 5.5.1, we need to understand the precise structure of $H_0$. For easy reference, we record this information here.

**Lemma 5.5.3** *Let* $T = \mathrm{P\Omega}_n^+(q)$ *and let* $H$ *be a* $\mathscr{C}_5$*-subgroup of type* $\mathrm{O}_n^+(q_0)$ *as in Case III of Table 5.5.1, with* $q = q_0^2$. *Then*

$$H_0 = \begin{cases} \Omega_n^+(q_0) & p = 2 \\ \mathrm{PSO}_n^+(q_0) & n \equiv 2 \ (\mathrm{mod}\ 4),\ q_0 \equiv 1 \ (\mathrm{mod}\ 4) \\ \mathrm{Inndiag}(\mathrm{P\Omega}_n^+(q_0)) & otherwise \end{cases}$$

*Proof* The result follows immediately from [86, Proposition 4.5.10], unless $q$ is odd and either $n \equiv 0 \ (\mathrm{mod}\ 4)$, or $n \equiv 2 \ (\mathrm{mod}\ 4)$ and $q_0 \equiv 3 \ (\mathrm{mod}\ 4)$. Here

$$H_0 = \mathrm{PSO}_n^+(q_0).2 \leqslant \mathrm{PGO}_n^+(q_0)$$

and $H_0$ does not contain an element that fuses the two $T$-orbits on maximal totally singular subspaces of the natural $T$-module. We conclude that $H_0 = \mathrm{Inndiag}(\mathrm{P\Omega}_n^+(q_0))$ is the only possibility. $\square$

**Lemma 5.5.4** *Let* $T = P\Omega_n^+(q)$ *and let* $H$ *be a* $\mathscr{C}_5$-*subgroup of type* $O_n^-(q_0)$ *as in Case IV of Table 5.5.1, so* $q = q_0^2$. *Then*

$$H_0 = \begin{cases} \Omega_n^-(q_0) & p = 2 \\ \mathrm{Inndiag}(P\Omega_n^-(q_0)) & n \equiv 2 \ (\mathrm{mod}\ 4),\ q_0 \equiv 1 \ (\mathrm{mod}\ 4) \\ \mathrm{PSO}_n^-(q_0) & otherwise \end{cases}$$

*Proof* As in the proof of the previous lemma, the result follows immediately from [86, Proposition 4.5.10], unless $n \equiv 2 \ (\mathrm{mod}\ 4)$ and $q_0 \equiv 1 \ (\mathrm{mod}\ 4)$, in which case

$$H_0 = \mathrm{PSO}_n^-(q_0).2 \leqslant \mathrm{PGO}_n^-(q_0)$$

Clearly, no element of $H_0$ can interchange the two $T$-orbits of maximal totally singular subspaces of $V$. Therefore, from the definition of $\mathrm{Inndiag}(P\Omega_n^-(q_0))$ given in Construction 2.5.12, we deduce that $H_0 \leqslant \mathrm{Inndiag}(P\Omega_n^-(q_0))$ and the result follows.                                                                    □

Finally, let us turn to Case VI in Table 5.5.1. Following [86, p.142], we first describe the construction of a natural embedding $O_n^\varepsilon(q) \leqslant GU_n(q)$, where $q$ is odd.

**Construction 5.5.5** *The embedding* $O_n^\varepsilon(q) \leqslant GU_n(q)$, $q$ *odd.*

Let $V$ be an $n$-dimensional vector space over $\mathbb{F}_{q^2}$, where $q$ is odd, and let $B$ be a nondegenerate hermitian form on $V$. Let $\beta = \{v_1, \ldots, v_n\}$ be an orthonormal basis for $V$ and let $V_\#$ denote the $\mathbb{F}_q$-span of $\beta$. Then $B(v, w) \in \mathbb{F}_q$ for all $v, w \in V_\#$, so the restriction of $B$ to $V_\#$ is a nondegenerate symmetric bilinear form, which corresponds in the usual way to a quadratic form $Q_\#$ on $V_\#$. Note that $D(Q_\#) = \square$. To obtain a quadratic form $Q_\#$ with $D(Q_\#) = \boxtimes$, we can repeat the above construction, replacing $v_1$ by $\mu v_1$, where $\mathbb{F}_{q^2}^\times = \langle \mu \rangle$. In particular, if $n$ is even then this construction yields an embedding of both $O_n^+(q)$ and $O_n^-(q)$ in $GU_n(q)$. Also note that we obtain an embedding of $GO_n^\varepsilon(q)$ in $\Delta U_n(q)$ such that the similarity map $\tau : GO_n^\varepsilon(q) \to \mathbb{F}_q^\times$ is the restriction of the corresponding map for $\Delta U_n(q)$.                                                          □

In the statement of the next lemma, if $n \equiv 2 \ (\mathrm{mod}\ 4)$ then $\gamma'_{(n+2)/4}$ denotes the involutory graph automorphism of $P\Omega_n^\varepsilon(q)$ discussed in Section 3.5.2.15.

**Lemma 5.5.6** *Let* $T = PSU_n(q)$ *and let* $H$ *be a* $\mathscr{C}_5$-*subgroup of type* $O_n^\varepsilon(q)$ *as in Case VI in Table 5.5.1, so* $q$ *is odd. Then*

$$H_0 = \begin{cases} \mathrm{PSO}_n(q) & n \text{ odd} \\ \mathrm{PO}_n^\varepsilon(q) & (q+1)_2 > (n)_2 > 1 \\ \mathrm{PSO}_n^\varepsilon(q).\langle \gamma'_{(n+2)/4} \rangle & n \equiv 2 \ (\mathrm{mod}\ 4),\ q \equiv 1 \ (\mathrm{mod}\ 4) \\ \mathrm{PSO}_n^\varepsilon(q).2 & q \equiv 3 \ (\mathrm{mod}\ 4),\ (q+1)_2 = (n)_2 \\ \mathrm{Inndiag}(\mathrm{P\Omega}_n^\varepsilon(q)) & \text{otherwise} \end{cases}$$

where $\mathrm{PSO}_n^\varepsilon(q).2 \leqslant \mathrm{PGO}_n^\varepsilon(q)$ is the unique nonsplit extension of $\mathrm{PSO}_n^\varepsilon(q)$ by $C_2$.

*Proof*   According to [86, Proposition 4.5.5] we have

$$H_0 = \mathrm{PSO}_n^\varepsilon(q).(2,n) \leqslant \mathrm{PGO}_n^\varepsilon(q)$$

and so the result follows for $n$ odd. For the remainder, assume $n$ is even. Let $\xi \in \mathbb{F}_{q^2}$ be an element of order $2(q-1)_2$.

If $(q+1)_2 > (n)_2$ then $T$ contains involutions of type $t_1 = [-I_1, I_{n-1}]$, whence $H_0 = \mathrm{PO}_n^\varepsilon(q)$ as claimed.

Next assume $n \equiv 2 \ (\mathrm{mod}\ 4)$ and $q \equiv 1 \ (\mathrm{mod}\ 4)$. Let $t = \hat{t}Z \in \mathrm{PGO}_n^\varepsilon(q)$ be an involution of type $\gamma'_{(n+2)/4}$ (where $Z = Z(\mathrm{GO}_n^\varepsilon(q))$). From the description of $t$ in Section 3.5.2.15, we deduce that $\det(\hat{t}) = (-\xi^2)^{n/2} = -\xi^n$ and $\tau(\hat{t}) = \xi^2$. Set $\mu = \lambda \xi \in \mathbb{F}_{q^2}$, where $\lambda \in \mathbb{F}_q$ has order 4. Then $\mu^{q+1} = \xi^2$ and $\mu^n = -\xi^n$, so $t \in T$ by Lemma 2.3.1 and thus $H_0 = \mathrm{PSO}_n^\varepsilon(q).\langle \gamma'_{(n+2)/4} \rangle$.

Now assume $n \equiv 0 \ (\mathrm{mod}\ 4)$ and $q \equiv 1 \ (\mathrm{mod}\ 4)$. Let

$$t = \hat{t}Z \in \mathrm{Inndiag}(\mathrm{P\Omega}_n^\varepsilon(q)) \setminus \mathrm{PSO}_n^\varepsilon(q)$$

be an involution of type $t''_{n/4}$ if $\varepsilon = +$, and type $t'_{n/4}$ if $\varepsilon = -$, so $\det(\hat{t}) = (-\xi^2)^{n/2} = \xi^n$ and $\tau(\hat{t}) = \xi^2$ (see Sections 3.5.2.7 and 3.5.2.9). Again, if we set $\mu = \lambda \xi \in \mathbb{F}_{q^2}$, where $\lambda \in \mathbb{F}_q$ has order 4, then $\mu^{q+1} = \xi^2$ and $\mu^n = \xi^n$, so $t \in T$ and thus $H_0 = \mathrm{Inndiag}(\mathrm{P\Omega}_n^\varepsilon(q))$.

Finally, suppose $n \equiv 0 \ (\mathrm{mod}\ 4)$ and $q \equiv 3 \ (\mathrm{mod}\ 4)$. Define $t$ as in the previous paragraph, so $t \in \mathrm{Inndiag}(\mathrm{P\Omega}_n^\varepsilon(q)) \setminus \mathrm{PSO}_n^\varepsilon(q)$ is an involution of type $t''_{n/4}$ if $\varepsilon = +$, and type $t'_{n/4}$ if $\varepsilon = -$. Then $\det(\hat{t}) = \xi^n = 1$ and $\tau(\hat{t}) = \xi^2 = -1$ since $\xi$ has order 4. By Lemma 2.3.1, we have $t \in T$ (and thus $H_0 = \mathrm{Inndiag}(\mathrm{P\Omega}_n^\varepsilon(q))$) if and only if there exists a scalar $\mu \in \mathbb{F}_{q^2}$ such that $\mu^{q+1} = -1$ and $\mu^n = 1$. Clearly, such a scalar exists if and only if $(q+1)_2 < (n)_2$. In particular, if $(q+1)_2 = (n)_2$ then $H_0 = \mathrm{PSO}_n^\varepsilon(q).2$ is a nonsplit extension. Moreover, this is the unique nonsplit extension of $\mathrm{PSO}_n^\varepsilon(q)$ by $C_2$ in $\mathrm{PGO}_n^\varepsilon(q)$ (see Sections 3.5.9.4 and 3.5.9.7). $\quad\square$

### 5.5.2 Proof of Theorem 5.5.1

**Proposition 5.5.7** *Theorem 5.5.1 holds in Case I of Table 5.5.1.*

*Proof* Here $T = \mathrm{PSL}_n^\varepsilon(q)$ and $H$ is of type $\mathrm{GL}_n^\varepsilon(q_0)$, where $q = q_0^k$ and $k$ is a prime (with $k > 2$ if $\varepsilon = -$). Set $H_0 = H \cap T$ and $\mathbb{F} = \mathbb{F}_{q^u}$, where $u = 1$ if $\varepsilon = +$ and $u = 2$ if $\varepsilon = -$. According to [86, Proposition 4.5.3] we have

$$H_0 = \mathrm{PSL}_n^\varepsilon(q_0).d \leqslant \mathrm{PGL}_n^\varepsilon(q_0)$$

where

$$d = \frac{(n, q_0 - \varepsilon)\left(q_0 - \varepsilon, \frac{q-\varepsilon}{(n,q-\varepsilon)}\right)}{q_0 - \varepsilon}$$

By Lemmas A.5 and A.6 (see Appendix A), we have $d \in \{1, k\}$, with $d = k$ if and only if $(q - \varepsilon)_k > (n)_k > 1$. Let $r$ be a prime divisor of $|\Omega|$.

*Case 1. $r = p$.* If $p = 2$ then every involution in $T$ has Jordan form $[J_2^m, J_1^{n-2m}]$ for some positive integer $m$, and by applying Propositions 3.2.7 and 3.3.10 we see that there is a unique class of such elements in both $T$ and $H_0$. Therefore, $T$ is 2-elusive.

Now assume $r = p > 2$. Let $\lambda = (p^{a_p}, \ldots, 1^{a_1})$ be a nontrivial partition of $n$ and set

$$v = \gcd\{j \mid a_j \neq 0\}$$

By Propositions 3.2.7 and 3.3.10, $T$ has precisely $(v, q - \varepsilon)$ distinct $T$-classes of unipotent elements with Jordan form $[J_p^{a_p}, \ldots, J_1^{a_1}]$, and there are precisely $(v, q_0 - \varepsilon)$ such $\mathrm{PSL}_n^\varepsilon(q_0)$-classes in $\mathrm{PSL}_n^\varepsilon(q_0)$. Note that $(v, q_0 - \varepsilon) \leqslant (v, q - \varepsilon)$. Of course, if $(v, q - \varepsilon) = 1$ then there is a unique class in both $T$ and $H_0$, so these unipotent elements have fixed points. On the other hand, if $(v, q_0 - \varepsilon) < (v, q - \varepsilon)$ then $T$ contains derangements of order $p$ since there are more $T$-classes of elements in $T$ with Jordan form $[J_p^{a_p}, \ldots, J_1^{a_1}]$ than there are $H_0$-classes in $H_0$ of this form. It remains to consider the case where $(v, q_0 - \varepsilon) = (v, q - \varepsilon) > 1$.

Let $\delta = [\mu, I_{n-1}]Z \in \mathrm{PGL}_n^\varepsilon(q)$, where $\mu \in \mathbb{F}$ has order $q - \varepsilon$, and let $x \in \mathrm{PSL}_n^\varepsilon(q_0)$ be a unipotent element with Jordan form $[J_p^{a_p}, \ldots, J_1^{a_1}]$. By Lemmas 3.2.8 and 3.3.11, the $(v, q_0 - \varepsilon)$ distinct $\mathrm{PSL}_n^\varepsilon(q_0)$-classes of this form are represented by the elements

$$x^{\delta^{is}}, \quad 0 \leqslant i < (v, q_0 - \varepsilon) \tag{5.5.1}$$

where $s = (q - \varepsilon)/(q_0 - \varepsilon)$. Suppose two of these $\mathrm{PSL}_n^\varepsilon(q_0)$-classes are fused in $T$. Then according to Lemmas 3.2.8 and 3.3.11, there exist integers $i, j$ such that $0 \leqslant j < i < (v, q_0 - \varepsilon)$ and $(v, q - \varepsilon)$ divides $s(i - j)$. Since $(v, q - \varepsilon) =$

$(v, q_0 - \varepsilon)$ it follows that $(v, q - \varepsilon)$ does not divide $i - j$, hence there is a prime divisor $\ell$ of $(v, q_0 - \varepsilon)$ that also divides $s$. But $k$ is the only prime dividing both $q_0 - \varepsilon$ and $s$ (see Lemma A.4 in Appendix A), so $\ell = k$ is the only possibility. In particular, it follows that none of the $\mathrm{PSL}_n^{\varepsilon}(q_0)$-classes are fused in $T$ if either $k \geqslant p$ or $(n, q_0 - \varepsilon)$ is indivisible by $k$, so in this situation every element in $T$ with Jordan form $[J_p^{a_p}, \ldots, J_1^{a_1}]$ has fixed points.

Finally, suppose $k < p$ and $k$ divides $(n, q_0 - \varepsilon)$. Consider the unipotent elements in $\mathrm{PSL}_n^{\varepsilon}(q_0)$ with Jordan form $[J_k^{n/k}]$. As before, the distinct $\mathrm{PSL}_n^{\varepsilon}(q_0)$-classes are represented by the elements in (5.5.1). Since $k = (v, q - \varepsilon) = (v, q_0 - \varepsilon)$ divides $s$, Lemmas 3.2.8 and 3.3.11 imply that all of these $\mathrm{PSL}_n^{\varepsilon}(q_0)$-classes are fused in $T$, so $T$ contains derangements of order $p$.

*Case 2.* $r > 2$, $r \neq p$. Set $i = \Phi(r, q)$ and $i_0 = \Phi(r, q_0)$ (see (3.1.2)), and observe that

$$ i = \begin{cases} i_0/k & \text{if } k \text{ divides } i_0 \\ i_0 & \text{otherwise} \end{cases} \tag{5.5.2} $$

As in Section 3.1.1, let $\mathscr{S}_r$ be the set of nontrivial $r$th roots of unity in $\mathbb{F}_{q^i}$, and let $\sigma$ and $\sigma_0$ be the permutations of $\mathscr{S}_r$ defined by $\lambda \mapsto \lambda^q$ and $\lambda \mapsto \lambda^{q_0}$, respectively. Note that if $k$ divides $i_0$ then each $\sigma_0$-orbit on $\mathscr{S}_r$ is a union of $k$ distinct $\sigma$-orbits. Define the integer $c = c(i, \varepsilon)$ as in (5.3.6). Let $x = \hat{x} Z \in T$ be an element of order $r$.

First assume $k$ divides $i_0$. Note that if $k$ is odd, then either $i \equiv i_0 \pmod{4}$, or both $i$ and $i_0$ are odd. If $c > 1$ then

$$ \hat{x} = \begin{cases} [\Lambda, \Lambda^{-q}, I_{n-c}] & \varepsilon = - \text{ and } i \not\equiv 2 \pmod{4} \\ [\Lambda, I_{n-c}] & \text{otherwise} \end{cases} $$

defines a derangement in $T$ of order $r$. Now assume $c = 1$ (so $r$ divides $q - \varepsilon$). If $n$ is indivisible by $r$ then set $\hat{x} = [\Lambda, I_{n-1}]$. Similarly, if $r$ divides $n$ and $r < n$ then take $\hat{x} = [\Lambda^r, I_{n-r}]$, and define $\hat{x} = [\Lambda, \Lambda^{-1}, I_{n-2}]$ if $r = n$ and $k > 2$. In each of these cases it is easy to check that $x$ is a derangement in $T$, so we have reduced to the case where $k = 2$, $r = n$ and $c = 1$. Note that $\varepsilon = +$ since $k = 2$, so $i = 1$ and $i_0 = 2$.

Let $\mu \in \mathbb{F}_q$ be a primitive $r$th root of unity. First observe that if $r \geqslant 7$ then $[\mu, \mu^2, \mu^{r-3}, I_{n-3}]$ is a derangement of order $r$. Next assume that $n = r = 5$, in which case every element $x \in T$ of order 5 is diagonalisable over $\mathbb{F}_q$ (see Proposition 3.2.2). Recall that $i_0 = 2$, so $q_0 \equiv -1 \pmod 5$. In particular, if $y \in H_0$ has order 5 then $\det(\hat{y}) = 1$. Therefore, if $q_0 \equiv -1 \pmod{25}$ then $[\mu, I_4]$ is a derangement in $T$ of order 5 (by Lemma 2.1.1, such an element is contained in $T$). On the other hand, if $q_0 \not\equiv -1 \pmod{25}$ then $\det(\hat{x}) = 1$ for all $x = \hat{x} Z \in T$ of order 5. Moreover, by choosing an appropriate scalar multiple, it is not difficult to see that we may assume

$$\hat{x} = [\mu I_a, \mu^4 I_a, \mu^2 I_b, \mu^3 I_b, I_c]$$

for some non-negative integers $a, b$ and $c$ with $2(a+b) + c = 5$. Therefore, $x$ has fixed points and thus $T$ is 5-elusive. Similarly, if $n = r = 3$ and $k = i_0 = 2$ then $T$ is 3-elusive if and only if $q_0 \not\equiv -1 \pmod 9$.

Now suppose $k$ does not divide $i_0$, so $i = i_0$. We claim that $r = k$. To see this, first assume $\varepsilon = +$ and write $n = mi + t$, where $0 \leqslant t < i$. Note that $r$ divides $|\Omega|$ if and only if $r$ divides

$$\prod_{\ell=1}^{m} \frac{(q_0^{\ell k i} - 1)}{(q_0^{\ell i} - 1)}$$

By Lemma A.4 (see Appendix A) we have

$$(q_0^{\ell k i} - 1)_r = (q_0^{\ell i} - 1)_r (k)_r$$

so $r$ divides $|\Omega|$ if and only if $r = k$. Similar reasoning applies when $\varepsilon = -$. For example, suppose $\varepsilon = -$ and $i \equiv 2 \pmod 4$. Then $r$ divides

$$\prod_{\ell=1}^{m'} \frac{(q_0^{\ell k i/2} - (-1)^\ell)}{(q_0^{\ell i/2} - (-1)^\ell)}$$

where $n = m'i/2 + t'$ and $0 \leqslant t' < i/2$. If $r \neq k$ then Lemma A.4 implies that

$$(q_0^{\ell k i/2} - (-1)^\ell)_r = (q_0^{\ell i/2} - (-1)^\ell)_r$$

whence $r$ divides $|\Omega|$ if and only if $r = k$. This establishes the claim. Since $i = i_0$, we conclude that every element in $T$ of order $r = k$ has fixed points. In addition, observe that the condition $r = k$ implies that $i = i_0$, since $r$ divides $q_0^{r-1} - 1$ by Fermat's Little Theorem.

*Case 3. $r = 2$, $p > 2$.* Finally, let us assume $r = 2$ and $p > 2$. First we claim that $k = 2$. To see this, suppose first that $k > 2$ and $\varepsilon = +$. Since $r$ divides $|\Omega|$ it follows that $r$ must also divide

$$N := \prod_{\ell=2}^{n} \frac{(q_0^{\ell k} - 1)}{(q_0^{\ell} - 1)}$$

But Lemma A.4 states that $(q_0^{\ell k} - 1)_2 = (q_0^{\ell} - 1)_2$ for each $\ell$, hence $N$ is odd. Similarly, if $\varepsilon = -$ then Lemma A.4 implies that $(q_0^{\ell k} + 1)_2 = (q_0^{\ell} + 1)_2$ and the claim follows. In particular, for the remainder of the proof we may assume that $k = 2$, so $\varepsilon = +$ and $q \equiv 1 \pmod 4$.

Recall that $H_0 = \mathrm{PSL}_n(q_0).d$, where $d \in \{1,2\}$. More precisely, $d = 2$ if and only if $(q-1)_2 > (n)_2 > 1$. In particular, if $(q-1)_2 > (n)_2$ then either $(q_0-1)_2 > (n)_2$ or $d = 2$. This implies that every involution $x \in T$ with $\nu(x) < n/2$ has fixed points.

Now assume $n$ is even and $x \in T$ is an involution such that $\nu(x) = n/2$. If $(n)_2 \leqslant (q-1)_2$ then $T$ has a unique conjugacy class of such elements and thus $T$ is 2-elusive (note that there are involutions $y \in H_0$ with $\nu(y) = n/2$). Finally, let us assume $(n)_2 > (q-1)_2$, in which case $T$ and $H_0 = \mathrm{PSL}_n(q_0)$ both contain two classes of involutions $x$ with $\nu(x) = n/2$. Clearly, $[-I_{n/2}, I_{n/2}] \in H_0$ is $T$-conjugate to $t_{n/2}$. The other class in $H_0$ is represented by a block-diagonal matrix of the form $\hat{x} = [A^{n/2}]$, where

$$A = \begin{pmatrix} 0 & 1 \\ \xi^2 & 0 \end{pmatrix}$$

and $\xi \in \mathbb{F}_q \setminus \mathbb{F}_{q_0}$ has order $2(q_0-1)_2$ (see Section 3.2.2.2). In particular, the eigenvalues of $\hat{x}$ belong to $\mathbb{F}_q$, and thus $x$ is $T$-conjugate to $t_{n/2}$. We conclude that involutions of type $t'_{n/2}$ are derangements. □

**Proposition 5.5.8** *Theorem 5.5.1 holds in Case II of Table 5.5.1.*

*Proof* Here $T = \mathrm{PSp}_n(q)$ and $H$ is of type $\mathrm{Sp}_n(q_0)$, where $q = q_0^k$ for some prime $k$. Set $H_0 = H \cap T$. By [86, Proposition 4.5.4] we have

$$H_0 = \mathrm{PSp}_n(q_0).(2, q-1, k) \leqslant \mathrm{PGSp}_n(q_0)$$

Let $r$ be a prime divisor of $|\Omega|$.

First assume $r = 2$. From the description of unipotent involutions in Section 3.4.4, it is clear that every involution in $T$ has fixed points when $p = 2$. Now assume $p > 2$. As in the proof of the previous proposition, we may assume that $k = 2$ since $r$ divides $|\Omega|$, so $H_0 = \mathrm{PGSp}_n(q_0)$ and $q \equiv 1 \pmod 4$. Visibly, any involution $x \in T$ with $\nu(x) < n/2$ has fixed points, so let us assume that $\nu(x) = n/2$. If $n \equiv 2 \pmod 4$ then $T$ contains a unique class of such involutions, so $T$ is 2-elusive in this situation. Now assume $n \equiv 0 \pmod 4$, so the relevant classes in $T$ are represented by the elements $t_{n/4}$ and $t_{n/2}$ (see Section 3.4.2). Clearly $[-I_{n/2}, I_{n/2}] \in H_0$ is $T$-conjugate to $t_{n/4}$ if the two eigenspaces are nondegenerate, and it is $T$-conjugate to $t_{n/2}$ if they are totally isotropic. We conclude that $T$ is 2-elusive.

Next suppose $r = p > 2$. If $k = 2$ then $H_0 = \mathrm{PGSp}_n(q_0)$ and Proposition 3.4.12 implies that there are two $T$-classes of unipotent elements with Jordan form $[J_2, J_1^{n-2}]$, but only one such class in $H_0$, so $T$ contains derangements of order $p$.

Now assume $k > 2$, so $H_0 = \mathrm{PSp}_n(q_0)$. Here $T$ and $H_0$ have the same number of conjugacy classes of elements of order $p$, so in order to deduce that $T$ is $p$-elusive we need to show that none of the relevant $H_0$-classes are fused in $T$.

In [93], Liebeck and O'Brien present explicit representatives of the unipotent classes in $\mathrm{Sp}_n(q)$. In particular, they prove that each class can be represented by a matrix with entries in $\{0, \pm 1, \alpha\}$, where $\alpha \in \mathbb{F}_q$ is a fixed nonsquare. Since $k$ is odd, the scalar $\alpha$ can be chosen in the subfield $\mathbb{F}_{q_0}$, which implies that every unipotent class in $\mathrm{Sp}_n(q)$ has a representative in $\mathrm{Sp}_n(q_0)$. We conclude that none of the $H_0$-classes of elements of order $p$ in $H_0$ are fused in $T$.

Finally, let us assume $r > 2$ and $r \ne p$. Set $i = \Phi(r, q)$, $i_0 = \Phi(r, q_0)$ and note that (5.5.2) holds. Also observe that $i_0 \equiv 0 \pmod 4$ if $k = 2$ (since $r$ divides $|\Omega|$), so $i$ is even in this situation. Let $x = \hat{x}Z \in T$ be an element of order $r$. If $k$ divides $i_0$ then

$$
\hat{x} = \begin{cases} [\Lambda, \Lambda^{-1}, I_{n-2i}] & i \text{ odd} \\ [\Lambda, I_{n-i}] & i \text{ even} \end{cases} \tag{5.5.3}
$$

is a derangement. Finally, suppose $i = i_0$. Since $r$ divides $|\Omega|$, we deduce that $r = k$, and it is straightforward to check that all elements in $T$ of order $r$ have fixed points. $\qquad\square$

**Proposition 5.5.9** *Theorem 5.5.1 holds in Case III of Table 5.5.1.*

*Proof* Here $T = \mathrm{P\Omega}_n^{\varepsilon}(q)$ and $H$ is of type $\mathrm{O}_n^{\varepsilon}(q_0)$, where $q = q_0^k$ for a prime $k$ (with $k > 2$ if $\varepsilon = -$). Recall that $n \geqslant 7$.

Let $Q$ be the underlying quadratic form on $V$ and let $\{v_1, \ldots, v_n\}$ be an $\mathbb{F}_q$-basis for $V$ with the property that $Q(v_i) \in \mathbb{F}_{q_0}$ for all $i$. If $n$ is odd or $\varepsilon = +$ then such a basis is presented in Proposition 2.2.7, and we refer the reader to the basis in Remark 2.2.8 when $\varepsilon = -$ (note that if $\varepsilon = -$ then $k$ is odd, so an element of $\mathbb{F}_{q_0}$ is a square in $\mathbb{F}_q$ if and only if it is a square in $\mathbb{F}_{q_0}$). Let $V_\#$ be the $\mathbb{F}_{q_0}$-span of this basis and let $Q_\#$ denote the restriction of $Q$ to $V_\#$.

Let $r$ be a prime divisor of $|\Omega|$ and set $H_0 = H \cap T$. By [86, Propositions 4.5.8 and 4.5.10] we have $H_0 = \mathrm{P\Omega}_n^{\varepsilon}(q_0)$ if $k > 2$ or $p = 2$, and $H_0 = \mathrm{PSO}_n(q_0)$ if $k = 2$ and $n$ is odd. If $k = 2$ and $n$ is even (so $\varepsilon = +$), then the structure of $H_0$ is given in Lemma 5.5.3.

*Case 1.* $r = p$. First assume $r = p = 2$, so $n$ is even and $H_0 = \Omega_n^{\varepsilon}(q_0)$. From the description of the involution classes in $T$ and $H_0$ given in Section 3.5.4, we immediately deduce that every involution in $T$ has fixed points.

Now assume $r = p > 2$. First we deal with the case $k > 2$, so $H_0 = \mathrm{P}\Omega_n^\varepsilon(q_0)$. By appealing to Propositions 3.5.12 and 3.5.14, we deduce that $H_0$ and $T$ have the same number of classes of elements of order $p$ (note that $q \equiv q_0 \pmod 4$ since $k > 2$). More precisely, there are the same number of classes of elements with a fixed Jordan form $[J_p^{a_p}, \dots, J_1^{a_1}]$. To see that $T$ is $p$-elusive, we need to show that none of the relevant $H_0$-classes are fused in $T$.

First we claim that none of the $\mathrm{PSO}_n^\varepsilon(q_0)$-classes are fused in $\mathrm{PSO}_n^\varepsilon(q)$. To see this, we can argue as in the proof of Proposition 5.5.8, using the explicit class representatives given in [93]. Indeed, every unipotent class in $\mathrm{SO}_n^\varepsilon(q)$ can be represented by a matrix with entries in $\{0, \pm 1, -\alpha, -2\alpha\}$ for some fixed nonsquare $\alpha \in \mathbb{F}_q$. As before, $\alpha$ can be taken to be in the subfield $\mathbb{F}_{q_0}$ and the claim quickly follows.

We can also give an alternative argument based on the description of unipotent classes in Proposition 3.5.12. Let $x, y \in H_0$ be representatives of two distinct $\mathrm{PO}_n^\varepsilon(q_0)$-classes corresponding to the labelled partitions

$$\rho_1 = (\delta_p p^{a_p}, (p-1)^{a_{p-1}}, \dots, 2^{a_2}, \delta_1 1^{a_1})$$

and

$$\rho_2 = (\delta_p' p^{a_p}, (p-1)^{a_{p-1}}, \dots, 2^{a_2}, \delta_1' 1^{a_1})$$

respectively, where $\delta_j, \delta_j' \in \{\square, \boxtimes\}$ for all odd $j$. Since $\rho_1 \neq \rho_2$, it follows that $a_i > 0$ for at least one odd integer $i$.

As explained in Proposition 3.5.12, $x$ and $y$ fix certain orthogonal decompositions of $V_\#$. In terms of this decomposition, let $U_\#$ be the unique nondegenerate subspace of $V_\#$ such that $\hat{x}|_{U_\#}$ has Jordan form $[J_\ell^{a_\ell}]$, where $\ell$ is a fixed odd integer such that $a_\ell > 0$. Similarly, define $W_\# \subseteq V_\#$ so that $\hat{y}|_{W_\#}$ has Jordan form $[J_\ell^{a_\ell}]$. By definition of the labels in $\rho_1$ and $\rho_2$, $\delta_\ell$ is the discriminant of $Q_\#|_{U_\#}$, and similarly $\delta_\ell'$ is the discriminant of $Q_\#|_{W_\#}$. Recall that an element of $\mathbb{F}_{q_0}$ is a square in $\mathbb{F}_q$ if and only if it is a square in $\mathbb{F}_{q_0}$ (since $k$ is odd). Therefore, if we define nondegenerate $\mathbb{F}_q$-spaces $U = U_\# \otimes \mathbb{F}_q$ and $W = W_\# \otimes \mathbb{F}_q$ then the discriminants of $Q|_U$ and $Q|_W$ are also $\delta_\ell$ and $\delta_\ell'$, respectively. It follows that the $\mathrm{PO}_n^\varepsilon(q)$-class of $x$ corresponds to the labelled partition $\rho_1$, and similarly the $\mathrm{PO}_n^\varepsilon(q)$-class of $y$ corresponds to $\rho_2$. But $\rho_1 \neq \rho_2$, so $x$ and $y$ are not $\mathrm{PO}_n^\varepsilon(q)$-conjugate.

Next we claim that none of the relevant $\mathrm{PSO}_n^\varepsilon(q_0)$-classes are fused in $\mathrm{PSO}_n^\varepsilon(q)$. To reach this conclusion, it suffices to show that if the $\mathrm{PO}_n^\varepsilon(q_0)$-class of $x$ splits into two $\mathrm{PSO}_n^\varepsilon(q_0)$-classes, say

$$x^{\mathrm{PO}_n^\varepsilon(q_0)} = x^{\mathrm{PSO}_n^\varepsilon(q_0)} \cup u^{\mathrm{PSO}_n^\varepsilon(q_0)}$$

then $x$ and $u$ are not $\mathrm{PSO}_n^\varepsilon(q)$-conjugate. Here $n$ must be even (since $\mathrm{PO}_n(q_0) = \mathrm{PSO}_n(q_0)$ when $n$ is odd) and Proposition 3.5.14 implies that the $\mathrm{PO}_n^\varepsilon(q)$-class of $x$ splits into two $\mathrm{PSO}_n^\varepsilon(q)$-classes, whence

$$C_{\mathrm{PO}_n^\varepsilon(q)}(x) = C_{\mathrm{PSO}_n^\varepsilon(q)}(x) \qquad (5.5.4)$$

Write $u = x^a$ for some $a \in \mathrm{PO}_n^\varepsilon(q_0) \setminus \mathrm{PSO}_n^\varepsilon(q_0)$. Note that $a \in \mathrm{PO}_n^\varepsilon(q) \setminus \mathrm{PSO}_n^\varepsilon(q)$. If $x$ and $u$ are $\mathrm{PSO}_n^\varepsilon(q)$-conjugate, say $u = x^b$ for some $b \in \mathrm{PSO}_n^\varepsilon(q)$, then

$$ba^{-1} \in C_{\mathrm{PO}_n^\varepsilon(q)}(x) \setminus C_{\mathrm{PSO}_n^\varepsilon(q)}(x)$$

and this contradicts (5.5.4). This justifies the claim.

Finally, let us prove that $T$ is $p$-elusive. To do this, it suffices to show that if the $\mathrm{PSO}_n^\varepsilon(q_0)$-class of $x$ splits into two $H_0$-classes, then these classes are not fused in $T$. We can essentially repeat the argument in the previous paragraph. Firstly, Proposition 3.5.14(i) implies that the $\mathrm{PSO}_n^\varepsilon(q)$-class of $x$ splits into two $T$-classes, so

$$C_{\mathrm{PSO}_n^\varepsilon(q)}(x) = C_T(x) \qquad (5.5.5)$$

Write $x^{\mathrm{PSO}_n^\varepsilon(q_0)} = x^{H_0} \cup v^{H_0}$, where $v = y^c$ for some $c \in \mathrm{PSO}_n^\varepsilon(q_0) \setminus H_0$. Note that $c \in \mathrm{PSO}_n^\varepsilon(q) \setminus T$. If $v = y^d$ for some $d \in T$ then

$$dc^{-1} \in C_{\mathrm{PSO}_n^\varepsilon(q)}(x) \setminus C_T(x)$$

which contradicts (5.5.5). We conclude that $T$ is $p$-elusive if $k > 2$.

Now assume $r = p > 2$ and $k = 2$. For now let us also assume $n$ is odd, so $H_0 = \mathrm{PSO}_n(q_0)$. Define a partition $\lambda$ of $n$ as follows:

$$\lambda = \begin{cases} (2^{(n-1)/2}, 1) & n \equiv 1 \ (\mathrm{mod}\ 4) \\ (3, 2^{(n-3)/2}) & n \equiv 3 \ (\mathrm{mod}\ 4) \end{cases}$$

Now Propositions 3.5.12 and 3.5.14 imply that $H_0$ has a unique class of unipotent elements whose Jordan form corresponds to $\lambda$, but $T$ has two such classes. It follows that $T$ contains derangements of order $p$.

Finally, suppose that $n$ is even and $k = 2$, so $\varepsilon = +$ and Lemma 5.5.3 gives

$$H_0 = \begin{cases} \mathrm{PSO}_n^+(q_0) & n \equiv 2 \ (\mathrm{mod}\ 4),\ q_0 \equiv 1 \ (\mathrm{mod}\ 4) \\ \mathrm{Inndiag}(\mathrm{P}\Omega_n^+(q_0)) & \text{otherwise} \end{cases}$$

By applying Propositions 3.5.12 and 3.5.14 we deduce that there are precisely two $T$-classes and two $\mathrm{PSO}_n^+(q_0)$-classes of elements of order $p$ with Jordan form $[J_3^2, J_1^{n-6}]$. Since $D(Q) = \square$ (see Lemma 2.2.9), the two $T$-classes correspond to the labelled partitions

$$(\square 3^2, \square 1^{n-6}), \quad (\boxtimes 3^2, \boxtimes 1^{n-6})$$

Similarly, the two $\mathrm{PSO}_n^+(q_0)$-classes correspond to labelled partitions

$$(\delta_3 3^2, \delta_1 1^{n-6}), \quad (\delta_3' 3^2, \delta_1' 1^{n-6})$$

where $\delta_3 \delta_1 = \delta_3' \delta_1' = D(Q_\#)$. Let $x$ be in the $\mathrm{PSO}_n^+(q_0)$-class corresponding to $(\delta_3 3^2, \delta_1 1^{n-6})$, and let $U_\#$ be the unique nondegenerate subspace of $V_\#$ such that $\hat{x}|_{U_\#}$ has Jordan form $[J_3^2]$. Then $Q_\#|_{U_\#}$ has discriminant $\delta_3$. However, every scalar $\mu \in \mathbb{F}_{q_0}$ is a square in $\mathbb{F}_q$, so if we set $U = U_\# \otimes \mathbb{F}_q$ then the discriminant of $Q|_U$ is a square. Therefore, the two $\mathrm{PSO}_n^+(q_0)$-classes are fused in $T$, and we deduce that the unipotent elements in $T$ with labelled partition $(\boxtimes 3^2, \boxtimes 1^{n-6})$ are derangements.

*Case 2. $r = 2$, $p > 2$.* In the usual manner, we reduce to the case $k = 2$ since $r$ divides $|\Omega|$. In particular, note that $q \equiv 1 \pmod 8$. If $n$ is odd then $H_0 = \mathrm{PSO}_n(q_0)$ and for each integer $1 \leqslant \ell < n/2$ there is a unique $T$-class of involutions $x \in T$ with $v(x) = \ell$. Consequently, every involution in $T$ has fixed points.

Now assume $n$ is even, so $\varepsilon = +$. Since $q \equiv 1 \pmod 8$, the only involutions $x \in T$ with $v(x) < n/2$ are of type $t_i$ with $1 \leqslant i < n/4$, and they clearly have fixed points. To complete the analysis of involutions, it remains to show that every involution $x \in T$ with $v(x) = n/2$ has fixed points.

First assume $n \equiv 2 \pmod 4$, so $H_0 = \mathrm{PSO}_n^+(q_0)$ if $q_0 \equiv 1 \pmod 4$, and $H_0 = \mathrm{Inndiag}(\mathrm{P\Omega}_n^+(q_0))$ if $q_0 \equiv 3 \pmod 4$ (see Lemma 5.5.3). Here $T$ has a unique class of involutions $x$ with $v(x) = n/2$ (labelled $t_{n/2}$), which clearly have fixed points. Therefore, $T$ is 2-elusive in this situation.

Now assume that $k = 2$ and $n \equiv 0 \pmod 4$, so $H_0 = \mathrm{Inndiag}(\mathrm{P\Omega}_n^+(q_0))$. Here $T$ contains three classes of involutions with $v(x) = n/2$, labelled $t_{n/4}$, $t_{n/2}$ and $t_{n/2-1}$. Clearly, involutions of type $t_{n/4}$ have fixed points, so let us turn to the involutions of type $t_{n/2}$ and $t_{n/2-1}$. Fix a decomposition $V = U \oplus W$, where $U$ and $W$ are maximal totally singular subspaces of $V$.

If $q_0 \equiv 1 \pmod 4$ then take $x = [\lambda I_{n/2}, \lambda^{-1} I_{n/2}] \in H_0$ with respect to this decomposition, where $\lambda \in \mathbb{F}_{q_0}$ has order 4. Similarly, take $x = [I_{n/2}, -I_{n/2}] \in H_0$ if $q_0 \equiv 3 \pmod 4$. In both cases, $x$ has eigenvalues in $\mathbb{F}_q$ and totally singular eigenspaces $U$ and $W$, so $x$ embeds in $T$ as an involution of type $t_{n/2}$ or $t_{n/2-1}$. As explained in Section 3.5.2.10, to obtain an involution in the other $T$-class of this type we use a decomposition $V = U' \oplus W'$, where $U'$ and $W'$ are maximal totally singular subspaces of $V$ such that $\dim U - \dim(U \cap U') = 1$ and $\dim W - \dim(W \cap W') = 1$. We conclude that $T$ is 2-elusive, as claimed.

*Case 3. $r > 2$, $r \neq p$.* As usual, set $i = \Phi(r, q)$, $i_0 = \Phi(r, q_0)$ and note that (5.5.2) holds. First assume $k$ divides $i_0$. If we set $x = \hat{x} Z \in T$, where $\hat{x}$ is defined in (5.5.3), then $x$ is a derangement unless $i$ is odd and $k = 2$. Suppose $k = 2$ (so $\varepsilon = +$ if $n$ is even). If $n \not\equiv 2 \pmod 4$ then $i_0 \equiv 0 \pmod 4$ (since $r$ divides $|\Omega|$), so $i$ is always even in this situation. Similarly, if $n \equiv 2 \pmod 4$ then $i = n/2$ is the only possibility with $i$ odd, and the same element $x$ is a derangement (note that $|H_0|$ is indivisible by $r$). Finally, if $i = i_0$ then $r = k$ (since $r$ divides $|\Omega|$), so $k > 2$ and it is easy to check that $T$ is $r$-elusive.    □

**Proposition 5.5.10** *Theorem 5.5.1 holds in Case IV of Table 5.5.1.*

*Proof*  Here $T = \mathrm{P}\Omega_n^+(q)$ and $H$ is of type $\mathrm{O}_n^-(q_0)$, where $q = q_0^2$. Set $H_0 = H \cap T$ and note that the structure of $H_0$ is given in Lemma 5.5.4. Let $r$ be a prime divisor of $|\Omega|$.

First assume $r = p = 2$, so $H_0 = \Omega_n^-(q_0)$. If $n \equiv 0 \pmod 4$ then there are three $T$-classes of involutions with $v(x) = n/2$, but there is only one in $H_0$, so $T$ contains derangements of order 2 (more precisely, involutions of type $a_{n/2}$ and $a'_{n/2}$ are derangements). However, if $n \equiv 2 \pmod 4$ then the description of unipotent involutions in Section 3.5.4 implies that $T$ is 2-elusive.

Next suppose $r = p > 2$. Recall that the $\mathrm{PO}_n^+(q)$-classes of elements of order $p$ in $T$ are parameterised by labelled partitions (see Proposition 3.5.12). Let $x \in T$ be an element of order $p$ corresponding to the labelled partition $(\boxtimes 3^2, \boxtimes 1^{n-6})$ (note that $D(Q) = \square$ since $q \equiv 1 \pmod 4$). There are precisely two $\mathrm{PSO}_n^-(q_0)$-classes of elements with Jordan form $[J_3^2, J_1^{n-6}]$ and by repeating the argument in the proof of Proposition 5.5.9 (see Case 1, where $n$ is even and $k = 2$) we deduce that both of these classes fuse to the $\mathrm{PO}_n^+(q)$-class corresponding to the labelled partition $(\square 3^2, \square 1^{n-6})$. Therefore, $x$ is a derangement of order $p$.

Next suppose $r = 2$ and $p$ is odd. Let $x \in T$ be an involution and note that $q \equiv 1 \pmod 8$. Visibly, $x$ has fixed points if $v(x) < n/2$, so let us assume $v(x) = n/2$. If $n \equiv 0 \pmod 4$ then there are three such classes in $T$, but only one in $H_0 = \mathrm{PSO}_n^-(q_0)$, so in this case $T$ contains derangements of order 2. Now assume $n \equiv 2 \pmod 4$, so there is a unique class of involutions $x \in T$ with $v(x) = n/2$. If $q_0 \equiv 3 \pmod 4$ then there are involutions $y \in \mathrm{PSO}_n^-(q_0)$ with $v(y) = n/2$ (see Section 3.5.2.13), so in this case $T$ is 2-elusive. Similarly, if $q_0 \equiv 1 \pmod 4$ then $H_0 = \mathrm{Inndiag}(\mathrm{P}\Omega_n^-(q_0))$ and once again we can find involutions $y \in H_0$ with $v(y) = n/2$. Therefore $T$ is 2-elusive if $n \equiv 2 \pmod 4$.

Finally, let us assume $r \neq p$ and $r > 2$. As before, set $i = \Phi(r, q)$, $i_0 = \Phi(r, q_0)$ and note that (5.5.2) holds (with $k = 2$). If $i_0 \equiv 0 \pmod 4$ then $i$ is even and $[\Lambda, I_{n-i}]$ is a derangement. Now assume $i_0 \not\equiv 0 \pmod 4$. Here $i$ is odd and $r$ divides $q_0^{n/2} - 1$ (since $r$ divides $|\Omega|$), so $i_0$ divides $n/2$ and thus $n \equiv 0$

(mod $2i$). In view of Remark 3.5.5(iii), it is easy to see that $[(\Lambda, \Lambda^{-1})^{n/2i}]$ is a derangement in $T$ of order $r$. □

**Proposition 5.5.11** *Theorem 5.5.1 holds in Case V of Table 5.5.1.*

*Proof* Here $T = \mathrm{PSU}_n(q)$ and $H$ is of type $\mathrm{Sp}_n(q)$. Set $H_0 = H \cap T$ and observe that

$$H_0 = \begin{cases} \mathrm{PGSp}_n(q) & (q+1)_2 < (n)_2 \\ \mathrm{PSp}_n(q) & \text{otherwise} \end{cases}$$

(see [86, Proposition 4.5.6]). Let $r$ be a prime divisor of $|\Omega|$.

If $r = p > 2$ then $[J_3, J_1^{n-3}] \in T$ is a derangement of order $r$ since odd size blocks must have even multiplicity in the Jordan form of any unipotent element in $H_0$. On the other hand, if $p = 2$ then $T$ is clearly 2-elusive.

Next suppose $r \neq p$ and $r > 2$. Set $i = \Phi(r, q)$ as in (3.1.2) and note that $i \equiv 2$ (mod 4) since $r$ divides $|\Omega|$. If $i > 2$ then $[\Lambda, I_{n-i/2}]$ is a derangement since $\Lambda \neq \Lambda^{-1}$. Similarly, if $i = 2$ and $(n, r) = 1$ then $[\Lambda, I_{n-1}]$ is a derangement. Finally, if $i = 2$ and $r$ divides $n$ then $[\Lambda^r, I_{n-r}]$ is a derangement.

For the remainder, let us assume $r = 2$ and $p > 2$. If $(q+1)_2 > (n)_2$ then $t_1 \in T$ is a derangement, so let us assume that $(q+1)_2 \leqslant (n)_2$. We claim that $T$ is 2-elusive. If $x = t_i \in T$ with $1 \leqslant i < n/2$ then $i$ is even and thus $x$ has fixed points. It remains to show that all involutions $x \in T$ with $\nu(x) = n/2$ also have fixed points. If $n \equiv 2$ (mod 4) then $T$ has a unique class of involutions with this property (namely, the involutions of type $t'_{n/2}$), so the claim is clear in this case since we can always find an involution $y \in H_0$ with $\nu(y) = n/2$.

Now assume $n \equiv 0$ (mod 4). Clearly, involutions of type $t_{n/4}$ in $\mathrm{PSp}_n(q)$ are $T$-conjugate to $t_{n/2}$, so these elements have fixed points. If $(q+1)_2 = (n)_2$ then these are the only involutions $x \in T$ with $\nu(x) = n/2$, so we may assume that $(q+1)_2 < (n)_2$, in which case $H_0 = \mathrm{PGSp}_n(q)$. Here $H_0$ contains involutions of type $t'_{n/4}$ with eigenvalues in $\mathbb{F}_{q^2} \setminus \mathbb{F}_q$, so these elements embed in $T$ as involutions of type $t'_{n/2}$. We conclude that $T$ is 2-elusive. □

**Proposition 5.5.12** *Theorem 5.5.1 holds in Case VI of Table 5.5.1.*

*Proof* Here $T = \mathrm{PSU}_n(q)$ and $H$ is of type $\mathrm{O}_n^\varepsilon(q)$, where $q$ is odd. Set $H_0 = H \cap T$. The structure of $H_0$ is given in Lemma 5.5.6, and a description of the relevant embedding $\mathrm{O}_n^\varepsilon(q) \leqslant \mathrm{GU}_n(q)$ is given in Construction 5.5.5.

Let $r$ be a prime divisor of $|\Omega|$. If $r = p$ then $[J_2, J_1^{n-2}]$ is a derangement of order $r$ (recall that every even size block in the Jordan form of a unipotent element in $H_0$ must occur with even multiplicity).

Next suppose $r \neq p$ and $r$ is odd. Set $i = \Phi(r, q)$ and let $x = \hat{x}Z \in T$ be an element of order $r$. First assume $i = 2$. If $(n, r) = 1$ then $[\Lambda, I_{n-1}]$ is a derangement. Similarly, if $r$ divides $n$ and $r < n$ then $[\Lambda^r, I_{n-r}]$ is a derangement. Now assume $i = 2$ and $r = n$. Let $\mu \in \mathbb{F}_{q^2}$ be a primitive $r$th root of unity and observe that every element $x \in T$ of order $r$ is diagonalisable over $\mathbb{F}_{q^2}$ (see Proposition 3.3.3).

If $r \geqslant 7$ then $\hat{x} = [\mu, \mu^2, \mu^{r-3}, I_{n-3}]$ is a derangement of order $r$. Next assume that $n = r = 5$. If $q \equiv -1 \pmod{25}$ then $[\mu, I_4]$ is a derangement in $T$ of order 5 (note that such an element belongs to $T$ by Lemma 2.3.1). On the other hand, if $q \not\equiv -1 \pmod{25}$ then $\det(\hat{x}) = 1$ for all $x = \hat{x}Z \in T$ of order 5, and by choosing an appropriate scalar multiple we may assume that

$$\hat{x} = [\mu I_a, \mu^4 I_a, \mu^2 I_b, \mu^3 I_b, I_c]$$

for some non-negative integers $a, b$ and $c$ with $2(a + b) + c = 5$. Therefore $x$ has fixed points and we conclude that $T$ is 5-elusive. Similarly, if $n = r = 3$ and $i = 2$ then $T$ is 3-elusive if and only if $q \not\equiv -1 \pmod{9}$.

Now assume $i \neq 2$. If $n$ is odd then $i \equiv 2 \pmod{4}$ (since $r$ divides $|\Omega|$) and thus $[\Lambda, I_{n-i/2}]$ is a derangement (since $\Lambda \neq \Lambda^{-1}$, there are no such elements in $H_0$). Next suppose $n$ is even and $\varepsilon = +$, in which case either $i \equiv 2 \pmod{4}$, or $i$ divides $n$ and $n/i$ is odd. Set $\hat{x} = [\Lambda, I_{n-i/2}]$ if $i \equiv 2 \pmod{4}$, otherwise set $\hat{x} = [(\Lambda, \Lambda^{-q})^{n/i}]$. Then $x$ is a derangement in $T$ of order $r$ (note that if $i = n$ then $r$ does not divide $|H_0|$). Finally, suppose $n$ is even and $\varepsilon = -$, so either $i \equiv 2 \pmod{4}$ or $i$ divides $n/2$ (since $r$ divides $|\Omega|$). Set

$$\hat{x} = \begin{cases} [\Lambda, I_{n-i/2}] & i \equiv 2 \pmod{4} \\ [(\Lambda, \Lambda^{-q})^{n/2i}] & i \text{ divides } n/2 \text{ and } i \text{ is odd} \\ [(\Lambda, \Lambda^{-q})^{n/i}] & i \text{ divides } n/2 \text{ and } i \equiv 0 \pmod{4} \end{cases}$$

In each case, it is easy to see that $x$ is a derangement (see Remark 3.5.5(iii)).

Finally, let us assume $r = 2$. If $n$ is odd then $H_0 = \mathrm{PSO}_n(q)$ and it is clear that each involution in $T$ has fixed points. For the remainder, let us assume $n$ is even, so $|H_0 : \mathrm{PSO}_n^\varepsilon(q)| = 2$.

First assume $(q + 1)_2 \leqslant (n)_2$. Note that $T$ does not contain any involutions of type $t_i$ with $i \leqslant n/2$ odd (see Section 3.3.2.1). Clearly, if $x = t_i$ and $i < n/2$ is even then $x$ has fixed points, so it remains to consider the involutions labelled $t_{n/2}$ and $t'_{n/2}$. From the structure of $H_0$ in Lemma 5.5.6 it is easy to see that $H_0$ always contains involutions $y$ with $\nu(y) = n/2$, so $T$ is 2-elusive if it has a unique class of such involutions. Therefore, we may assume that $T$ contains involutions of type $t_{n/2}$ and $t'_{n/2}$, in which case $n \equiv 0 \pmod{4}$ and $(q + 1)_2 < (n)_2$. In particular, $H_0 = \mathrm{Inndiag}(\mathrm{P}\Omega_n^\varepsilon(q))$ (see Lemma 5.5.6).

Now, if $x \in H_0$ is an involution of type $[-I_{n/2}, I_{n/2}]$ with nondegenerate eigenspaces, then $x$ is clearly $T$-conjugate to $t_{n/2}$, so it remains to determine whether or not $H_0$ contains involutions of type $t'_{n/2}$. From the description of semisimple involutions in Section 3.5.2, it follows that there are involutions $y \in H_0$ with $v(y) = n/2$ and eigenvalues in $\mathbb{F}_{q^2} \setminus \mathbb{F}_q$ (also see Table B.9 in Appendix B), which must embed in $T$ as involutions of type $t'_{n/2}$. Therefore, $T$ is 2-elusive.

Finally, let us assume $(q+1)_2 > (n)_2 > 1$, so $H_0 = \mathrm{PO}_n^\varepsilon(q)$ (see Lemma 5.5.6). As before, any involution in $T$ of type $t_i$ with $i < n/2$ has fixed points. The condition $(q+1)_2 > (n)_2$ implies that $T$ has a unique class of involutions $x$ with $v(x) = n/2$, so these elements also have fixed points and once again we conclude that $T$ is 2-elusive. $\qquad\square$

This completes the proof of Theorem 5.5.1.

## 5.6 $\mathscr{C}_6$: Local subgroups

Let $k$ be a prime number. A finite $k$-group $R$ is said to be of *symplectic-type* if every characteristic abelian subgroup of $R$ is cyclic. As noted in Section 2.6.2.6, the structure of such a group is closely related to the extraspecial $k$-groups (see [116, p. 75–76], for example).

Let $R$ be a symplectic-type $k$-group of exponent $k(k,2)$, and fix a prime $p \neq k$. It is well known that $R$ has precisely $|Z(R)| - 1$ inequivalent faithful absolutely irreducible representations over an algebraically closed field of characteristic $p$ (see [86, Proposition 4.6.3]). Furthermore, each of these representations has degree $k^m$ for some fixed integer $m \geqslant 1$, and the smallest field over which they can be realised is $\mathbb{F}_{p^e}$, where

$$e = \min\{z \in \mathbb{N} \mid p^z \equiv 1 \ (\mathrm{mod}\ |Z(R)|)\} \qquad (5.6.1)$$

This allows us to embed $R$ into certain classical groups $J$ of dimension $n = k^m$ over $\mathbb{F}_q$, where $L = J^\infty$ is quasisimple and

$$q = \begin{cases} p^{e/2} & L \text{ is unitary} \\ p^e & \text{otherwise} \end{cases} \qquad (5.6.2)$$

Let $G$ be an almost simple classical group with socle $T = L/Z(L)$ and let $H_0 < T$ be the image (modulo scalars) of $N_L(R)$. Then the members of the collection $\mathscr{C}_6$ are the subgroups of the form $H = N_G(H_0)$ that arise in this way.

The above restriction on the underlying field $\mathbb{F}_q$ ensures that a subgroup in $\mathscr{C}_6$ is not contained in a member of the subfield subgroup collection $\mathscr{C}_5$. The full list of possibilities for $G$ and $H$ is given in Table 5.6.1 (see [86, Table 4.6.B]). Here $n = k^m$ and $q$ satisfies (5.6.2), with $e$ given in (5.6.1). As usual, the *type of H* provides an approximate description of the group-theoretic structure of $H \cap \mathrm{PGL}(V)$, where $V$ denotes the natural $T$-module.

**Theorem 5.6.1** *Let $G \leqslant \mathrm{Sym}(\Omega)$ be a primitive almost simple classical group with socle $T$ and point stabiliser $H \in \mathscr{C}_6$. Let $r$ be a prime divisor of $|\Omega|$. Then $T$ is r-elusive if and only if $(G, r)$ is one of the cases in Table 5.6.2.*

Table 5.6.1 *The collection $\mathscr{C}_6$*

| Case | $T$ | Type of $H$ | $|Z(R)|$ | Conditions |
|------|-----|-------------|----------|------------|
| I | $\mathrm{PSL}_n^\varepsilon(q)$ | $k^{2m}.\mathrm{Sp}_{2m}(k)$ | $k$ | $k$ odd, $\varepsilon = (-)^{e+1}$ |
| II | $\mathrm{PSL}_n^\varepsilon(q)$ | $2^{2m}.\mathrm{Sp}_{2m}(2)$ | $4$ | $k = 2, m \geqslant 2, q = p \equiv \varepsilon \ (\mathrm{mod}\ 4)$ |
| III | $\mathrm{PSL}_2(q)$ | $2^2.\mathrm{O}_2^-(2)$ | $2$ | $(k, m) = (2, 1), q = p \geqslant 5$ |
| IV | $\mathrm{PSp}_n(q)$ | $2^{2m}.\mathrm{O}_{2m}^-(2)$ | $2$ | $k = 2, m \geqslant 2, q = p \geqslant 3$ |
| V | $\mathrm{P\Omega}_n^+(q)$ | $2^{2m}.\mathrm{O}_{2m}^+(2)$ | $2$ | $k = 2, m \geqslant 3, q = p \geqslant 3$ |

Table 5.6.2 *r-elusive $\mathscr{C}_6$-actions*

| Case | r | Conditions |
|---|---|---|
| I | 2 | $(k,m) = (3,1)$ |
|   | 3 | $(k,m) = (3,1)$, $q \equiv \varepsilon \pmod 9$ |
| II | 2,3 | $m = 2$ |
|   | $2^m \pm 1, r \neq p$ | $c > n/2$ |
| III | 2,3 | always |
| IV | 2 | $m = 2$ |
|   | $2^m + 1, r \neq p$ | $c > n/2$ |
| V | $2^m - 1, r \neq p$ | $c > n/2$ |

**Remark 5.6.2** In Table 5.6.2, the integer $c$ is defined in Definition 5.1.1.

The following lemma will be useful in the proof of Theorem 5.6.1 (see [18, Lemma 6.3]).

**Lemma 5.6.3** *Let G be an almost simple classical group with socle T and natural module V with* $\dim V = n$. *Let H be a $\mathscr{C}_6$-subgroup of G. Then* $v(x) \geqslant n/4$ *for all nontrivial* $x \in H \cap \mathrm{PGL}(V)$.

**Proposition 5.6.4** *Theorem 5.6.1 holds in Case I of Table 5.6.1.*

*Proof* Here $T = \mathrm{PSL}_n^\varepsilon(q)$ and $H$ is of type $k^{2m}.\mathrm{Sp}_{2m}(k)$, where $k$ is odd, $\varepsilon = (-)^{e+1}$ and $n = k^m$ (here $e$ is the integer defined in (5.6.1), where $|Z(R)| = k$ in this case). We will assume $\varepsilon = +$ since a very similar argument applies when $\varepsilon = -$. Write $q = p^e$ and note that $e$ is odd and $k$ divides $q - 1$. Set $H_0 = H \cap T$. Then [86, Proposition 4.6.5] states that

$$H_0 = \begin{cases} 3^2.Q_8 & (k,m) = (3,1) \text{ and } q \equiv 4,7 \pmod 9 \\ k^{2m}.\mathrm{Sp}_{2m}(k) & \text{otherwise} \end{cases}$$

Let $r$ be a prime divisor of $|\Omega|$.

First assume $r = p$. In view of Lemma 5.6.3, we immediately deduce that $[J_2, J_1^{n-2}]$ is a derangement of order $r$ if $n \geqslant 5$. If $(n,p) = (3,2)$ then $e = 2$ is even, so we may assume $n = 3$ and $p \geqslant 5$ (note that $k = 3$ if $n = 3$, so $p \neq 3$). But $p$ does not divide $|H_0|$, so in this case every element in $T$ of order $p$ is a derangement.

Next suppose $r = 2$ and $p > 2$. Since $n$ is odd, there are involutions $x \in T$ of type $t_1$ with $v(x) = 1$, so as above we have reduced to the case $n = 3$ (with $p \geqslant 5$). Here $T$ contains a unique class of involutions, so $T$ is 2-elusive.

Now assume that $r = k$, so $r$ divides $q - 1$. Let $\mu \in \mathbb{F}_q$ be a primitive $r$th root of unity and set $\hat{x} = [\mu, \mu^{-1}, I_{n-2}]$, so $x = \hat{x}Z \in T$ has order $r$. Then $x$ is a derangement if $n \geqslant 9$ (see Lemma 5.6.3), so we may assume that $m = 1$ and $k \in \{3, 5, 7\}$. If $(k, m) = (7, 1)$ then [72, Theorem 7.1] implies that $v(y) \geqslant 3$ for all nontrivial $y \in H_0$, so the above element $x$ is a derangement in this case too.

Next suppose $(r, k, m) = (3, 3, 1)$, so $n = 3$, $q = p$ and $p \equiv 1 \pmod 3$. In fact, since $r$ divides $|\Omega|$ it follows that $p \equiv 1 \pmod 9$ and $H_0 = 3^2.\mathrm{Sp}_2(3)$. Observe that $H_0$ is the image (modulo scalars) of $N_{\mathrm{SL}_3(p)}(R)$, where $R = \langle a, b \rangle \cong 3^{1+2}$ is an extraspecial 3-group of order $3^3$ and exponent 3, and the generators $a$ and $b$ are as follows:

$$a = \begin{pmatrix} 0 & 1 & 0 \\ 0 & 0 & 1 \\ 1 & 0 & 0 \end{pmatrix}, \quad b = \begin{pmatrix} 1 & & \\ & \mu & \\ & & \mu^2 \end{pmatrix}$$

(see [86, p. 151]). Now $T$ has precisely two classes of subgroups of order 3, with representatives generated by the diagonal matrices $x_1 = [1, 1, \mu]$ and $x_2 = [1, \mu, \mu^2]$ (modulo scalars). Both $x_1$ and $x_2$ clearly normalise $R$, so all elements in $T$ of order 3 have fixed points and thus $T$ is 3-elusive.

Now consider the case $(r, k, m) = (5, 5, 1)$. Here $n = 5$, $q = p$, $p \equiv 1 \pmod 5$ and $H_0 = 5^2.\mathrm{Sp}_2(5)$. If $p \equiv 1 \pmod{25}$ then Lemma 5.6.3 implies that $[\mu, I_4] \in T$ is a derangement, so let us assume that $p \not\equiv 1 \pmod{25}$. Note that there are three conjugacy classes of subgroups of order 5 in $5^2.\mathrm{Sp}_2(5)$ and the same number in $T$. We claim that $v(x) \geqslant 3$ for all $x \in H_0$ of order 5. To see this, we first use MAGMA [11] to construct $R = 5^{1+2} < \mathrm{SL}_5(11)$ and the (ordinary) character table of $K = N_{\mathrm{SL}_5(11)}(R)$, where $R = \langle a, b \rangle$ with

$$a = \begin{pmatrix} 0 & 1 & 0 & 0 & 0 \\ 0 & 0 & 1 & 0 & 0 \\ 0 & 0 & 0 & 1 & 0 \\ 0 & 0 & 0 & 0 & 1 \\ 1 & 0 & 0 & 0 & 0 \end{pmatrix}, \quad b = \begin{pmatrix} 1 & & & & \\ & \mu & & & \\ & & \mu^2 & & \\ & & & \mu^3 & \\ & & & & \mu^4 \end{pmatrix}$$

Note that $H_0 = K/Z$, where $Z \cong C_5$ is the centre of $K$. The character table indicates that $K$ has precisely four faithful irreducible ordinary characters of degree 5, one of which must be the Brauer character $\chi$ corresponding to the relevant irreducible representation $\rho : K \to \mathrm{SL}_5(p)$. By inspecting the values of $\chi$, we quickly deduce that $v(x) \geqslant 3$ for all $x \in K$ of order 5. This justifies the claim and we conclude that $[\mu, \mu^{-1}, I_3]$ is a derangement in $T$ of order 5.

To complete the proof of the proposition, we may assume that $r > 2$ and $r \neq p, k$. Since $r$ divides $|T|$, we may define $i = \Phi(r, q)$ as in (3.1.2) (so $i \geqslant 1$ is minimal such that $r$ divides $q^i - 1$). If $1 < i < n/4$ then $[\Lambda, I_{n-i}]$ is a derangement by Lemma 5.6.3. Next suppose $i = 1$. Here a combination of Lemma 5.6.3 and [72, Theorem 7.1] implies that $[\Lambda, \Lambda^{-1}, I_{n-2}]$ is a derangement if $n \geqslant 7$, so we may assume that $n \in \{3, 5\}$. If $n = 3$ then $k = 3$, $q = p$ and we can discard this case since 2 and 3 are the only prime divisors of $|H_0|$. For $n = 5$, the only possibility is $r = 3$. Here $H_0 = 5^2.\mathrm{Sp}_2(5)$ contains a unique conjugacy class of subgroups of order 3, but there are at least two $T$-classes of such subgroups in $T$. Therefore, $T$ contains derangements of order 3.

For the remainder we may assume that $i \geqslant n/4$ and $r$ divides both $|\Omega|$ and $|H_0|$. Since $(r, q) = 1$, Fermat's Little Theorem implies that $r$ divides $q^{r-1} - 1$, so $i$ divides $r - 1$. Furthermore, since $r \neq k$ and $r$ divides $|H_0|$, it follows that $r$ divides $\prod_{\ell=1}^m (k^{2\ell} - 1)$. In particular, $r \leqslant (k^m + 1)/2$ (we may divide by 2 since $r$ and $k$ are both odd) and thus

$$i \leqslant (k^m - 1)/2 < n/2.$$

Since $i \geqslant n/4$ and $i$ divides $r - 1$, we have $r \geqslant k^m/4 + 1$. This implies that $r$ divides $k^{2m} - 1$; if not then $r \leqslant (k^{m-1} + 1)/2$, which is incompatible with the previous lower bound. More precisely, if we set $i_k = \Phi(r, k)$ then either $i_k = 2m$, or $m$ is odd and $i_k = m$. In particular, there is a unique $H_0$-class of subgroups of order $r$ in $H_0$. However, the condition $i < n/2$ implies that there are at least two $T$-classes of such subgroups in $T$, so $T$ contains derangements of order $r$.  $\square$

**Proposition 5.6.5** *Theorem 5.6.1 holds in Case II of Table 5.6.1.*

*Proof* Here $T = \mathrm{PSL}_n^\varepsilon(q)$ and $H$ is of type $2^{2m}.\mathrm{Sp}_{2m}(2)$, where $n = 2^m$, $m \geqslant 2$ and $q = p \equiv \varepsilon \pmod 4$. We will assume that $\varepsilon = -$ since the case $\varepsilon = +$ is very similar. Set $H_0 = H \cap T$. By [86, Proposition 4.6.6] we have

$$H_0 = \begin{cases} 2^4.A_6 & n = 4, \ p \equiv 3 \pmod 8 \\ 2^{2m}.\mathrm{Sp}_{2m}(2) & \text{otherwise} \end{cases}$$

Let $r$ be a prime divisor of $|\Omega|$.

First assume $r = p$. If $m \geqslant 3$ then Lemma 5.6.3 implies that $[J_2, J_1^{n-2}]$ is a derangement, so let us assume $m = 2$. Clearly, if $r$ does not divide $|H_0|$ then $T$ contains derangements of order $r$, so we may assume that $r$ divides $|H_0|$, in which case $p = 3$ is the only possibility. There are precisely two $H_0$-classes of elements of order 3 in $H_0$, but there are four such classes in $T$. We conclude that $T$ always contains derangements of order $p$.

Next suppose $r = 2$. A combination of Lemma 5.6.3 and [72, Theorem 7.1] implies that $t_2$ involutions are derangements if $m \geqslant 3$, so we reduce to the case $m = 2$. If $p \equiv 3 \pmod 8$ then $T$ contains a unique class of involutions, so $T$ is

2-elusive in this situation. Now assume $p \equiv 7 \pmod 8$, so $H_0 = 2^4.\mathrm{Sp}_4(2) = K/Z$, where $Z \cong C_4$ is the centre of $K = N_{\mathrm{SU}_4(p)}(R)$ and

$$R = \langle a,b,c,d,e \rangle \cong C_4 \circ D_8 \circ D_8 = 4 \circ 2^{1+4}$$

is a symplectic-type 2-group of order $2^6$, with generators

$$a = \mu I_4, \quad b = [\mu I_2, \mu^{-1} I_2], \quad c = [\mu, \mu^{-1}, \mu, \mu^{-1}]$$

$$d = \begin{pmatrix} & & 1 & 0 \\ & & 0 & 1 \\ -1 & 0 & & \\ 0 & -1 & & \end{pmatrix}, \quad e = \begin{pmatrix} 0 & 1 & & \\ -1 & 0 & & \\ & & 0 & 1 \\ & & -1 & 0 \end{pmatrix} \qquad (5.6.3)$$

in terms of an orthonormal basis of the natural module for $\mathrm{SU}_4(p)$, with $\mu \in \mathbb{F}_{p^2}$ of order 4 (see [86, p. 152]). Now $T$ contains precisely two classes of involutions, with representatives labelled $t_1$ and $t_2$ (see Section 3.3.2). It is easy to check that $R$ is normalised by both elements, so all involutions in $T$ have fixed points.

Finally, suppose $r > 2$ and $r \ne p$. We may assume that $|H_0|$ is divisible by $r$. Set $i = \Phi(r,q)$ and define the integer $c = c(i)$ as in Definition 5.1.1, so

$$c = \begin{cases} i & i \equiv 0 \pmod 4 \\ i/2 & i \equiv 2 \pmod 4 \\ 2i & i \text{ odd} \end{cases}$$

It is easy to see that $T$ contains derangements of order $r$ if $1 < c < n/4$. For example, if $i \not\equiv 2 \pmod 4$ then Lemma 5.6.3 implies that $[\Lambda, \Lambda^{-q}, I_{n-c}]$ is a derangement. Next suppose $c = 1$, so $i = 2$. By applying Lemma 5.6.3 and [72, Theorem 7.1], we deduce that $[\Lambda, \Lambda^{-1}, I_{n-2}]$ is a derangement if $m \geqslant 3$. Now assume $m = 2$, so $r = 3$ or 5 (since $r$ divides $|H_0|$). There is a unique $H_0$-class of subgroups of order 5 in $H_0$, but there are at least two such classes in $T$, so $T$ contains derangements of order 5. Similarly, we can find derangements of order 3 since $H_0$ has two classes of subgroups of order 3, but there are three in $T$.

For the remainder we may assume that $c \geqslant n/4$. First assume $c > n/2$. Here $T$ contains a unique class of subgroups of order $r$, so $T$ is $r$-elusive in this situation. More precisely, we claim that $r = 2^m \pm 1$ is the only possibility (so $r$ is a Fermat or Mersenne prime). To see this, first note that $i$ divides $r - 1$ (by Fermat's Little Theorem), and $r$ divides $\prod_{\ell=1}^{m} (2^{2\ell} - 1)$. In particular, if $i$ is even then $r \geqslant i + 1 \geqslant c + 1 > 2^{m-1} + 1$, so $r \geqslant 2^{m-1} + 3$ and thus $r = 2^m \pm 1$ as claimed. Similarly, if $i$ is odd then $i$ divides $(r-1)/2$, so $r \geqslant 2i + 1 = c + 1 > 2^{m-1} + 1$ and the same conclusion follows. We note that examples do arise. For instance, suppose $G = \mathrm{PSU}_{16}(131)$, $H = 2^8.\mathrm{Sp}_8(2)$ and $r = 17$. Then $r$ divides

$|H_0|$ and $|\Omega|$, and we calculate that $c = 16$ so $T$ has a unique class of elements of order 17, and all such elements have fixed points.

Finally, let us assume that $n/4 \leqslant c \leqslant n/2$. As before, $i$ divides $r - 1$, and $r$ divides $\prod_{\ell=1}^{m} (2^{2\ell} - 1)$. Let $i_2 = \Phi(r, 2)$, so $i_2 \geqslant 1$ is minimal such that $r$ divides $2^{i_2} - 1$.

First assume $i \equiv 0 \pmod 4$, so $c = i$. Then $r \geqslant 2^{m-2} + 1$ and we deduce that either $i_2 \in \{2m - 4, 2m - 2, 2m\}$, or $m$ is even and $i_2 = m - 1$, or $m$ is odd and $i_2 = m$. In particular, if $m \geqslant 5$ then $H_0$ contains a unique conjugacy class of subgroups of order $r$, but the condition $c \leqslant n/2$ implies that there are at least two such classes in $T$. Therefore $T$ contains derangements of order $r$ if $m \geqslant 5$. If $m = 4$ then we may as well assume $i_2 = 4$ (otherwise $H_0$ has a unique class of subgroups of order $r$), so $r = 5$ and $i = 4$. Here $H_0$ has two classes of subgroups of order 5, and $T$ has at least four, so once again $T$ contains derangements. Similarly, if $m = 3$ then we may as well assume $i_2 = 2$, so $r = 3$ but this is incompatible with the condition $i \equiv 0 \pmod 4$. Finally, the case $m = 2$ does not arise since we are assuming $n/4 \leqslant c \leqslant n/2$.

Next suppose $i \equiv 2 \pmod 4$, so $c = i/2$. Then $r \geqslant i + 1 \geqslant 2^{m-1} + 1$, so either $i_2 \in \{2m - 2, 2m\}$, or $i_2 = m$ is odd. If $(m, r) = (2, 3)$ then $i = i_2 = 2$ and we calculate that $H_0$ has exactly two classes of subgroups of order 3, but $T$ has three, so $T$ contains derangements of order 3. In every other case, $H_0$ contains a unique class of subgroups of order $r$, so once again $T$ is not $r$-elusive.

Finally, let us assume $i$ is odd, so $c = 2i$. Here $i$ divides $(r - 1)/2$, so $r \geqslant 2i + 1 \geqslant 2^{m-2} + 1$ and as before we quickly reduce to the case $m \leqslant 4$. If $m = 4$ then we may as well assume $i_2 = 4$, so $r = 5$. But this is not possible since $c = 2i \geqslant n/4 = 4$ and $i$ is odd. If $m = 3$ then we may assume $i_2 = 2$, so $r = 3$ and $i = 1$. Here $T$ contains four classes of subgroups of order 3, but $H_0$ only contains three. Finally, suppose $m = 2$, so $i = 1$ and $r \in \{3, 5\}$. If $r = 5$ then $i_2 = 4$ and $H_0$ has a unique class of subgroups of order $r$, so $T$ contains derangements of order $r$. On the other hand, if $r = 3$ then $T$ and $H_0$ both contain two classes of subgroups of order 3, and we claim that all elements in $T$ of order 3 have fixed points.

To see this, we first use MAGMA and the generators in (5.6.3) to construct $R \cong 4 \circ 2^{1+4}$ as a subgroup of $\mathrm{SU}_4(7)$, so $H_0 \leqslant K/Z$ where $K = N_{\mathrm{SU}_4(7)}(R)$ and $Z \cong C_4$ is the centre of $K$. From the character table, we observe that $K$ has two ordinary irreducible characters of degree 4, one of which must be the Brauer character $\chi$ of the relevant irreducible representation $\rho : K \to \mathrm{SU}_4(p)$. Now $K$ has precisely two classes of elements of order 3, with representatives $x_1$ and $x_2$, say, and we note that $T$ also has two such classes. The character table of $K$ indicates that $\chi(x_1) \neq \chi(x_2)$, so the corresponding classes in $H_0$ are not fused in $T$. We conclude that $T$ is 3-elusive. $\qquad\square$

**Proposition 5.6.6** *Theorem 5.6.1 holds in Case III of Table 5.6.1.*

*Proof* Here $T = \mathrm{PSL}_2(q)$ and $H$ is of type $2^2.O_2^-(2)$, with $q = p \geqslant 5$. Set $H_0 = H \cap T$. By [86, Proposition 4.6.7] we have

$$H_0 = \begin{cases} A_4 & p \equiv \pm 3 \ (\mathrm{mod}\ 8) \\ S_4 & \text{otherwise} \end{cases}$$

We may as well assume $r$ divides $|H_0|$, so $r = 2$ or $3$. Now $T$ has a unique class of involutions, so $T$ is 2-elusive. Similarly, $T$ contains a unique class of subgroups of order 3, so $T$ is also 3-elusive. $\qquad\qquad\square$

**Proposition 5.6.7** *Theorem 5.6.1 holds in Case IV of Table 5.6.1.*

*Proof* Here $T = \mathrm{PSp}_n(q)$ and $H$ is of type $2^{2m}.O_{2m}^-(2)$, where $q = p \geqslant 3$ and $n = 2^m$ with $m \geqslant 2$. Set $H_0 = H \cap T$. According to [86, Proposition 4.6.9] we have

$$H_0 = \begin{cases} 2^{2m}.\Omega_{2m}^-(2) & p \equiv \pm 3 \ (\mathrm{mod}\ 8) \\ 2^{2m}.O_{2m}^-(2) & \text{otherwise} \end{cases}$$

Let $r$ be a prime divisor of $|\Omega|$. We may assume that $r$ also divides $|H_0|$.

First assume $r = p$. If $m \geqslant 3$ then $[J_2, J_1^{n-2}]$ is a derangement (see Lemma 5.6.3) so we may assume that $m = 2$, in which case $r \in \{3, 5\}$. If $p = 3$ then $H_0$ has a unique class of elements of order 3, but $T$ has four. Similarly, if $p = 5$ then $H_0$ has two classes of elements of order 5, but there are six in $T$. We conclude that $T$ always contains derangements of order $p$.

Next suppose $r = 2$. If $x \in T$ is a $t_1$ involution then $v(x) = 2$ and thus $x$ is a derangement if $m \geqslant 3$ (in view of Lemma 5.6.3 and [72, Theorem 7.1]), so once again we reduce to the case $m = 2$. Here we claim that $T$ is 2-elusive. To see this, first observe that $H_0 \leqslant K/Z$, where $K = N_{\mathrm{Sp}_4(p)}(R)$, $Z \cong C_2$ is the centre of $K$ and $R$ is the subgroup of $\mathrm{Sp}_4(p)$ generated by the matrices

$$a = \begin{pmatrix} 0 & 1 & & \\ -1 & 0 & & \\ & & 0 & 1 \\ & & -1 & 0 \end{pmatrix}, \quad b = \begin{pmatrix} 1 & & & \\ & -1 & & \\ & & 1 & \\ & & & -1 \end{pmatrix}$$

$$c = \begin{pmatrix} & & 1 & 0 \\ & & 0 & 1 \\ -1 & 0 & & \\ 0 & -1 & & \end{pmatrix}, \quad d = \begin{pmatrix} u & 0 & v & 0 \\ 0 & u & 0 & v \\ v & 0 & -u & 0 \\ 0 & v & 0 & -u \end{pmatrix}$$

where $u, v \in \mathbb{F}_p$ satisfy the equation $u^2 + v^2 = -1$ (see [86, p. 153–154]). Here the matrices are written with respect to a symplectic basis $\{e_1, e_2, f_1, f_2\}$ of the natural $\mathrm{Sp}_4(p)$-module. In particular, if $p \equiv 1 \pmod 4$ then we may assume that $u \in \mathbb{F}_p$ has order 4 and $v = 0$. The desired result now follows since $b$ is a $t_1$-type involution, $d$ is of type $t_2$ if $p \equiv 1 \pmod 4$, and $a$ is of type $t_2'$ if $p \equiv 3 \pmod 4$.

Finally, let us assume $r > 2$ and $r \neq p$. Set $i = \Phi(r, q)$. First assume $i$ is even. If $i < n/4$ then Lemma 5.6.3 implies that $[\Lambda, I_{n-i}]$ is a derangement, so let us assume $i \geqslant n/4$. If $i > n/2$ then $T$ has a unique class of subgroups of order $r$, so $T$ is $r$-elusive. Moreover, $r = 2^m + 1$ is a Fermat prime. To see this, observe that $i$ divides $r - 1$ by Fermat's Little Theorem, so $r \geqslant 2^{m-1} + 3$ and $r$ divides

$$(2^m + 1) \prod_{\ell=1}^{m-1} (2^{2\ell} - 1)$$

so $r = 2^m + 1$ is the only possibility. Note that examples do exist. For instance, take $G = \mathrm{PSp}_{16}(653)$, $H = 2^8.\Omega_8^-(2)$ and $r = 17$.

Next assume $i$ is even and $n/4 \leqslant i \leqslant n/2$. Set $i_2 = \Phi(r, 2)$. Then $r \geqslant n/4 + 1 = 2^{m-2} + 1$ and

$$i_2 \in \{m - 1, 2m - 4, 2m - 2, 2m\}$$

where $i_2 = m - 1$ only if $m$ is even. In particular, $H_0$ has a unique class of subgroups of order $r$, but $T$ has at least two, so $T$ contains derangements of order $r$.

Finally, assume $i$ is odd. If $i < n/8$ then $[\Lambda, \Lambda^{-1}, I_{n-2i}]$ is a derangement by Lemma 5.6.3. Next suppose $i > n/4$. As above, we see that $i$ divides $r - 1$, so $i$ divides $(r - 1)/2$ since $ir$ is odd. It follows that $r \geqslant 2i + 1 \geqslant 2^{m-1} + 3$ and again we deduce that $r = 2^m + 1$ is the only possibility. Examples do exist. For instance, take $G = \mathrm{PSp}_{16}(3469)$, $H = 2^8.\Omega_8^-(2)$ and $r = 17$. Finally, suppose $i$ is odd and $n/8 \leqslant i \leqslant n/4$. Then $r \geqslant 2i + 1 \geqslant 2^{m-2} + 1$ and as before we deduce that $H_0$ has a unique class of subgroups of order $r$, but there are at least two classes in $T$. Therefore, $T$ contains derangements of order $r$.                   $\square$

**Proposition 5.6.8** *Theorem 5.6.1 holds in Case V of Table 5.6.1.*

*Proof* This is similar to the proof of the previous proposition. Here $T = \mathrm{P}\Omega_n^+(q)$ and $H$ is of type $2^{2m}.O_{2m}^+(2)$, where $q = p \geqslant 3$ and $n = 2^m$ with $m \geqslant 3$. Set $H_0 = H \cap T$ and observe that

$$H_0 = \begin{cases} 2^{2m}.\Omega_{2m}^+(2) & p \equiv \pm 3 \pmod 8 \\ 2^{2m}.O_{2m}^+(2) & \text{otherwise} \end{cases}$$

(see [86, Proposition 4.6.8]). Let $r$ be a prime divisor of $|\Omega|$. By applying Lemma 5.6.3 and [72, Theorem 7.1], we see that $[J_2^2, J_1^{n-4}]$ is a derangement of order $p$, and involutions of type $t_1$ and $t_1'$ are also derangements.

For the remainder, we may assume that $r > 2$, $r \neq p$ and $r$ divides $|H_0|$. Set $i = \Phi(r, q)$ and assume that $i$ is even for now. If $i < n/4$ then Lemma 5.6.3 implies that $[\Lambda, I_{n-i}]$ is a derangement, so let us assume $i \geqslant n/4$. If $i > n/2$ then $T$ has a unique class of subgroups of order $r$, so $T$ is $r$-elusive. Moreover, we claim that $r = 2^m - 1$ is a Mersenne prime. Indeed, $i$ divides $r - 1$ by Fermat's Little Theorem, so $r \geqslant 2^{m-1} + 3$ and the claim follows since $r$ divides

$$(2^m - 1) \prod_{\ell=1}^{m-1} (2^{2\ell} - 1)$$

It is easy to see that examples do exist. For instance, take $G = \mathrm{P}\Omega_8^+(19)$, $H = 2^6.\Omega_6^+(2)$ and $r = 7$.

Next assume $i$ is even and $n/4 \leqslant i \leqslant n/2$. Set $i_2 = \Phi(r, 2)$ and exclude the case $m = 4$ for now. Then $r \geqslant n/4 + 1 = 2^{m-2} + 1$ and we deduce that

$$i_2 \in \{m-1, m, 2m-4, 2m-2\}$$

where $i_2 = m - 1$ only if $m$ is even, and $i_2 = m$ only if $m$ is odd. In particular, $H_0$ contains a unique class of subgroups of order $r$, but there are at least two such classes in $T$, so $T$ contains derangements of order $r$. Finally, suppose $m = 4$, so $i_2 \in \{3, 4, 6\}$. If $i_2 = 3$ or $6$ then once again $H_0$ has a unique class of subgroups of order $r$ and the result follows. Similarly, if $i_2 = 4$ then $r = 5$ and $i = 4$ (since we are assuming $i \geqslant n/4 = 4$), so $T$ contains derangements of order $5$ since there are more classes of subgroups of order $5$ in $T$ than there are in $H_0$.

To complete the proof of the proposition, let us assume $i$ is odd. If $i < n/8$ then $[\Lambda, \Lambda^{-1}, I_{n-2i}]$ is a derangement by Lemma 5.6.3. Next suppose $i > n/4$. As before, we note that $i$ divides $r - 1$, but $ir$ is odd so $i$ divides $(r-1)/2$. Therefore $r \geqslant 2i + 1 \geqslant 2^{m-1} + 3$ and again we deduce that $r = 2^m - 1$ is the only possibility. Here $T$ has a unique class of subgroups of order $r$, so $T$ is $r$-elusive in this situation. Moreover, examples do exist. For instance, take $G = \mathrm{P}\Omega_8^+(67)$, $H = 2^6.\Omega_6^+(2)$ and $r = 7$. Finally, suppose $i$ is odd and $n/8 \leqslant i \leqslant n/4$. Then $r \geqslant 2i + 1 \geqslant 2^{m-2} + 1$ and as before we deduce that $H_0$ contains a unique class of subgroups of order $r$, so $T$ contains derangements of order $r$. $\square$

This completes the proof of Theorem 5.6.1.

## 5.7 $\mathscr{C}_7$: Tensor product subgroups, II

The subgroups $H$ in the collection $\mathscr{C}_7$ are the stabilisers in $G$ of appropriate tensor product decompositions

$$V = V_1 \otimes \cdots \otimes V_t$$

of the natural $T$-module $V$, where each $V_i$ is $a$-dimensional and $t \geqslant 2$, so $\dim V = n = a^t$. The specific cases we need to consider are listed in Table 5.7.1 (see [86, Tables 3.5.B–E, 4.7.A]). Note that if $T = \mathrm{P}\Omega_8^+(q)$ and $H$ is of type $\mathrm{Sp}_2(q) \wr S_3$ (with $q$ even), then $H$ is non-maximal (see [85, p. 194]), so this case is excluded.

**Theorem 5.7.1** *Let $G \leqslant \mathrm{Sym}(\Omega)$ be a primitive almost simple classical group with socle $T$ and point stabiliser $H \in \mathscr{C}_7$. Let $r$ be a prime divisor of $|\Omega|$. Then $T$ is $r$-elusive if and only if $T = \mathrm{PSp}_8(q)$, $H$ is of type $\mathrm{Sp}_2(q) \wr S_3$ and $r = 2$.*

We begin with a preliminary lemma, which can be viewed as a tensor product analogue of Lemma 5.2.6.

**Lemma 5.7.2** *Let $x \in \mathrm{GL}_n(\mathbb{F})$ be an element of prime order $r$, where $\mathbb{F}$ is an algebraically closed field of characteristic $p$. Suppose $x$ cyclically permutes the subspaces $\{V_1, \ldots, V_r\}$ in a decomposition*

$$V = V_1 \otimes \cdots \otimes V_r$$

*of the natural $\mathrm{GL}_n(\mathbb{F})$-module, where each $V_i$ is an $a$-dimensional space. Set $b = (a^r - a)/r$.*

(i) *If $r \neq p$ then $\dim C_V(x) = a + b$ and every nontrivial $r$th root of unity occurs as an eigenvalue of $x$ with multiplicity $b$.*

(ii) *If $r = p$ then $x$ has Jordan form $[J_p^b, J_1^a]$.*

Table 5.7.1 *The collection $\mathscr{C}_7$*

| Case | $T$ | Type of $H$ | Conditions |
|------|-----|-------------|------------|
| I | $\mathrm{PSL}_n^\varepsilon(q)$ | $\mathrm{GL}_a^\varepsilon(q) \wr S_t$ | $a \geqslant 3$ |
| II | $\mathrm{PSp}_n(q)$ | $\mathrm{Sp}_a(q) \wr S_t$ | $a$ even, $qt$ odd |
| III | $\Omega_n(q)$ | $\mathrm{O}_a(q) \wr S_t$ | $aq$ odd, $a \geqslant 3$ |
| IV | $\mathrm{P}\Omega_n^+(q)$ | $\mathrm{Sp}_a(q) \wr S_t$ | $a, qt$ even, $(a,t) \neq (2,3)$ |
| V | $\mathrm{P}\Omega_n^+(q)$ | $\mathrm{O}_a^\varepsilon(q) \wr S_t$ | $a$ even, $a \geqslant 5 + \varepsilon$, $q$ odd |

*Proof* This follows immediately from the fact that $(t-1)^a(t^r-1)^b$ is the characteristic polynomial of $x$. Alternatively, let $\{e_{1,1},\ldots,e_{a,1}\}$ be a basis for $V_1$ and set $e_{i,j+1} = (e_{i,1})^{x^j}$ for all $1 \leqslant i \leqslant a$ and $1 \leqslant j < r$. Then $x$ induces a permutation of cycle-shape $(r^b, 1^a)$ on the vectors in the basis

$$\{e_{i_1,1} \otimes \cdots \otimes e_{i_r,r} \mid i_j \in \{1,\ldots,a\}\}$$

of $V$, and the fixed points are the elements $e_{i,1} \otimes \cdots \otimes e_{i,r}$ with $1 \leqslant i \leqslant a$. The result follows. $\qquad\qquad\square$

We will also need the following result, which follows from [18, Lemma 7.1].

**Lemma 5.7.3** *Let $G$ be an almost simple classical group and let $H$ be a maximal $\mathscr{C}_7$-subgroup of $G$ that stabilises a decomposition $V = V_1 \otimes \cdots \otimes V_t$ with $\dim V_i = a$. Suppose $x \in H \cap \mathrm{PGL}(V)$ has prime order $r$. Then either $\nu(x) \geqslant a^{t/2}$, or $(a,t,r) = (2,3,2)$ and $\nu(x) = 2$.*

**Proposition 5.7.4** *Theorem 5.7.1 holds in Case I of Table 5.7.1.*

*Proof* Here $T = \mathrm{PSL}_n^\varepsilon(q)$ and $H$ is the stabiliser in $G$ of a decomposition

$$V = V_1 \otimes \cdots \otimes V_t \tag{5.7.1}$$

where each $V_i$ is $a$-dimensional, so $n = a^t$, $a \geqslant 3$ and $H$ is of type $\mathrm{GL}_a^\varepsilon(q) \wr S_t$. Set $H_0 = H \cap T$ and observe that

$$H_0 \leqslant \mathrm{PGL}_a^\varepsilon(q) \wr S_t$$

(see [86, Proposition 4.7.3]). Let $r$ be a prime divisor of $|\Omega|$.

By Lemma 5.7.3, $[J_2, J_1^{n-2}]$ is a derangement of order $p$. Similarly, if $p > 2$ then $t_2$ involutions are derangements. For the remainder we may assume that $r \neq p$ and $r > 2$. Set $i = \Phi(r,q)$ and define the integer $c = c(i)$ as in Definition 5.1.1. As usual, we may as well assume that $r$ divides $|H_0|$.

If $c = 1$ then Lemma 5.7.3 implies that $[\Lambda, \Lambda^{-1}, I_{n-2}]$ is a derangement. Now assume $c \geqslant 2$. Let $x = \hat{x}Z \in T$ be an element of order $r$, where

$$\hat{x} = \begin{cases} [\Lambda, \Lambda^{-q}, I_{n-c}] & \varepsilon = - \text{ and } i \not\equiv 2 \pmod 4 \\ [\Lambda, I_{n-c}] & \text{otherwise} \end{cases}$$

Suppose $c > a$. Here $|\mathrm{PGL}_a^\varepsilon(q)|$ is indivisible by $r$, so if $x$ stabilises the decomposition in (5.7.1) then it must induce a nontrivial permutation on the $V_i$, so $r$ divides $t!$ and Lemma 5.7.2 implies that every primitive $r$th root of unity has multiplicity at least $a$ as an eigenvalue of $\hat{x}$ on $V \otimes \overline{\mathbb{F}}_q$. But $\hat{x}$ does not have this form, so $x$ is a derangement.

Finally, let us assume that $2 \leqslant c \leqslant a$. Since $v(x) = c$, $x$ is a derangement if $(t,c) \neq (2,a)$. However, if $(t,c) = (2,a)$ then Lemma 5.4.2 implies that $v(y) \geqslant a(a-1)$ for all $y \in H_0$ of order $r$, hence $x$ is also a derangement in this case. $\quad\square$

**Proposition 5.7.5** *Theorem 5.7.1 holds in Case II of Table 5.7.1.*

*Proof* Here $T = \mathrm{PSp}_n(q)$ and $H$ is of type $\mathrm{Sp}_a(q) \wr S_t$, where $a$ is even and $qt$ is odd. Set $H_0 = H \cap T$. By arguing as in the proof of the previous proposition, using Lemmas 5.7.2 and 5.7.3, we quickly reduce to the case $(a,t) = (2,3)$. Here $T = \mathrm{PSp}_8(q)$ and

$$\mathrm{PSp}_2(q) \wr S_3 \leqslant H_0 \leqslant \mathrm{PGSp}_2(q) \wr S_3$$

(see [86, Proposition 4.7.4]). Let $r$ be a prime divisor of $|\Omega|$.

By Lemma 5.7.3, $[J_2, J_1^6]$ is a derangement of order $p$. Similarly, if $r \neq p$, $r > 2$ and $i = \Phi(r,q)$, then either $i > 2$ and every element in $T$ of order $r$ is a derangement (since $|H_0|$ is indivisible by $r$), or $x = \hat{x}Z \in T$ is a derangement, where

$$\hat{x} = \begin{cases} [\Lambda, \Lambda^{-1}, I_6] & i = 1 \\ [\Lambda, I_6] & i = 2 \end{cases}$$

Finally, let us assume $r = 2$. We claim that $T$ is 2-elusive. In terms of the notation introduced in Section 3.4.2, representatives of the three involution classes in $T$ are labelled $t_1, t_2$ and either $t_4$ (if $q \equiv 1 \pmod 4$) or $t_4'$ (if $q \equiv 3 \pmod 4$). Let us denote a typical element $x \in \mathrm{GSp}_2(q) \wr S_3$ by writing $x = (x_1 \otimes x_2 \otimes x_3)\pi$, where $x_i \in \mathrm{GSp}_2(q)$ and $\pi \in S_3$.

It is easy to see that

$$(I_2 \otimes I_2 \otimes I_2)(1,2), \quad [-1,1] \otimes [-1,1] \otimes I_2$$

are involutions in $H_0$ of type $t_1$ and $t_2$, respectively. Similarly, if $q \equiv 1 \pmod 4$ and $\lambda \in \mathbb{F}_q$ has order 4, then $[\lambda, \lambda^{-1}] \otimes I_2 \otimes I_2$ is of type $t_4$, and if $q \equiv 3 \pmod 4$ then $A \otimes I_2 \otimes I_2$ is of type $t_4'$, where

$$A = \begin{pmatrix} 0 & 1 \\ -1 & 0 \end{pmatrix}$$

This justifies the claim. $\quad\square$

**Proposition 5.7.6** *Theorem 5.7.1 holds in Cases III–V in Table 5.7.1.*

*Proof* Consider Case V, so $T = \mathrm{P}\Omega_n^+(q)$ and $H$ is of type $\mathrm{O}_a^\varepsilon(q) \wr S_t$, where $q$ is odd and $a \geqslant 5 + \varepsilon$ is even. Set $H_0 = H \cap T$ and note that $n \equiv 0 \pmod 4$. By

Lemma 5.7.3, it follows that $[J_2^2, J_1^{n-4}]$ is a derangement of order $p$, and $t_1$ and $t_1'$ involutions are also derangements.

Now assume $r \neq p$ is an odd prime that divides $|\Omega|$ and $|H_0|$. Set $i = \Phi(r, q)$ and let $x = \hat{x}Z \in T$ be an element of order $r$, where

$$\hat{x} = \begin{cases} [\Lambda, \Lambda^{-1}, I_{n-2i}] & i \text{ odd} \\ [\Lambda, I_{n-i}] & i \text{ even} \end{cases}$$

Suppose $i > a$. If $x$ stabilises an appropriate decomposition of $V$, then it must induce a nontrivial permutation on the tensor factors, but this is ruled out by Lemma 5.7.2 and thus $x$ is a derangement.

Now assume $i \leqslant a$. If $i$ is even and $(t, i) \neq (2, a)$ then $v(x) = i < a^t/2$ and thus $x$ is a derangement. In the exceptional case $(t, i) = (2, a)$, we can argue as in the proof of Proposition 5.7.4 (using Lemma 5.4.2) to deduce that $x$ is a derangement. Now assume $i$ is odd. Here $v(x) = 2i \leqslant 2(a-1)$ and thus $v(x) < a^t/2$ if $t \geqslant 3$. Therefore, we can assume that $t = 2$, in which case $r$ divides $|O_a^\varepsilon(q)|$. If $i < a/2$ then $x$ is a derangement since $v(x) < a$. On the other hand, if $i \geqslant a/2$ then $i = a/2$ and $\varepsilon = +$ is the only possibility (since we are assuming that $O_a^\varepsilon(q)$ contains an element of order $r$), so $v(x) = a$. However, Lemma 5.4.2 implies that $v(y) \geqslant a(a-1)$ for all elements $y \in H_0$ of order $r$, so once again we deduce that $x$ is a derangement.

Cases III and IV in Table 5.7.1 are very similar and we ask the reader to check the details. □

This completes the proof of Theorem 5.7.1.

## 5.8 $\mathscr{C}_8$: Classical subgroups

The subgroups in the collection $\mathscr{C}_8$ are the stabilisers of appropriate nondegenerate forms $f$ on the natural $T$-module $V$. The four cases that arise are listed in Table 5.8.1. In Cases I, II and III we have $T = \mathrm{PSL}_n(q)$ and $f$ is respectively an alternating, quadratic and hermitian form. Recall that if a subgroup $H$ preserves a nondegenerate quadratic form on $V$, then it also preserves a symmetric bilinear form $B$ on $V$; if $p = 2$ then $B$ is alternating and thus $H$ is contained in the corresponding symplectic group on $V$. This explains how Case IV arises.

**Theorem 5.8.1** *Let $G \leqslant \mathrm{Sym}(\Omega)$ be a primitive almost simple classical group with socle $T$ and point stabiliser $H \in \mathscr{C}_8$. Let $r$ be a prime divisor of $|\Omega|$. Then $T$ is $r$-elusive if and only if $(G, r)$ is one of the cases in Table 5.8.2.*

**Remark 5.8.2** In Case III, if $p$ is odd then $T$ is $p$-elusive if and only if $n$ is odd and $(v, q_0 + 1) = (v, q - 1)$ for every nontrivial partition $(p^{a_p}, \ldots, 1^{a_1})$ of $n$, where $v = \gcd\{j \mid a_j \neq 0\}$. In particular, if $p = 3$ then $T$ is $p$-elusive if and only if $n$ is odd.

Set $H_0 = H \cap T$. In order to study Case II in Table 5.8.1, we need to determine the precise structure of $H_0$. In the following lemma, if $n \equiv 2 \pmod 4$ then $\gamma_{(n+2)/4}$ denotes the involutory graph automorphism of $\mathrm{P}\Omega_n^\varepsilon(q)$ discussed in Section 3.5.2.15.

Table 5.8.1 *The collection $\mathscr{C}_8$*

| Case | $T$ | Type of $H$ | Conditions |
|------|-----|-------------|------------|
| I | $\mathrm{PSL}_n(q)$ | $\mathrm{Sp}_n(q)$ | $n \geqslant 4$ even |
| II | $\mathrm{PSL}_n(q)$ | $\mathrm{O}_n^\varepsilon(q)$ | $n \geqslant 3$, $q$ odd |
| III | $\mathrm{PSL}_n(q)$ | $\mathrm{GU}_n(q_0)$ | $n \geqslant 3$, $q = q_0^2$ |
| IV | $\mathrm{PSp}_n(q)$ | $\mathrm{O}_n^\varepsilon(q)$ | $p = 2$ |

Table 5.8.2 *$r$-elusive $\mathscr{C}_8$-actions*

| Case | $r$ | Conditions |
|------|-----|------------|
| I | 2 | $(q-1)_2 \leqslant (n)_2$ |
| II | 2 | always |
|  | 3, 5 | $n = r$, $(q-1)_r = r$ |
| III | 2 | $p = 2$ or $(n)_2 \leqslant (q-1)_2$ |
|  | $p > 2$ | see Remark 5.8.2 |
|  | 3, 5 | $n = r$, $(q_0 - 1)_r = r$ |
| IV | 2 | $\varepsilon = +$ or $n \equiv 2 \pmod 4$ |

**Lemma 5.8.3** *Let* $T = \mathrm{PSL}_n(q)$ *and let* $H$ *be a* $\mathscr{C}_8$*-subgroup of type* $\mathrm{O}_n^\varepsilon(q)$ *as in Case II of Table 5.8.1, so* $n \geqslant 3$ *and* $q$ *is odd. Then*

$$
H_0 = \begin{cases}
\mathrm{PSO}_n(q) & n \text{ odd} \\
\mathrm{PO}_n^\varepsilon(q) & (q-1)_2 > (n)_2 > 1 \\
\mathrm{PSO}_n^\varepsilon(q).\langle \gamma'_{(n+2)/4} \rangle & n \equiv 2 \ (\mathrm{mod}\ 4),\ q \equiv 3 \ (\mathrm{mod}\ 4) \\
\mathrm{PSO}_n^\varepsilon(q).2 & q \equiv 1 \ (\mathrm{mod}\ 4),\ (q-1)_2 = (n)_2 > 1 \\
\mathrm{Inndiag}(\mathrm{P}\Omega_n^\varepsilon(q)) & \text{otherwise}
\end{cases}
$$

*where* $\mathrm{PSO}_n^\varepsilon(q).2 \leqslant \mathrm{PGO}_n^\varepsilon(q)$ *is the unique nonsplit extension of* $\mathrm{PSO}_n^\varepsilon(q)$ *by* $C_2$.

*Proof* By [86, Proposition 4.8.4] we have

$$
H_0 = \mathrm{PSO}_n^\varepsilon(q).(2,n) \leqslant \mathrm{PGO}_n^\varepsilon(q)
$$

so the $n$ odd case is clear. Now assume $n$ is even. Let $\xi \in \mathbb{F}_{q^2}$ be an element of order $2(q-1)_2$. With minor modifications, we can argue as in the proof of Lemma 5.5.6.

For example, suppose $n \equiv 2 \ (\mathrm{mod}\ 4)$ and $q \equiv 3 \ (\mathrm{mod}\ 4)$. Let $t = \hat{t}Z \in \mathrm{PGO}_n^\varepsilon(q)$ be an involution of type $\gamma'_{(n+2)/4}$ (where $Z = Z(\mathrm{GO}_n^\varepsilon(q))$). Then

$$
\det(\hat{t}) = (-\xi^2)^{n/2} = -\xi^n = 1
$$

(see Section 3.5.2.15), so $t \in T$ and thus $H_0 = \mathrm{PSO}_n^\varepsilon(q).\langle \gamma'_{(n+2)/4} \rangle$ as claimed.

Similarly, suppose $q \equiv 1 \ (\mathrm{mod}\ 4)$ and $(q-1)_2 < (n)_2$, so $n \equiv 0 \ (\mathrm{mod}\ 8)$. Let

$$
t = \hat{t}Z \in \mathrm{Inndiag}(\mathrm{P}\Omega_n^\varepsilon(q)) \setminus \mathrm{PSO}_n^\varepsilon(q)
$$

be an involution of type $t''_{n/4}$ if $\varepsilon = +$, and type $t'_{n/4}$ if $\varepsilon = -$. From the description of these involutions given in Sections 3.5.2.7 and 3.5.2.9, we see that $\det(\hat{t}) = \xi^n = 1$ (note that $2(q-1)_2$ divides $n$) and thus $t \in T$, so in this situation we have $H_0 = \mathrm{Inndiag}(\mathrm{P}\Omega_n^\varepsilon(q))$. The other cases are similar, and the reader can check the details. $\qquad\square$

**Proposition 5.8.4** *Theorem 5.8.1 holds in Cases I and II of Table 5.8.1.*

*Proof* Case I is entirely similar to the proof of Proposition 5.5.11, and we ask the reader to make the necessary minor adjustments. (This case is discussed in [86, Proposition 4.8.3].)

Now consider Case II. Here $T = \mathrm{PSL}_n(q)$, $n \geqslant 3$, $q$ is odd and the structure of $H_0$ is given in Lemma 5.8.3. We can proceed as in the proof of Proposition 5.5.12. In particular, we find that $T$ is $r$-elusive if and only if $r = 2$, or $r \in \{3,5\}$, $n = r$ and $(q-1)_r = r$. $\qquad\square$

**Proposition 5.8.5** *Theorem 5.8.1 holds in Case III of Table 5.8.1.*

*Proof* This is very similar to the proof of Proposition 5.5.7. Here $T = \mathrm{PSL}_n(q)$, $q = q_0^2$ and [86, Proposition 4.8.5] states that

$$H_0 = \mathrm{PSU}_n(q_0).d \leqslant \mathrm{PGU}_n(q_0)$$

where

$$d = \frac{(n, q_0 + 1)\left(q_0 + 1, \frac{q-1}{(n,q-1)}\right)}{q_0 + 1}$$

By Lemma A.5 in Appendix A, we have $d \in \{1, 2\}$, with $d = 2$ if and only if $(q - 1)_2 > (n)_2 > 1$. Let $r$ be a prime divisor of $|\Omega|$.

*Case 1.* $r = p > 2$. First assume $r = p$ is odd. Let $\lambda = (p^{a_p}, \ldots, 1^{a_1})$ be a nontrivial partition of $n$ and let $v = \gcd\{j \mid a_j \neq 0\}$. By Propositions 3.2.7 and 3.3.10, there are precisely $(v, q - 1)$ distinct $T$-classes of unipotent elements in $T$ with Jordan form $[J_p^{a_p}, \ldots, J_1^{a_1}]$, and there are $(v, q_0 + 1)$ such classes in $\mathrm{PSU}_n(q_0)$.

Suppose $n$ is even and let $\lambda = (2^{n/2})$, so $(v, q_0 + 1) = (v, q - 1) = 2$ and thus there are two $T$-classes of elements $x \in T$ of this form, and there are also two $\mathrm{PSU}_n(q_0)$-classes. We claim that the two $\mathrm{PSU}_n(q_0)$-classes are fused in $T$, which implies that $T$ contains derangements of order $p$. To see this, let $\delta = [\mu, I_{n-1}]Z \in \mathrm{PGL}_n(q)$, where $\mu \in \mathbb{F}_q$ has order $q - 1$. By Lemma 3.3.11, the two $\mathrm{PSU}_n(q_0)$-classes are represented by the elements $x$ and $x^{\delta^{q_0-1}}$, which are $T$-conjugate if and only if $q_0 - 1$ is divisible by $(v, q - 1) = 2$ (see Lemma 3.2.8). The latter condition clearly holds, so the two $\mathrm{PSU}_n(q_0)$-classes are indeed fused in $T$.

Now assume $n$ is odd, so $H_0 = \mathrm{PSU}_n(q_0)$. If $(v, q - 1) = 1$ then both $T$ and $H_0$ contain a unique class of such elements, so they all have fixed points. On the other hand, if $(v, q_0 + 1) < (v, q - 1)$ then $T$ contains derangements of order $p$ since there are more $T$-classes of unipotent elements in $T$ with Jordan form $[J_p^{a_p}, \ldots, J_1^{a_1}]$ than there are $H_0$-classes in $H_0$ of this form. Now assume $(v, q_0 + 1) = (v, q - 1) > 1$. We claim that none of the relevant $H_0$-classes are fused in $T$.

Let $x \in \mathrm{PSU}_n(q_0)$ be a unipotent element with Jordan form $[J_p^{a_p}, \ldots, J_1^{a_1}]$. By Lemma 3.3.11, the $(v, q_0 + 1)$ distinct $\mathrm{PSU}_n(q_0)$-classes of unipotent elements of this form are represented by

$$x^{\delta^{is}}, \ 0 \leqslant i < (v, q_0 + 1)$$

where $s = q_0 - 1$. Suppose two of these $\mathrm{PSU}_n(q_0)$-classes are fused in $T$. Then Lemma 3.2.8 implies that there exist integers $i, j$ such that $0 \leqslant j < i < (v, q_0 + \varepsilon)$ and $(v, q - 1)$ divides $s(i - j)$. Since $(v, q - 1) = (v, q_0 + 1)$ it follows that $i - j$ is indivisible by $(v, q - 1)$, hence there is a prime divisor $\ell$ of $(v, q_0 + 1)$ that also divides $s$. Therefore $\ell$ divides both $q_0 + 1$ and $q_0 - 1$, so $\ell = 2$. However, $\ell$ divides $v$, and $v$ divides $n$, which is odd, so this is not possible. We conclude that none of the relevant $\mathrm{PSU}_n(q_0)$-classes are fused in $T$.

To summarise, if $p$ is odd then $T$ is $p$-elusive if and only if $n$ is odd and $(v, q_0 + 1) = (v, q - 1)$ for every nontrivial partition $(p^{a_p}, \ldots, 1^{a_1})$ of $n$.

*Case 2. $r \neq p$, $r > 2$.* Next suppose $r \neq p$ and $r > 2$. Let $x = \hat{x}Z \in T$ be an element of order $r$. Set $i = \Phi(r, q)$ and $i_0 = \Phi(r, q_0)$, so $i = i_0/2$ if $i_0$ is even, otherwise $i = i_0$. Since $r$ divides $|\Omega|$, it follows that $i_0 \not\equiv 2 \pmod 4$. In particular, if $i > 1$, or if $n$ is indivisible by $r$, then $\hat{x} = [\Lambda, I_{n-i}]$ is a derangement. For instance, if $i > 1$ is odd then $i_0 = i$ and thus Proposition 3.3.2 implies that every element in $H_0$ of order $r$ is of the form $[(\Lambda_1, \Lambda_1^{-q_0})^{a_1}, \ldots, (\Lambda_k, \Lambda_k^{-q_0})^{a_k}, I_e]$ for some $a_j, k \geqslant 1$, where $|\Lambda_j| = i$, but $\hat{x}$ does not have this form.

Now assume $i = 1$ and $r$ divides $n$, in which case $i_0 = 1$ since we have already noted that $i_0 \not\equiv 2 \pmod 4$. Let $\mu \in \mathbb{F}_q$ be a primitive $r$th root of unity. If $r < n$ then $[\mu I_r, I_{n-r}]$ is a derangement, so assume $n = r$. Note that if $x = \hat{x}Z \in T$ has order $r$, then $\hat{x}$ is diagonalisable over $\mathbb{F}_q$ (see Proposition 3.2.2). If $r \geqslant 7$ then $\hat{x} = [\mu, \mu^2, \mu^{r-3}, I_{n-3}]$ is a derangement. Next assume that $n = r = 5$, so $q_0 \equiv 1 \pmod 5$. If $q_0 \equiv 1 \pmod{25}$ then $[\mu, I_4]$ is a derangement in $T$ of order 5 (by Lemma 2.1.1, such an element is contained in $T$). On the other hand, if $q_0 \not\equiv 1 \pmod{25}$ then $\det(\hat{x}) = 1$ for all $x = \hat{x}Z \in T$ of order 5. By choosing an appropriate scalar multiple, we may assume that

$$\hat{x} = [\mu I_a, \mu^4 I_a, \mu^2 I_b, \mu^3 I_b, I_c]$$

for some non-negative integers $a, b$ and $c$ with $2(a + b) + c = 5$. It follows that $x$ has fixed points and thus $T$ is 5-elusive. Similarly, if $n = r = 3$ and $i_0 = 1$ then $T$ is 3-elusive if and only if $q_0 \not\equiv 1 \pmod 9$.

*Case 3. $r = 2$.* To complete the proof, we may assume that $r = 2$. If $p = 2$ then it is clear that every involution $x \in T$ has fixed points since there is a unique class of elements with Jordan form $[J_2^\ell, J_1^{n-2\ell}]$ in both $T$ and $H_0$, for all possible $\ell$.

Now assume $p > 2$. Recall that $H_0 = \mathrm{PSU}_n(q_0).d$, where $d \in \{1, 2\}$. More precisely, $d = 2$ if and only if $(q - 1)_2 > (n)_2 > 1$. In particular, if $n$ is odd then $d = 1$ and $T$ is 2-elusive (both $T$ and $H_0$ contain a unique class of involutions of type $[-I_i, I_{n-i}]$, for all possible $i$). Now assume $n$ is even. We use the notation for involutions introduced in Section 3.2.2.

If $(q-1)_2 > (n)_2$ then $d = 2$ and it is clear that each involution in $T$ of type $t_i$ (with $i \leqslant n/2$) has fixed points. Now assume that $(q-1)_2 \leqslant (n)_2$. Again, it is clear that all $t_i$ involutions in $T$ have fixed points. Finally, suppose that $x \in T$ is an involution of type $t'_{n/2}$, so $(q-1)_2 < (n)_2$ (see Section 3.2.2.2) and thus $H_0 = \mathrm{PSU}_n(q_0)$. Visibly, $[-I_{n/2}, I_{n/2}] \in H_0$ is $T$-conjugate to $t_{n/2}$. There is an additional class of involutions $y = \hat{y}Z \in H_0$ such that $\hat{y} = [\xi I_{n/2}, \xi^{-q_0} I_{n/2}]$, where $\xi \in \mathbb{F}_q$ has order $2(q_0+1)_2$ (see Section 3.3.2.2). The eigenvalues of $\hat{y}$ are in $\mathbb{F}_q$, so $y$ must be $T$-conjugate to $t_{n/2}$. We conclude that involutions of type $t'_{n/2}$ are derangements. □

**Proposition 5.8.6** *Theorem 5.8.1 holds in Case IV of Table 5.8.1.*

*Proof* Here $T = \mathrm{PSp}_n(q)'$ and $H$ is of type $\mathrm{O}_n^\varepsilon(q)$, where $p = 2$ and $n \geqslant 4$ is even. By [86, Proposition 4.8.6],

$$H_0 = \mathrm{O}_n^\varepsilon(q) = \Omega_n^\varepsilon(q).2$$

and thus $|\Omega| = \frac{1}{2} q^{n/2}(q^{n/2} + \varepsilon)$. Let $r$ be a prime divisor of $|\Omega|$.

First assume $r = 2$. From the description of the involutions in $T$ and $H_0$ (see Sections 3.4.4 and 3.5.4) we immediately deduce that $T$ is 2-elusive if $\varepsilon = +$ or $n \equiv 2 \pmod 4$. However, if $\varepsilon = -$ and $n \equiv 0 \pmod 4$ then involutions in $T$ of type $a_{n/2}$ are derangements.

Now assume $r > 2$ and set $i = \Phi(r, q)$. First assume $\varepsilon = +$, so $r$ divides $q^{n/2} + 1$ and thus $i$ divides $n$, but not $n/2$. In particular, $n/i$ is odd and $i$ is even, so $\hat{x} = [\Lambda^{n/i}]$ is a derangement (see Remark 3.5.5(ii)). The case $\varepsilon = -$ is entirely similar. Here $r$ divides $q^{n/2} - 1$, so $i$ divides $n/2$. Set

$$\hat{x} = \begin{cases} [(\Lambda, \Lambda^{-1})^{n/2i}] & i \text{ odd} \\ [\Lambda^{n/i}] & i \text{ even} \end{cases}$$

In both cases, it is easy to check that $x = \hat{x}Z \in T$ is a derangement. □

This completes the proof of Theorem 5.8.1.

## 5.9 Novelty subgroups

Let $G \leqslant \mathrm{Sym}(\Omega)$ be a primitive almost simple classical group over $\mathbb{F}_q$ with socle $T$ and point stabiliser $H$. Let $V$ be the natural $T$-module. Observe that either $G \leqslant \mathrm{P\Gamma L}(V)$, or one of the following holds:

(i) $T = \mathrm{PSL}_n(q)$, $n \geqslant 3$ and $G$ contains graph or graph-field automorphisms;
(ii) $T = \mathrm{Sp}_4(q)'$, $p = 2$ and $G$ contains graph-field automorphisms;
(iii) $T = \mathrm{P\Omega}_8^+(q)$ and $G$ contains triality automorphisms.

If $G \leqslant \mathrm{P\Gamma L}(V)$ then Aschbacher's main theorem [3] states that $H$ is contained in one of nine subgroup collections, labelled $\mathscr{C}_1, \ldots, \mathscr{C}_8, \mathscr{S}$. The case $H \in \mathscr{C}_1$ was investigated in Chapter 4, and we handled the remaining geometric actions with $H \in \mathscr{C}_i$ (for $2 \leqslant i \leqslant 8$) in Section 5.$i$.

In this section, we will consider the special cases (i), (ii) and (iii) above. In order to deal with the primitive groups arising in (i), Aschbacher introduces an extra family of subgroups of $G$, denoted by $\mathscr{C}_1'$, and he establishes a suitable extension of his main theorem in this case (see [3, Section 13]). Following [86], we have included the $\mathscr{C}_1'$ collection in our definition of $\mathscr{C}_1$; the extra cases are labelled I$'$ and I$''$ in Table 4.1.1 in Chapter 4 (see Propositions 4.3.4 and 4.3.6). Similarly, in [3, Section 14] Aschbacher describes the subgroup structure of the groups in (ii) by modifying his original definition of the $\mathscr{C}_i$ collections. Partial information in case (iii) is given in [3, Section 15]; a complete classification of the maximal subgroups of $G$ was obtained later by Kleidman in [85].

Therefore, in order to complete the proof of Theorem 1.5.1, it remains to deal with the groups arising in cases (ii) and (iii) above. Let $G$ be such a group. Set $H_0 = H \cap T$ and assume that $H \notin \mathscr{S}$. If $H_0$ is a maximal subgroup of $T$ then $H_0$ belongs to one of the collections $\mathscr{C}_i$ with $1 \leqslant i \leqslant 8$, and so we have already handled this situation. However, there are a few cases in which $H_0$ is non-maximal; following Wilson [121], in this situation we say that $H$ is a *novelty subgroup* of $G$, and we write $\mathscr{N}$ to denote the collection of subgroups that arises in this way. The specific cases to be considered are listed in Table 5.9.1; see [3, Section 14] and [85] (also see [13, Tables 8.14 and 8.50]). Note that the collection $\mathscr{N}$ is empty unless $T = \mathrm{Sp}_4(q)'$ (with $p = 2$) or $\mathrm{P\Omega}_8^+(q)$.

**Theorem 5.9.1** *Let $G \leqslant \mathrm{Sym}(\Omega)$ be a primitive almost simple classical group with socle $T$ and point stabiliser $H \in \mathscr{N}$. Let $r$ be a prime divisor of $|\Omega|$. Then $T$ is $r$-elusive if and only if $(G, H, r)$ is one of the cases in Table 5.9.2.*

Table 5.9.1 *The collection $\mathcal{N}$*

| Case | $T$ | Type of $H$ | Conditions |
|------|-----|-------------|------------|
| I | $\mathrm{Sp}_4(q)'$ | $\mathrm{O}_2^\varepsilon(q) \wr S_2$ | $p = 2$ |
| II | | $\mathrm{O}_2^-(q^2).2$ | $p = 2$ |
| III | | $P_{1,2} = [q^4].\mathrm{GL}_1(q)^2$ | $p = 2$ |
| IV | $\mathrm{P}\Omega_8^+(q)$ | $\mathrm{GL}_1^\varepsilon(q) \times \mathrm{GL}_3^\varepsilon(q)$ | |
| V | | $\mathrm{O}_2^-(q^2) \times \mathrm{O}_2^-(q^2)$ | |
| VI | | $G_2(q)$ | |
| VII | | $[2^9].\mathrm{SL}_3(2)$ | $q = p > 2$ |
| VIII | | $P_{1,3,4} = [q^{11}].\mathrm{GL}_2(q)\mathrm{GL}_1(q)^2$ | |

Table 5.9.2 *r-elusive $\mathcal{N}$-actions*

| Case | $r$ | Conditions |
|------|-----|------------|
| I | 2 | always |
| IV | 2 | $q \equiv \varepsilon \pmod 4$ |
| V | 2 | $q$ odd |
| VII | 2 | $q \equiv \pm 1 \pmod 8$ |
| | 7 | $q^2 \not\equiv 1 \pmod 7$, $q^3 \equiv \pm 1 \pmod{49}$ |
| VIII | 2 | $q \equiv 1 \pmod 4$ |
| | 3 | $q \equiv 1 \pmod 3$ |

### 5.9.1 4-dimensional symplectic groups

**Proposition 5.9.2** *Theorem 5.9.1 holds in Cases I–III of Table 5.9.1.*

*Proof*  Let $H_0 = H \cap T$. In Case III we note that $|H_0| = q^4(q-1)^2$ and $|\Omega| = (q+1)^2(q^2+1)$ are coprime, so we may assume that we are in one of the first two cases. Here $r = 2$ is the only common prime divisor of $|H_0|$ and $|\Omega|$. We will use the notation for involutions introduced in Section 3.4.4.

In Case I, $H_0$ is the stabiliser in $\mathrm{O}_4^+(q) < \mathrm{Sp}_4(q)$ of an orthogonal decomposition $V = V_1 \perp V_2$, where $V_1$ and $V_2$ are nondegenerate 2-spaces of type $\varepsilon$. The involutions $(J_2, I_2)$ and $(J_2, J_2)$ in $\mathrm{O}_2^\varepsilon(q) \times \mathrm{O}_2^\varepsilon(q) < H_0$ are of type $b_1$ and $c_2$, respectively, and any involution in $H_0$ that interchanges $V_1$ and $V_2$ is of type $a_2$. Therefore, $T$ is 2-elusive.

Case II is similar. Here $H_0$ is a field extension subgroup of type $\mathrm{O}_2^-(q^2)$ in $\mathrm{O}_4^-(q) < \mathrm{Sp}_4(q)$. As noted in Section 3.5.4, the group $\mathrm{O}_4^-(q)$ does not contain any involutions of type $a_2$, so these elements are derangements. (In fact, every involution in $H_0$ is $T$-conjugate to $c_2$.) $\qquad\square$

### 5.9.2 8-dimensional orthogonal groups

We begin with a preliminary lemma, which provides a convenient summary of some of the information on involutions in $T = \mathrm{P}\Omega_8^+(q)$ presented in Section 3.5. We use the notation for involutions introduced in Sections 3.5.2 and 3.5.4.

**Lemma 5.9.3** *Let* $T = \mathrm{P}\Omega_8^+(q)$, *let* $x \in T$ *be an involution and let* $\tau \in \mathrm{Aut}(T)$ *be a triality graph automorphism.*

(i) *If* $q \equiv 1 \pmod 4$ *then* $x$ *is* $T$*-conjugate to* $t_1$, $t_2$, $t_3$ *or* $t_4$, *and* $\tau$ *cyclically permutes the* $T$*-classes represented by* $t_1$, $t_3$ *and* $t_4$.

(ii) *If* $q \equiv 3 \pmod 4$ *then* $x$ *is* $T$*-conjugate to* $t_1'$, $t_2$, $t_3'$ *or* $t_4'$, *and* $\tau$ *cyclically permutes the* $T$*-classes represented by* $t_1'$, $t_3'$ *and* $t_4'$.

(iii) *If* $p = 2$ *then* $x$ *is* $T$*-conjugate to* $a_2$, $c_2$, $a_4$ *or* $a_4'$, *and* $\tau$ *cyclically permutes the* $T$*-classes represented by* $c_2$, $a_4$ *and* $a_4'$.

*Proof* For $T$-class representatives, see Proposition 3.5.10 if $p \neq 2$ and Proposition 3.5.16 (and also Remark 3.5.18) if $p = 2$. The fusion of classes under a triality graph automorphism is described in Proposition 3.5.24. $\square$

**Proposition 5.9.4** *Theorem 5.9.1 holds in Case IV of Table 5.9.1.*

*Proof* Here $H_0 = H \cap T$ is the image (modulo scalars) of the subgroup

$$\left( C_{(q-\varepsilon)/d} \times \frac{1}{d}\mathrm{GL}_3^\varepsilon(q) \right).2^d < \Omega_8^+(q)$$

where $d = (2, q-1)$ (see [85, Propositions 3.2.2, 3.2.3]). (Note that if $q$ is odd then $\frac{1}{2}\mathrm{GL}_3^\varepsilon(q)$ denotes the unique index-two subgroup of $\mathrm{GL}_3^\varepsilon(q)$.) If $\varepsilon = +$ then $H_0$ is the intersection of the stabiliser in $T$ of a hyperbolic 2-space and the stabiliser of a decomposition $V = U \oplus W$, where $U$ and $W$ are totally singular 4-spaces. Moreover, if

$$\{e_1, \ldots, e_4, f_1, \ldots, f_4\}$$

is a hyperbolic basis for $V$, then we may assume that $H_0$ is the stabiliser in $T$ of the specific decomposition

$$V = \langle e_1, f_1 \rangle \oplus \langle e_2, e_3, e_4 \rangle \oplus \langle f_2, f_3, f_4 \rangle \tag{5.9.1}$$

On the other hand, if $\varepsilon = -$ then $H_0$ is the intersection of the stabiliser of an elliptic 2-space and a $\mathscr{C}_3$-subgroup of type $\mathrm{GU}_4(q)$.

Let $r$ be a prime divisor of $|H_0|$ and $|\Omega|$, and let $x = \hat{x}Z \in T$ be an element of order $r$. If $r = p > 2$ then any element $x \in T$ with Jordan form $[J_2^4]$ is a derangement of order $r$ since $x$ does not fix a nondegenerate 2-space (of any isometry type). Similarly, if $r = p = 2$ then $a_4$-type involutions are derangements.

Next assume $r \neq p$ and $r > 2$. Set $i = \Phi(r,q)$ and note that $i = \frac{1}{2}(3 + \varepsilon)$ since $r$ divides $|H_0|$ and $|\Omega|$. If $\varepsilon = +$ then $\hat{x} = [\Lambda, I_6]$ does not stabilise a decomposition of $V$ into maximal totally singular subspaces, whence $x$ is a derangement. Similarly, if $\varepsilon = -$ then $[\Lambda, \Lambda^{-1}, I_6]$ is a derangement since every eigenvalue of an element of order $r$ in a $\mathcal{C}_3$-subgroup of $T$ of type $GU_4(q)$ has even multiplicity in its action on $V$ (see Lemma 5.3.11(ii)).

Finally, let us assume $r = 2$ and $p \neq 2$. We claim that $T$ is 2-elusive if and only if $q \equiv \varepsilon \pmod 4$.

First assume $\varepsilon = +$ and $q \equiv 1 \pmod 4$. Since $H_0$ is normalised by a triality graph automorphism of $T$, in order to deduce that $T$ is 2-elusive it suffices to show that there are involutions of type $t_1$ and $t_2$ in $H_0$ (see Lemma 5.9.3). This is very straightforward. Indeed, in terms of the hyperbolic basis

$$\{e_1, f_1, e_2, f_2, e_3, f_3, e_4, f_4\}$$

we observe that $[-I_2, I_6] \in H_0$ is an involution of type $t_1$ and $[-I_4, I_4] \in H_0$ is of type $t_2$ (both elements clearly stabilise the decomposition in (5.9.1)).

Next suppose $\varepsilon = +$ and $q \equiv 3 \pmod 4$. Let $x \in T$ be an involution of type $t_1'$. The $(-1)$-eigenspace of $\hat{x}$ is an elliptic 2-space (see Section 3.5.2.4), so $x$ does not fix a maximal totally singular subspace of $V$. Moreover, $x$ does not stabilise a decomposition $V = U \oplus W$ into maximal totally singular subspaces (since $v(x) = 2$, $x$ does not interchange $U$ and $W$). We conclude that $x$ is a derangement.

Now assume $\varepsilon = -$ and $q \equiv 1 \pmod 4$. Let $x \in T$ be an involution of type $t_1$, so the $(-1)$-eigenspace of $\hat{x}$ is a hyperbolic 2-space. Now, if $y = [-I_1, I_3]$ is an involution in a $\mathcal{C}_3$-subgroup of $T$ of type $GU_4(q)$, then the $(-1)$-eigenspace of $y$ on $V$ is an elliptic 2-space (see Construction 2.5.14). Therefore, there are no $t_1$ involutions in a subgroup of type $GU_4(q)$ (indeed, we observed this in the proof of Proposition 5.3.13), whence $x$ is a derangement.

Finally, suppose $\varepsilon = -$ and $q \equiv 3 \pmod 4$. In view of Lemma 5.9.3, to see that $T$ is 2-elusive it suffices to show that $H_0$ contains involutions of type $t_1'$ and $t_2$. Clearly, a $t_1'$ involution fixes an elliptic 2-space, and in the previous paragraph we observed that such elements are in $\mathcal{C}_3$-subgroups of type $GU_4(q)$. Therefore, $t_1'$ involutions have fixed points. Similarly, a $t_2$ involution fixes an elliptic 2-space (both eigenspaces are hyperbolic 4-spaces), and we can also find $t_2$ involutions in $GU_4(q)$ (as the image of $[-I_2, I_2] \in GU_4(q)$). Therefore, these involutions also have fixed points, and we conclude that $T$ is 2-elusive. $\qquad\square$

**Proposition 5.9.5** *Theorem 5.9.1 holds in Case V of Table 5.9.1.*

*Proof*  Let $R$ be a Sylow $r$-subgroup of $T$, where $r$ is an odd prime divisor of $q^2 + 1$. Then [85, Proposition 3.3.1] states that

$$H_0 = H \cap T = N_T(R) = (D_{2(q^2+1)/d} \times D_{2(q^2+1)/d}).2^2$$

where $d = (2, q - 1)$ and $D_m$ denotes the dihedral group of order $m$. In particular, 2 is the only common prime divisor of $|\Omega|$ and $|H_0|$. The proof of [85, Proposition 3.3.1] also indicates that $H_0 \leqslant L$, where $L \cong (\Omega_4^-(q) \times \Omega_4^-(q)).2^2$ is the stabiliser of an orthogonal decomposition $V = V_1 \perp V_2$, where $V_1$ and $V_2$ are elliptic 4-spaces. More precisely, $H_0$ is the intersection of such a subgroup $L$ with a $\mathscr{C}_3$-subgroup $K$ of type $O_4^+(q^2)$ (see [85, Table I]).

If $p = 2$ then every $a_2$-type involution is a derangement since there are no such elements in the overgroup $L$. For the remainder, let us assume $p \neq 2$. We claim that $T$ is 2-elusive. Let $x = \hat{x}Z \in T$ be an involution.

First assume $q \equiv 1 \pmod 4$. As in the proof of the previous proposition, it suffices to show that there are $t_1$ and $t_2$ involutions in $H_0$. Let $\hat{x} = [-I_2, I_6]$ be a $t_1$ involution, so both eigenspaces are hyperbolic. Clearly, $L$ contains such elements (take an element that centralises $V_2$ and acts on $V_1$ as $[-I_2, I_2]$, where the $(-1)$-eigenspace on $V_1$ is hyperbolic). They also arise in $K$ as the image of a reflection $[-I_1, I_3]$ with a nonsquare-parabolic $(-1)$-eigenspace over $\mathbb{F}_{q^2}$ (see Construction 2.5.13). Therefore, $t_1$ involutions have fixed points. Similarly, $t_2$ involutions also have fixed points; in $L$, take $(z, z) \in \Omega_4^-(q) \times \Omega_4^-(q)$, where $z = [-I_2, I_2]$ has an elliptic $(-1)$-eigenspace; in $K$, take the image of $[-I_2, I_2]$, where both eigenspaces are hyperbolic $\mathbb{F}_{q^2}$-spaces.

The case $q \equiv 3 \pmod 4$ is very similar and we omit the details.  □

**Proposition 5.9.6** *Theorem 5.9.1 holds in Case VI of Table 5.9.1.*

*Proof*  Here $H_0 = H \cap T = C_T(\tau) = G_2(q)$, where $\tau \in \mathrm{Aut}(T)$ is an appropriate $G_2$-type triality graph automorphism (see Section 3.5.7). We can also view $H_0$ as the intersection of the stabiliser in $T$ of a nonsingular 1-space (a reducible subgroup of type $O_7(q)$), and the image of this stabiliser under $\tau$, which is an irreducible subgroup of type $O_7(q)$ (obtained by restricting one of the 8-dimensional spin representations of $\Omega_8^+(q)$). Let $r$ be a prime divisor of $|H_0|$ and $|\Omega|$, and let $x = \hat{x}Z \in T$ be an element of order $r$.

If $r = p > 2$ then any element $x \in T$ with Jordan form $[J_2^4]$ is a derangement since $x$ does not fix a nonsingular 1-space. For the same reason, if $p = 2$ then $a_4$-type involutions are derangements, and similarly if $p \neq 2$ then involutions of type $t_3$ (if $q \equiv 1 \pmod 4$) and $t_3'$ (if $q \equiv 3 \pmod 4$) are derangements. Finally, suppose $r > 2$ and $r \neq p$. Set $i = \Phi(r,q)$ and note that $i \in \{1,2\}$. If $i = 1$ then set $\hat{x} = [(\Lambda, \Lambda^{-1})^4]$, and if $i = 2$ set $\hat{x} = [\Lambda^4]$. In both cases, $x$ fails to fix a nonsingular 1-space, so $x$ is a derangement.                                    □

**Proposition 5.9.7** *Theorem 5.9.1 holds in Case VII of Table 5.9.1.*

*Proof* Here $q = p > 2$ and $H_0 = H \cap T = [2^9].\mathrm{SL}_3(2)$, so

$$|\Omega| = \frac{q^{12}(q^2-1)(q^4-1)^2(q^6-1)}{2^{14}.3.7}$$

Let $r$ be a prime divisor of $|H_0|$ and $|\Omega|$, so $r \in \{2,3,7\}$. As explained in the proof of [85, Proposition 3.4.2], $H_0$ is contained in the stabiliser $N$ of an orthogonal decomposition

$$V = \langle v_1 \rangle \perp \ldots \perp \langle v_8 \rangle \tag{5.9.2}$$

where the $\langle v_i \rangle$ are isometric nondegenerate 1-spaces (that is, the $\langle v_i \rangle$ are either all square-parabolic spaces, or all nonsquare-parabolic spaces). In particular, we note that $H_0$ contains the centraliser $C$ of this decomposition (that is, $C \cong 2^6$ is the subgroup of $T$ that fixes the above decomposition componentwise). Let $x = \hat{x}Z \in T$ be an element of order $r$.

If $r = p$ then the elements in $T$ with Jordan form $[J_2^2, J_1^4]$ are derangements since there are no such elements in $N$ (note that any unipotent element $y \in N$ must induce a nontrivial permutation of the $\langle v_i \rangle$, in which case the Jordan form of $y$ will involve Jordan blocks of size $p$ – see Lemma 5.2.6).

Next suppose $r \in \{3,7\}$ and $r \neq p$. Define $i = \Phi(r,q)$ as before. If $i = 2$ then set $\hat{x} = [\Lambda^4]$, and if $i = 1$ let $\hat{x} = [(\Lambda, \Lambda^{-1})^4]$. Since $C$ is a 2-group, every element $y \in N$ of order $r$ must induce a nontrivial permutation of the $\langle v_j \rangle$, so Lemma 5.2.6 implies that $C_V(\hat{y})$ is nontrivial. It follows that $x$ is a derangement. Finally, if $r = 7$ and $i \in \{3,6\}$ then $T$ has a unique class of elements of order $r$, so $T$ is $r$-elusive. Here the condition $i \in \{3,6\}$ is equivalent to the pair of congruences $q^2 \not\equiv 1 \pmod 7$ and $q^3 \equiv \pm 1 \pmod 7$. In fact, we need the stronger condition $q^3 \equiv \pm 1 \pmod{49}$ in Table 5.9.2 to ensure that $|\Omega|$ is divisible by 7.

To complete the proof, we may assume that $r = 2$. We claim that $T$ is 2-elusive. First assume $q \equiv 1 \pmod 4$. As before, since $H_0$ is normalised by a triality graph automorphism of $T$, it suffices to show that $H_0$ contains $t_1$ and $t_2$ involutions. Let $\hat{x} = [-I_2, I_6]$ be a $t_1$ involution. The discriminant of the restriction of $Q$ (the nondegenerate quadratic form on $V$ defining $T$) to both eigenspaces of $\hat{x}$ is a square (see Lemma 2.2.9), so Lemma 2.2.11 implies that an appropriate conjugate of $x$ centralises the decomposition of $V$ in (5.9.2) (recall that the $\langle v_i \rangle$ are either all square-parabolic spaces or all nonsquare-parabolic spaces) and thus $H_0$ contains $t_1$ involutions. The same argument shows that $t_2 = [-I_4, I_4]$ also centralises the above decomposition, hence $T$ is 2-elusive. An entirely similar argument applies when $q \equiv 3 \pmod{4}$.

Finally, note that if $q \equiv \pm 3 \pmod{8}$ then $(q^2 - 1)_2 = (q^6 - 1)_2 = 2^3$ and $(q^4 - 1)_2 = 2^4$, so $|\Omega|$ is odd. This explains the additional condition $q \equiv \pm 1 \pmod{8}$ in Table 5.9.2 (for $r = 2$). $\qquad\qquad\square$

**Proposition 5.9.8** *Theorem 5.9.1 holds in Case VIII of Table 5.9.1.*

*Proof* Here $H_0 = H \cap T = [q^{11}].\mathrm{GL}_2(q)\mathrm{GL}_1(q)^2$ is a parabolic subgroup of $T$, which can be constructed by intersecting the stabiliser of a totally singular 1-space $U$ and the stabilisers of two totally singular 4-spaces $W_1, W_2$ such that $U \leqslant W_1 \cap W_2$ and $\dim(W_1 \cap W_2) = 3$ (so $W_1$ and $W_2$ are representatives of the two $T$-orbits on maximal totally singular subspaces; see Proposition 2.5.4). In particular, if $x \in T$ does not fix a totally singular 1-space, then it is a derangement. Therefore, by applying Proposition 4.6.2, we may assume that one of the following holds:

(i) $r = 2$ and $q \equiv 1 \pmod{4}$, or
(ii) $r$ is odd and $r$ divides $q - 1$.

In fact, if (ii) holds then $r = 3$ is the only possibility since

$$|\Omega| = \frac{1}{d}(q+1)^3(q^2+1)^2 \left( \frac{q^6 - 1}{q^2 - 1} \right)$$

where $d = 4$ if $q$ is odd, otherwise $d = 1$.

Consider (i). Here the involutions in $T$ are of type $t_1, t_2, t_3$ and $t_4$, and it is easy to see that $t_1 = [-I_2, I_6]$ and $t_2 = [-I_4, I_4]$ fix a totally singular 1-space $U$ and totally singular 4-spaces $W_1$ and $W_2$, with $\dim(W_1 \cap W_2) = 3$ and $U \leqslant W_1 \cap W_2$. As before, this is sufficient to deduce that $T$ is 2-elusive.

Now let us turn to (ii), so $r = 3$ as noted above. Let

$$\{e_1, \ldots, e_4, f_1, \ldots, f_4\}$$

be a hyperbolic basis. In terms of this basis, Proposition 3.5.4(i) implies that every element $x \in T$ of order 3 is of the form $x = \hat{x}Z$, where

$$\hat{x} = [\lambda_1, \lambda_2, \lambda_3, \lambda_4, \lambda_1^{-1}, \lambda_2^{-1}, \lambda_3^{-1}, \lambda_4^{-1}]$$

and each $\lambda_i \in \mathbb{F}_q$ satisfies $\lambda_i^3 = 1$, with $\lambda_i \neq 1$ for some $i$. Clearly, such an element fixes a totally singular 1-space $U$ and a pair of totally singular 4-spaces

$W_1$ and $W_2$ such that $\dim(W_1 \cap W_2) = 3$ and $U \leqslant W_1 \cap W_2$. We conclude that $T$ is $r$-elusive. $\qquad \square$

This completes the proof of Theorem 5.9.1.

In view of our earlier work in Chapter 4 and Sections 5.2–5.8, we have now completed the proof of Theorem 1.5.1.

# 6

# Low-dimensional classical groups

In this final chapter we use our earlier work to determine precise results on the $r$-elusivity of primitive actions of the low-dimensional classical groups. In particular, we prove Theorem 1.5.5.

## 6.1 Introduction

As before, let $G \leqslant \mathrm{Sym}(\Omega)$ be a primitive almost simple classical group over $\mathbb{F}_q$ with socle $T$ and point stabiliser $H$. Recall that $H$ is a maximal subgroup of $G$ and let $n$ be the dimension of the natural $T$-module $V$. As described in Section 2.6, Aschbacher's theorem [3] implies that either $H$ is a *geometric* subgroup in one of the eight $\mathscr{C}_i$ collections, or $H$ is a *non-geometric* subgroup with a simple socle that acts irreducibly on $V$ (we use $\mathscr{S}$ to denote the latter subgroup collection). Also recall that a small additional collection of *novelty* subgroups (denoted by $\mathscr{N}$) arises if $T = \mathrm{Sp}_4(q)'$ (with $p = 2$) or $\mathrm{P}\Omega_8^+(q)$.

In general, it is very difficult to give a complete description of the maximal subgroups of $G$, and in particular those in the collection $\mathscr{S}$. However, the maximal subgroups in the low-dimensional classical groups with $n \leqslant 12$ are determined in the recent book [13] by Bray, Holt and Roney-Dougal. By applying our earlier work in Chapters 4 and 5 on geometric actions, we can use the information in [13] to obtain precise results on the $r$-elusivity of all primitive actions of the low-dimensional classical groups (including non-geometric actions). We anticipate that this sort of information may be useful in applications.

The main result of this chapter is Theorem 6.1.1 below. For convenience we will assume that $n \leqslant 5$, although it would be feasible to extend the analysis to higher dimensions.

**Theorem 6.1.1** *Let $G \leqslant \mathrm{Sym}(\Omega)$ be a primitive almost simple classical group with socle $T$ and point stabiliser $H$, where*

$$T \in \{\mathrm{PSL}_2(q), \mathrm{PSL}_3^\varepsilon(q), \mathrm{PSL}_4^\varepsilon(q), \mathrm{PSp}_4(q)', \mathrm{PSL}_5^\varepsilon(q)\}$$

*Let r be a prime. Then T is r-elusive if and only if $(G, H, r)$ is one of the cases recorded in Tables 6.4.1–6.4.8.*

The proof of Theorem 6.1.1 for geometric actions is essentially a straightforward application of our earlier work in Chapters 4 and 5. Now assume $H$ is a maximal non-geometric subgroup of $G$, so $H$ is almost simple, with socle $S$ say. The embedding of $H$ in $G$ corresponds to an absolutely irreducible representation

$$\rho : \hat{S} \to \mathrm{GL}(V)$$

of an appropriate covering group $\hat{S}$ of $S$. The possibilities for $H$ are listed in [13, Chapter 8] and we can use the representation $\rho$ (and its Brauer character $\chi$) to investigate the fusion of $H$-classes in $G$. For example, if $\chi(\hat{x}) \neq \chi(\hat{y})$ then the elements $x = \hat{x}Z$ and $y = \hat{y}Z$ in $S$ (where $Z = Z(\hat{S})$) are not $T$-conjugate.

We partition the proof of Theorem 6.1.1 into two parts. The cases where $H \in \bigcup_i \mathscr{C}_i \cup \mathscr{N}$ are handled in Section 6.2, and the remaining non-geometric actions are studied in Section 6.3. The tables referred to in the statement of Theorem 6.1.1 are presented in Section 6.4.

## 6.2 Geometric actions

**Proposition 6.2.1** *Theorem 6.1.1 holds if $H \notin \mathscr{S}$.*

*Proof* We consider each possibility for $T$ in turn, using our earlier results in Chapters 4 and 5. In several cases, some additional work is required to determine precise conditions for $r$-elusivity. To illustrate the general approach, we will give details for the case $T = \mathrm{PSL}_3(q)$. Let $V$ be the natural $T$-module and set $H_0 = H \cap T$. Let $r$ be a prime divisor of $|\Omega|$.

*Case 1. Subspace actions.* To begin with, we will assume that $H \in \mathscr{C}_1$ is a subspace subgroup, so there are three cases to consider (see Table 4.1.1). First assume that $H_0 = P_1$ is the stabiliser of a 1-dimensional subspace of $V$, so $|\Omega| = (q^3 - 1)/(q - 1)$. By Proposition 4.3.2, $T$ is $r$-elusive if and only if $r$ divides $q - 1$. Note that $|\Omega|$ is odd, so $r$ is also odd. Now, if $r$ is an odd prime divisor of $q - 1$ then $r$ divides $|\Omega|$ if and only if $r = 3$ (see Lemma A.4). Therefore, $T$ is $r$-elusive if and only if $r = 3$ and $q \equiv 1 \pmod{3}$, which is the condition recorded in Table 6.4.2.

Next suppose that $H$ is a $\mathcal{C}_1$-subgroup of type $\mathrm{GL}_1(q) \times \mathrm{GL}_2(q)$ (this corresponds to Case I' in Table 4.1.1, with $m = 1$). Here $|\Omega| = q^2(q^3 - 1)/(q - 1)$ and by applying Proposition 4.3.4 we deduce that $T$ is 2-elusive if and only if $p = 2$ (note that $|\Omega|$ is odd if $q$ is odd). If $r = p > 2$ then $T$ is $r$-elusive if and only if every partition of 3 contains a partition of 1, which is clearly false (take the partition $\lambda = (3)$). Finally, if $r > 2$ and $r \neq p$, then as in the previous case we find that $T$ is $r$-elusive if and only if $r = 3$ and $q \equiv 1 \pmod{3}$.

To complete the analysis of subspace actions, suppose that $H = P_{1,2}$ is the stabiliser of a pair of subspaces $U$ and $W$ of $V$, where $U < W$, $\dim U = 1$ and $\dim W = 2$. Here $|\Omega| = (q+1)(q^3 - 1)/(q - 1)$ and by arguing as before (using Proposition 4.3.6) it is easy to check that $T$ is $r$-elusive if and only if $r \in \{2, 3\}$ and $q \equiv 1 \pmod{r}$.

*Case 2. Non-subspace actions.* Now let us consider the non-subspace actions. First observe that the collections $\mathcal{C}_4$ and $\mathcal{C}_7$ are empty. If $H$ is a $\mathcal{C}_3$-subgroup of type $\mathrm{GL}_1(q^3)$, then Proposition 5.3.7 implies that $T$ is $r$-elusive if and only if $r = 3$ and $q \equiv 4, 7, 8 \pmod{9}$.

Next assume that $H$ is a $\mathcal{C}_2$-subgroup of type $\mathrm{GL}_1(q) \wr S_3$, in which case

$$|\Omega| = \frac{1}{6}q^3(q+1)\left(\frac{q^3 - 1}{q - 1}\right)$$

By Proposition 5.2.12, if $r = p$ then $T$ is $r$-elusive if and only if $r = 2$, so let us assume that $r \neq p$. Set $i = \Phi(r, q)$ (see (3.1.2)) and write $3 \equiv k \pmod{i}$ with $0 \leqslant k < i$. Note that $i \in \{1, 2, 3\}$. According to Proposition 5.2.20, $T$ is $r$-elusive if and only if one of the following holds:

(a) $i = r - 1 > 1$ and $k > 0$;
(b) $i = 1$.

If (a) holds, then $i = 2$ is the only possibility, so $r = 3$ and we need $(q+1)_3 \geqslant 9$ to ensure that $|\Omega|$ is divisible by 3. This is equivalent to the condition $q \equiv -1 \pmod{9}$. Now assume that (b) holds. If $r = 2$ and $q$ is odd, then $|\Omega|$ is even if and only if $(q+1)_2 \geqslant 4$, so we need $q \equiv -1 \pmod{4}$. If $r > 2$ then $r$ must divide $(q^3 - 1)/(q - 1)$, so $r = 3$ is the only possibility. But $(q^3 - 1)_3 = 3(q - 1)_3$, so $|\Omega|$ is indivisible by 3. We conclude that $T$ is $r$-elusive if and only if $r = 2$ and $q \not\equiv 1 \pmod{4}$, or $r = 3$ and $q \equiv -1 \pmod{9}$.

Now suppose $H$ is a $\mathcal{C}_5$-subgroup of type $\mathrm{GL}_3(q_0)$, where $q = q_0^k$ and $k$ is a prime. First assume that $r = p$. By Proposition 5.5.7, if $p = 2$ then $T$ is 2-elusive, so let us assume that $r = p > 2$. As noted in Remark 5.5.2(ii), $T$ is $r$-elusive if and only if every nontrivial partition $\lambda = (p^{a_p}, \ldots, 1^{a_1})$ of 3 satisfies the following conditions:

(a) $(v, q-1) = (v, q_0-1)$, where $v = \gcd\{j \mid a_j \neq 0\}$, and
(b) if $(v, q-1) = (v, q_0-1) > 1$ then either $k \geqslant p$, or $(3, q_0-1)$ is indivisible by $k$.

Clearly, if $\lambda = (2,1)$ then $v = 1$, so (a) and (b) are satisfied. Now assume that $\lambda = (3)$, so $v = 3$. If $p = 3$ then $(v, q-1) = (v, q_0-1) = 1$ and thus $T$ is 3-elusive, so let us assume that $p > 3$. If $q_0 \equiv -1 \pmod 3$ then $(v, q_0-1) = 1$, and we have $(v, q-1) = 1$ if and only if $k > 2$, so in this situation $T$ is $p$-elusive if and only if $k > 2$. Now assume that $q_0 \equiv 1 \pmod 3$, so $(v, q_0-1) = 3$. Here (a) holds in all cases, whereas (b) holds if and only if $k \geqslant p$ or $k \neq 3$, so $T$ is $p$-elusive if and only if $k \neq 3$. We conclude that $T$ is $p$-elusive if and only if one of the following holds:

(i) $p \in \{2,3\}$;
(ii) $p > 3$ and $k > 3$;
(iii) $p > 3$, $k \in \{2,3\}$ and $q_0 \equiv (-1)^k \pmod 3$.

Finally, if $r \neq p$ then Proposition 5.5.7 implies that $T$ is $r$-elusive if and only if $r = k$, or $r = 3$, $k = 2$ and $(q_0 + 1)_3 = 3$.

The analysis of the remaining cases with $H \in \mathcal{C}_6 \cup \mathcal{C}_8$ is very similar, and we omit the details (see Tables 5.6.2 and 5.8.2).      □

## 6.3 Non-geometric actions

**Proposition 6.3.1** *Theorem 6.1.1 holds if $H \in \mathcal{S}$.*

The possibilities for $H$ are given in [13, Chapter 8], and we record the relevant cases in Table 6.3.1, where $S$ denotes the socle of $H$ (note that the conditions recorded in the final column of Table 6.3.1 are necessary for the existence and maximality of $H$ in $G$, but they are not always sufficient for maximality; see the relevant tables in [13, Chapter 8] for further details). We consider each of these cases in turn. Set $H_0 = H \cap T$ and let $r$ be a prime that divides $|\Omega|$ and $|H_0|$.

**Lemma 6.3.2** *Proposition 6.3.1 holds in Case I of Table 6.3.1.*

*Proof* Here $T = \mathrm{PSL}_5^\varepsilon(q)$ and $H_0 = \mathrm{PSU}_4(2)$, where $q = p \equiv \varepsilon \pmod 6$. Since $r$ divides $|H_0|$, we have $r \in \{2,3,5\}$. There are two classes of involutions in both $H_0$ and $T$, and the character table of $H_0$ (see [41]) indicates that the two $H_0$-classes are not fused in $T$, so $T$ is 2-elusive (note that $|\Omega|$ is even). On the

Table 6.3.1 *The collection $\mathscr{S}$, $n \leqslant 5$*

| Case | $T$ | $S$ | Conditions |
|------|-----|-----|------------|
| I | $\mathrm{PSL}_5^{\varepsilon}(q)$ | $\mathrm{PSU}_4(2)$ | $q = p \equiv \varepsilon \pmod 6$ |
| II | | $\mathrm{PSL}_2(11)$ | $q = p \equiv \varepsilon, 3\varepsilon, 4\varepsilon, 5\varepsilon, 9\varepsilon \pmod{11}$ |
| III | | $\mathrm{M}_{11}$ | $(\varepsilon, q) = (+, 3)$ |
| IV | $\mathrm{PSL}_4^{\varepsilon}(q)$ | $\mathrm{PSU}_4(2)$ | $q = p \equiv \varepsilon \pmod 6$ |
| V | | $A_7$ | $q = p \equiv \varepsilon, 2\varepsilon, 4\varepsilon \pmod 7$ |
| VI | | $\mathrm{PSL}_2(7)$ | $q = p \equiv \varepsilon, 2\varepsilon, 4\varepsilon \pmod 7$, $q \neq 2, 3$ |
| VII | | $\mathrm{PSL}_3(4)$ | $(\varepsilon, q) = (-, 3)$ |
| VIII | $\mathrm{PSp}_4(q)'$ | $\mathrm{Sz}(q)$ | $p = 2$, $\log_2 q > 1$ odd |
| IX | | $\mathrm{PSL}_2(q)$ | $p \geqslant 5$, $q \neq 5$ |
| X | | $A_6$ | $q = p \equiv 1, 5, 7, 11 \pmod{12}$, $q \neq 7$ |
| XI | | $A_7$ | $q = 7$ |
| XII | $\mathrm{PSL}_3^{\varepsilon}(q)$ | $\mathrm{PSL}_2(7)$ | $q = p \equiv \varepsilon, 2\varepsilon, 4\varepsilon \pmod 7$, $q \neq 2$ |
| XIII | | $A_6$ | $q = p \equiv \varepsilon, 4\varepsilon \pmod{15}$, |
| | | | or $\varepsilon = +, q = p^2, p \equiv 2, 3 \pmod 5, p \neq 3$, |
| | | | or $(\varepsilon, q) = (-, 5)$ |
| XIV | | $A_7$ | $(\varepsilon, q) = (-, 5)$ |
| XV | $\mathrm{PSL}_2(q)$ | $A_5$ | $q = p \equiv \pm 1 \pmod{10}$, |
| | | | or $q = p^2, p \equiv \pm 3 \pmod{10}$ |

other hand, $T$ is not 3-elusive since there are four $H_0$-classes of elements of order 3, but there are six such classes in $T$ (see Propositions 3.2.2 and 3.3.3, for example).

Finally, suppose $r = 5$. Note that $H_0$ contains a unique class of elements of order 5, so $T$ is not 5-elusive if $p = 5$. Now assume $p \neq 5$. Set $i = \Phi(r, q)$ as in (3.1.2), so $i \in \{1, 2, 4\}$. If $i = 4$ then Propositions 3.2.1 and 3.3.2 imply that $T$ also has a unique class of elements of order 5, so $T$ is 5-elusive. In addition, it is easy to check that $|\Omega|$ is divisible by 5 if and only if $(q^2 + 1)_5 \geqslant 25$, so we include the condition $q^2 \equiv -1 \pmod{25}$ in Tables 6.4.7 and 6.4.8. Finally, if $i = 1$ or 2 then $T$ has at least two classes of elements of order 5, so $T$ is not 5-elusive in this situation. □

**Lemma 6.3.3** *Proposition 6.3.1 holds in Case II of Table 6.3.1.*

*Proof*  Here $T = \mathrm{PSL}_5^{\varepsilon}(q)$, $H_0 = \mathrm{PSL}_2(11)$ and $r \in \{2, 3, 5, 11\}$, where $q = p$ satisfies the congruence condition in Table 6.3.1. Now $H_0$ has a unique conjugacy class of subgroups of order $r$, so $T$ is $r$-elusive if and only if $T$ also has a unique class of subgroups of order $r$. For example, if $\varepsilon = +$ then it is easy to see that $T$ has this property if and only if $r = 5$ and $\Phi(5, q) = 4$, or $r = 11$ and $\Phi(11, q) = 5$. The case $\varepsilon = -$ is entirely similar. Finally, for divisibility we require $q^2 \equiv -1 \pmod{25}$ if $r = 5$, and $q^5 \equiv \varepsilon \pmod{121}$ if $r = 11$. □

**Lemma 6.3.4** *Proposition 6.3.1 holds in Case IV of Table 6.3.1.*

*Proof* Here $T = \mathrm{PSL}_4^\varepsilon(q)$ and $H_0 = \mathrm{PSU}_4(2)$, where $q = p \equiv \varepsilon \pmod 6$. Since $r$ divides $|H_0|$ we have $r \in \{2,3,5\}$. As noted in the proof of Lemma 6.3.2, $H_0$ has a unique class of elements of order 5 and we deduce that $T$ is 5-elusive if and only if $q^2 \equiv -1 \pmod{25}$. Similarly, both $H_0$ and $T$ contain four classes of elements of order 3, and the character table of $2.\mathrm{PSU}_4(2) \cong \mathrm{Sp}_4(3)$ indicates that none of the relevant $H_0$-classes are fused in $T$. Therefore $T$ is 3-elusive, and we note that the condition $q \equiv \varepsilon \pmod 9$ is needed for divisibility.

Now assume $r = 2$. Note that $|\Omega|$ is even. Now $H_0$ has two classes of involutions, and from the character table of $\mathrm{Sp}_4(3)$ we deduce that $\nu(x) = 2$ for all $x \in H_0$ of order 2. In particular, if $q \equiv \varepsilon \pmod 8$ then $t_1$-type involutions in $T$ are derangements (see Table B.1). On the other hand, if $q \equiv 5\varepsilon \pmod 8$ then $T$ has a unique class of involutions, so in this situation $T$ is 2-elusive.

Finally, suppose that $r = 2$ and $q \equiv -\varepsilon \pmod 4$. Here $T$ has two classes of involutions, with representatives $t_2$ and $t_2'$ (see Table B.1), and we claim that $T$ is 2-elusive. Let

$$\rho : \mathrm{Sp}_4(3) \to \mathrm{SL}_4^\varepsilon(q)$$

be the corresponding irreducible representation, and note that this map is injective (here $\rho$ is a *Weil representation* of $\mathrm{Sp}_4(3)$). Let $x_1, x_2$ be representatives of the two classes of involutions in $\mathrm{PSp}_4(3)$, where $x_1$ lifts to an involution in $\mathrm{Sp}_4(3)$, and $x_2$ lifts to an element of order 4. Then $\rho(\hat{x}_1)$ has order 2, so $x_1$ is $T$-conjugate to $t_2$. However, $\rho(\hat{x}_2)$ has order 4 and thus $x_2$ is $T$-conjugate to $t_2'$. This justifies the claim, and we conclude that $T$ is 2-elusive if and only if $q \not\equiv \varepsilon \pmod 8$. $\square$

**Lemma 6.3.5** *Proposition 6.3.1 holds in Case V of Table 6.3.1.*

*Proof* Here $T = \mathrm{PSL}_4^\varepsilon(q)$ and $H_0 = A_7$, where $q = p \equiv \varepsilon, 2\varepsilon, 4\varepsilon \pmod 7$. Since $r$ divides $|H_0|$ we have $r \in \{2,3,5,7\}$. If $r \in \{2,5,7\}$ then $H_0$ has a unique conjugacy class of subgroups of order $r$ and the result quickly follows. In particular, note that $T$ has a unique class of involutions if and only if $q \equiv 5\varepsilon \pmod 8$. Also note that $T$ has a unique class of subgroups of order $r \in \{5,7\}$ if and only if $r \neq p$ and $q^2 \not\equiv 1 \pmod r$. In Tables 6.4.4 and 6.4.5, we need the additional conditions $q^2 \equiv -1 \pmod{25}$ (for $r = 5$) and $q^3 \equiv \varepsilon \pmod{49}$ (for $r = 7$) for divisibility. Note that for $r = 7$, the congruences $q^2 \not\equiv 1 \pmod 7$ and $q^3 \equiv \varepsilon \pmod{49}$ are equivalent to the single congruence $q(q+\varepsilon) \equiv -1 \pmod{49}$.

Now assume $r = 3$ and note that $H_0$ contains two classes of elements of order 3. The character table of $2.A_7$ indicates that these two classes are not fused in $T$, whence $T$ is 3-elusive if and only if $q \equiv -\varepsilon \pmod 3$. In fact, it is easy to check that we need the stronger condition $q \equiv -\varepsilon \pmod 9$ to ensure that $|\Omega|$ is divisible by 3. $\qquad\square$

**Lemma 6.3.6** *Proposition 6.3.1 holds in Case VI of Table 6.3.1.*

*Proof* Here $T = \mathrm{PSL}_4^{\varepsilon}(q)$ and $H_0 = \mathrm{PSL}_2(7)$, where $q = p \equiv \varepsilon, 2\varepsilon, 4\varepsilon$ (mod 7) and $q \neq 2,3$. Note that $r \in \{2,3,7\}$. Now $H_0$ has a unique class of elements of order 2 and 3. Therefore, $T$ is not 3-elusive (since $T$ has at least two classes of elements of order 3), and we deduce that $T$ is 2-elusive if and only if $q \equiv 5\varepsilon \pmod 8$. Similarly, $H_0$ has a unique class of subgroups of order 7, so $T$ is 7-elusive if and only if $q^2 \not\equiv 1 \pmod 7$ and $q^3 \equiv \varepsilon \pmod 7$ (in fact, we need $q^3 \equiv \varepsilon \pmod{49}$ for divisibility, so the required condition in Tables 6.4.4 and 6.4.5 is $q(q+\varepsilon) \equiv -1 \pmod{49}$). $\qquad\square$

**Lemma 6.3.7** *Proposition 6.3.1 holds in Case VIII of Table 6.3.1.*

*Proof* Here $T = \mathrm{PSp}_4(q)'$ and $H_0 = \mathrm{Sz}(q) = {}^2B_2(q)$, where $p = 2$ and $\log_2 q > 1$ is odd. We may view $H_0$ as the centraliser $C_T(\tau)$ of an involutory graph-field automorphism $\tau$ of $T$ (see Section 3.4.6). It is well known that $H_0$ has a unique class of involutions, so $T$ is not 2-elusive (note that $T$ has three classes of involutions, with representatives labelled $b_1, a_2$ and $c_2$ in Section 3.4.4).

Now assume $r$ is odd and set $i = \Phi(r, q)$. Since $r$ divides

$$|H_0| = q^2(q^2+1)(q-1), \quad |\Omega| = q^2(q+1)(q^2-1)$$

it follows that $i = 1$. Here [18, Proposition 3.52] implies that $x = [\Lambda, \Lambda^{-1}, I_2]$ $\in T$ is not centralised by $\tau$, so $x$ is a derangement. $\qquad\square$

**Lemma 6.3.8** *Proposition 6.3.1 holds in Case IX of Table 6.3.1.*

*Proof* Here we have $T = \mathrm{PSp}_4(q)$ and $H_0 = \mathrm{PSL}_2(q)$, where $p \geqslant 5$ and $q \neq 5$. Now $H_0$ has a unique conjugacy class of involutions, but there are two classes in $T$, so $T$ is not 2-elusive. Similarly, it is easy to check that $T$ is not $p$-elusive. Now assume that $r \neq p$ and $r > 2$. Set $i = \Phi(r, q)$ and note that $i \in \{1, 2\}$ since $r$ divides $|H_0|$. Then $H_0$ has a unique conjugacy class of subgroups of order $r$, but $T$ has at least two, so we conclude that $T$ is not $r$-elusive. $\qquad\square$

**Lemma 6.3.9** *Proposition 6.3.1 holds in Case X of Table 6.3.1.*

*Proof* Here $T = \mathrm{PSp}_4(q) \cong \Omega_5(q)$ and $S = A_6$, where $q = p \equiv 1,5,7,11$ (mod 12) and $q \neq 7$. More precisely, [13, Table 8.13] indicates that $H_0 = A_6$ if $q \equiv 5,7$ (mod 12), otherwise $H_0 = S_6$. Note that $r \in \{2,3,5\}$ since $r$ divides $|H_0|$.

Both $T$ and $H_0$ contain two classes of elements of order 3, and by inspecting the character table of $2.A_6$ (or $2.S_6$) we deduce that the two $H_0$-classes are not fused in $T$, so $T$ is 3-elusive (we need the condition $q^2 \equiv 1$ (mod 9) in Table 6.4.6 for divisibility). Since $H_0$ has a unique class of subgroups of order 5, it follows that $T$ is 5-elusive if and only if $q^2 \equiv -1$ (mod 5) (and in fact, we need $q^2 \equiv -1$ (mod 25) for divisibility).

Finally let us assume $r = 2$. Note that $T$ has two classes of involutions. If $H_0 = A_6$ then $H_0$ has a unique class of involutions, so $T$ is not 2-elusive. Now assume that $H_0 = S_6$, so $H_0$ has three classes of involutions. The involution classes in $T$ are represented by the elements $t_1$ and $t_2$ (or $t_2'$ if $q \equiv 3$ (mod 4)). We claim that $T$ is 2-elusive. To see this, it is convenient to work with $T = \Omega_5(q)$ since this allows us to identify the natural $T$-module $V$ with the fully deleted permutation module for $S_6$ over $\mathbb{F}_q$ (see [86, p. 185–187]). An easy calculation with this module shows that a transposition in $S_6$ is a $t_1$ involution, and the other two classes of involutions in $S_6$ are of type $t_2$ or $t_2'$. Therefore $T$ is 2-elusive, as claimed.                     □

**Lemma 6.3.10** *Proposition 6.3.1 holds in Case XII of Table 6.3.1.*

*Proof* Here $T = \mathrm{PSL}_3^\varepsilon(q)$ and $H_0 = \mathrm{PSL}_2(7)$, where $q = p \equiv \varepsilon, 2\varepsilon, 4\varepsilon$ (mod 7) and $q \neq 2$. Now $r \in \{2,3,7\}$ since $r$ divides $|H_0|$, and we note that $H_0$ has a unique class of subgroups of order $r$. Therefore, $T$ is $r$-elusive if and only if it also has a unique class of such subgroups. We immediately deduce that $T$ is 2-elusive. Similarly, $T$ is 7-elusive if and only if $c \in \{2,3\}$, where $c$ is the integer defined in Definition 5.1.1 (with $i = \Phi(7,q)$). Here the condition $c = 2$ is equivalent to the congruence $q \equiv -\varepsilon$ (mod 7) (and in fact, we require $q \equiv -\varepsilon$ (mod 49) for divisibility). Similarly, the condition $c = 3$ is equivalent to the pair of congruences $q^2 \not\equiv 1$ (mod 7) and $q^3 \equiv \varepsilon$ (mod 7) (which in turn is equivalent to the single congruence $q(q+\varepsilon) \equiv -1$ (mod 7)). Moreover, we need $q(q+\varepsilon) \equiv -1$ (mod 49) for divisibility.

Finally, suppose that $r = 3$. If $p = 3$ then $T$ has two classes of elements of order 3, so $T$ is not 3-elusive. Now assume $p \neq 3$. If $q \equiv -\varepsilon$ (mod 3) then $T$ has a unique class of elements of order 3, so $T$ is 3-elusive (we need $q \equiv -\varepsilon$ (mod 9) for divisibility). Now assume $q \equiv \varepsilon$ (mod 3), so $\mathrm{PGL}_3^\varepsilon(q)$ has five classes of elements of order 3 (this can be computed from Propositions 3.2.2 and 3.3.3). If $\varepsilon = +$ then either $T$ has a unique class of such elements, or there exists a scalar $\xi \in \mathbb{F}_q$ such that $\xi^3$ is a primitive cube root of unity (in which case $T$ has three such classes). Therefore, $T$ is 3-elusive if and only if $(q - 1)_3 = 3$ (that is, if and only if $q \equiv 4, 7 \pmod 9$). Similarly, if $\varepsilon = -$ then $(q + 1)_3 = 3$ is a necessary and sufficient condition for 3-elusivity. To summarise, we have shown that $T$ is 3-elusive if and only if $q \equiv 4\varepsilon, 7\varepsilon, 8\varepsilon \pmod 9$, which is the condition recorded in Tables 6.4.2 and 6.4.3.     $\Box$

**Lemma 6.3.11** *Proposition 6.3.1 holds in Case XIII of Table 6.3.1.*

*Proof* Here $T = \mathrm{PSL}_3^\varepsilon(q)$ and $S = A_6$, with the conditions on $p$ and $q$ given in Table 6.3.1. Moreover, either $H_0 = A_6$, or $(\varepsilon, q) = (-, 5)$ and $H_0 = \mathrm{M}_{10} = A_6.2_3$ (see [13, Tables 8.4, 8.6]). We may assume $r \in \{2, 3, 5\}$.

If $(\varepsilon, q) = (-, 5)$ then $H_0 = \mathrm{M}_{10}$ and $r = 5$ is the only prime that divides both $|\Omega|$ and $|H_0|$. Here $T$ has four classes of elements of order 5, but there is only one class in $H_0$, so $T$ contains derangements of order 5.

For the remainder, we may assume that $p \ne 5$ and $H_0 = A_6$. Note that $T$ is 2-elusive since it has a unique class of involutions. Now assume $r = 3$ and note that the conditions on $q$ in Table 6.3.1 imply that $q \equiv \varepsilon \pmod 3$. If $(q - \varepsilon)_3 = 3$ then $|\Omega|$ is indivisible by 3, so we may assume that $q \equiv \varepsilon \pmod 9$. Here $T$ has three classes of elements of order 3, but $H_0$ has only two, so $T$ is not 3-elusive. Finally, suppose that $r = 5$. If $q \equiv \varepsilon \pmod 5$ then $T$ has six classes of elements of order 5, so in this case $T$ is not 5-elusive. However, if $q \equiv -\varepsilon \pmod 5$ then $T$ has a unique class of subgroups of order 5, so $T$ is 5-elusive (and we need the condition $q \equiv -\varepsilon \pmod{25}$ for divisibility).     $\Box$

**Lemma 6.3.12** *Proposition 6.3.1 holds in Case XV of Table 6.3.1.*

*Proof* Here $T = \mathrm{PSL}_2(q)$ and $H_0 = A_5$, where $p$ and $q$ satisfy the conditions in Table 6.3.1 (in particular, $q$ is odd). We may assume that $r$ divides $|H_0|$, so $r \in \{2, 3, 5\}$.

Since $T$ has a unique class of involutions, it follows that $T$ is 2-elusive, and one can check that we require the condition $q \equiv \pm 1 \pmod 8$ for divisibility. Similarly, $T$ is 3-elusive if $r = 3$ and $p \ne 3$ (with the additional condition $q \equiv \pm 1 \pmod 9$ for divisibility). However, if $p = 3$ then $q = 9$ and there are two classes in $T$, but only one in $H_0$, so $T$ is not 3-elusive in this situation. Finally, we observe that both $T$ and $H_0$ have a unique class of subgroups of order 5, so $T$ is 5-elusive. The condition $q \equiv \pm 1 \pmod{25}$ in Table 6.4.1 is required for divisibility.     $\Box$

**Lemma 6.3.13** *Proposition 6.3.1 holds in the remaining cases in Table 6.3.1.*

*Proof* It remains to deal with the cases labelled III, VII, XI and XIV in Table 6.3.1. It is easy to check that $T$ is $r$-elusive if and only if $r$ is one of the following:

Case III, $r = 11$;
Case VII, $r = 2$;
Case XI, $r = 5$;
Case XIV, $r = 2$

We leave the reader to check the details. □

This completes the proof of Proposition 6.3.1.

## 6.4 Tables

In this final section, we present the tables referred to in the statement of Theorem 6.1.1, which record the $r$-elusive primitive actions of the low-dimensional classical groups.

In the first column of each table we record the Aschbacher collection that $H$ belongs to. The type of $H$ is given in the next column; as before, this provides an approximate description of the group-theoretic structure of $H \cap \mathrm{PGL}(V)$, where $V$ is the natural $T$-module. Note that the type of a subgroup $H$ in the collection $\mathscr{S}$ is simply the socle of $H$. Following [86], we write $P_m$ for the stabiliser of a totally singular $m$-space, and $P_{m,n-m}$ for the stabiliser of a pair of spaces $\{U, W\}$, where $U < W$ with $\dim U = m$ and $\dim W = n - m$.

Finally, we present necessary and sufficient conditions for $r$-elusivity in the final column of the tables. These are additional conditions to those needed for the existence and maximality of $H$ in $G$, which can be read off from the appropriate tables in Chapters 4 and 5 (together with Table 6.3.1 for subgroups in the collection $\mathscr{S}$), or from the relevant tables in [13, Chapter 8]. For instance, Table 6.4.7 indicates that if $T = \mathrm{PSL}_5(q)$ and $H$ is a subfield subgroup of type $\mathrm{GL}_5(q_0)$, where $q = q_0^k$ with $k$ a prime, then $T$ is $p$-elusive unless $p > 5$, $k \in \{2, 5\}$ and $q_0 \equiv (-1)^{k+1} \pmod 5$.

Table 6.4.1 *r-elusive actions,* $T = \mathrm{PSL}_2(q)$

|  | Type of $H$ | Conditions |
|---|---|---|
| $\mathscr{C}_1$ | $P_1$ | $r = 2, q \equiv 1 \pmod 4$ |
| $\mathscr{C}_2$ | $\mathrm{GL}_1(q) \wr S_2$ | $r = 2, q \not\equiv 1 \pmod 4$ |
| $\mathscr{C}_3$ | $\mathrm{GL}_1(q^2)$ | $r = 2, q \not\equiv 3 \pmod 4$ |
| $\mathscr{C}_5$ | $\mathrm{GL}_2(q_0), q = q_0^k$ | $r = p$, with $k > 2$ if $p > 2$ |
|  |  | $r = k$ |
| $\mathscr{C}_6$ | $2^2.\mathrm{O}_2^-(2)$ | $r = 2, q \equiv \pm 1 \pmod{16}$ |
|  |  | $r = 3, q \equiv \pm 1 \pmod 9$ |
| $\mathscr{S}$ | $A_5$ | $r = 2, q \equiv \pm 1 \pmod 8$ |
|  |  | $r \in \{3,5\}, q \equiv \pm 1 \pmod{r^2}$ |

Table 6.4.2 *r-elusive actions,* $T = \mathrm{PSL}_3(q)$

|  | Type of $H$ | Conditions |
|---|---|---|
| $\mathscr{C}_1$ | $P_1$ | $r = 3, q \equiv 1 \pmod 3$ |
|  | $P_{1,2}$ | $r \in \{2,3\}, q \equiv 1 \pmod r$ |
|  | $\mathrm{GL}_1(q) \times \mathrm{GL}_2(q)$ | $r = p = 2$ |
|  |  | $r = 3, q \equiv 1 \pmod 3$ |
| $\mathscr{C}_2$ | $\mathrm{GL}_1(q) \wr S_3$ | $r = 2, q \not\equiv 1 \pmod 4$ |
|  |  | $r = 3, q \equiv -1 \pmod 9$ |
| $\mathscr{C}_3$ | $\mathrm{GL}_1(q^3)$ | $r = 3, q \equiv 4,7,8 \pmod 9$ |
| $\mathscr{C}_5$ | $\mathrm{GL}_3(q_0), q = q_0^k$ | $r = p \in \{2,3\}$ |
|  |  | $r = p > 3$, with $q_0 \equiv (-1)^k \pmod 3$ if $k \in \{2,3\}$ |
|  |  | $r = 3, k = 2, (q_0+1)_3 = 3$ |
|  |  | $r \neq p, r = k$ |
| $\mathscr{C}_6$ | $3^2.\mathrm{Sp}_2(3)$ | $r = 2$ |
|  |  | $r = 3, q \equiv 1 \pmod 9$ |
| $\mathscr{C}_8$ | $\mathrm{O}_3(q)$ | $r = 2$ |
|  |  | $r = 3, (q-1)_3 = 3$ |
|  | $\mathrm{GU}_3(q_0), q = q_0^2$ | $r = 2$ |
|  |  | $r = 3, (q_0-1)_3 = 3$ |
|  |  | $r = p > 2$, with $q_0 \equiv -1 \pmod 3$ if $p > 3$ |
| $\mathscr{S}$ | $\mathrm{PSL}_2(7)$ | $r = 2$ |
|  |  | $r = 3, q \equiv 4,7,8 \pmod 9$ |
|  |  | $r = 7, q \equiv -1 \pmod{49}$ or $q(q+1) \equiv -1 \pmod{49}$ |
|  | $A_6$ | $r = 2$ |
|  |  | $r = 5, q \equiv -1 \pmod{25}$ |

Table 6.4.3  *r-elusive actions*, $T = \mathrm{PSU}_3(q)$

|        | Type of $H$ | Conditions |
|--------|-------------|------------|
| $\mathscr{C}_1$ | $P_1$ | $r = 2, p > 2$ |
|        | $\mathrm{GU}_1(q) \times \mathrm{GU}_2(q)$ | $r = p = 2$ |
|        |             | $r = 3, q \equiv -1 \pmod 3$ |
| $\mathscr{C}_2$ | $\mathrm{GU}_1(q) \wr S_3$ | $r = 2, q \not\equiv -1 \pmod 4$ |
|        |             | $r = 3, q \equiv 1 \pmod 9$ |
| $\mathscr{C}_3$ | $\mathrm{GU}_1(q^3)$ | $r = 3, q \equiv 1, 2, 5 \pmod 9$ |
| $\mathscr{C}_5$ | $\mathrm{GU}_3(q_0), q = q_0^k$ | $r = p$, with $k > 3$ if $p > 2$ and $q_0 \equiv -1 \pmod 3$ |
|        |             | $r \neq p, r = k$ |
|        | $O_3(q)$ | $r = 2$ |
|        |             | $r = 3, (q+1)_3 = 3$ |
| $\mathscr{C}_6$ | $3^2.\mathrm{Sp}_2(3)$ | $r = 2$ |
|        |             | $r = 3, q \equiv -1 \pmod 9$ |
| $\mathscr{S}$ | $\mathrm{PSL}_2(7)$ | $r = 2$ |
|        |             | $r = 3, q \equiv 1, 2, 5 \pmod 9$ |
|        |             | $r = 7, q \equiv 1 \pmod{49}$ or $q(q-1) \equiv -1 \pmod{49}$ |
|        | $A_6$ | $r = 2, q \neq 5$ |
|        |             | $r = 5, q \equiv 1 \pmod{25}$ |
|        | $A_7$ | $r = 2, q = 5$ |

Table 6.4.4  *r-elusive actions*, $T = \mathrm{PSL}_4(q)$

|        | Type of $H$ | Conditions |
|--------|-------------|------------|
| $\mathscr{C}_1$ | $P_1$ | $r = 2, q \equiv 1 \pmod 4$ |
|        | $P_2$ | $r \in \{2,3\}, q \equiv 1 \pmod r$ |
|        | $P_{1,3}$ | $r = 2, q \equiv 1 \pmod 4$ |
|        |             | $r = 3, q \equiv 1 \pmod 3$ |
|        | $\mathrm{GL}_1(q) \times \mathrm{GL}_3(q)$ | $r = 2, q \equiv 1 \pmod 4$ |
| $\mathscr{C}_2$ | $\mathrm{GL}_1(q) \wr S_4$ | $r = 2, q \not\equiv 1 \pmod 4$ |
|        | $\mathrm{GL}_2(q) \wr S_2$ | $r = p = 2$ |
|        |             | $r = 3, q \equiv 1 \pmod 3$ |
| $\mathscr{C}_3$ | $\mathrm{GL}_2(q^2)$ | $r = 2, p > 2, q \not\equiv 1 \pmod 8$ |
| $\mathscr{C}_5$ | $\mathrm{GL}_4(q_0), q = q_0^k$ | $r = p$, with $k > 2$ if $p > 2$ |
|        |             | $r = k$ |
| $\mathscr{C}_6$ | $2^4.\mathrm{Sp}_4(2)$ | $r = 2, q \equiv 1 \pmod 8$ |
|        |             | $r = 3, q \equiv -1 \pmod 9$ |
|        |             | $r = 5, q^2 \equiv -1 \pmod{25}$ |
| $\mathscr{C}_8$ | $\mathrm{Sp}_4(q)$ | $r = 2, q \not\equiv 1, 3, 7 \pmod 8$ |
|        | $O_4^\varepsilon(q)$ | $r = 2$ |
|        | $\mathrm{GU}_4(q_0), q = q_0^2$ | $r = 2$ |

Table 6.4.4 (Continued)

|   | Type of $H$ | Conditions |
|---|---|---|
| $\mathscr{S}$ | $\mathrm{PSU}_4(2)$ | $r = 2, q \not\equiv 1 \pmod 8$ |
|   |   | $r = 3, q \equiv 1 \pmod 9$ |
|   |   | $r = 5, q^2 \equiv -1 \pmod{25}$ |
|   | $A_7$ | $r = 2, q \equiv 5 \pmod 8$ |
|   |   | $r = 3, q \equiv -1 \pmod 9$ |
|   |   | $r = 5, q^2 \equiv -1 \pmod{25}$ |
|   |   | $r = 7, q(q+1) \equiv -1 \pmod{49}$ |
|   | $\mathrm{PSL}_2(7)$ | $r = 2, q \equiv 5 \pmod 8$ |
|   |   | $r = 7, q(q+1) \equiv -1 \pmod{49}$ |

Table 6.4.5  *r-elusive actions,* $T = \mathrm{PSU}_4(q)$

|   | Type of $H$ | Conditions |
|---|---|---|
| $\mathscr{C}_1$ | $P_1$ | $r \in \{2,3\}, q \equiv -1 \pmod r$ |
|   | $P_2$ | $r = 2, p > 2, q \not\equiv -1 \pmod 8$ |
|   | $\mathrm{GU}_1(q) \times \mathrm{GU}_3(q)$ | $r = 2, q \equiv 3 \pmod 4$ |
| $\mathscr{C}_2$ | $\mathrm{GU}_1(q) \wr S_4$ | $r = 2, q \not\equiv -1 \pmod 4$ |
|   | $\mathrm{GU}_2(q) \wr S_2$ | $r = p = 2$ |
|   |   | $r = 3, q \equiv -1 \pmod 3$ |
|   | $\mathrm{GL}_2(q^2)$ | $r = 2, p > 2, q \not\equiv -1 \pmod 8$ |
| $\mathscr{C}_5$ | $\mathrm{GU}_4(q_0), q = q_0^k$ | $r \in \{p, k\}$ |
|   | $\mathrm{Sp}_4(q)$ | $r = 2, q \not\equiv 1, 5, 7 \pmod 8$ |
|   | $\mathrm{O}_4^\varepsilon(q)$ | $r = 2$ |
| $\mathscr{C}_6$ | $2^4.\mathrm{Sp}_4(2)$ | $r = 2, q \equiv -1 \pmod 8$ |
|   |   | $r = 3, q \equiv 1 \pmod 9$ |
|   |   | $r = 5, q^2 \equiv -1 \pmod{25}$ |
| $\mathscr{S}$ | $\mathrm{PSU}_4(2)$ | $r = 2, q \not\equiv -1 \pmod 8$ |
|   |   | $r = 3, q \equiv -1 \pmod 9$ |
|   |   | $r = 5, q^2 \equiv -1 \pmod{25}$ |
|   | $A_7$ | $r = 2, q \equiv 3 \pmod 8$ |
|   |   | $r = 3, q \equiv 1 \pmod 9$ |
|   |   | $r = 5, q^2 \equiv -1 \pmod{25}$ |
|   |   | $r = 7, q(q-1) \equiv -1 \pmod{49}$ |
|   | $\mathrm{PSL}_2(7)$ | $r = 2, q \equiv 3 \pmod 8$ |
|   |   | $r = 7, q(q-1) \equiv -1 \pmod{49}$ |
|   | $\mathrm{PSL}_3(4)$ | $r = 2, q = 3$ |

Table 6.4.6 *r-elusive actions,* $T = \mathrm{PSp}_4(q)'$

|  | Type of $H$ | Conditions |
|---|---|---|
| $\mathscr{C}_1$ | $P_1$ | $r = 2, q \equiv 1 \ (\mathrm{mod}\ 4)$ |
|  | $P_2$ | $r = 2, p > 2$ |
| $\mathscr{C}_2$ | $\mathrm{Sp}_2(q) \wr S_2$ | $r = p \in \{2,3\}$ |
|  | $\mathrm{GL}_2(q)$ | $r = 2$ |
| $\mathscr{C}_3$ | $\mathrm{Sp}_2(q^2)$ | $r = 2, p > 2$ |
|  | $\mathrm{GU}_2(q)$ | $r = 2$ |
| $\mathscr{C}_5$ | $\mathrm{Sp}_4(q_0), q = q_0^k$ | $r = p$, with $k > 2$ if $p > 2$ |
|  |  | $r = k$ |
| $\mathscr{C}_6$ | $2^4.\mathrm{O}_4^-(2)$ | $r = 2, q \equiv \pm 1 \ (\mathrm{mod}\ 8)$ |
|  |  | $r = 5, q^2 \equiv -1 \ (\mathrm{mod}\ 25)$ |
| $\mathscr{C}_8$ | $\mathrm{O}_4^+(q)$ | $r = 2$ |
| $\mathscr{N}$ | $\mathrm{O}_2^\varepsilon(q) \wr S_2$ | $r = 2$ |
| $\mathscr{S}$ | $A_6$ | $r = 2, q \equiv \pm 1 \ (\mathrm{mod}\ 12)$ |
|  |  | $r = 3, q^2 \equiv 1 \ (\mathrm{mod}\ 9)$ |
|  |  | $r = 5, q^2 \equiv -1 \ (\mathrm{mod}\ 25)$ |
|  | $A_7$ | $r = 5, q = 7$ |

Table 6.4.7 *r-elusive actions,* $T = \mathrm{PSL}_5(q)$

|  | Type of $H$ | Conditions |
|---|---|---|
| $\mathscr{C}_1$ | $P_1$ | $r = 5, q \equiv 1 \ (\mathrm{mod}\ 5)$ |
|  | $P_2$ | $r \in \{2,5\}, q \equiv 1 \ (\mathrm{mod}\ r)$ |
|  | $P_{1,4}$ | $r \in \{2,5\}, q \equiv 1 \ (\mathrm{mod}\ r)$ |
|  | $P_{2,3}$ | $r \in \{2,3,5\}, q \equiv 1 \ (\mathrm{mod}\ r)$ |
|  | $\mathrm{GL}_1(q) \times \mathrm{GL}_4(q)$ | $r = p = 2$ |
|  |  | $r = 5, q \equiv 1 \ (\mathrm{mod}\ 5)$ |
|  | $\mathrm{GL}_2(q) \times \mathrm{GL}_3(q)$ | $r = 2$ |
|  |  | $r = p = 3$ |
|  |  | $r = 5, q \equiv 1 \ (\mathrm{mod}\ 5)$ |
| $\mathscr{C}_2$ | $\mathrm{GL}_1(q) \wr S_5$ | $r = 2, q \not\equiv 1 \ (\mathrm{mod}\ 4)$ |
|  |  | $r = 5, q^2 \equiv -1 \ (\mathrm{mod}\ 25)$ |
| $\mathscr{C}_3$ | $\mathrm{GL}_1(q^5)$ | $r = 5, q^2 \equiv -1 \ (\mathrm{mod}\ 25)$ |

Table 6.4.7 (Continued)

| | Type of $H$ | Conditions |
|---|---|---|
| $\mathscr{C}_5$ | $\mathrm{GL}_5(q_0), q = q_0^k$ | $r = p \in \{2,3,5\}$ |
| | | $r = p > 5$, with $q_0 \not\equiv (-1)^{k+1} \pmod 5$ if $k \in \{2,5\}$ |
| | | $r = 5, k = 2, (q_0 + 1)_5 = 5$ |
| | | $r \neq p, r = k$ |
| $\mathscr{C}_8$ | $\mathrm{O}_5^\varepsilon(q)$ | $r = 2$ |
| | | $r = 5, (q-1)_5 = 5$ |
| | $\mathrm{GU}_5(q_0), q = q_0^2$ | $r = 2$ |
| | | $r = 5, (q_0 - 1)_5 = 5$ |
| | | $r = p > 2$, with $q_0 \not\equiv 1 \pmod 5$ if $p > 5$ |
| $\mathscr{S}$ | $\mathrm{PSU}_4(2)$ | $r = 2$ |
| | | $r = 5, q^2 \equiv -1 \pmod{25}$ |
| | $\mathrm{PSL}_2(11)$ | $r = 5, q^2 \equiv -1 \pmod{25}$ |
| | | $r = 11, q \not\equiv 1 \pmod{11}, q^5 \equiv 1 \pmod{121}$ |
| | $\mathrm{M}_{11}$ | $r = 11, q = 3$ |

Table 6.4.8 *r-elusive actions, $T = \mathrm{PSU}_5(q)$*

| | Type of $H$ | Conditions |
|---|---|---|
| $\mathscr{C}_1$ | $P_1$ | $r \in \{2,3\}, q \equiv -1 \pmod r$ |
| | $P_2$ | $r = 2, p > 2$ |
| | $\mathrm{GU}_1(q) \times \mathrm{GU}_4(q)$ | $r = p = 2$ |
| | | $r = 5, q \equiv -1 \pmod 5$ |
| | $\mathrm{GU}_2(q) \times \mathrm{GU}_3(q)$ | $r = 2$ |
| | | $r = p = 3$ |
| | | $r = 5, q \equiv -1 \pmod 5$ |
| $\mathscr{C}_2$ | $\mathrm{GU}_1(q) \wr S_5$ | $r = 2, q \not\equiv -1 \pmod 4$ |
| | | $r = 5, q^2 \equiv -1 \pmod{25}$ |
| $\mathscr{C}_3$ | $\mathrm{GU}_1(q^5)$ | $r = 5, q^2 \equiv -1 \pmod{25}$ |
| $\mathscr{C}_5$ | $\mathrm{GU}_5(q_0), q = q_0^k$ | $r = p$, with $k \neq 5$ if $p > 5$ and $q_0 \equiv -1 \pmod 5$ |
| | | $r \neq p, r = k$ |
| | $\mathrm{O}_5(q)$ | $r = 2$ |
| | | $r = 5, (q+1)_5 = 5$ |
| $\mathscr{S}$ | $\mathrm{PSU}_4(2)$ | $r = 2$ |
| | | $r = 5, q^2 \equiv -1 \pmod{25}$ |
| | $\mathrm{PSL}_2(11)$ | $r = 5, q^2 \equiv -1 \pmod{25}$ |
| | | $r = 11, q \not\equiv -1 \pmod{11}, q^5 \equiv -1 \pmod{121}$ |

# Appendix A
## Number-theoretic miscellanea

In this short appendix we present some elementary number-theoretic results that are required throughout the text. In the lemmas, $q$ is a $p$-power, where $p$ is a prime number. In addition, if $r \neq p$ is a prime then we define $\Phi(r,q)$ as in (3.1.2), so $\Phi(r,q)$ is the smallest integer $i \geqslant 1$ such that $r$ divides $q^i - 1$. Set

$$c = c(i) = \begin{cases} i & i \equiv 0 \ (\mathrm{mod}\ 4) \\ i/2 & i \equiv 2 \ (\mathrm{mod}\ 4) \\ 2i & i \ \mathrm{odd} \end{cases} \tag{A.1}$$

and

$$d = d(i) = \begin{cases} i & i \ \mathrm{even} \\ 2i & i \ \mathrm{odd} \end{cases} \tag{A.2}$$

Also recall that $(a,b)$ denotes the greatest common divisor of the positive integers $a$ and $b$.

**Lemma A.1** *Let $r \neq p$ be an odd prime and set $i = \Phi(r,q)$. Let $m$ be a positive integer. Then the following hold:*

(i) *$r$ divides $q^m - 1$ if and only if $i$ divides $m$;*
(ii) *if $r$ divides $q^m + 1$ and $m$ is odd then $i \equiv 2 \ (\mathrm{mod}\ 4)$;*
(iii) *if $r$ divides $q^m + 1$ and $m$ is even then $i \equiv 0 \ (\mathrm{mod}\ 4)$.*

*Proof* First consider (i). If $i$ divides $m$, say $m = \ell i$, then

$$q^m - 1 = (q^i - 1)(1 + q^\ell + q^{2\ell} + \cdots + q^{(i-1)\ell})$$

and thus $r$ divides $q^m - 1$. Conversely, suppose $r$ divides $q^m - 1$. Write $m = \ell i + s$ with $0 \leqslant s < i$. Then $r$ divides $(q^m - 1) - (q^{\ell i} - 1) = q^{\ell i}(q^s - 1)$, so $r$ divides $q^s - 1$ and thus the minimality of $i$ implies that $s = 0$.

Next consider (ii). Here $i$ divides $2m$, but not $m$ (since $r > 2$). Therefore $2m/i$ is odd and we deduce that $i \equiv 2 \pmod 4$ since $m$ is odd. Part (iii) is entirely similar. $\qquad\square$

**Lemma A.2** *Let $r \neq p$ be an odd prime, set $i = \Phi(r,q)$ and define $d$ as in (A.2). Let $n$ be a positive integer and write $n \equiv k \pmod d$ with $0 \leqslant k < d$. Suppose that $j$ is the largest even integer such that $j \leqslant n$ and $r$ divides $q^j - 1$. Then $j = n - k$ is divisible by $d$.*

*Proof* By Lemma A.1(i), $r$ divides $q^m - 1$ if and only if $i$ divides $m$, so $i$ divides $j$. Moreover, since $j$ is even, it follows that $2i$ divides $j$ if $i$ is odd and thus $d$ divides $j$. Now $r$ also divides $q^{j+d} - 1$, so the maximality of $j$ implies that $j + d > n$, that is $0 \leqslant n - j < d$ and we quickly deduce that $n - j = k$, as required. $\qquad\square$

**Lemma A.3** *Let $r \neq p$ be an odd prime, set $i = \Phi(r,q)$ and define $c$ as in (A.1). Let $n$ be a positive integer and write $n \equiv k \pmod c$ with $0 \leqslant k < c$. Suppose that $j$ is the largest integer such that $j \leqslant n$ and $r$ divides $q^j - (-1)^j$. Then $j = n - k$ is divisible by $c$.*

*Proof* First assume $i$ is odd. By Lemma A.1, $r$ does not divide $q^m + 1$ for any $m$, so $j$ is even. The same lemma also implies that $c = 2i$ divides $j$. Moreover, since $j + i$ is odd it follows that $r$ does not divide $q^{j+i} - (-1)^{j+i}$, but it does divide $q^{j+c} - (-1)^{j+c}$. Therefore $j = n - k$ by the maximality of $j$.

The case $i \equiv 0 \pmod 4$ is very similar. Here Lemma A.1(ii) implies that $j$ is even, so $c = i$ divides $j$. Since $j + i$ is also even we deduce that $r$ divides $q^{j+i} - (-1)^{j+i}$, so once again $j = n - k$ by the maximality of $j$.

Finally, let us assume $i \equiv 2 \pmod 4$, so $c = i/2$ is odd. By Lemma A.1, $r$ divides $q^m - (-1)^m$ if and only if $m$ is even and $i$ divides $m$, or $m$ is odd and $i$ divides $2m$. It follows that $c$ divides $j$. Now, if $j$ is even then $r$ divides $q^{j+c} + 1$. On the other hand, if $j$ is odd then $j + c$ is even and divisible by $i$, so $r$ divides $q^{j+c} - 1$. In both cases, the maximality of $j$ implies that $j = n - k$. $\qquad\square$

Recall that if $n$ is a positive integer and $r$ is a prime, then $(n)_r$ denotes the highest power of $r$ dividing $n$.

**Lemma A.4** *Let $n$ be a positive integer and let $r$ be a prime divisor of $q - \varepsilon$, where $\varepsilon = \pm$. Then*

$$(q^n - \varepsilon)_r = \begin{cases} (q-\varepsilon)_r(n)_r & n \text{ odd, or } r \text{ odd and } \varepsilon = + \\ (q^2-1)_2(n)_2/2 & n \text{ even, } r = 2, \varepsilon = + \\ (r,2) & n \text{ even, } \varepsilon = - \end{cases}$$

*Proof* The case $\varepsilon = +$ follows from [63, Lemmas 3.3 and 3.4], so we may assume that $\varepsilon = -$.

First assume $n$ is even. If $r$ is odd then $\Phi(r,q) = 2$ and thus Lemma A.1 implies that $r$ divides $q^n - 1$, so $r$ does not divide $q^n + 1$. On the other hand, if $r = 2$ then $q^2 \equiv 1 \pmod 4$, so $q^n - 1$ is divisible by 4 and thus $(q^n + 1)_2 = 2$. We conclude that $(q^n + 1)_r = (r, 2)$ if $n$ is even.

Now assume $n$ is odd, so

$$q^n + 1 = (q+1)(q^{n-1} - q^{n-2} + q^{n-3} - q^{n-4} + \cdots - q + 1)$$

Since $q \equiv -1 \pmod r$ we have

$$q^{n-1} - q^{n-2} + q^{n-3} - q^{n-4} + \cdots - q + 1 \equiv n \pmod r$$

Therefore, if $(n)_r = 1$ then $(q^n + 1)_r = (q+1)_r$ as required. Now assume that $r$ divides $n$, say $(n)_r = r^\ell$ for some $\ell \geqslant 1$. Note that $r$ is odd. Then

$$q^n + 1 = (q+1)\frac{q^r + 1}{q+1}\frac{q^{r^2}+1}{q^r+1}\cdots\frac{q^{r^\ell}+1}{q^{r^{\ell-1}}+1}\frac{(q^{r^\ell})^{n/r^\ell}+1}{q^{r^\ell}+1}$$

and for each positive integer $j$ we have

$$\frac{q^{r^{j+1}}+1}{q^{r^j}+1} = q^{(r-1)r^j} - q^{(r-2)r^j} + q^{(r-3)r^j} - q^{(r-4)r^j} + \cdots - q^{r^j} + 1$$

Since $r$ is odd we have $q^{r^j} \equiv -1 \pmod{r^2}$, so

$$\frac{q^{r^{j+1}}+1}{q^{r^j}+1} \equiv r \pmod{r^2}$$

and thus

$$\left(\frac{q^r + 1}{q+1}\frac{q^{r^2}+1}{q^r+1}\cdots\frac{q^{r^\ell}+1}{q^{r^{\ell-1}}+1}\right)_r = r^\ell = (n)_r$$

Moreover, since $n/r^\ell$ is coprime to $r$, we have

$$\left(\frac{(q^{r^\ell})^{n/r^\ell}+1}{q^{r^\ell}+1}\right)_r = 1$$

The result follows.                                                      □

Our final two results are technical lemmas that are useful for determining the precise structure of subfield subgroups of linear and unitary groups (see Propositions 5.5.7 and 5.8.5).

**Lemma A.5** *Suppose $q = q_0^2$, $n \geqslant 2$ is an integer and $\varepsilon = \pm$. Set*

$$d = \frac{(n, q_0 - \varepsilon)\left(q_0 - \varepsilon, \frac{q-1}{(n,q-1)}\right)}{q_0 - \varepsilon}$$

*Then $d = 1$ or $2$. Moreover, $d = 2$ if and only if $(q-1)_2 > (n)_2 > 1$.*

*Proof*   For any positive integer $m$ we may write $m = (m)_2 m'$, where $m' \geqslant 1$ is odd and $(m)_2 = 2^i$ for some integer $i \geqslant 0$. Set $a = (q_0 + \varepsilon)'$ and observe that

$$(n, q-1) = (n', a)(n', q_0 - \varepsilon)((n)_2, (q-1)_2)$$

Then

$$\frac{q-1}{(n, q-1)} = \frac{b(q_0 - \varepsilon)'(q-1)_2}{(n', q_0 - \varepsilon)((n)_2, (q-1)_2)}$$

for some odd integer $b$ coprime to $q_0 - \varepsilon$, and thus

$$\left(q_0 - \varepsilon, \frac{q-1}{(n, q-1)}\right) = \begin{cases} \dfrac{(q_0 - \varepsilon)'}{(n', q_0 - \varepsilon)} & (q-1)_2 \leqslant (n)_2 \\[3mm] \dfrac{2^i(q_0 - \varepsilon)'}{(n', q_0 - \varepsilon)} & (q-1)_2 > (n)_2 \end{cases}$$

where $2^i = ((q_0 - \varepsilon)_2, (q-1)_2 / ((n)_2, q-1))$.

If $(q-1)_2 \leqslant (n)_2$ then

$$d = \frac{(n, q_0 - \varepsilon)(q_0 - \varepsilon)'}{(n', q_0 - \varepsilon)(q_0 - \varepsilon)} = \frac{((n)_2, q_0 - \varepsilon)}{(q_0 - \varepsilon)_2} = 1$$

so for the remainder we may assume that $(q-1)_2 > (n)_2$, in which case $q$ is odd and

$$d = \frac{((n)_2, q_0 - \varepsilon)}{(q_0 - \varepsilon)_2}\left((q_0 - \varepsilon)_2, \frac{(q-1)_2}{((n)_2, q-1)}\right)$$

$$= \begin{cases} \left((q_0 - \varepsilon)_2, \frac{(q-1)_2}{(n)_2}\right) & (q_0 - \varepsilon)_2 \leqslant (n)_2 \\[3mm] \dfrac{(n)_2}{(q_0 - \varepsilon)_2}\left((q_0 - \varepsilon)_2, \frac{(q-1)_2}{(n)_2}\right) & (q_0 - \varepsilon)_2 > (n)_2 \end{cases}$$

First consider the case $(q_0 - \varepsilon)_2 \leqslant (n)_2$. If $q_0 \equiv -\varepsilon \pmod 4$ then

$$d = \left((q_0 - \varepsilon)_2, \frac{(q-1)_2}{(n)_2}\right) = 2$$

Similarly, if $q_0 \equiv \varepsilon \pmod 4$ then $(q_0 - \varepsilon)_2 = (n)_2$ and once again we conclude that $d = 2$.

Finally, let us assume $(q_0 - \varepsilon)_2 > (n)_2$. If $q_0 \equiv -\varepsilon \pmod 4$ then $(n)_2 = 1$ and

$$d = \frac{(n)_2}{(q_0 - \varepsilon)_2}\left((q_0 - \varepsilon)_2, \frac{(q-1)_2}{(n)_2}\right) = 1$$

Similarly, if $q_0 \equiv \varepsilon \pmod 4$ then

$$d = \frac{(n)_2}{(q_0 - \varepsilon)_2}\left((q_0 - \varepsilon)_2, \frac{2(q_0 - \varepsilon)_2}{(n)_2}\right)$$

and thus $d = 2$ if $n$ is even, otherwise $d = 1$.

To summarise, we have now shown that $d \in \{1, 2\}$, with $d = 2$ if and only if $(q-1)_2 > (n)_2$ and one of the following holds:

(i) $(q_0 - \varepsilon)_2 \leqslant (n)_2$;
(ii) $n$ is even and $q_0 \equiv \varepsilon \pmod 4$.

The condition $(q-1)_2 > (n)_2$ implies that $q$ is odd, so $n$ is even if (i) holds. Similarly, if $n$ is even and $(q_0 - \varepsilon)_2 > (n)_2$, then $q_0 \equiv \varepsilon \pmod 4$. We conclude that $d = 2$ if and only if $(q-1)_2 > (n)_2 > 1$. $\qquad\square$

**Lemma A.6** *Suppose $q = q_0^k$, where $k$ is an odd prime, $\varepsilon = \pm$ and $n \geqslant 2$ is an integer. Set*

$$d = \frac{(n, q_0 - \varepsilon)\left(q_0 - \varepsilon, \frac{q-\varepsilon}{(n,q-\varepsilon)}\right)}{q_0 - \varepsilon}$$

*Then $d = 1$ or $k$. Moreover, $d = k$ if and only if $(q - \varepsilon)_k > (n)_k > 1$.*

*Proof* Set $s = (q - \varepsilon)/(q_0 - \varepsilon)$. By Lemma A.4, if $r$ is a prime that divides $q_0 - \varepsilon$ and $s$, then $r = k$, and $r^2$ does not divide $s$. Therefore

$$(n, q - \varepsilon) = \begin{cases} ak(n, q_0 - \varepsilon) & (q_0 - \varepsilon)_k < (n)_k \text{ and } (s)_k > 1 \\ a(n, q_0 - \varepsilon) & \text{otherwise} \end{cases}$$

for some integer $a$ coprime to $q_0 - \varepsilon$. Hence

$$\frac{q - \varepsilon}{(n, q - \varepsilon)} = \begin{cases} \frac{bk(q_0-\varepsilon)}{(n,q_0-\varepsilon)} & (q_0 - \varepsilon)_k \geqslant (n)_k \text{ and } (s)_k > 1 \\ \frac{b(q_0-\varepsilon)}{(n,q_0-\varepsilon)} & \text{otherwise} \end{cases}$$

for some integer $b$ coprime to $q_0 - \varepsilon$. Therefore, if $(q_0 - \varepsilon)_k \geqslant (n)_k$ and $(s)_k > 1$ then

$$\left(q_0 - \varepsilon, \frac{q-\varepsilon}{(n,q-\varepsilon)}\right) = \begin{cases} \frac{k(q_0-\varepsilon)}{(n,q_0-\varepsilon)} & (n, q_0 - \varepsilon)_k > 1 \\ \frac{q_0-\varepsilon}{(n,q_0-\varepsilon)} & (n, q_0 - \varepsilon)_k = 1 \end{cases}$$

and similarly, if $(q_0 - \varepsilon)_k < (n)_k$ or $(s)_k = 1$, then

$$\left( q_0 - \varepsilon, \frac{q - \varepsilon}{(n, q - \varepsilon)} \right) = \frac{q_0 - \varepsilon}{(n, q_0 - \varepsilon)}$$

Finally, observe that if $(q_0 - \varepsilon)_k > 1$ then Lemma A.4 implies that $(s)_k > 1$ and $(q - \varepsilon)_k = k(q_0 - \varepsilon)_k$. We conclude that

$$d = \begin{cases} k & (q - \varepsilon)_k > (n)_k > 1 \\ 1 & \text{otherwise} \end{cases}$$

as required.                                                                    □

# Appendix B

## Tables

Let $T$ be a finite simple classical group over $\mathbb{F}_q$ with natural module $V$. Here we provide a convenient summary of some of the information on the conjugacy classes of elements of prime order in $\mathrm{Aut}(T)$ given in Chapter 3.

For semisimple involutions, we use the notation for $\mathrm{Inndiag}(T)$-class representatives introduced in Sections 3.2.2, 3.3.2, 3.4.2 and 3.5.2 (recall that this is consistent with the notation used by Gorenstein, Lyons and Solomon in [67, Table 4.5.1]). In the third column of Tables B.1, B.5, B.8, B.10 and B.11 we give necessary and sufficient conditions for the given involution $x$ to lie in $T$. Similarly, in the fourth column of Tables B.10 and B.11 we give the precise conditions needed for $x$ to be in $\mathrm{PSO}_n^\varepsilon(q)$.

Further information on semisimple involutions is presented in Tables B.2, B.6 and B.9. For instance, consider Table B.2. Here $T = \mathrm{PSL}_n^\varepsilon(q)$ and $x = \hat{x}Z \in \mathrm{PGL}_n^\varepsilon(q)$ is an involution, where $\hat{x} \in \mathrm{GL}_n^\varepsilon(q) \leqslant \mathrm{GL}(V)$ and $Z = Z(\mathrm{GL}_n^\varepsilon(q))$. In the fourth column, we write $i_2(\hat{x}Z)$ to denote the number of involutions in the coset $\hat{x}Z$. The next column gives the splitting field $K$ of $\hat{x}$, which is the smallest extension of $\mathbb{F}_q$ containing all the eigenvalues of $\hat{x}$. In the sixth column, we describe the eigenspaces of $\hat{x}$ on $V$, using the standard abbreviations $t.i.$ and $n.d.$ for the terms *totally isotropic* and *nondegenerate*, respectively (in Table B.9, we also use $t.s.$ for *totally singular*). Of course, this only makes sense when the eigenspaces of $\hat{x}$ are defined over the appropriate underlying field $\mathbb{F}_{q^u}$, where $u = 1$ if $\varepsilon = +$ and $u = 2$ if $\varepsilon = -$. In the last column of the table we state whether or not $\hat{x}$ is $\mathrm{SL}_n^\varepsilon(q)$-conjugate to its negative $-\hat{x}$.

Finally, in Tables B.3, B.4, B.7 and B.12 we provide a more general summary of our earlier work on conjugacy classes of automorphisms $x \in \mathrm{Aut}(T)$ of order $r$, where $r$ is an arbitrary prime (in Table B.12 we assume $r > 2$). Note that in the second column of these tables we use the notation u, s, etc. to denote the type of $x$; this is explained in Table 3.2.2 (see Section 3.2.7).

# B.1 Linear and unitary groups

Table B.1 *Involutions in $G = \mathrm{PGL}_n^{\varepsilon}(q)$, $p \neq 2$, part I*

| $x$ | Conditions | Inner | $|C_G(x)|$ |
|---|---|---|---|
| $t_i$ | $1 \leqslant i < n/2$ | $i$ even or $(q-\varepsilon)_2 > (n)_2$ | $\alpha\,|\mathrm{GL}_i^{\varepsilon}(q)|\,|\mathrm{GL}_{n-i}^{\varepsilon}(q)|$ |
| $t_{n/2}$ | $n$ even | $n \equiv 0 \pmod 4$ or $q \equiv \varepsilon \pmod 4$ | $2\alpha\,|\mathrm{GL}_{n/2}^{\varepsilon}(q)|^2$ |
| $t'_{n/2}$ | $n$ even | $q \equiv -\varepsilon \pmod 4$ or $(n)_2 > (q-\varepsilon)_2$ | $2\alpha\,|\mathrm{GL}_{n/2}(q^2)|$ |

$\alpha = (q-\varepsilon)^{-1}$

Table B.2 *Involutions in $G = \mathrm{PGL}_n^{\varepsilon}(q)$, $p \neq 2$, part II*

| $\varepsilon$ | $x$ | Conditions | $i_2(\hat{x}Z) > 0$? | $K$ | Eigenspaces | $\hat{x} \sim -\hat{x}$? |
|---|---|---|---|---|---|---|
| $+$ | $t_i$ | $1 \leqslant i < n/2$ | yes | $\mathbb{F}_q$ | t.i. | no |
|  | $t_{n/2}$ | $n$ even | yes | $\mathbb{F}_q$ | t.i. | yes |
|  | $t'_{n/2}$ | $n$ even | no | $\mathbb{F}_{q^2}$ | — | yes |
| $-$ | $t_i$ | $1 \leqslant i < n/2$ | yes | $\mathbb{F}_q$ | n.d. | no |
|  | $t_{n/2}$ | $n$ even | yes | $\mathbb{F}_q$ | n.d. | yes |
|  | $t'_{n/2}$ | $n$ even | no | $\mathbb{F}_{q^2}$ | t.i. | yes |

$x = \hat{x}Z \in \mathrm{PGL}_n^{\varepsilon}(q)$, $Z = Z(\mathrm{GL}_n^{\varepsilon}(q))$
$\sim$ denotes $\mathrm{SL}_n^{\varepsilon}(q)$-conjugacy

Table B.3 *Elements of prime order r in* $\mathrm{Aut}(\mathrm{PSL}_n(q))$, $n \ge 2$

| $x$ | Type | Conditions | $|C_G(x)|$ | References |
|---|---|---|---|---|
| $[J_p^{a_p}, \ldots, J_1^{a_1}]$ | u | $r = p$ | $\alpha q^\beta \prod_j |\mathrm{GL}_{a_j}(q)|$ | 3.2.6 |
| $[\Lambda_1^{a_1}, \ldots, \Lambda_t^{a_t}, I_e]$ | s | $i > 1$ | $\alpha |\mathrm{GL}_e(q)| \prod_j |\mathrm{GL}_{a_j}(q)|$ | 3.2.1 |
| $[I_{a_0}, \lambda I_{a_1}, \ldots, \lambda^{r-1} I_{a_{r-1}}]$ | s | $i = 1$ | $\alpha \prod_j |\mathrm{GL}_{a_j}(q)|$ | 3.2.2(i), 3.2.5 |
| $[I_{n/r}, \lambda I_{n/r}, \ldots, \lambda^{r-1} I_{n/r}]$ | s | $i = 1,\ r \mid n$ | $r\alpha |\mathrm{GL}_{n/r}(q)|^r$ | 3.2.2(i), 3.2.5 |
| $[\Lambda^{n/r}]$ | s | $i = 1,\ r \mid n$ | $r\alpha |\mathrm{GL}_{n/r}(q^r)|$ | 3.2.2(ii), 3.2.5 |
| $\phi^{fj/r},\ 1 \le j < r$ | f | $r \mid f$ | $|\mathrm{PGL}_n(q^{1/r})|$ | 3.2.9 |
| $\phi^{f/2}\iota$ | gf | $r = 2,\ f$ even, $n \ge 3$ | $|\mathrm{PGU}_n(q^{1/2})|$ | 3.2.15 |
| $\gamma_1$ | g | $r = 2,\ n$ odd | $|\mathrm{Sp}_{n-1}(q)|$ | 3.2.11 |
| $\gamma_1$ | g | $r = 2,\ n \ge 4$ even | $|\mathrm{PGSp}_n(q)|$ | 3.2.11 |
| $\gamma_2$ | g | $r = 2,\ n \ge 4$ even, $p \ne 2$ | $|\mathrm{PGO}_n^+(q)|$ | 3.2.11 |
| $\gamma_2'$ | g | $r = 2,\ n \ge 4$ even, $p \ne 2$ | $|\mathrm{PGO}_n^-(q)|$ | 3.2.11 |
| $\gamma_3$ | g | $r = p = 2,\ n \ge 4$ even | $q^{n^2/4} \prod_{i=1}^{n/2-1}(q^{2i} - 1)$ | 3.2.11 |

$G = \mathrm{Inndiag}(\mathrm{PSL}_n(q)) = \mathrm{PGL}_n(q)$, $q = p^f$

$\alpha = (q-1)^{-1}$

In the first row, $\beta = 2\sum_{i<j} ia_i a_j + \sum_i (i-1)a_i^2$

For semisimple $x$, $i = \Phi(r; q)$ is the smallest positive integer such that $r$ divides $q^i - 1$

In the third row, $a_j \ne n/r$ for some $j$

In the third and fourth rows, $\lambda \in \mathbb{F}_q$ is a nontrivial $r$th root of unity

Table B.4 *Elements of prime order $r$ in* $\mathrm{Aut}(\mathrm{PSU}_n(q)), n \geqslant 3$

| $x$ | Type | Conditions | $|C_G(x)|$ | References |
|---|---|---|---|---|
| $[J_p^{a_p}, \ldots, J_1^{a_1}]$ | u | $r = p$ | $\alpha q^\beta \prod_j |\mathrm{GU}_{a_j}(q)|$ | 3.3.7 |
| $[(\Lambda_1, \Lambda_1^{-q})^{a_1}, \ldots, (\Lambda_{s/2}, \Lambda_{s/2}^{-q})^{a_{s/2}}, I_e]$ | s | $i$ odd, $r > 2$ | $\alpha |\mathrm{GU}_e(q)| \prod_j |\mathrm{GL}_{a_j}(q^{2i})|$ | 3.3.2(i) |
| $[(\Lambda_1, \Lambda_1^{-q})^{a_1}, \ldots, (\Lambda_{s/2}, \Lambda_{s/2}^{-q})^{a_{s/2}}, I_e]$ | s | $i \equiv 0 \pmod 4$ | $\alpha |\mathrm{GU}_e(q)| \prod_j |\mathrm{GL}_{a_j}(q^i)|$ | 3.3.2(i) |
| $[\Lambda_1^{a_1}, \ldots, \Lambda_s^{a_s}, I_e]$ | s | $i \equiv 2 \pmod 4, i > 2$ | $\alpha |\mathrm{GU}_e(q)| \prod_j |\mathrm{GU}_{a_j}(q^{i/2})|$ | 3.3.2(ii) |
| $[I_{a_0}, \lambda I_{a_1}, \ldots, \lambda^{r-1} I_{a_{r-1}}]$ | s | $i = 2$ | $\alpha \prod_j |\mathrm{GU}_{a_j}(q)|$ | 3.3.3(i) |
| $[I_{n/r}, \lambda I_{n/r}, \ldots, \lambda^{r-1} I_{n/r}]$ | s | $i = 2, r \mid n$ | $r\alpha |\mathrm{GU}_{n/r}(q)|^r$ | 3.3.3(i) |
| $[\Lambda^{n/r}]$ | s | $i = 2, r > 2, r \mid n$ | $r\alpha |\mathrm{GU}_{n/r}(q^r)|$ | 3.3.3(ii) |
| $t_j, 1 \leqslant j < n/2$ | s | $r = 2, p \neq 2$ | $\alpha |\mathrm{GU}_j(q)| |\mathrm{GU}_{n-j}(q)|$ | 3.3.6 |
| $t_{n/2}$ | s | $r = 2, p \neq 2, n$ even | $2\alpha |\mathrm{GU}_{n/2}(q)|^2$ | 3.3.6 |
| $t'_{n/2}$ | s | $r = 2, p \neq 2, n$ even | $2\alpha |\mathrm{GL}_{n/2}(q^2)|$ | 3.3.6 |
| $\phi^2 f j/r, 1 \leqslant j < r$ | f | $r > 2, r \mid f$ | $|\mathrm{PGU}_n(q^{1/r})|$ | 3.3.12 |
| $\gamma_1$ | g | $r = 2, n$ odd | $|\mathrm{Sp}_{n-1}(q)|$ | 3.3.15 |
| $\gamma_1$ | g | $r = 2, n \geqslant 4$ even | $|\mathrm{PGSp}_n(q)|$ | 3.3.15 |
| $\gamma_2$ | g | $r = 2, n \geqslant 4$ even, $p \neq 2$ | $|\mathrm{PGO}_n^+(q)|$ | 3.3.15 |
| $\gamma_2'$ | g | $r = 2, n \geqslant 4$ even, $p \neq 2$ | $|\mathrm{PGO}_n^-(q)|$ | 3.3.15 |
| $\gamma_3$ | g | $r = p = 2, n \geqslant 4$ even | $q^{n^2/4} \prod_{i=1}^{n/2-1} (q^{2i} - 1)$ | 3.3.15 |

$G = \mathrm{Inndiag}(\mathrm{PSU}_n(q)) = \mathrm{PGU}_n(q), q = p^f$

$\alpha = (q+1)^{-1}$

In the first row, $\beta = 2\sum_{i<j} j a_i a_j + \sum_i (i-1) a_i^2$

For semisimple $x$, $i = \Phi(r, q)$

In the fifth row, $a_j \neq n/r$ for some $j$

In the fifth and sixth rows, $\lambda \in \mathbb{F}_{q^2}$ is a nontrivial $r$th root of unity

## B.2 Symplectic groups

Table B.5 *Involutions in $G = \mathrm{PGSp}_n(q)$, $p \neq 2$, part I*

| $x$ | Conditions | Inner | $|C_G(x)|$ |
|---|---|---|---|
| $t_i$ | $1 \leqslant i < n/4$ | always | $|\mathrm{Sp}_{2i}(q)||\mathrm{Sp}_{n-2i}(q)|$ |
| $t_{n/4}$ | $n \equiv 0 \pmod 4$ | always | $2|\mathrm{Sp}_{n/2}(q)|^2$ |
| $t'_{n/4}$ | $n \equiv 0 \pmod 4$ | never | $2|\mathrm{Sp}_{n/2}(q^2)|$ |
| $t_{n/2}$ | | $q \equiv 1 \pmod 4$ | $2|\mathrm{GL}_{n/2}(q)|$ |
| $t'_{n/2}$ | | $q \equiv 3 \pmod 4$ | $2|\mathrm{GU}_{n/2}(q)|$ |

Table B.6 *Involutions in $G = \mathrm{PGSp}_n(q)$, $p \neq 2$, part II*

| $x$ | Conditions | $i_2(\hat{x}Z) > 0$? | $K$ | Eigenspaces | $\hat{x} \sim -\hat{x}$? |
|---|---|---|---|---|---|
| $t_i$ | $1 \leqslant i < n/4$ | yes | $\mathbb{F}_q$ | n.d. | no |
| $t_{n/4}$ | $n \equiv 0 \pmod 4$ | yes | $\mathbb{F}_q$ | n.d. | yes |
| $t'_{n/4}$ | $n \equiv 0 \pmod 4$ | no | $\mathbb{F}_{q^2}$ | — | yes |
| $t_{n/2}$ | | yes | $\mathbb{F}_q$ | t.i. | yes |
| $t'_{n/2}$ | | no | $\mathbb{F}_{q^2}$ | — | yes |

$x = \hat{x}Z \in \mathrm{PGSp}_n(q)$, $Z = Z(\mathrm{GSp}_n(q))$
$\sim$ denotes $\mathrm{Sp}_n(q)$-conjugacy

Table B.7 *Elements of prime order r in* Aut(PSp$_n$(q)), $n \geq 4$

| $x$ | Type | Conditions | $|C_G(x)|$ | References |
|---|---|---|---|---|
| $[J_p^{a_p}, J_{p-1}^{a_{p-1}, \varepsilon_{p-1}}, \ldots, J_2^{a_2, \varepsilon_2}, J_1^{a_1}]$ | u | $r = p > 2$ | $2^{-\beta} q^\gamma \prod_{i \text{ even}} |O_{a_i}^{\varepsilon_i}(q)| \prod_{i \text{ odd}} |\mathrm{Sp}_{a_i}(q)|$ | 3.4.10 |
| $a_s, 2 \leq s \leq n/2$ even | u | $r = p = 2$ | $q^{ns - 3s^2/2 + s/2} |\mathrm{Sp}_s(q)| |\mathrm{Sp}_{n-2s}(q)|$ | 3.4.13 |
| $b_s, 1 \leq s \leq n/2$ odd | u | $r = p = 2$ | $q^{ns - 3s^2/2 + s/2} |\mathrm{Sp}_{s-1}(q)| |\mathrm{Sp}_{n-2s}(q)|$ | 3.4.13 |
| $c_s, 2 \leq s \leq n/2$ even | u | $r = p = 2$ | $q^{ns - 3s^2/2 + 3s/2 - 1} |\mathrm{Sp}_{s-2}(q)| |\mathrm{Sp}_{n-2s}(q)|$ | 3.4.13 |
| $[(\Lambda_1, \Lambda_1^{-1})^{a_1}, \ldots, (\Lambda_{t/2}, \Lambda_{t/2}^{-1})^{a_{t/2}}, I_e]$ | s | $ir$ odd | $|\mathrm{Sp}_e(q)| \prod_j |\mathrm{GL}_{a_j}(q^j)|$ | 3.4.3(i) |
| $[\Lambda_1^{a_1}, \ldots, \Lambda_r^{a_r}, I_e]$ | s | $i$ even, $r > 2$ | $|\mathrm{Sp}_e(q)| \prod_j |\mathrm{GU}_{a_j}(q^{i/2})|$ | 3.4.3(ii) |
| $t_j, 1 \leq j < n/4$ | s | $r = 2, p \neq 2$ | $|\mathrm{Sp}_{2j}(q)| |\mathrm{Sp}_{n-2j}(q)|$ | 3.4.6 |
| $t_{n/4}$ | s | $r = 2, p \neq 2, n \equiv 0 \pmod 4$ | $2|\mathrm{Sp}_{n/2}(q)|^2$ | 3.4.6 |
| $t'_{n/4}$ | s | $r = 2, p \neq 2, n \equiv 0 \pmod 4$ | $2|\mathrm{Sp}_{n/2}(q^2)|$ | 3.4.6 |
| $t_{n/2}$ | s | $r = 2, p \neq 2$ | $2|\mathrm{GL}_{n/2}(q)|$ | 3.4.6 |
| $t'_{n/2}$ | s | $r = 2, p \neq 2$ | $2|\mathrm{GU}_{n/2}(q)|$ | 3.4.6 |
| $\phi^{fj/r}, 1 \leq j < r$ | f | $r \mid f$ | $|\mathrm{PGSp}_n(q^{1/r})|$ | 3.4.15 |
| $\tau$ | gf | $(n, r, p) = (4, 2, 2), f$ odd | $|\mathrm{Sz}(q)| = q^2(q^2 + 1)(q - 1)$ | 3.4.16 |

$G = \text{Inndiag}(\mathrm{PSp}_n(q)) = \mathrm{PGSp}_n(q), q = p^f$

In the first row, $\beta = 1$ if $a_j$ is odd for some $j$, otherwise $\beta = 0$

Also in the first row, $\gamma = \sum_{i<j} ia_i a_j + \frac{1}{2}\sum_i (i-1)a_i^2 + \frac{1}{2}\sum_{i \text{ even}} a_i$

For semisimple $x, i = \Phi(r, q)$

# B.3 Orthogonal groups

Table B.8 *Involutions in $G = \mathrm{PGO}_n(q)$, $nq$ odd*

| $x$ | Conditions | Inner | $|C_G(x)|$ |
|---|---|---|---|
| $t_i$ | $1 \leqslant i \leqslant (n-1)/2$ | $q^i \equiv 1 \pmod 4$ | $2|\mathrm{SO}_{2i}^+(q)||\mathrm{SO}_{n-2i}(q)|$ |
| $t_i'$ | $1 \leqslant i \leqslant (n-1)/2$ | $q^i \equiv 3 \pmod 4$ | $2|\mathrm{SO}_{2i}^-(q)||\mathrm{SO}_{n-2i}(q)|$ |

Table B.9 *Involutions in $G = \mathrm{Inndiag}(\mathrm{P}\Omega_n^\varepsilon(q))$, $p \neq 2$*

| $\varepsilon$ | $x$ | Conditions | $i_2(\hat{x}Z) > 0$? | $K$ | Eigenspaces | $\hat{x} \sim -\hat{x}$? |
|---|---|---|---|---|---|---|
| $+$ | $t_i$ | $1 \leqslant i < n/4$ | yes | $\mathbb{F}_q$ | n.d. | no |
|  | $t_i'$ | $1 \leqslant i < n/4$ | yes | $\mathbb{F}_q$ | n.d. | no |
|  | $t_{n/4}$ | $n \equiv 0 \pmod 4$ | yes | $\mathbb{F}_q$ | n.d. | yes |
|  | $t_{n/4}'$ | $n \equiv 0 \pmod 4$ | yes | $\mathbb{F}_q$ | n.d. | yes |
|  | $t_{n/4}'', t_{n/4}'''$ | $n \equiv 0 \pmod 4$ | no | $\mathbb{F}_{q^2}$ | — | yes |
|  | $t_{n/2}, t_{n/2-1}$ | $n \equiv 0 \pmod 4$ | yes[†] | $\mathbb{F}_q$ | t.s. | yes |
|  | $t_{n/2}', t_{n/2-1}'$ | $n \equiv 0 \pmod 4$ | no | $\mathbb{F}_{q^2}$ | — | yes |
|  | $t_{n/2}$ | $n \equiv 2 \pmod 4$ | yes[†] | $\mathbb{F}_q$ | t.s. | no[‡] |
| $-$ | $t_i$ | $1 \leqslant i < n/4$ | yes | $\mathbb{F}_q$ | n.d. | no |
|  | $t_i'$ | $1 \leqslant i < n/4$ | yes | $\mathbb{F}_q$ | n.d. | no |
|  | $t_{n/4}$ | $n \equiv 0 \pmod 4$ | yes | $\mathbb{F}_q$ | n.d. | no |
|  | $t_{n/4}'$ | $n \equiv 0 \pmod 4$ | no | $\mathbb{F}_{q^2}$ | — | yes |
|  | $t_{n/2}$ | $n \equiv 2 \pmod 4$ | no | $\mathbb{F}_{q^2}$ | — | yes |
| $\circ$ | $t_i$ | $1 \leqslant i \leqslant (n-1)/2$ | yes | $\mathbb{F}_q$ | n.d. | no |
|  | $t_i'$ | $1 \leqslant i \leqslant (n-1)/2$ | yes | $\mathbb{F}_q$ | n.d. | no |

$x = \hat{x}Z \in \mathrm{Inndiag}(\mathrm{P}\Omega_n^\varepsilon(q))$, $Z = Z(\mathrm{GO}_n^\varepsilon(q))$
$\sim$ denotes $\Omega_n^\varepsilon(q)$-conjugacy
[†] $\hat{x}Z$ contains involutions in $\mathrm{GO}_n^+(q)$, but not in $\mathrm{O}_n^+(q)$
[‡] $\hat{x}$ is $\mathrm{O}_n^+(q)$-conjugate to $-\hat{x}$, but not $\mathrm{SO}_n^+(q)$-conjugate

Table B.10 *Inner-diagonal and graph involutions in* $\mathrm{Aut}(\mathrm{P}\Omega_n^+(q))$, $p \neq 2$

| $x$ | Conditions | Inner | $\mathrm{PSO}_n^+(q)$ | $|C_G(x)|$ |
|---|---|---|---|---|
| $t_i$ | $1 \leqslant i < n/4$ | $q^i \equiv 1 \pmod 4$ or $n/2 - i$ even | always | $2|\mathrm{SO}_{2i}^+(q)||\mathrm{SO}_{n-2i}^+(q)|$ |
| $t_i'$ | $1 \leqslant i < n/4$ | $q^i \equiv 3 \pmod 4$, or $q \equiv 3 \pmod 4$ and $n/2 - i$ odd | always | $2|\mathrm{SO}_{2i}^-(q)||\mathrm{SO}_{n-2i}^-(q)|$ |
| $t_{n/4}$ | $n \equiv 0 \pmod 4$ | $q^{n/4} \equiv 1 \pmod 4$ | always | $|\mathrm{O}_{n/2}^+(q)|^2$ |
| $t_{n/4}'$ | $n \equiv 0 \pmod 4$ | $q^{n/4} \equiv 3 \pmod 4$ | always | $|\mathrm{O}_{n/2}^-(q)|^2$ |
| $t_{n/4}'', t_{n/4}'''$ | $n \equiv 0 \pmod 4$ | never | never | $2|\mathrm{O}_{n/2}^+(q^2)|$ |
| $t_{n/2}, t_{n/2-1}$ | $n \equiv 0 \pmod 4$ | $q \equiv 1 \pmod 4$ | $q \equiv 1 \pmod 4$ | $2|\mathrm{GL}_{n/2}(q)|$ |
| $t_{n/2}', t_{n/2-1}'$ | $n \equiv 0 \pmod 4$ | $q \equiv 3 \pmod 4$ | $q \equiv 3 \pmod 4$ | $2|\mathrm{GU}_{n/2}(q)|$ |
| $t_{n/2}$ | $n \equiv 2 \pmod 4$ | $q \equiv 1 \pmod 8$ | $q \equiv 1 \pmod 4$ | $|\mathrm{GL}_{n/2}(q)|$ |
| $\gamma_i$ | $1 \leqslant i < (n+2)/4$ | never | never | $|\mathrm{SO}_{2i-1}(q)||\mathrm{SO}_{n-2i+1}(q)|$ |
| $\gamma_{(n+2)/4}$ | $n \equiv 2 \pmod 4$ | never | never | $2|\mathrm{SO}_{n/2}(q)|^2$ |
| $\gamma_{(n+2)/4}'$ | $n \equiv 2 \pmod 4$ | never | never | $2|\mathrm{SO}_{n/2}(q^2)|$ |
| $\gamma_1^*, \gamma_1^{**}$ | $n = 8$ | never | never | $|\mathrm{SO}_7(q)|$ |
| $\gamma_2^*, \gamma_2^{**}$ | $n = 8$ | never | never | $|\mathrm{SO}_3(q)||\mathrm{SO}_5(q)|$ |

$G = \mathrm{Imdiag}(\mathrm{P}\Omega_n^+(q))$

Table B.11 *Involutions in* $\mathrm{PGO}_n^-(q)$, $p \neq 2$

| $x$ | Conditions | Inner | $\mathrm{PSO}_n^-(q)$ | $|C_G(x)|$ |
|---|---|---|---|---|
| $t_i$ | $1 \leq i < n/4$ | $q^i \equiv 1 \pmod 4$ or $n/2 - i$ odd | always | $2|\mathrm{SO}_{2i}^+(q)||\mathrm{SO}_{n-2i}^-(q)|$ |
| $t_i'$ | $1 \leq i < n/4$ | $q^i \equiv 3 \pmod 4$ or $n/2 - i$ even or $q \equiv 1 \pmod 4$ | always | $2|\mathrm{SO}_{2i}^-(q)||\mathrm{SO}_{n-2i}^+(q)|$ |
| $t_{n/4}$ | $n \equiv 0 \pmod 4$ | always | always | $2|\mathrm{SO}_{n/2}^+(q)||\mathrm{SO}_{n/2}^-(q)|$ |
| $t_{n/4}'$ | $n \equiv 0 \pmod 4$ | never | never | $2|\mathrm{SO}_{n/2}^-(q^2)|$ |
| $t_{n/2}$ | $n \equiv 2 \pmod 4$ | $q \equiv 7 \pmod 8$ | $q \equiv 3 \pmod 4$ | $|\mathrm{GU}_{n/2}(q)|$ |
| $\gamma_i$ | $1 \leq i < (n+2)/4$ | never | never | $|\mathrm{SO}_{2i-1}(q)||\mathrm{SO}_{n-2i+1}(q)|$ |
| $\gamma_{(n+2)/4}$ | $n \equiv 2 \pmod 4$ | never | never | $2|\mathrm{SO}_{n/2}(q)|^2$ |
| $\gamma_{(n+2)/4}'$ | $n \equiv 2 \pmod 4$ | never | never | $2|\mathrm{SO}_{n/2}(q^2)|$ |

$G = \mathrm{Inndiag}(P\Omega_n^-(q))$

Table B.12 *Elements of odd prime order $r$ in $\mathrm{Aut}(\mathrm{P}\Omega_n^\varepsilon(q))$, $n \geqslant 7$*

| $x$ | Type | Conditions | $\|C_G(x)\|$ | References |
|---|---|---|---|---|
| $[J_p^{a_p,\delta_p}, J_{p-1}^{a_{p-1}}, \ldots, J_1^{a_1,\delta_1}]$ | u | $r = p$ | $2^{-\beta} q^\gamma \prod_{i \text{ odd}} \|O_{a_i}^{\varepsilon_i}(q)\| \prod_{i \text{ even}} \|\mathrm{Sp}_{a_i}(q)\|$ | 3.5.12, 3.5.14 |
| $[(\Lambda_1,\Lambda_1^{-1})^{a_1}, \ldots, (\Lambda_{t/2},\Lambda_{t/2}^{-1})^{a_{t/2}}, I_e]$ | s | $i$ odd | $2^{-\delta}\|O_e^\varepsilon(q)\| \prod_j \|\mathrm{GL}_{a_j}(q^i)\|$ | 3.5.4(i) |
| $[\Lambda_1^{a_1}, \ldots, \Lambda_t^{a_t}, I_e]$ | s | $i$ even | $2^{-\delta}\|O_e^{\varepsilon'}(q)\| \prod_j \|\mathrm{GU}_{a_j}(q^{i/2})\|$ | 3.5.4(ii) |
| $\phi^{fj/r}, \; 1 \leqslant j < r$ | f | $r \mid f$ | $\|\mathrm{SO}_n^\varepsilon(q^{1/r})\|$ | 3.5.20 |
| $\tau_1^{\pm 1}$ | g | $(n,\varepsilon,r) = (8,+,3)$ | $\|G_2(q)\|$ | 3.5.23 |
| $\tau_2^{\pm 1}$ | g | $\begin{cases}(n,\varepsilon,r)=(8,+,3),\, q \equiv \varepsilon' \text{ (mod 3)}\\ (n,\varepsilon,r)=(8,+,3),\, p=3\end{cases}$ | $\begin{cases}\|\mathrm{PGL}_3^{\varepsilon'}(q)\|\\ q^5\|\mathrm{SL}_2(q)\|\end{cases}$ | 3.5.23 |
| $\tau_1^{\pm 1}\phi^{f/3}, \; \tau_1^{\pm 1}\phi^{2f/3}$ | gf | $(n,\varepsilon,r) = (8,+,3),\, q = q_0^3$ | $\|{}^3D_4(q_0)\|$ | 3.5.23 |

$G = \mathrm{Imdiag}(\mathrm{P}\Omega_n^\varepsilon(q))$, $q = p^f$

In the first row, if $i$ is odd and $a_i$ is odd, then $\varepsilon_i = \circ$

Similarly, if $i$ is odd and $a_i$ is even, then $\varepsilon_i = (-)^{ia_i(q-1)/4}$ if $\delta_i = \square$ and $\varepsilon_i = (-)^{1+ia_i(q-1)/4}$ if $\delta_i = \boxtimes$

Also $\beta = 0$ if $a_i = 0$ for all odd $i$, $\beta = 2$ if $n$ is even and $a_i$ is odd for some $i$, otherwise $\beta = 1$

Also in the first row, $\gamma = \sum_{i<j} ia_i a_j + \frac{1}{2}\sum_i (i-1)a_i^2 - \frac{1}{2}\sum_{i \text{ even}} a_i$

In the first row, if $n$ is even then there are two $G$-classes of this form if and only if $a_i = 0$ for all odd $i$

In the first row, if $a_i$ is odd for some odd $i$ then $[J_p^{a_p,\delta_p}, J_{p-1}^{a_{p-1}}, \ldots, J_1^{a_1,\delta_1}]$ and $[J_p^{a_p,\boxtimes\delta_p}, J_{p-1}^{a_{p-1}}, \ldots, J_1^{a_1,\boxtimes\delta_1}]$ are $G$-conjugate

In the second and third rows, $\delta = 1$ if $e \neq 0$, otherwise $\delta = 0$

For semisimple $x$, $i = \Phi(r,q)$

In the second row, if $e = 0$ then $\varepsilon = +$ and there are two $G$-classes of this type

In the third row, if $e = 0$ then $\varepsilon = (-)^{\sum_j a_j}$ and there are two $G$-classes of this type

In the third row, $\varepsilon' = \varepsilon$ if and only if $n$ is odd, or if there is an even number of odd terms $a_j$

# References

[1]  Alspach, B. 1989. Lifting Hamilton cycles of quotient graphs. *Discrete Math.*, **78**, 25–36.

[2]  Arvind, V. 2013. The parameterized complexity of fixpoint free elements and bases in permutation groups. Pages 4–15 of: Gutin, G., and Szeider, S. (eds), *Parameterized and Exact Computation.* Lecture Notes in Computer Science, vol. 8246. Springer, Switzerland.

[3]  Aschbacher, M. 1984. On the maximal subgroups of the finite classical groups. *Invent. Math.*, **76**, 469–514.

[4]  Aschbacher, M. 2000. *Finite Group Theory* (Second edition). Cambridge Studies in Advanced Mathematics, vol. 10. Cambridge University Press, Cambridge.

[5]  Aschbacher, M., and Seitz, G. M. 1976. Involutions in Chevalley groups over fields of even order. *Nagoya Math. J.*, **63**, 1–91.

[6]  Bamberg, J., Giudici, M., Liebeck, M. W., Praeger, C. E., and Saxl, J. 2013. The classification of almost simple $\frac{3}{2}$-transitive groups. *Trans. Am. Math. Soc.*, **365**, 4257–4311.

[7]  Bang, A. S. 1886. Taltheoretiske undersølgelser. *Tidskrifft Math.*, **5**, 70–80, 130–137.

[8]  Bereczky, Á. 1995. Fixed-point-free $p$-elements in transitive permutation groups. *Bull. London Math. Soc.*, **27**, 447–452.

[9]  Bienert, R., and Klopsch, B. 2010. Automorphism groups of cyclic codes. *J. Algebraic Combin.*, **31**, 33–52.

[10] Biggs, N. 1973. Three remarkable graphs. *Can. J. Math.*, **25**, 397–411.

[11] Bosma, W., Cannon, J., and Playoust, C. 1997. The Magma algebra system I: The user language. *J. Symbolic Comput.*, **24**, 235–265.

[12] Boston, N., Dabrowski, W., Foguel, T., Gies, P. J., Jackson, D. A., Leavitt, J., and Ose, D. T. 1993. The proportion of fixed-point-free elements of a transitive permutation group. *Commun. Algebra*, **21**, 3259–3275.

[13] Bray, J. N., Holt, D. F., and Roney-Dougal, C. M. 2013. *The Maximal Subgroups of the Low-Dimensional Finite Classical Groups.* London Mathematical Society Lecture Note Series, vol. 407. Cambridge University Press, Cambridge.

[14] Breuer, T., Guralnick, R. M., and Kantor, W. M. 2008. Probabilistic generation of finite simple groups, II. *J. Algebra*, **320**, 443–494.

[15]  Britnell, J. R., and Maróti, A. 2013. Normal coverings of linear groups. *Algebra Number Theory*, **7**, 2085–2102.

[16]  Bubboloni, D., Praeger, C. E., and Spiga, P. 2013. Normal coverings and pair-wise generation of finite alternating and symmetric groups. *J. Algebra*, **390**, 199–215.

[17]  Burness, T. C. 2007a. Fixed point ratios in actions of finite classical groups, I. *J. Algebra*, **309**, 69–79.

[18]  Burness, T. C. 2007b. Fixed point ratios in actions of finite classical groups, II. *J. Algebra*, **309**, 80–138.

[19]  Burness, T. C. 2007c. Fixed point ratios in actions of finite classical groups, III. *J. Algebra*, **314**, 693–748.

[20]  Burness, T. C. 2007d. Fixed point ratios in actions of finite classical groups, IV. *J. Algebra*, **314**, 749–788.

[21]  Burness, T. C. 2007e. On base sizes for actions of finite classical groups. *J. London Math. Soc.*, **75**, 545–562.

[22]  Burness, T. C., and Giudici, M. On $2'$-elusive biquasiprimitive permutation groups. In preparation.

[23]  Burness, T. C., Giudici, M., and Wilson, R. A. 2011b. Prime order derangements in primitive permutation groups. *J. Algebra*, **341**, 158–178.

[24]  Burness, T. C., and Guest, S. 2013. On the uniform spread of almost simple linear groups. *Nagoya Math. J.*, **209**, 35–109.

[25]  Burness, T. C., Guralnick, R. M., and Saxl, J. 2011a. On base sizes for symmetric groups. *Bull. London Math. Soc.*, **43**, 386–391.

[26]  Burness, T. C., Guralnick, R. M., and Saxl, J. 2014. Base sizes for $\mathscr{S}$-actions of finite classical groups. *Isr. J. Math.*, **199**, 711–756.

[27]  Burness, T. C., Liebeck, M. W., and Shalev, A. 2009. Base sizes for simple groups and a conjecture of Cameron. *Proc. London Math. Soc.*, **98**, 116–162.

[28]  Burness, T. C., O'Brien, E. A., and Wilson, R. A. 2010. Base sizes for sporadic simple groups. *Isr. J. Math.*, **177**, 307–333.

[29]  Burness, T. C., Praeger, C. E., and Seress, Á. 2012a. Extremely primitive classical groups. *J. Pure Appl. Algebra*, **216**, 1580–1610.

[30]  Burness, T. C., Praeger, C. E., and Seress, Á. 2012b. Extremely primitive sporadic and alternating groups. *Bull. London Math. Soc.*, **44**, 1147–1154.

[31]  Burness, T. C., and Tong-Viet, H. P. 2015. Derangements in primitive permutation groups, with an application to character theory. *Q. J. Math.*, **66**, 63–96.

[32]  Burness, T. C., and Tong-Viet, H. P. Primitive permutation groups and derangements of prime power order. Manuscripta Math. In press.

[33]  Burnside, W. 1911. *Theory of Groups of Finite Order* (Second edition). Cambridge University Press, Cambridge.

[34]  Cameron, P. J. (ed.). 1997. Research problems from the Fifteenth British Combinatorial Conference (Stirling, 1995). *Discrete Math.*, **167/168**, 605–615.

[35]  Cameron, P. J. 1999. *Permutation Groups*. London Mathematical Society Student Texts, vol. 45. Cambridge University Press, Cambridge.

[36]  Cameron, P. J. 2000. *Notes on Classical Groups*. Unpublished lecture notes, available at www.maths.qmul.ac.uk/~pjc/class_gps/cg.pdf.

[37] Cameron, P. J., and Cohen, A. M. 1992. On the number of fixed point free elements in a permutation group. *Discrete Math.*, **106/107**, 135–138.

[38] Cameron, P. J., Frankl, P., and Kantor, W. M. 1989. Intersecting families of finite sets and fixed-point-free 2-elements. *Eur. J. Combin.*, **10**, 149–160.

[39] Cameron, P. J., Giudici, M., Jones, G. A., Kantor, W. M., Klin, M. H., Marušič, D., and Nowitz, L. A. 2002. Transitive permutation groups without semiregular subgroups. *J. London Math. Soc.*, **66**, 325–333.

[40] Carter, R. W. 1989. *Simple Groups of Lie Type*. Wiley Classics Library. John Wiley & Sons, New York.

[41] Conway, J. H., Curtis, R. T., Norton, S. P., Parker, R. A., and Wilson, R. A. 1985. *Atlas of Finite Groups*. Oxford University Press, Eynsham.

[42] Crestani, E., and Lucchini, A. 2012. Normal coverings of solvable groups. *Arch. Math. (Basel)*, **98**, 13–18.

[43] Crestani, E., and Spiga, P. 2010. Fixed-point-free elements in $p$-groups. *Isr. J. Math.*, **180**, 413–424.

[44] Diaconis, P., Fulman, J., and Guralnick, R. 2008. On fixed points of permutations. *J. Algebraic Combin.*, **28**, 189–218.

[45] Dickson, L. E. 1901. *Linear Groups, with an Exposition of the Galois Field Theory*. B. G. Teubner, Leipzig.

[46] Dieudonné, J. 1951. On the automorphisms of the classical groups. *Mem. Am. Math. Soc.*, **2**.

[47] Dieudonné, J. 1955. *La Géométrie des Groupes Classiques*. Springer, Berlin.

[48] Dixon, J. D., and Mortimer, B. 1996. *Permutation Groups*. Graduate Texts in Mathematics, vol. 163. Springer, New York.

[49] Dobson, E., Malnič, A., Marušič, D., and Nowitz, L. A. 2007a. Minimal normal subgroups of transitive permutation groups of square-free degree. *Discrete Math.*, **307**, 373–385.

[50] Dobson, E., Malnič, A., Marušič, D., and Nowitz, L. A. 2007b. Semiregular automorphisms of vertex-transitive graphs of certain valencies. *J. Combin. Theory Ser. B*, **97**, 371–380.

[51] Dobson, E., and Marušič, D. 2011. On semiregular elements of solvable groups. *Commun. Algebra*, **39**, 1413–1426.

[52] Fein, B., Kantor, W. M., and Schacher, M. 1981. Relative Brauer groups, II. *J. Reine Angew. Math.*, **328**, 39–57.

[53] Feit, W. 1980. Some consequences of the classification of finite simple groups. Pages 175–181 of: *The Santa Cruz Conference on Finite Groups, 1979*. Proceeding of Symposia in Pure Mathematics, vol. 37. American Mathematical Society, Providence, RI.

[54] Frucht, R. 1970. How to describe a graph. *Ann. N. Y. Acad. Sci.*, **175**, 159–167.

[55] Fulman, J., and Guralnick, R. M. 2003. Derangements in simple and primitive groups. Pages 99–121 of: *Groups, Combinatorics & Geometry (Durham, 2001)*. World Scientific, River Edge, NJ.

[56] Fulman, J., and Guralnick, R. M. 2012. Bounds on the number and sizes of conjugacy classes in finite Chevalley groups with applications to derangements. *Trans. Am. Math. Soc.*, **364**, 3023–3070.

[57] Fulman, J., and Guralnick, R. M. Derangements in finite classical groups for actions related to extension field and imprimitive subgroups and the solution of the Boston-Shalev conjecture. Submitted (arxiv:1508.00039).

[58] Fulman, J., and Guralnick, R. M. Derangements in subspace actions of finite classical groups. *Trans. Am. Math. Soc.*, to appear.

[59] Galois, E. 1846. Oeuvres mathématiques: Lettre de Galois à M. Auguste Chevalier (29 Mai 1832). *J. Math. Pures Appl. (Liouville)*, **11**, 408–415.

[60] Gill, N. 2007. Polar spaces and embeddings of classical groups. *N. Z. J. Math.*, **36**, 175–184.

[61] Giudici, M. 2003. Quasiprimitive groups with no fixed point free elements of prime order. *J. London Math. Soc.*, **67**, 73–84.

[62] Giudici, M. 2007. New constructions of groups without semiregular subgroups. *Commun. Algebra*, **35**, 2719–2730.

[63] Giudici, M., and Kelly, S. 2009. Characterizing a family of elusive permutation groups. *J. Group Theory*, **12**, 95–105.

[64] Giudici, M., Morgan, L., Potočnik, P., and Verret, G. 2015. Elusive groups of automorphisms of digraphs of small valency. *Eur. J. Combin.*, **46**, 1–9.

[65] Giudici, M., and Xu, J. 2007. All vertex-transitive locally-quasiprimitive graphs have a semiregular automorphism. *J. Algebraic Combin.*, **25**, 217–232.

[66] Gorenstein, D., and Lyons, R. 1983. The local structure of finite groups of characteristic 2 type. *Mem. Am. Math. Soc.*, **276**.

[67] Gorenstein, D., Lyons, R., and Solomon, R. 1998. *The Classification of the Finite Simple Groups. Number 3.* Mathematical Surveys and Monographs, vol. 40. American Mathematical Society, Providence, RI.

[68] Guralnick, R. M. 1990. Zeroes of permutation characters with applications to prime splitting and Brauer groups. *J. Algebra*, **131**, 294–302.

[69] Guralnick, R. M. Conjugacy classes of derangements in finite transitive groups. *Proc. Steklov Inst. Math.* In press.

[70] Guralnick, R. M., and Kantor, W. M. 2000. Probabilistic generation of finite simple groups. *J. Algebra*, **234**, 743–792.

[71] Guralnick, R. M., Müller, P., and Saxl, J. 2003. The rational function analogue of a question of Schur and exceptionality of permutation representations. *Mem. Am. Math. Soc.*, **773**.

[72] Guralnick, R. M., and Saxl, J. 2003. Generation of finite almost simple groups by conjugates. *J. Algebra*, **268**, 519–571.

[73] Guralnick, R. M., and Wan, D. 1997. Bounds for fixed point free elements in a transitive group and applications to curves over finite fields. *Isr. J. Math.*, **101**, 255–287.

[74] Hartley, R. W. 1925. Determination of the ternary collineation groups whose coefficients lie in the $GF(2^n)$. *Ann. Math.*, **27**, 140–158.

[75] Herstein, I. N. 1975. *Topics in Algebra* (Second edition). John Wiley & Sons, New York.

[76] Hiss, G., and Malle, G. 2001. Low-dimensional representations of quasi-simple groups. *LMS J. Comput. Math.*, **4**, 22–63.

[77] Isbell, J. R. 1957. Homogeneous games. *Math. Student*, **25**, 123–128.

[78] Isbell, J. R. 1960. Homogeneous games. II. *Proc. Am. Math. Soc.*, **11**, 159–161.

[79] Isbell, J. R. 1964. Homogeneous games. III. Pages 255–265 of: *Advances in Game Theory.* Princeton University Press, Princeton, NJ.

[80] Jones, G. A. 2002. Cyclic regular subgroups of primitive permutation groups. *J. Group Theory*, **5**, 403–407.

[81] Jones, J. W., and Roberts, D. P. 2014. The tame-wild principle for discriminant relations for number fields. *Algebra Number Theory*, **8**, 609–645.

[82] Jordan, C. 1872. Recherches sur les substitutions. *J. Math. Pures Appl. (Liouville)*, **17**, 351–367.

[83] Jordan, D. 1988. Eine Symmetrieeigenschaft von Graphen. Pages 17–20 of: *Graphentheorie und ihre Anwendungen (Stadt Wehlen, 1988)*. Dresdner Reihe Forsch., vol. 9. Päd. Hochsch., Dresden.

[84] Khukhro, E. I., and Mazurov, V. D. (eds.). 2014. *The Kourovka Notebook: Unsolved Problems in Group Theory* (Eighteenth edition). Institute of Mathematics, Novosibirsk.

[85] Kleidman, P. B. 1987. The maximal subgroups of the finite 8-dimensional orthogonal groups $P\Omega_8^+(q)$ and of their automorphism groups. *J. Algebra*, **110**, 173–242.

[86] Kleidman, P., and Liebeck, M. 1990. *The Subgroup Structure of the Finite Classical Groups*. London Mathematical Society Lecture Note Series, vol. 129. Cambridge University Press, Cambridge.

[87] Knapp, A. W. 2007. *Advanced Algebra*. Cornerstones. Birkhäuser, Boston, MA.

[88] Kutnar, K., and Šparl, P. 2010. Distance-transitive graphs admit semiregular automorphisms. *Eur. J. Combin.*, **31**, 25–28.

[89] Leighton, F. T. 1983. On the decomposition of vertex-transitive graphs into multicycles. *J. Res. Natl. Bur. Stand.*, **88**, 403–410.

[90] Li, C. H. 2003. The finite primitive permutation groups containing an abelian regular subgroup. *Proc. London Math. Soc.*, **87**, 725–747.

[91] Li, C. H. 2005. Permutation groups with a cyclic regular subgroup and arc transitive circulants. *J. Algebraic Combin.*, **21**, 131–136.

[92] Li, C. H. 2006. Finite edge-transitive Cayley graphs and rotary Cayley maps. *Trans. Am. Math. Soc.*, **358**, 4605–4635.

[93] Liebeck, M. W., and O'Brien, E. A. Conjugacy classes in finite groups of Lie type: representatives, centralizers and algorithms. In preparation.

[94] Liebeck, M. W., Praeger, C. E., and Saxl, J. 1988. On the O'Nan–Scott theorem for finite primitive permutation groups. *J. Aust. Math. Soc. Ser. A*, **44**, 389–396.

[95] Liebeck, M. W., Praeger, C. E., and Saxl, J. 2000. Transitive subgroups of primitive permutation groups. *J. Algebra*, **234**, 291–361.

[96] Liebeck, M. W., and Seitz, G. M. 2012. *Unipotent and Nilpotent Classes in Simple Algebraic Groups and Lie Algebras*. Mathematical Surveys and Monographs, vol. 180. American Mathematical Society, Providence, RI.

[97] Liebeck, M. W., and Shalev, A. 1999. Simple groups, permutation groups, and probability. *J. Am. Math. Soc.*, **12**, 497–520.

[98] Lübeck, F. 2001. Small degree representations of finite Chevalley groups in defining characteristic. *LMS J. Comput. Math.*, **4**, 135–169.

[99] Malle, G., and Testerman, D. 2011. *Linear Algebraic Groups and Finite Groups of Lie Type*. Cambridge Studies in Advanced Mathematics, vol. 133. Cambridge University Press, Cambridge.

[100] Malnič, A., Marušič, D., Šparl, P., and Frelih, B. 2007. Symmetry structure of bicirculants. *Discrete Math.*, **307**, 409–414.

[101] Marušič, D. 1981. On vertex symmetric digraphs. *Discrete Math.*, **36**, 69–81.

[102] Marušič, D., and Scapellato, R. 1998. Permutation groups, vertex-transitive digraphs and semiregular automorphisms. *Eur. J. Combin.*, **19**, 707–712.

[103] McKay, B. D., and Royle, G. F. 1990. The transitive graphs with at most 26 vertices. *Ars Combin.*, **30**, 161–176.

[104] Mitchell, H. H. 1911. Determination of the ordinary and modular ternary linear groups. *Trans. Am. Math. Soc.*, **12**, 207–242.

[105] Mitchell, H. H. 1914. The subgroups of the quaternary abelian linear group. *Trans. Am. Math. Soc.*, **15**, 379–396.

[106] Montmort, P. R. de. 1708. *Essay d'analyse sur les Jeux de Hazard*. Quillau, Paris.

[107] Neumann, P. M., and Praeger, C. E. 1998. Derangements and eigenvalue-free elements in finite classical groups. *J. London Math. Soc.*, **58**, 564–586.

[108] Pless, V. 1964. On Witt's theorem for nonalternating symmetric bilinear forms over a field of characteristic 2. *Proc. Am. Math. Soc.*, **15**, 979–983.

[109] Praeger, C. E., Li, C. H., and Niemeyer, A. C. 1997. Finite transitive permutation groups and finite vertex-transitive graphs. Pages 277–318 of: *Graph Symmetry (Montreal, 1996)*. NATO Advanced Science Institutes Series C: Mathematical and Physical Sciences, vol. 497. Kluwer, Dordrecht.

[110] Sabidussi, G. 1958. On a class of fixed-point-free graphs. *Proc. Am. Math. Soc.*, **9**, 800–804.

[111] Saxl, J., and Seitz, G. M. 1997. Subgroups of algebraic groups containing regular unipotent elements. *J. London Math. Soc.*, **55**, 370–386.

[112] Schur, I. 1933. Zur Theorie der einfach transitiven Permutationsgruppen. *S. B. Preuss. Akad. Wiss., Phys.-Math. Kl.*, 598–623.

[113] Serre, J.-P. 2003. On a theorem of Jordan. *Bull. Am. Math. Soc.*, **40**, 429–440.

[114] Spiga, P. 2013. Permutation 3-groups with no fixed-point-free elements. *Algebra Colloq.*, **20**, 383–394.

[115] Steinberg, R. 1968. *Lectures on Chevalley Groups*. Department of Mathematics, Yale University.

[116] Suzuki, M. 1986. *Group Theory II*. Springer, New York.

[117] Takács, L. 1979/1980. The problem of coincidences. *Arch. Hist. Exact Sci.*, **21**, 229–244.

[118] Taylor, D. E. 1992. *The Geometry of the Classical Groups*. Sigma Series in Pure Mathematics, vol. 9. Heldermann Verlag, Berlin.

[119] Wall, G. E. 1963. On the conjugacy classes in the unitary, symplectic and orthogonal groups. *J. Aust. Math. Soc.*, **3**, 1–62.

[120] Wielandt, H. 1964. *Finite Permutation Groups*. Academic Press, New York.

[121] Wilson, R. A. 1985. Maximal subgroups of automorphism groups of simple groups. *J. London Math. Soc.*, **32**, 460–466.

[122] Wilson, R. A. 2009. *The Finite Simple Groups*. Graduate Texts in Mathematics, vol. 251. Springer, London.

[123] Zsigmondy, K. 1892. Zur Theorie der Potenzreste. *Monatsh. Math. Phys.*, **3**, 265–284.

# Index

Printed in the United States
By Bookmasters